Introduction to Stochastic Models in Operations Research

Physics Assignment #7
2.35
2.44
2.46
2.48
Text ch 24/8

Introduction to Stochastic Models in Operations Research

Frederick S. Hillier
Professor of Operations Research
Stanford University

Gerald J. Lieberman
Professor of Operations Research and Statistics
Stanford University

McGraw-Hill Publishing Company

New York St. Louis San Francisco Auckland Bogotá
Caracas Hamburg Lisbon London Madrid Mexico
Milan Montreal New Delhi Oklahoma City Paris
San Juan São Paulo Singapore Sydney Tokyo Toronto

Introduction to Stochastic Models in Operations Research

2 3 4 5 6 7 8 9 0 DOH DOH 9 5 4 3 2 1 0

P/N 028938-7

This book was set in Times Roman by Waldman Graphics, Inc.
The editors were Eric M. Munson and Scott Amerman;
the production supervisor was Janelle S. Travers.
Design was done by Caliber Design Planning, Inc.
R. R. Donnelley & Sons Company was printer and binder.

Library of Congress Cataloging-in-Publication Data

Hillier, Frederick S.
Introduction to stochastic models in operations research / Frederick S. Hillier, Gerald J. Lieberman.
p. cm.
This volume is derived from the authors' Introduction to operations research, 5th. ed., and is intended for the first part of the course required of industrial engineering majors.
P/N 028938-7
1. Operations research. 2. Stochastic programming. 3. Stochastic processes.
I. Lieberman, Gerald J. II. Hillier, Frederick S. Introduction to operations research.
III. Title.
T57.6.H55 1990
003'.76—dc20 89-49757

About the Authors

Frederick S. Hillier was born and raised in Aberdeen, Washington, where he was an award winner in statewide high school contests in essay writing, mathematics, debate, and music. As an undergraduate at Stanford University, he ranked first in his engineering class, won the McKinsey Prize for technical writing, and won the Hamilton Award for combining excellence in engineering with notable achievements in the humanities and social sciences. Upon his graduation with a B.S. degree in Industrial Engineering, he was awarded three national fellowships (National Science Foundation, Tau Beta Pi, and Danforth) for graduate study at Stanford with specialization in operations research. After receiving his Ph.D. degree, he joined the faculty of Stanford University, where he is now Professor of Operations Research.

Dr. Hillier's research has extended into a variety of areas, including integer programming, queueing theory and its application, statistical quality control, and the application of operations research to the design of production systems and to capital budgeting. He has published widely, and his seminal papers have been selected for republication in books of selected readings at least ten times. He was the first-prize winner of a research contest on "Capital Budgeting of Interrelated Projects" sponsored by The Institute of Management Sciences and the U.S. Office of Naval Research. He also has served as Treasurer of the Operations Research Society of America, Vice President for Meetings of The Institute of Management Sciences (TIMS), and Co-General Chairman of the 1989 TIMS International Meeting in Osaka, Japan.

In addition to *Introduction to Stochastic Models in Operations Research* and the two companion volumes, *Introduction to Operations Research* and *Introduction to Mathematical Programming*, his books are *The Evaluation of Risky Interrelated Investments* (North-Holland, 1969) and *Queueing Tables and Graphs* (Elsevier North-Holland, 1981, co-authored by O. S. Yu, with D. M. Avis, L. D. Fossett, F. D. Lo, and M. I. Reiman).

Gerald J. Lieberman is Professor of Operations Research and Statistics at Stanford University. He has served as Vice Provost and Dean of Graduate Studies and Research at Stanford and was the founding chair of the Department of Operations Research. He is both an engineer (having received an undergraduate degree in mechanical engineering from Cooper Union) and an operations research statistician (with an A.M. from Columbia University in mathematical statistics and a Ph.D. from Stanford University in statistics).

His research interests have been in the stochastic areas of operations research, often at the interface of applied probability and statistics. He has extensive publications in the areas of reliability and quality control, and in the modeling of complex systems, including their optimal design, when resources are limited.

Dr. Lieberman's professional honors include being elected to the National Academy of Engineering, receiving the Shewhart Medal of the American Society for Quality Control, receiving the Cuthbertson Award for exceptional service to Stanford University, and serving as a fellow at the Center for Advanced Study in the Behavioral Sciences. He also served as President of The Institute of Management Sciences.

In addition to *Introduction to Stochastic Models in Operations Research* and the two companion volumes, *Introduction to Operations Research* and *Introduction to Mathematical Programming*, his books are *Handbook of Industrial Statistics* (Prentice-Hall, 1955, co-authored by A. H. Bowker), *Tables of the Non-Central t-Distribution* (Stanford University Press, 1957, co-authored by G. J. Resnikoff), *Tables of the Hypergeometric Probability Distribution* (Stanford University Press, 1961, co-authored by D. Owen), and *Engineering Statistics*, Second Edition (Prentice-Hall, 1972, co-authored by A. H. Bowker).

To Our Parents

Contents

Preface

During the 22 years since the initial publication of our textbook *Introduction to Operations Research* we have witnessed many new developments and considerable growth in the field. As subsequent editions expanded to keep up with this growth, many survey courses in operations research using this book were split into more specialized courses. Frequently, the new courses cover either mathematical programming (perhaps entitled ''Deterministic Models in Operations Research'') or stochastic models in operations research.

To accommodate this evolution, the current publication of the even bigger fifth edition of *Introduction to Operations Research* (nearly 1,000 pages) is being accompanied by the publication of two more-compact split volumes, namely, *Introduction to Stochastic Models in Operations Research* and *Introduction to Mathematical Programming*. All of the material in the fifth edition has been duplicated in one or the other of the split volumes (and occasionally in both). In particular, this book, *Introduction to Stochastic Models in Operations Research*, includes Part 1 (Introduction) with slight alterations, Part 4 (Probabilistic Models), and the appendices from the fifth edition. Because of their relevance for a course on stochastic models in operations research, slightly modified versions of the fifth edition's Chap. 11 (Dynamic Programming) and Chap. 12 (Game Theory) also have been included here as Chaps. 8 and 13, respectively.

In addition, this book includes some material from previous editions of *Introduction to Operations Research* that were subsequently deleted for space reasons. This material appears here as Chap. 3 (Probability Theory), Chap. 4 (Statistical Inference), Sec. 7.6 (Estimating Model Parameters) in Chap. 7 (The Application of Queueing Theory), and Sec. 9.4 (a more mathematical treatment of stochastic models) in Chap. 9 (Inventory Theory).

Those familiar with the fourth edition of *Introduction to Operations Research* will find that new material has been added to the chapters appearing in both this book and the fifth edition. Three of the chapters have been thoroughly revised. Chapter 5 (Markov Chains) has been rewritten, and now includes more examples and applications. Much of Chap. 10 (Forecasting) has been rewritten so as to present an elementary, but rigorous, treatment of a subject that is becoming increasingly important. Chapter 15 (Simulation) also received a major revision, with considerable updating and an increased emphasis on the major building blocks of a simulation model. The other chapters received a careful review (updated as needed) with the objective to increase the clarity of the presentation. Our goal has been to make this a real ''student's book''—clear, interesting, and well organized, with lots of helpful examples and illustrations, good motivation and perspective, easy-to-find important material, enjoyable homework, etc.—without too much notation, terminology, and dense mathematics.

One notable feature in striving toward this goal is the incorporation of software that has been specifically designed to be a teaching supplement to this book. This software is packaged in the back of the book for use on either an IBM (or IBM-compatible) personal computer with a graphics card or a Macintosh. This is very different from the usual software now widely available for running the algorithms of operations research on microcomputers. Instead, it is true tutorial software for helping to learn the material in this book (or the two companion volumes). The demonstration examples, the routines for interactive execution of algorithms, and the routines for post-optimality analysis should make the learning process far more efficient and effective, as well as more stimulating and enjoyable. We are convinced that this type of innovative software will usher in a new era for operations research education.

We feel very fortunate to have had the services of a talented and creative operations research student (and one with good family connections), Mark Hillier, to develop this software.

The prerequisites for a course using this book can be relatively modest, because the mathematics has been kept at a relatively elementary level. A previous introduction to probability theory is assumed. However, an extensive review chapter (Chapter 3) is included. Calculus is used in a few places. However, the mathematical maturity that a student achieves through taking an elementary calculus course is useful throughout.

The content of the book is aimed largely at the upper division undergraduate level and at first-year (Master's level) graduate students. There is ample material for a one-quarter or one-semester course with some selectivity of topics, or for a longer course.

The book has been organized to provide great flexibility in the selection of topics. Chapters 1 and 2 provide a general introduction. Chapter 3 reviews elementary probability theory, and Chap. 4 briefly reviews some of the important concepts in statistics. Chapter 5 (Markov Chains) should be covered before the chapters on queueing theory (Chaps. 6 and 7) and Markovian decision processes and applications (Chapter 11) are studied. The order of presentation of the other chapters are at the instructor's discretion.

We thank our wives, Ann and Helen, for their editorial and word processing assistance, as well as their encouragement and understanding when we devoted too many evenings and weekends to preparing this book and the companion volumes. Our children, David, John, and Mark Hillier, Janet Lieberman Argyres, and Joanne, Michael, and Diana Lieberman also have provided strong support in a variety of ways. Most of them have used *Introduction to Operations Research* as a text in their own college courses, given considerable advice, and even (in the case of Mark Hillier) become a full-fledged collaborator. It is a joy to have them on the team.

Acknowledgments

We are deeply indebted to many people for their part in making possible the preparation of this book and the two companion volumes. Those making helpful comments are literally too numerous to mention. However, we would like to acknowledge particularly the advice and support of Stephen Argyres, John Birge, Richard Cottle, Hans

Daellenbach, George Dantzig, Cyrus Derman, B. Curtis Eaves, Philip Gill, Peter Glynn, David Hillier, Mark Hillier, Donald Iglehart, Carolyn Kinser, Steve Love, Tom Magnanti, Alan Manne, Douglas Montgomery, Walter Murray, Stephen Robinson, Matthew Rosenshine, Sheldon Ross, Michael Saunders, Siegfried Schaible, Gerald Shedler, N. D. Singpurwalla, John Tomlin, and Arthur Veinott, Jr. We also thank dozens of Stanford students who gave us helpful written suggestions at our request.

Ann Hillier devoted numerous long days and nights to sitting with a Macintosh, doing word processing, and constructing many figures and tables. She was a vital member of the team.

We are indebted to the Office of Naval Research and the National Science Foundation for supporting some of our research that led to results included in the chapters on queueing theory and reliability.

It was a real pleasure working with McGraw-Hill's thoroughly professional editorial and production staff, including Eric Munson, John Corrigan, and Scott Amerman.

FREDERICK S. HILLIER
GERALD J. LIEBERMAN

Introduction to Stochastic Models in Operations Research

1

The Nature of Operations Research

1.1 The Origins of Operations Research

Since the advent of the industrial revolution, the world has seen a remarkable growth in the size and complexity of organizations. The artisans' small shops of an earlier era have evolved into the billion-dollar corporations of today. An integral part of this revolutionary change has been a tremendous increase in the division of labor and segmentation of management responsibilities in these organizations. The results have been spectacular. However, along with its blessings, this increasing specialization has created new problems, problems that are still occurring in many organizations. One problem is a tendency for the many components of an organization to grow into relatively autonomous empires with their own goals and value systems, thereby losing sight of how their activities and objectives mesh with those of the overall organization. What is best for one component frequently is detrimental to another, so they may end up working at cross purposes. A related problem is that as the complexity and specialization in an organization increase, it becomes more and more difficult to allocate its available resources to its various activities in a way that is most effective for the organization as a whole. These kinds of problems and the need to find a better way to resolve them provided the environment for the emergence of operations research.

The roots of operations research can be traced back many decades, when early attempts were made to use a scientific approach in the management of organizations. However, the beginning of the activity called *operations research* has generally been attributed to the military services early in World War II. Because of the war effort, there was an urgent need to allocate scarce resources to the various military operations and to the activities within each operation in an effective manner. Therefore the British and then the American military management called upon a large number of scientists to apply a scientific approach to dealing with this and other strategic and tactical problems. In effect they were asked to do research on (military) operations. These teams of scientists were the first operations research teams. Their efforts allegedly were instrumental in winning the Air Battle of Britain, the Island Campaign in the Pacific, the Battle of the North Atlantic, and others.

Spurred on by the apparent success of operations research in the military, industry gradually became interested in this new field. As the industrial boom following the war was running its course, the problems caused by the increasing complexity and specialization in organizations were again coming to the forefront. It was becoming apparent to a growing number of people, including business consultants who had served on or with the operations research teams during the war, that these were basically the same problems that had been faced by the military but in a different context. In this way operations research began to creep into industry, business, and civil government. By 1951, it had already taken hold in Great Britain and was in the process of doing so in the United States. Since then the field has developed very rapidly, as will be described further in Sec. 1.3.

At least two other factors that played a key role in the rapid growth of operations research during this period can be identified. One was the substantial progress that was made early in improving the techniques available to operations research. After the war, many of the scientists who had participated on operations research teams or who had heard about this work were motivated to pursue research relevant to the field; important advancements in the state of the art resulted. A prime example is the simplex method for solving linear programming problems, developed by George Dantzig in 1947. Many of the standard tools of operations research, e.g., linear programming, dynamic programming, queueing theory, and inventory theory, were relatively well developed before the end of the 1950s. In addition to this rapid advancement in the theory of operations research, a second factor that gave great impetus to the growth of the field was the onslaught of the computer revolution. A large amount of computation is usually required to deal most effectively with the complex problems typically considered by operations research. Doing this by hand would often be out of the question. Therefore, the development of electronic digital computers, with their ability to perform arithmetic calculations thousands or even millions of times faster than a human being can, was a tremendous boon to operations research. Today, mainframe computers, minicomputers, or microcomputers are essential for solving real-world operations research problems.

1.2 The Nature of Operations Research

What is operations research? One way of trying to answer this question is to give a definition. For example, operations research may be described as a scientific approach to decision making that involves the operations of organizational systems. However, this description, like earlier attempts at a definition, is so general that it is equally applicable to many other fields as well. Therefore, perhaps the best way of grasping the unique nature of operations research is to examine its outstanding characteristics.

As its name implies, operations research involves "research on operations." This says something about both the approach and the area of application of the field. Thus operations research is applied to problems that concern how to conduct and coordinate the operations or activities within an organization. The nature of the organization is essentially immaterial, and, in fact, operations research has been applied extensively in business, industry, the military, civil government and agencies, hospitals, and so forth. Therefore, the breadth of application is unusually wide. The approach of operations research is that of the scientific method. In particular, the process begins by carefully observing and formulating the problem and then constructing a scientific (typically mathematical) model that attempts to abstract the essence of the real problem. It is then hypothesized that this model is a sufficiently precise representation of the essential features of the situation, so that the conclusions (solutions) obtained from the model are also valid for the real problem. This hypothesis is then modified and verified by suitable experimentation. Thus in a certain sense operations research involves creative scientific research into the fundamental properties of operations. However, there is more to it than this. Specifically, operations research is also concerned with the practical management of the organization. Therefore, to be successful it must also provide positive, understandable conclusions to the decision maker(s) when they are needed.

Still another characteristic of operations research is its broad viewpoint. As implied in the preceding section, operations research adopts an organizational point of view. Thus it attempts to resolve the conflicts of interest among the components of the organization in a way that is best for the organization as a whole. This does not imply that the study of each problem must give explicit consideration to all aspects of the organization; rather, the objectives being sought must be consistent with those of the overall organization. An additional characteristic that was mentioned in passing is that operations research attempts to find the best or optimal solution to the problem under consideration. Rather than being content with merely improving the status quo, the goal is to identify the best possible course of action. Although it must be interpreted carefully, this "search for optimality" is a very important theme in operations research.

All these characteristics lead quite naturally to still another one. It is evident that no single individual should be expected to be an expert on all the many aspects of operations research work or the problems typically considered; this would require a group of individuals having diverse backgrounds and skills. Therefore, when undertaking a full-fledged operations research study of a new problem, it is usually necessary to use a team approach. Such an operations research team typically needs to include individuals who collectively are highly trained in mathematics, statistics and probability theory, economics, business administration, electronic computing, engineering and the physical sciences, the behavioral sciences, and the special techniques of operations research. The team also needs to have the necessary experience and variety of skills to give appropriate consideration to the many ramifications of the problem throughout the organization and to execute effectively all the diverse phases of the operations research study.

In summary, operations research is concerned with optimal decision making in, and modeling of, deterministic and probabilistic systems that originate from real life.

These applications, which occur in government, business, engineering, economics, and the natural and social sciences, are characterized largely by the need to allocate limited resources. In these situations, considerable insight can be obtained from scientific analysis such as that provided by operations research. The contribution from the operations research approach stems primarily from:

1. Structuring the real-life situation into a mathematical model, abstracting the essential elements so that a solution relevant to the decision maker's objectives can be sought. This involves looking at the problem in the context of the entire system.
2. Exploring the structure of such solutions and developing systematic procedures for obtaining them.
3. Developing a solution, including the mathematical theory, if necessary, that yields an optimal value of the system measure of desirability (or possibly comparing alternative courses of action by evaluating their measure of desirability).

1.3 The Impact of Operations Research

Operations research has had an increasingly great impact on the management of organizations in recent years. Both the number and the variety of its applications continue to grow rapidly, and no slowdown is in sight. In fact, with the exception of the advent of the electronic computer, the extent of this impact seems to be unrivaled by that of any other recent development.

After their success with operations research during World War II, the British and American military services continued to have active operations research groups, often at different levels of command. As a result, there now exists a large number of people called ''military operations researchers'' who are applying an operations research approach to problems of national defense. For example, they engage in tactical planning for requirements and use of weapon systems as well as consider the larger problems of the allocation and integration of effort. Some of their techniques involve quite sophisticated ideas in political science, mathematics, economics, probability theory, and statistics.

Operations research is also being used widely in other types of organizations, including business and industry. Sometimes the term ''management science'' is used as a designation for this activity. Almost all the dozen or so largest corporations in the world, and a sizable proportion of the small industrial organizations, either have well-established operations research groups or have integrated the activity into the regular components of the organization. Many industries, including aircraft and missile, automobile, communication, computer, electric power, electronics, food, metallurgy, mining, paper, petroleum, and transportation, have made widespread use of operations research. Financial institutions, governmental agencies, and hospitals are rapidly increasing their use of operations research.

As an example of the impact of operations research in government, the President's Commission on Aviation Safety issued its report in April 1988 on the strengths, weaknesses, and problems of the airspace system along with 15 recommendations for change. One of these recommendations stated that ''Operations research (applied mathematics methods and models for solving complex operations problems) should be recognized as a standard approach for problem solving in the FAA.''

To be more specific, consider some of the problems that have been solved by particular techniques of operations research. Linear programming has been used successfully in the solution of problems concerned with assignment of personnel, blending of materials, distribution and transportation, and investment portfolios. Dynamic programming has been applied successfully to such areas as planning advertising expenditures, distributing sales effort, and production scheduling. Queueing theory has had application in solving problems concerned with traffic congestion, servicing machines subject to breakdown, determining the level of a service force, air traffic scheduling, design of dams, production scheduling, and hospital operation. Other techniques of operations research, such as inventory theory, game theory, and simulation, also have been successfully applied to a variety of contexts.

The extent of operations research activities, and the profile of operations research practitioners, in U.S. corporations have been frequently surveyed. Among the surveys that emphasized the companies in *Fortune's* list of the top 500 corporations are the following:

1. In 1972, Turban[1] reported on a survey of operations research activities that provide a snapshot of activities in 1969. Mail questionnaires were sent to the directors of operations research/management science of 475 companies. These companies were selected from *Fortune's* list of the top 500, using the 300 largest industrial corporations, 50 industrial corporations drawn from the companies ranking between 300 and 500, and the 25 largest companies in each of the service categories, banks, utilities, merchandising, life insurance, and transportation. There were 107 questionnaires returned.
2. In 1977, Ledbetter and Cox[2] published the results of a survey of *Fortune's* 500 firms (1975 listing) concerning utilization of operations research techniques in their firms. There were 176 respondents.
3. In 1979, Thomas and DaCosta[3] reported on a survey of operations research activities in 1977. Mail questionnaires were sent to 420 individual corporations, including 260 firms from *Fortune's* 1975 list of the top 500, the largest 100 industrial firms in California, and the balance from California financial institutions. There were 150 questionnaires returned.
4. Finally, in 1983, Forgionne[4] published the results of a survey of corporate usage of operations research/management science. A questionnaire was mailed in 1982 to a random sample of 500 corporations drawn from the 1,500 largest U.S. corporations. There were 125 respondents.

Both Turban and Thomas and DaCosta indicated that nearly half of the companies reporting had a special department that was engaged mainly in operations research/management science (OR/MS) activities. While this ratio has remained virtually constant, and since all of the respondents used OR/MS techniques, Thomas and DaCosta concluded that "management science is becoming a part of the everyday

[1] Turban, E.: "A Sample Survey of Operations Research Activities at the Corporate Level," *Operations Research*, **20**:708–721, 1972.

[2] Ledbetter, W. N., and J. F. Cox: "Are OR Techniques Being Used," *Industrial Engineering*, pp. 19–21, February 1977.

[3] Thomas, G., and J. DaCosta: "A Sample Survey of Corporate Operations Research," *Interfaces*, **9**:102–111, 1979.

[4] Forgionne, G. A.: "Corporate Management Science Activities: An Update," *Interfaces*, **13**:20–23, 1983.

Table 1.1 **Ranking of the Techniques of Operations Research/Management Science***

	Turban (1969)	Ledbetter and Cox (1975)	Thomas and DaCosta (1977)	Forgionne (1982)
Bayesian decision analysis	—	—	9	—
Delphi	—	—	13.5	—
Dynamic programming	6	6	10	7
Financial methods	—	—	13.5	—
Game theory	—	7	—	8
Heuristic programming	8.5	—	8	—
Integer and mixed programming	—	—	12	—
Inventory theory	4	—	5	—
Linear programming	3	2	3	4
Network models	—	4	—	—
Nonlinear programming	7	—	7	6
PERT/CPM	5	—	4	3
Risk analysis	—	—	11	—
Queueing theory	8.5	5	6	5
Simulation	2	3	2	2
Statistical analysis	1	1	1	1

* Rank 1 denotes the most frequently used technique.

activities of the modern firm and therefore is no longer a specialized function to be undertaken by a separate specialized department.'' An interesting finding of the Turban survey was that almost all of the specialized departments reported to the company president, vice-president, or controller.

All of the surveys attempted to find which techniques of OR/MS were most frequently used. Table 1.1 presents a ranking of the techniques of operations research that are being applied in U.S. corporations based upon the aforementioned surveys. It is interesting to note that all four surveys, representing a span of 14 years, are in basic agreement. Statistical analysis, simulation, and linear programming were the most widely used techniques, although PERT/CPM, inventory theory, and queueing theory were not far behind.

As noted earlier, OR/MS techniques are applied to a broad spectrum of corporate problem areas, and two of the surveys dealt with this issue. Table 1.2 presents a ranking of the application areas given in the Thomas and DaCosta and Forgionne surveys. The classification of application areas differs somewhat in these two surveys, but it is evident that capital budgeting, forecasting, inventory control, production planning, and project planning are the most popular application areas.

Because of the great impact of operations research, professional societies devoted to this field and related activities have been founded in a number of countries throughout the world. In the United States, the Operations Research Society of America (ORSA), established in 1952, and The Institute of Management Sciences (TIMS), founded in 1953, each has over 6,000 members. ORSA publishes the journal *Operations Research* and TIMS *Management Science*. The two societies also jointly publish *Mathematics of Operations Research* and *Interfaces*. These four journals contain well over 3,000 pages per year reporting new research and applications in the field. In addition, there are many other similar journals published in such countries as the United States, England, France, India, Japan, Canada, and West Germany. Indeed, there are 32 member countries (including the United States) in the International Fed-

Table 1.2 **Ranking of the Application Areas of Operations Research/Management Science***

	Thomas and DaCosta (1977)	Forgionne (1982)
Accounting	11	5
Advertising and sales research	8	—
Capital budgeting	4	2
Equipment replacement	9	—
Forecasting—market planning	1	6
Inventory control	2.5	4
Maintenance	10	9
Packaging	12	—
Personnel management	—	10
Plant location	6	8
Production planning and scheduling	2.5	3
Project planning	—	1
Quality control	7	7
Transportation	5	—

* Rank 1 denotes the most frequent application area.

eration of Operational Research Societies (IFORS), with each country having a national operations research society.

Operations research also has had considerable impact in colleges and universities. Today most of the major American universities offer courses in this field, and many offer advanced degrees that are either in or with specialization in operations research. As a result, there are now thousands of students taking at least one course in operations research each year. Much of the basic research in the field is also being done in the universities.

1.4 Training for a Career in Operations Research

Because of the great growth of operations research, career opportunities in this field appear to be outstanding. The demand for trained people continues to far exceed the supply, and both attractive starting positions and rapid advancement are readily available. Because of the nature of their work, operations research groups tend to have a prominent staff position, with access to higher-level management in the organization. The problems they work on tend to be important, challenging, and interesting. Therefore, any individual with a mathematics and science orientation who is also interested in the practical management of organizations is likely to find a career in operations research very rewarding.

Three complementary types of academic training are particularly relevant for a career in operations research. The first is basic training in the fundamentals upon which operations research is based. This includes the basic methodology of mathematics and science as well as such topics as linear algebra and matrix theory, probability theory, statistical inference, stochastic processes, computer science, microeconomics, accounting and business administration, organization theory, and the behavioral sciences.

A second important type of training is in operations research per se, including special techniques of the field such as linear and nonlinear programming, dynamic programming, inventory theory, network flow theory, queueing models, reliability, game theory, and simulation. It should also include an introduction to the methodology of operations research, where the various techniques and their role in an operations research study involving specific problem areas would be placed in perspective. Often courses covering certain of these topics are offered in more than one department within a university, including departments of business, industrial engineering, mathematics, statistics, computer science, economics, and electrical engineering. This is a natural reflection of the broad scope of application of the field. Since it does spread across traditional disciplinary lines, separate programs or departments in operations research also are being established in some universities.

Finally, it is also good to have specialized training in some field other than operations research, for example, mathematics, statistics, industrial engineering, business, or economics. This additional training provides one with an area of special competence for applying operations research, and it should make that person a more valuable member of an operations research team.

The early operations researchers were people whose primary training and work had been in some traditional field, such as physics, chemistry, mathematics, engineering, or economics. They tended to have little or no formal education in operations research per se. However, as the body of special knowledge has expanded, it has become increasingly more difficult to enter the field without considerable prior education in this area. As a result, although it is still common for new operations researchers to have their college degree(s) in a traditional field, they generally have specialized too in operations research as part of their academic program. The traditional fields that have most commonly served as a vehicle into operations research are indicated in Table 1.3, which is based on the 1972 survey by Turban described in the preceding section. The Thomas and DaCosta survey confirms the diversity of the educational backgrounds of OR/MS practitioners. They noted that the percentage of Ph.D.'s has decreased from 20 percent to 13 percent during the time interval of the two surveys, which they speculate may be due to the "maturity" of the OR/MS techniques in industry and to the lack of need for separate specialized OR/MS departments.

Finally, in 1982[1] a survey of TIMS membership provided a profile of its membership, including information on educational training, job activity, and compensation

Table 1.3 **Educational Background of Operations Research Personnel (Turban Survey)**

	Percentage of Total at Degree Level			
Major Field of Study	Bachelor's	Master's	Doctorate	All Degree Levels
Operations research and management science	3	24	32	12
Mathematics and statistics	26	16	21	22
Business administration	20	27	2	22
Engineering	34	17	29	28
Other	17	16	16	16
Percentage of total	27	53	20	

[1] Hall, J. R., Jr.: "Career Paths and Compensation in Management Science: Results of a TIMS Membership Survey," *Interfaces,* **14**(3):15–23, May–June 1984.

for professionals in industry, government, universities, and consulting. This report reinforced the point that, as the profession has matured, fewer people with formal training in non–operations research fields are entering the profession than occurred in the previous decades.

1.5 The Road Ahead

As an introduction to stochastic models in operations research, this book is designed to acquaint students with the formulation, solution, and implementation of stochastic operations research models for analyzing complex systems problems in industry or government. This chapter and the next introduce the reader to the field of operations research. The next chapter also provides an overview of the operations research modeling approach and describes the major phases of an operations research study. Chapters 3 through 5 *review* certain of the fundamentals upon which the analysis of stochastic models in operations research is based. These fundamentals (namely, elementary probability theory, statistical inference, and Markov chains), are particularly important for an understanding of subsequent portions of the book, so they are summarized in Chapters 3 through 5 for convenient reference by the reader. Finally, the remainder of the book is devoted to the more important stochastic models in operations research. Much of the material presented in these chapters consists of typical examples of situations that are encountered in practice. Synopses of several such examples are presented first, and detailed solutions are given later. A *queueing theory* model is illustrated by a hospital emergency room. The emergency room provides quick medical care for emergency cases that are brought to the hospital by ambulance or private automobile. At any hour there is always one doctor on duty in the emergency room. However, because of a growing tendency for emergency cases to use these facilities rather than go to a private physician, the hospital has been experiencing a continuing increase in the number of emergency room visits each year. As a result, when patients arrive during peak usage hours (the early evening), they have to wait until it is their turn to be treated by the doctor. Therefore, a proposal has been made that a second doctor should be assigned to the emergency room during these hours so that two emergency cases can be treated simultaneously. By recognizing that the emergency room is a queueing system, several alternative queueing theory models can be applied to predict the waiting characteristics of the system with one doctor and with two doctors. These models will aid the hospital in its evaluation of the proposal to add a second physician.

A similar queueing example in a very different context concerns determination of the *optimal number of repairers* for a group of machines. A company uses 10 identical machines in its production facility. However, because these machines break down and require repair frequently, the company has only enough operators to operate *eight* machines at a time, so two machines are available on a standby basis for use while other machines are down. Thus eight machines are always operating whenever no more than two machines are waiting to be repaired, but the number of operating machines is reduced by one for each additional machine waiting to be repaired. The probability distribution of the time until any given operating machine breaks down and the probability distribution of the time required to repair a machine are known from past history. Up until now the company has had just *one* repairer to repair these machines. However, this has frequently resulted in reduced productivity by having *fewer than eight* operating machines. Therefore, consideration is being given to hiring a *second* repairer so that *two* machines can be repaired simultaneously. Thus the queueing system to be studied has the repairers as its servers and the machines requiring repair as its customers, where the problem is to choose between having *one* or *two* servers (or possibly more). Given the cost of each repairer and the cost of inoperable machines, the optimal number of repairers is to be determined.

Inventory theory is illustrated by a television manufacturing company that produces its own speakers, which are used in the production of its television sets. The television sets are assembled on a continuous production line at a known monthly rate. The speakers are produced in batches because they do not warrant setting up a continuous production line and because relatively large quantities can be produced in a short time. The company is interested in determining when and how many to produce. Several costs must be considered. (1) Each time a batch is produced, a setup cost is incurred. This cost includes the cost of "tooling up," administrative costs, record keeping, and so on. (2) The production of speakers in large batch sizes leads to a large inventory, resulting in a monthly cost for keeping a speaker in stock. This cost includes the cost of capital tied up, storage space, insurance, taxes, protection, and so forth. (3) A cost of producing a single speaker (excluding the setup cost) is incurred. (4) Company policy prohibits deliberately planning for shortages of any of its components. However, a shortage of speakers occasionally occurs, resulting in a monthly cost for each speaker unavailable when required. This cost includes the cost of installing speakers after the television set is fully assembled, storage space, delayed revenue, record keeping, and so on. Given data on these costs, the optimal batch size (and period between production) is to be determined.

The use of *Markovian decision processes* can be described in terms of a production process that contains a machine that deteriorates rapidly in both quality and output under heavy usage, so that it is inspected at the end of each day. Immediately after inspection, the condition of the machine is noted and classified into one of four possible states: 0 (as good as new), 1 (operable—minor deterioration), 2 (operable—major deterioration), and 3 (inoperable—output of unacceptable quality). The state of the system is assumed to evolve according to some known probabilistic "laws of motion." At the end of each day, one of three decisions can be made: (1) leave the machine alone; (2) overhaul the machine, which results in leaving it operable with minor deterioration; and (3) replace it, which results in a new machine. As a result of the state of the system found at the end of the day and the decision taken, a cost is incurred. Given these costs and a description of the probabilistic "laws of motion," an optimal maintenance policy is to be found.

An important supplement to this book is the personal software (entitled PROBMOD) which is packaged in the back of the book. It has been designed to be the student's personal tutor, illustrating and illuminating key concepts in a manner that cannot be duplicated on a printed page. It also should save considerable time on homework by enabling the student to focus on concepts while the computer does the number crunching. Although use of this software will result in an enhancement of the textbook, it is not essential for the student to have access to a microcomputer to comprehend the material presented in the book.

2

Overview of the Operations Research Modeling Approach

The bulk of this book is devoted to the mathematical methods of stochastic models in operations research. This is quite appropriate because these quantitative techniques form the main part of what is known about operations research. However, it does not imply that practical operations research studies are primarily mathematical exercises. As a matter of fact, the mathematical analysis often represents only a relatively small part of the total effort required. The purpose of this chapter is to place things into better perspective by describing all the major phases of a typical operations research study.

One way of summarizing the usual phases of an operations research study is the following[1]:

1. Formulating the problem.
2. Constructing a mathematical model to represent the system under study.
3. Deriving a solution from the model.
4. Testing the model and the solution derived from it.

[1] Ackoff, Russell L.: "The Development of Operations Research as a Science," *Operations Research*, **4**:265f, 1956.

5. Establishing controls over the solution.
6. Putting the solution to work: implementation.

Each of these phases will be discussed in turn in the following sections.

2.1 Formulating the Problem

In contrast to textbook examples, most practical problems are initially communicated to an operations research team in a vague, imprecise way. Therefore, the first order of business is to study the relevant system and develop a well-defined statement of the problem to be considered. This includes determining such things as the appropriate objectives, the constraints on what can be done, interrelationships between the area to be studied and other areas of the organization, the possible alternative courses of action, time limits for making a decision, and so on. This process of problem formulation is a crucial one because it greatly affects how relevant the conclusions of the study will be. It is difficult to extract a "right" answer from the "wrong" problem! Consequently, this phase should be executed with considerable care, and the initial formulation should be continually reexamined in the light of new insights obtained during the later phases.

The first thing to recognize is that an operations research team is normally working in an *advisory capacity*. The team members are not just given a problem and told to solve it however they see fit. Instead, they are advising management (often one key decision maker). The team performs a detailed technical analysis of the problem and then presents its recommendations to management. Frequently, the report to management will identify a number of alternatives that are particularly attractive under different assumptions or over a different range of values of some policy parameter that can be evaluated only by management (e.g., the trade-off between *cost* and *benefits*). Management evaluates the study and its recommendations, takes into account a variety of intangible factors, and makes the final decision based on its best judgment. Consequently, it is vital for the operations research team to get on the same wavelength as management, including identifying the "right" problem from management's viewpoint, and to build the support of management for the course that the study is taking.

Determining the *appropriate objectives* is a very important aspect of problem formulation. To do this, it is necessary first to identify the member (or members) of management who actually will be making the decisions concerning the system under study and then to probe into this individual's thinking regarding the pertinent objectives. (Involving the decision maker from the outset also helps to build his or her support for the implementation of the study.) After the decision maker's objectives have been elicited, they should be analyzed and edited for the identification of the ultimate objectives that encompass the other objectives, the determination of the relative importance of these ultimate objectives, and the statement of them precisely in a way that does not eliminate worthwhile goals and alternatives.

By its nature, operations research is concerned with the welfare of the *entire organization* rather than that of only certain of its components. An operations research study seeks solutions that are optimal for the overall organization rather than suboptimal solutions that are best for only one component. Therefore, the objectives that are formulated should ideally be those of the entire organization. However, this is not always convenient to do. Many problems primarily concern only a portion of the organization, so the analysis would become unwieldy if the stated objectives were too general and if explicit consideration were given to all side effects on the rest of the organization. Granted that operations research takes the viewpoint of the overall organization, this does not imply that each problem should be broadened into a study of the entire organization. Instead, the objectives used in the study should be as specific as they can be while still encompassing the main goals of the decision maker and maintaining a reasonable degree of consistency with the higher-level objectives of the organization. Side effects on other segments of the organization must then be considered only to the extent that there are questions of consistency with these higher-level objectives.

For profit-making organizations, one possible approach to circumventing the problem of suboptimization is to use *long-run profit maximization* as the sole objective. The adjective *long-run* indicates that this objective provides the flexibility to consider activities that do not translate into profits *immediately* (e.g., research and development projects) but need to do so *eventually* in order to be worthwhile. At first glance, this approach appears to have considerable merit. In particular, this objective is specific enough to be used conveniently, and yet it seems to be broad enough to encompass the basic goal of profit-making organizations. In fact, some people believe that all other legitimate objectives can be translated into this one.

However, this is an oversimplification, and considerable caution is required! A number of studies of American corporations have found that the goal of *satisfactory profits,* combined with other objectives, is preferred over profit maximization. (In fact, inadequate consideration of long-run profits sometimes is cited as a major reason why American industry may be losing its competitive edge over that of other leading countries.) In particular, typical objectives might be to maintain stable profits, increase (or maintain) one's share of the market, provide for product diversification, maintain stable prices, improve worker morale, maintain family control of the business, and increase company prestige. Fulfilling these objectives might result in the achievement of long-run profit maximization, but the relationship is sufficiently obscure that it may not be convenient to incorporate them into this one objective.

Furthermore, there are additional considerations involving social responsibilities that are distinct from the profit motive. The five parties affected by a business firm located in a single country are: (1) the *owners* (stockholders), who desire profits (dividends, stock appreciation, etc.); (2) the *employees,* who desire steady employment at reasonable wages; (3) the *customers,* who desire a reliable product at a reasonable price; (4) the *vendors,* who desire integrity and a reasonable selling price for their goods; and (5) the *government* and, hence, the *nation,* which desires payment of fair taxes and consideration of the national interest. All five parties make essential contributions to the firm, and the firm should not be viewed as the exclusive servant of any one party for the exploitation of others. By the same token, international corporations acquire additional obligations to follow socially responsible practices. Therefore, although granting that management's prime responsibility is to make profits (which ultimately benefits all five parties), its broader social responsibilities also must be recognized.

2.2 Constructing a Mathematical Model

After formulating the decision maker's problem, the next phase is to reformulate this problem into a form that is convenient for analysis. The conventional operations research approach for doing this is to construct a mathematical model that represents the essence of the problem. Before discussing how to formulate such a model, let us first explore the nature of models in general and of mathematical models in particular.

Models, or idealized representations, are an integral part of everyday life. Common examples of models include model airplanes, portraits, globes, and so on. Similarly, models play an important role in science and business, as illustrated by models of the atom, models of genetic structure, mathematical equations describing physical laws of motion or chemical reactions, graphs, organization charts, and industrial accounting systems. Such models are invaluable for abstracting the essence of the subject of inquiry, showing interrelationships, and facilitating analysis.

Mathematical models are also idealized representations, but they are expressed in terms of mathematical symbols and expressions. Such laws of physics as $F = ma$ and $E = mc^2$ are familiar examples. Similarly, the mathematical model of a business problem is the system of equations and related mathematical expressions that describe the essence of the problem. Thus, if there are n related quantifiable decisions to be made, they are represented as **decision variables** (say, $x_1, x_2, \ldots, x_n$) whose respective values are to be determined. The appropriate measure of performance (e.g., profit) is then expressed as a mathematical function of these decision variables (e.g., $P = 3x_1 + 2x_2 + \cdots + 5x_n$). This function is called the **objective function.** Any restrictions on the values that can be assigned to these decision variables are also expressed mathematically, typically by means of inequalities or equations (e.g., $x_1 + 3x_1x_2 + 2x_2 \leq 10$). Such mathematical expressions for the restrictions often are called **constraints.** The constants (coefficients or right-hand sides) in the constraints and the objective function are called the **parameters** of the model. The mathematical model might then say that the problem is to choose the values of the decision variables so as to maximize the objective function, subject to the specified constraints. Such a model, and minor variations of it, typify the models used in operations research.

Mathematical models have many advantages over a verbal description of the problem. One obvious advantage is that a mathematical model describes a problem much more concisely. This tends to make the overall structure of the problem more comprehensible, and it helps to reveal important cause-and-effect relationships. In this way, it indicates more clearly what additional data are relevant to the analysis. It also facilitates dealing with the problem in its entirety and considering all its interrelationships simultaneously. Finally, a mathematical model forms a bridge to the use of high-powered mathematical techniques and computers to analyze the problem. Indeed, packaged software for both microcomputers and mainframe computers is becoming widely available for many mathematical models.

On the other hand, there are pitfalls to be avoided when using mathematical models. Such a model is necessarily an abstract idealization of the problem, so approximations and simplifying assumptions generally are required if the model is to be *tractable* (capable of being solved). Therefore, care must be taken to ensure that the model remains a valid representation of the problem. The proper criterion for judging the validity of a model is whether or not the model predicts the relative effects of the alternative courses of action with sufficient accuracy to permit a sound decision. Consequently, it is not necessary to include unimportant details or factors that have approximately the same effect for all the alternative courses of action considered. It is not even necessary that the absolute magnitude of the measure of performance be approximately correct for the various alternatives, provided that their relative values (i.e., the differences between their values) are sufficiently precise. Thus all that is required is that there be a high *correlation* between the prediction by the model and what would actually happen in the real world. To ascertain whether this requirement is satisfied, it is important to do considerable *testing* and consequent modifying of the model, which will be the subject of Sec. 2.4. Although this testing phase is placed later in the chapter, much of this *model validation* work actually is conducted during the model-building phase of the study to help guide the construction of the mathematical model.

When developing the model, a good approach is to begin with a very simple version and then move in evolutionary fashion toward more elaborate models that more nearly reflect the complexity of the real problem. This process of *model enrichment* continues only as long as the model remains tractable. The basic trade-off under consideration is between the *precision* and the *tractability* of the model. (See Selected Reference 6 for a detailed description of this process.)

A crucial step in formulating the mathematical model is constructing the objective function. This requires developing a quantitative measure of performance relative to each objective that has been formulated for the study. If there are multiple objectives, their respective measures commonly are then transformed and combined into a composite measure called the **overall measure of performance.** This overall measure might be something tangible (e.g., profit), corresponding to a higher goal of the organization, or it might be abstract (e.g., ''utility''). In the latter case, the task of developing this measure tends to be a complex one requiring a careful comparison of the objectives and their relative importance. After developing the overall measure of performance, the objective function is then obtained by expressing this measure as a mathematical function of the decision variables. Alternatively, there also are methods for explicitly considering multiple objectives simultaneously.

2.3 Deriving a Solution

After formulating a mathematical model for the problem under consideration, the next phase in an operations research study is to derive a solution from this model. You might think that this must be the major part of the study, but actually it is not in most cases. Sometimes, in fact, it is a relatively simple step, in which one of the standard **algorithms** (iterative solution procedures) of operations research is applied on a computer by using one of a number of readily available software packages. For experienced operations research practitioners, finding a solution is the ''fun part,'' whereas the real work comes in the preceding and following steps, including the *post-optimality analysis* discussed later in this section.

Since much of this book is devoted to the subject of how to obtain solutions for various important types of mathematical models, little needs to be said about it here. However, we do need to discuss the nature of such solutions.

A common theme in operations research is the search for an **optimal,** or best, **solution.** Indeed, many procedures have been developed, and are presented in this book, for finding such solutions for certain kinds of problems. However, it needs to be recognized that these solutions are optimal only with respect to the model being used. Since the model necessarily is an idealized rather than an exact representation of the real problem, there cannot be any Utopian guarantee that the optimal solution for the model will prove to be the best possible solution that could have been implemented for the real problem. There just are too many imponderables and uncertainties associated with real problems. However, if the model is well formulated and tested, the resulting solution should tend to be a good approximation to the ideal course of action for the real problem. Therefore, rather than be deluded into demanding the impossible, the test of the practical success of an operations research study should be whether it provides a better guide for action than can be obtained by other means.

The eminent management scientist and Nobel Laureate in Economics, Herbert Simon, points out that **satisficing** is much more prevalent than optimizing in actual practice. In coining the term *satisficing* as a combination of the words *satisfactory* and *optimizing,* Simon is describing the tendency of managers to seek a solution that is "good enough" for the problem at hand. Rather than trying to develop an overall measure of performance to optimally reconcile conflicts between various desirable objectives (including well-established criteria for judging the performance of different segments of the organization), a more pragmatic approach may be used. Goals may be set to establish minimum satisfactory levels of performance in various areas, based perhaps on past levels of performance or on what the competition is achieving. If a solution is found that enables all of these goals to be met, it is likely to be adopted without further ado. Such is the nature of satisficing.

The distinction between optimizing and satisficing reflects the difference between theory and the realities frequently faced in trying to implement that theory in practice. In the words of one of England's OR leaders, Samuel Eilon, "optimizing is the science of the ultimate; satisficing is the art of the feasible."[1]

Operations research teams attempt to bring as much of the "science of the ultimate" as possible to the decision-making process. However, the successful team does so in full recognition of the overriding need of the decision maker to obtain a satisfactory guide for action in a reasonable period of time. Therefore, the goal of an operations research study should be to conduct the study in an optimal manner, regardless of whether this involves finding an optimal solution to the model or not. Thus, in addition to pursuing the "science of the ultimate," the team should also consider the cost of the study and the disadvantages of delaying its completion, and then attempt to maximize the net benefits resulting from the study. In recognition of this concept, operations research teams occasionally use only **heuristic procedures** (i.e., intuitively designed procedures that do not guarantee an optimal solution) to find a good **suboptimal solution.** This is most often the case when the time or cost required to find an optimal solution for an adequate model of the problem would be very large.

The discussion thus far has implied that an operations research study seeks to find only *one* solution, which may or may not be required to be optimal. In fact, this usually is not the case. An optimal solution for the original model may be far from ideal for the real problem. Therefore, **post-optimality analysis** is a very important part of most operations research studies.

In part, post-optimality analysis involves conducting **sensitivity analysis** to determine which parameters of the model are most critical (the "sensitive parameters") in determining the solution. Some or all of the parameters generally are an estimate of some quantity (e.g., unit profit) whose exact value will become known only after the solution has been implemented. Therefore, after identifying the sensitive parameters, special attention is given to estimating each one more closely, or at least its

[1] Eilon, Samuel: "Goals and Constraints in Decision-making," *Operational Research Quarterly,* **23**:3–15, 1972; address given at the 1971 Annual Conference of the Canadian Operational Research Society.

range of likely values. One then seeks a solution that remains a particularly good one for all of the various combinations of likely values of the sensitive parameters.

In some cases, certain parameters of the model represent policy decisions (e.g., resource allocations). If so, there frequently is some flexibility in the values assigned to these parameters. Perhaps some can be increased by decreasing others. Post-optimality analysis includes the investigation of such trade-offs.

In conjunction with the study phase discussed in the next section (testing the model and the solution), post-optimality analysis also involves obtaining a sequence of solutions that comprises a series of improving approximations to the ideal course of action. Thus the apparent weaknesses in the initial solution are used to suggest improvements in the model, its input data, and perhaps the solution procedure. A new solution is then obtained, and the cycle is repeated. This process continues until the improvements in the succeeding solutions become too small to warrant continuation. Even then, a number of alternative solutions (perhaps solutions that are optimal for one of several plausible versions of the model and its input data) may be presented to management for the final selection. As suggested in Sec. 2.1, this presentation of alternative solutions would normally be done whenever the final choice among these alternatives should be based on considerations that are best left to the judgment of management.

Ways in which the model and its solution are evaluated and improved will be discussed in the next section.

2.4 Testing the Model and the Solution

One of the first lessons of operations research is that it is generally not sufficient to rely solely on one's intuition. This caution applies not only in obtaining a solution to a problem but also in evaluating the model that has been formulated to represent this problem. As indicated in Sec. 2.2, the proper criterion for judging the validity of a model is whether or not it predicts the *relative effects* of the alternative courses of action with sufficient accuracy to permit a sound decision. No matter how plausible the model may appear to be, it should not be accepted on faith that this condition is satisfied. Given the difficulty of communicating and understanding all the aspects and subtleties of a complex operational problem, there is a distinct possibility that the operations research team either has not been given all the true facts of the situation or has not interpreted them properly. For example, an important factor or interrelationship may not have been incorporated into the model or perhaps certain parameters have not been estimated accurately.

Before undertaking more elaborate tests, check for obvious errors or oversights in the model. Reexamining the formulation of the problem and comparing it with the model may help to reveal any such mistakes. Another useful check is to make sure that all the mathematical expressions are *dimensionally consistent* in the units they use. Additional insight into the validity of the model can sometimes be obtained by varying the parameters and/or the decision variables and checking to see whether the output from the model behaves in a plausible manner. This is often especially revealing when the parameters or variables are assigned extreme values near their maxima or minima.

A more systematic approach to testing the model is to use a **retrospective test.** When it is applicable, this test involves using historical data to reconstruct the past and then determining how well the model and the resulting solution would have performed if it had been used. Comparing the effectiveness of this hypothetical performance with what actually happened then indicates whether using this model tends to yield a significant improvement over current practice. It may also indicate areas where the model has shortcomings and requires modifications. Furthermore, by using alternative solutions from the model and determining their hypothetical historical performances, considerable evidence can be gathered regarding how well the model predicts the relative effects of alternative courses of action.

On the other hand, a disadvantage of retrospective testing is that it uses the same data that guided the formulation of the model. The crucial question is whether or not the past is truly representative of the future. If it is not, then the model might perform quite differently in the future than it would have in the past.

To circumvent this disadvantage of retrospective testing, it is sometimes useful to continue the status quo temporarily. This provides new data that were not available when the model was constructed. These data are then used in the same ways as those described here to evaluate the model.

If the final model is to be used repeatedly, it is important to continue checking the model and its solution after the initial implementation to make sure that they remain valid as conditions evolve over time. The establishment of such controls is the subject of the next section.

2.5 Establishing Control Over the Solution

What happens after the testing phase has been completed and an acceptable model has been developed? If the model is to be used repeatedly, the next step is to install a well-documented *system* for applying the model. This system would include the model, the solution procedure (including post-optimality analysis), and operating procedures for implementation. Then, even as personnel changes, the system can be called on at regular intervals to provide a specific numerical solution.

It is evident that this solution remains valid for the real problem only as long as this specific model remains valid. However, conditions are constantly changing in the real world. Therefore, changes might well occur that would invalidate this model; e.g., the values of the parameters might change significantly. If these values should change, it is vital that the change be detected as soon as possible so that the model, its solution, and the resulting course of action can be modified accordingly. A plan for detecting such changes and making the needed modifications should be part of the system for applying the model.

This plan will include provision for maintaining a general surveillance of the situation. In addition, it is often worthwhile to establish *systematic procedures* for detecting change and controlling the solution. To do this, it is necessary to identify the *sensitive parameters* of the model by sensitivity analysis, as discussed in Sec. 2.3. Next, a procedure is established for detecting *statistically significant changes* in each of these sensitive parameters. This procedure can sometimes be established by the process control charts used in statistical quality control. Finally, provision is made for adjusting the solution and consequent course of action whenever such a change is detected.

2.6 Implementation

The last phase of an operations research study is to implement the final solution as approved by the decision maker. This phase is a critical one because it is here, and only here, that the benefits of the study are reaped. Therefore, it is important for the operations research team to participate in launching this phase, both to make sure that the solution is accurately translated into an operating procedure and to rectify any flaws in the solution that are then uncovered.

The success of the implementation phase depends a great deal upon the support of both top management and operating management. Consequently, as mentioned in Secs. 2.1 and 2.4, the operations research team should encourage the active participation of management in formulating the problem and evaluating the solution. Obtaining the guidance of management is valuable in its own right for identifying relevant special considerations and thereby avoiding potential pitfalls during these phases. However, making management a party to the study also serves to enlist their active support for its implementation.

The implementation phase involves several steps. First, the operations research team gives operating management a careful explanation of the solution to be adopted and how it relates to operating realities. Next, these two parties share the responsibility for developing the procedures required to put this solution into operation. Operating management then sees that a detailed indoctrination is given to the personnel involved, and the new course of action is initiated. If successful, the model and the solution procedure may be used periodically to provide guidance to management. With this in mind, the operations research team monitors the initial experience with the course of action taken and seeks to identify any modifications that should be made in the future.

Upon culminating a study, it is appropriate for the operations research team to document its methodology clearly and accurately enough so that the work is *reproducible*. *Replicability* should be part of the professional ethical code of the operations researcher. This condition is especially crucial when controversial public policy issues are being studied.

2.7 Conclusions

Although the remainder of this book focuses primarily on *constructing* and *solving* mathematical stochastic models, we have tried to emphasize in the present chapter that this constitutes only a portion of the overall process involved in conducting a typical operations research study. The other phases described here also are very important to the success of the study. Try to keep in perspective the role of the model and the solution procedure in the overall process as you move through the subsequent chapters. Then, after gaining a deeper understanding of mathematical models, we suggest that you plan to return to review this chapter again in order to further sharpen this perspective.

Many of the phases discussed in this chapter entail the use of *software tools,* including *decision support systems*. Operations research is closely intertwined with the use of computers. Until recently, these have been almost exclusively mainframe computers, but now microcomputers also are being widely used for dealing with smaller problems.

In concluding this discussion of the major phases of an operations research study, it should be emphasized that there are many exceptions to the "rules" prescribed in this chapter. By its very nature, operations research requires considerable ingenuity and innovation, so it is impossible to write down any standard procedure that should always be followed by operations research teams. Rather, the preceding description may be viewed as a model that roughly represents how successful operations research studies are conducted.

SELECTED REFERENCES

1. Ackoff, Russell L., and Patrick Rivett: *A Manager's Guide to Operations Research,* Wiley, New York, 1963.
2. Ackoff, Russell L., and Maurice W. Sasieni: *Fundamentals of Operations Research,* Wiley, New York, 1968.
3. Churchman, C. West, Russell L. Ackoff, and E. L. Arnoff: *Introduction to Operations Research,* Wiley, New York, 1957.
4. Huysmans, Jan H. B. M.: *The Implementation of Operations Research,* Wiley, New York, 1970.
5. Miller, David W., and Martin K. Starr: *Executive Decisions and Operations Research,* 2d ed., Prentice-Hall, Englewood Cliffs, N.J., 1969.
6. Morris, William T.: "On the Art of Modeling," *Management Science,* **13**:B707–717, 1967.
7. Williams, H. P.: *Model Building in Mathematical Programming,* 2d ed., Wiley, New York, 1985.

3

Probability Theory

3.1 Introduction

In decision-making problems, one is often faced with making decisions based upon phenomena that have uncertainty associated with them. This uncertainty is caused by inherent variation due to sources of variation that elude control or the inconsistency of natural phenomena. Rather than treat this variability qualitatively, one can incorporate it into the mathematical model and thus handle it quantitatively. This generally can be accomplished if the natural phenomena exhibit some degree of regularity, so that their variation can be described by a probability model. The ensuing sections are concerned with methods for characterizing these probability models.

3.2 Sample Space

Suppose the demand for a product over a period of time, say a month, is of interest. From a realistic point of view, demand is not generally constant but exhibits the type of variation alluded to in the previous section. Suppose an experiment that will result in observing the demand for the product during the month is run. Whereas the outcome of the experiment cannot be predicted exactly, each *possible* outcome can be described. The demand during the period can be any one of the values 0, 1, 2, . . . , that is, the entire set of nonnegative integers. The set of all possible outcomes of the experiment is called the *sample space* and will be denoted by Ω. Each outcome in the sample space is called a *point* and will be denoted by ω. Actually, in the experiment just described, the possible demands may be bounded from above by N, where N would represent the size of the population that has any use for the product. Hence, the sample space would then consist of the set of the integers 0, 1, 2, . . . , N. Strictly speaking, the sample space is much more complex than just described. In fact, it may be extremely difficult to characterize precisely. Associated with this experiment are such factors as the dates and times that the demands occur, the prevailing weather, the disposition of the personnel meeting the demand, and so on. Many more factors could be listed, most of which are irrelevant. Fortunately, as noted in the next section, it is not necessary to describe completely the sample space, but only to record those factors that appear to be necessary for the purpose of the experiment.

Another experiment may be concerned with the time until the first customer arrives at a store. Since the first customer may arrive at any time until the store closes (assuming an 8-hour day), for the purpose of this experiment the sample space can be considered to be all points on the real line between 0 and 8 hours. Thus Ω consists of all points ω such that

$$0 \leq \omega \leq 8.^{\dagger}$$

Now consider a third example. Suppose that a modification of the first experiment is made by observing the demands during the first 2 months. The sample space Ω consists of all points (x_1, x_2), where x_1 represents the demand during the first month, $x_1 = 0, 1, 2, \ldots$, and x_2 represents the demand during the second month, $x_2 = 0, 1, 2, \ldots$. Thus Ω consists of the set of all possible points ω, where ω represents a pair of nonnegative integer values (x_1, x_2). The point $\omega = (3, 6)$ represents a possible outcome of the experiment where the demand in the first month is 3 units and the demand in the second month is 6 units. In a similar manner, the experiment can be extended to observing the demands during the first n months. In this situation Ω consists of all possible points $\omega = (x_1, x_2, \ldots, x_n)$, where x_i represents the demand during the ith month.

The experiment that is concerned with the time until the first arrival appears can also be modified. Suppose an experiment that measures the times of the arrival of the first customer on each of 2 days is performed. The set of all possible outcomes of the experiment Ω consists of all points (x_1, x_2), $0 \leq x_1, x_2 \leq 8$, where x_1 represents the time the first customer arrives on the first day, and x_2 represents the time the first customer arrives on the second day. Thus, Ω consists of the set of all possible points ω, where ω represents a point in two space lying in the square shown in Fig. 3.1.

This experiment can also be extended to observing the times of the arrival of the first customer on each of n days. The sample space Ω consists of all points $\omega = (x_1, x_2, \ldots, x_n)$, such that $0 \leq x_i \leq 8$ $(i = 1, 2, \ldots, n)$, where x_i represents the time the first customer arrives on the ith day.

An event is defined as a set of outcomes of the experiment. Thus, there are many events that can be of interest. For example, in the experiment concerned with

† It will be assumed that at least one customer arrives each day.

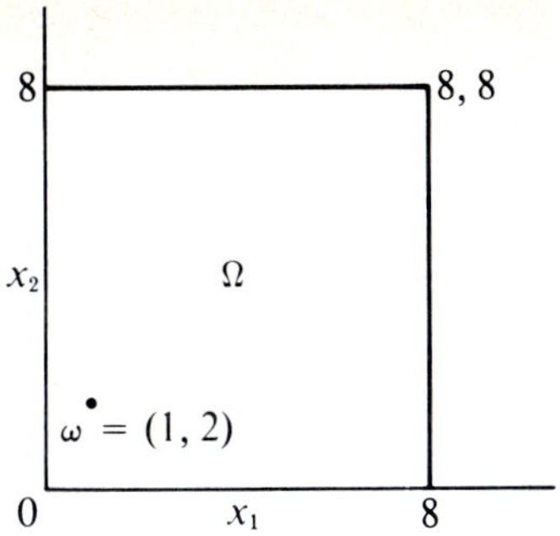

Figure 3.1 Sample space of the arrival-time experiment.

observation of the demand for a product in a given month, the set $\{\omega = 0, \omega = 1, \omega = 2, \ldots, \omega = 10\}$ is the event that the demand for the product does not exceed 10 units. Similarly, the set $\{\omega = 0\}$ denotes the event of no demand for the product during the month. In the experiment which measures the times of the arrival of the first customer on each of 2 days, the set $\{\omega = (x_1, x_2); x_1 < 1, x_2 < 1\}$, is the event that the first arrival on each day occurs before the first hour. It is evident that any subset of the sample space—for example, any point, collection of points, or the entire sample space—is an event.

Events may be combined, thereby resulting in the formation of new events. For any two events E_1 and E_2, the new event $E_1 \cup E_2$, referred to as the union of E_1 and E_2, is defined to contain all points in the sample space that are in either E_1 or E_2, or in both E_1 and E_2. Thus, the event $E_1 \cup E_2$ will occur if either E_1 or E_2 occurs. For example, in the demand experiment, let E_1 be the event of a demand in a single month of zero or 1 unit, and let E_2 be the event of a demand in a single month of 1 or 2 units. The event $E_1 \cup E_2$ is just $\{\omega = 0, \omega = 1, \omega = 2\}$, which is just the event of a demand of 0, 1, or 2 units.

The intersection of two events E_1 and E_2 is denoted by $E_1 \cap E_2$ (or equivalently by E_1E_2). This new event $E_1 \cap E_2$ is defined to contain all points in the sample space that are in both E_1 and E_2. Thus, the event $E_1 \cap E_2$ will occur only if both E_1 and E_2 occur. In the aforementioned example, the event $E_1 \cap E_2$ is $\{\omega = 1\}$, which is just the event of a demand of 1 unit.

Finally, the events E_1 and E_2 are said to be mutually exclusive (or disjoint) if their intersection does not contain any points. In the example, E_1 and E_2 are not disjoint. However, if the event E_3 is defined to be the event of a demand of 2 or 3 units, then $E_1 \cap E_3$ is disjoint. Events that do not contain any points, and therefore cannot occur, are called *null events*. (Of course, all these definitions can be extended to any finite number of events.)

3.3 Random Variables

It may occur frequently that in performing an experiment one is not interested directly in the entire sample space or in events defined over the sample space. For example, suppose that the experiment which measures the times of the first arrival on 2 days was performed to determine at what time to open the store. Prior to performing the experiment, the store owner decides that if the average of the arrival times is greater than an hour, thereafter he will not open his store until 10 A.M. (9 A.M. being the previous opening time). The average of x_1 and x_2 (the two arrival times) is not a point in the sample space, and hence he cannot make his decision by looking directly at the outcome of his experiment. Instead, he makes his decision according to the results of a rule that assigns the average of x_1 and x_2 to *each point* (x_1, x_2) in Ω. This resultant set is then partitioned into two parts: those points below 1 and those above 1. If the observed result of this rule (the average of the two arrival times) lies in the partition with points greater than 1, the store will be opened at 10 A.M.; otherwise, the store will continue to open at 9 A.M. The rule that assigns the average of x_1 and x_2 to each point in the sample space is called a random variable. Thus, a *random variable is a numerically valued function defined over the sample space*. Note that a function is, in a mathematical sense, just a rule that assigns a number to each value in the domain of definition, in this context the sample space.

Random variables play an extremely important role in probability theory. Experiments are usually very complex and contain information that may or may not be superfluous. For example, in measuring the arrival time of the first customer, the color of her shoes may be pertinent. Although this is unlikely, the prevailing weather may certainly be relevant. Hence, the choice of the random variable enables experimenters to describe the factors of importance to them and permits them to discard the superfluous characteristics that may be extremely difficult to characterize.

There is a multitude of random variables associated with each experiment. In the experiment concerning the arrival of the first customer on each of 2 days, it has been pointed out already that the average of the arrival times $\overline{X}$ is a random variable. Notationally, random variable will be characterized by capital letters, and the values the random variable takes on will be denoted by lowercase letters. Actually, to be precise, $\overline{X}$ should be written as $\overline{X}(\omega)$, where ω is any point shown in the square in Fig. 3.1 because $\overline{X}$ is a function. Thus, $\overline{X}(1, 2) = (1 + 2)/2 = 1.5$, $\overline{X}(1.6, 1.8) = (1.6 + 1.8)/2 = 1.7$, $\overline{X}(1.5, 1.5) = (1.5 + 1.5)/2 = 1.5$, $\overline{X}(8, 8) = (8 + 8)/2 = 8$. The values that the random variable $\overline{X}$ takes on are the set of values $\bar{x}$ such that $0 \leq \bar{x} \leq 8$. Another random variable, X_1, can be described as follows: For each ω in Ω, the random variable (the numerically valued function) disregards the x_2 coordinate and transforms the x_1 coordinate into itself. This random variable, then, represents the arrival time of the first customer on the first day. Hence, $X_1(1, 2) = 1$, $X_1(1.6, 1.8) = 1.6$, $X_1(1.5, 1.5) = 1.5$, $X_1(8, 8) = 8$. The values the random variable X_1 takes on are the set of values x_1 such that $0 \leq x_1 \leq 8$. In a similar manner, the random variable X_2 can be described as representing the arrival time of the first customer on the second day. A third random variable, S^2, can be described as follows: For each ω in Ω, the random variable computes the sum of squares of the deviations of the coordinates about their average; that is, $S^2(\omega) = S^2(x_1, x_2) = (x_1 - \bar{x})^2 + (x_2 - \bar{x})^2$. Hence, $S^2(1, 2) = (1 - 1.5)^2 + (2 - 1.5)^2 = 0.5$, $S^2(1.6, 1.8) = (1.6 - 1.7)^2 + (1.8 - 1.7)^2 = 0.02$, $S^2(1.5, 1.5) = (1.5 - 1.5)^2 + (1.5 - 1.5)^2 = 0$, $S^2(8, 8) = (8 - 8)^2 + (8 - 8)^2 = 0$. It is evident that the values the random variable S^2 takes on are the set of values s^2 such that $0 \leq s^2 \leq 32$.

All the random variables just described are called *continuous random variables* because they take on a continuum of values. Discrete random variables are those that take on a finite or countably infinite set of values.[1] An example of a discrete random

[1] A countably infinite set of values is a set whose elements can be put into one-to-one correspondence with the set of positive integers. The set of odd integers is countably infinite. The 1 can be paired with 1, 3 with 2, 5 with 3, . . . , $2n - 1$ with n. The set of all real numbers between 0 and $\frac{1}{2}$ is not countably infinite because there are too many numbers in the interval to pair with the integers.

variable can be obtained by referring to the experiment dealing with the measurement of demand. Let the discrete random variable X be defined as the demand during the month. (The experiment consists of measuring the demand for 1 month.) Thus $X(0) = 0$, $X(1) = 1$, $X(2) = 2, \ldots$, so that the random variable takes on the set of values consisting of the integers. Note that Ω and the set of values the random variable takes on are identical, so that this random variable is just the identity function.

From the above paragraphs it is evident that any function of a random variable is itself a random variable because a function of a function is also a function. Thus, in the previous examples $\bar{X} = (X_1 + X_2)/2$ and $S_2 = (X_1 - \bar{X})^2 + (X_2 - \bar{X})^2$ can also be recognized as random variables by noting that they are functions of the random variables X_1 and X_2.

This text is concerned with random variables that are real-valued functions defined over the real line or a subset of the real line.

3.4 Probability and Probability Distributions

Returning to the example of the demand for a product during a month, note that the actual demand is not a constant; instead, it can be expected to exhibit some "variation." In particular, this variation can be described by introducing the concept of probability defined over events in the sample space. For example, let E be the event $\{\omega = 0, \omega = 1, \omega = 2, \ldots, \omega = 10\}$. Then intuitively one can speak of $P\{E\}$, where $P\{E\}$ is referred to as the probability of having a demand of 10 or less units. Note that $P\{E\}$ can be thought of as a numerical value associated with the event E. If $P\{E\}$ is known for all events E in the sample space, then some "information" is available about the demand that can be expected to occur. Usually these numerical values are difficult to obtain, but nevertheless their existence can be postulated. To define the concept of probability rigorously is beyond the scope of this text. However, for most purposes it is sufficient to postulate the existence of numerical values $P\{E\}$ associated with events E in the sample space. The value $P\{E\}$ is the probability of the occurrence of the event E. Furthermore, it will be assumed that $P\{E\}$ satisfies the following reasonable properties:

1. $0 \le P\{E\} \le 1$. This implies that the probability of an event is always nonnegative and can never exceed 1.
2. If E_0 is an event that cannot occur (a null event) in the sample space, then $P\{E_0\} = 0$. Let E_0 denote the event of obtaining a demand of -7 units. Then $P\{E_0\} = 0$.
3. $P\{\Omega\} = 1$. If the event consists of obtaining a demand between 0 and N, that is, the entire sample space, the probability of having some demand between 0 and N is certain.
4. If E_1 and E_2 are disjoint (mutually exclusive) events in Ω, then $P\{E_1 \cup E_2\} = P\{E_1\} + P\{E_2\}$. Thus, if E_1 is the event of 0 or 1, and E_2 is the event of a demand of 4 or 5, then the probability of having a demand of 0, 1, 4, or 5, that is, $\{E_1 \cup E_2\}$, is given by $P\{E_1\} + P\{E_2\}$.

Although these properties are rather formal, they do conform to one's intuitive notion about probability. Nevertheless, these properties cannot be used to obtain values for $P\{E\}$. Occasionally the determination of exact values, or at least approximate values, is desirable. Approximate values, together with an interpretation of probability, can be obtained through a frequency interpretation of probability. This may be stated precisely as follows. Denote by n the number of times an experiment is performed and by m the number of successful occurrences of the event E in the n trials. Then $P\{E\}$ can be interpreted as

$$P\{E\} = \lim_{n\to\infty} \frac{m}{n},$$

assuming the limit exists for such a phenomena. The ratio m/n can be used to approximate $P\{E\}$. Furthermore, m/n satisfies the properties required of probabilities; that is,

1. $0 \le m/n \le 1$.
2. $0/n = 0$. (If the event E cannot occur, then $m = 0$.)
3. $n/n = 1$. (If the event E must occur every time the experiment is performed, then $m = n$.)
4. $(m_1 + m_2)/n = m_1/n + m_2/n$ if E_1 and E_2 are disjoint events. (If the event E_1 occurs m_1 times in the n trials and the event E_2 occurs m_2 times in the n trials, and E_1 and E_2 are disjoint, then the total number of successful occurrences of the event E_1 *or* E_2 is just $m_1 + m_2$.)

Since these properties are true for a finite n, it is reasonable to expect them to be true for

$$P\{E\} = \lim_{n\to\infty} \frac{m}{n}.$$

The trouble with the frequency interpretation as a definition of probability is that it is not possible to actually determine the probability of an event E because the question "How large must n be?" cannot be answered. Furthermore, such a definition does not permit a logical development of the theory of probability. However, a rigorous definition of probability, or finding methods for determining exact probabilities of events, is not of prime importance here.

The existence of probabilities, defined over events E in the sample space, has been described, and the concept of a random variable has been introduced. Finding the relation between probabilities associated with events in the sample space and "probabilities" associated with random variables is a topic of considerable interest.

Associated with every random variable is a *cumulative distribution function* (*CDF*). To define a CDF it is necessary to introduce some additional notation. Define the symbol $E_b^X = \{\omega | X(\omega) \le b\}$ (or equivalently, $\{X \le b\}$) as the set of outcomes ω in the sample space forming the event E_b^X such that the random variable X takes on values less than or equal to b.[1] Then $P\{E_b^X\}$ is just the probability of this event. Note that this probability is well defined because E_b^X is an event in the sample space, and this event depends upon both the random variable that is of interest and the value of b chosen. For example, suppose the experiment that measures the demand for a product during a month is performed. Let $N = 99$, and assume that the events $\{0\}$, $\{1\}$, $\{2\}$, . . . , $\{99\}$ each has probability equal to $\frac{1}{100}$; that is, $P\{0\} = P\{1\} = P\{2\} = \cdots = P\{99\} = \frac{1}{100}$. Let the random variable X be the square of the demand, and choose b equal to 150. Then

[1] The notation $\{X \le b\}$ suppresses the fact that this is really an event in the sample space. However, it is simpler to write, and the reader is cautioned to interpret it properly, that is, as the set of outcomes ω in the sample space, $\{\omega | X(\omega) \le b\}$.

$$E_{150}^X = \{\omega | X(\omega) \leq 150\} = \{X \leq 150\}$$

is the set $E_{150}^X = \{0, 1, 2, 3, 4, 5, 6, 7, 8, 9, 10, 11, 12\}$ (since the square of each of these numbers is less than 150). Furthermore,

$$P\{E_{150}^X\} = \tfrac{1}{100} + \tfrac{1}{100} + \tfrac{1}{100} + \tfrac{1}{100} + \tfrac{1}{100} + \tfrac{1}{100} + \tfrac{1}{100} + \tfrac{1}{100} + \tfrac{1}{100} + \tfrac{1}{100} + \tfrac{1}{100} + \tfrac{1}{100} + \tfrac{1}{100} = \tfrac{13}{100}$$

Thus $P\{E_{150}^X\} = P\{X \leq 150\} = \tfrac{13}{100}$.

For a given random variable X, $P\{X \leq b\}$, denoted by $F_X(b)$, is called the *CDF of the random variable* X and is defined for all real values of b. Where there is no ambiguity, the CDF will be denoted by $F(b)$; that is,

$$F(b) = F_X(b) = P\{E_b^X\} = P\{\omega | X(\omega) \leq b\} = P\{X \leq b\}.$$

Although $P\{X \leq b\}$ is defined through the event E_b^X in the sample space, it will often be read as the "probability" that the random variable X takes on a value less than or equal to b. The reader should interpret this statement properly, that is, in terms of the event E_b^X.

As mentioned, each random variable has a cumulative distribution function associated with it. This is not an arbitrary function but is induced by the probabilities associated with events of the form E_b^X defined over the sample space Ω. Furthermore, the CDF of a random variable is a numerically valued function defined for all b, $-\infty \leq b \leq \infty$, having the following properties:

1. $F_X(b)$ is a nondecreasing function of b,
2. $\lim_{b \to -\infty} F_X(b) = F_X(-\infty) = 0$,
3. $\lim_{b \to +\infty} F_X(b) = F_X(+\infty) = 1$.

The CDF is a versatile function. Events of the form

$$\{\omega | a < X(\omega) \leq b\},$$

that is, the set of outcomes ω in the sample space such that the random variable X takes on values greater than a but not exceeding b, can be expressed in terms of events of the form E_b^X. In particular, E_b^X can be expressed as the union of two disjoint sets; that is,

$$E_b^X = E_a^X \cup \{\omega | a < X(\omega) \leq b\}.$$

Thus, $P\{\omega | a < X(\omega) \leq b\} = P\{a < X \leq b\}$ can easily be seen to be

$$F_X(b) - F_X(a).$$

As another example, consider the experiment that measures the times of the arrival of the first customer on each of 2 days. Ω consists of all points (x_1, x_2) such that $0 \leq x_1, x_2 \leq 8$, where x_1 represents the times the first customer arrives on the first day, and x_2 represents the time the first customer arrives on the second day. Consider all events associated with this experiment, and assume that the probabilities of such events can be obtained. Suppose $\overline{X}$, the average of the two arrival times, is chosen as the random variable of interest and that $E_b^{\overline{X}}$ is the set of outcomes ω in the sample space forming the event $E_b^{\overline{X}}$ such that $\overline{X} \leq b$. Hence $F_{\overline{X}}(b) = P\{E_b^{\overline{X}}\} = P\{\overline{X} \leq b\}$. To illustrate how this can be evaluated, suppose that $b = 4$ hours. All the values of x_1, x_2 are sought such that $(x_1 + x_2)/2 \leq 4$ or $x_1 + x_2 \leq 8$. This is shown

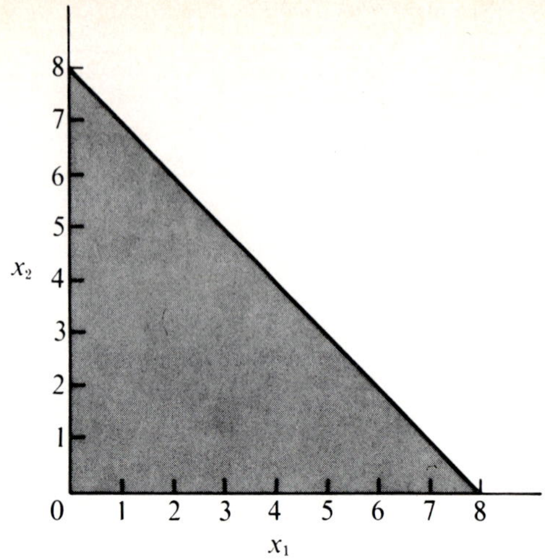

Figure 3.2 Shaded area represents the event $E_b^{\bar{x}} = \{\bar{X} \leq 4\}$.

by the shaded area in Fig. 3.2. Hence, $F_{\bar{X}}(b)$ is just the probability of a successful occurrence of the event given by the shaded area in Fig. 3.2. Presumably $F_{\bar{X}}(b)$ can be evaluated if probabilities of such events in the sample space are known.

Another random variable associated with this experiment is X_1, the time of the arrival of the first customer on the first day. Thus, $F_{X_1}(b) = P\{X_1 \leq b\}$, which can be obtained simply if probabilities of events over the sample space are given.

There is a simple frequency interpretation for the cumulative distribution function of a random variable. Suppose an experiment is repeated n times, and the random variable X is observed each time. Denote by $x_1, x_2, \ldots, x_n$ the outcomes of these n trials. Order these outcomes, letting $x_{(1)}$ be the smallest observation, $x_{(2)}$ the second smallest, . . . , $x_{(n)}$ the largest. Plot the following step function $F_n(x)$:

$$\text{For } x < x_{(1)}, \qquad \text{let } F_n(x) = 0.$$

$$\text{For } x_{(1)} \leq x < x_{(2)}, \qquad \text{let } F_n(x) = \frac{1}{n}.$$

$$\text{For } x_{(2)} \leq x < x_{(3)}, \qquad \text{let } F_n(x) = \frac{2}{n}.$$

$$\vdots$$

$$\text{For } x_{(n-1)} \leq x < x_{(n)}, \text{ let } F_n(x) = \frac{n-1}{n}.$$

$$\text{For } x \geq x_{(n)}, \qquad \text{let } F_n(x) = \frac{n}{n} = 1.$$

Such a plot is given in Fig. 3.3 and is seen to "jump" at the values that the random variable takes on.

$F_n(x)$ can be interpreted as the fraction of outcomes of the experiment less than or equal to x, and is called the *sample CDF*. It can be shown that as the number of

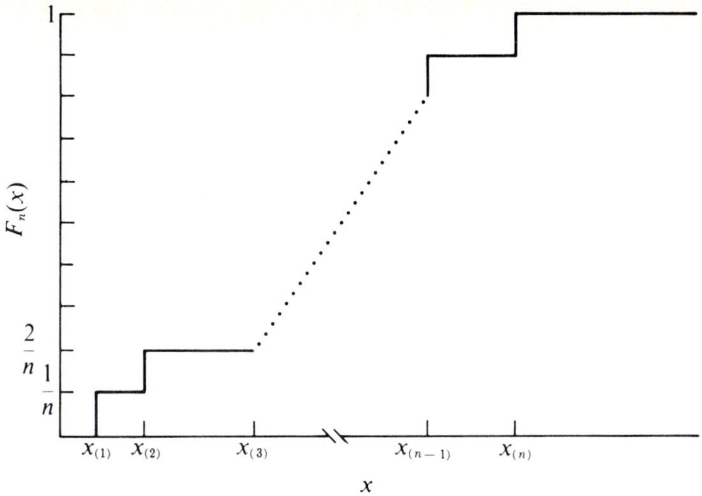

Figure 3.3 Sample cumulative distribution function.

repetitions n of the experiment gets large, the sample CDF approaches the CDF of the random variable X.

In most problems encountered in practice, one is not concerned with events in the sample space and their associated probabilities. Instead, interest is focused on random variables and their associated cumulative distribution functions. Generally, a random variable (or random variables) is chosen, and some assumption is made about the form of the CDF or about the random variable. For example, the random variable X_1, the time of the first arrival on the first day, may be of interest, and an assumption may be made that the form of its CDF is exponential. Similarly, the same assumption about X_2, the time of the first arrival on the second day, may also be made. If these assumptions are valid, then the CDF of the random variable $\overline{X} = (X_1 + X_2)/2$ can be derived. Of course, these assumptions about the form of the CDF are not arbitrary and really imply assumptions about probabilities associated with events in the sample space. One hopes that they can be substantiated by either empirical evidence or theoretical considerations.

3.5 Conditional Probability and Independent Events

Often experiments are performed so that some results are obtained early in time and some later in time. This is the case, for example, when the experiment consists of measuring the demand for a product during each of 2 months; the demand during the first month is observed at the end of the first month. Similarly, the arrival times of the first two customers on each of 2 days are observed sequentially in time. This early information can be useful in making predictions about the subsequent results of the experiment. Such information need not necessarily be associated with time. If the demand for two products during a month is investigated, knowing the demand of one may be useful in assessing the demand for the other. So that this information can be utilized, the concept of "conditional probability," defined over events occurring in the sample space, is introduced.

Consider two events in the sample space E_1 and E_2, where E_1 represents the event that has occurred, and E_2 represents the event whose occurrence or nonoccurrence is of interest. Furthermore, assume that $P\{E_1\} > 0$. The conditional probability of the occurrence of the events E_2, given that the event E_1 has occurred, $P\{E_2|E_1\}$, is defined to be

$$P\{E_2|E_1\} = \frac{P\{E_1 \cap E_2\}}{P\{E_1\}},$$

where $\{E_1 \cap E_2\}$ represents the event consisting of all points ω in the sample space common to *both* E_1 and E_2. For example, consider the experiment that consists of observing the demand for a product over each of 2 months. Suppose the sample space Ω consists of all points $\omega = (x_1, x_2)$, where x_1 represents the demand during the first month, and x_2 represents the demand during the second month, $x_1, x_2 = 0, 1, 2, \ldots, 99$. Furthermore, it is known that the demand during the first month has been 10. Hence the event E_1, which consists of the points $(10, 0), (10, 1), (10, 2), \ldots, (10, 99)$, has occurred. Consider the event E_2, which represents a demand for the product in the second month that does not exceed 1 unit. This event consists of the points $(0, 0), (1, 0), (2, 0), \ldots, (10, 0), \ldots, (99, 0), (0, 1), (1, 1), (2, 1), \ldots, (10, 1), \ldots, (99, 1)$. The event $\{E_1 \cap E_2\}$ consists of the points $(10, 0)$ and $(10, 1)$. Hence the probability of a demand which does not exceed 1 unit in the second month, given that a demand of 10 units occurred during the first month, that is, $P\{E_2|E_1\}$, is given by

$$\begin{aligned} P\{E_2|E_1\} &= \frac{P\{E_1 \cap E_2\}}{P\{E_1\}} \\ &= \frac{P\{\omega = (10, 0), \omega = (10, 1)\}}{P\{\omega = (10, 0), \omega = (10, 1), \ldots, \omega = (10, 99)\}}. \end{aligned}$$

The definition of conditional probability can be given a frequency interpretation. Denote by n the number of times an experiment is performed, and let n_1 be the number of times the event E_1 has occurred. Let n_{12} be the number of times that the event $\{E_1 \cap E_2\}$ has occurred in the n trials. The ratio n_{12}/n_1 is the proportion of times that the event E_2 occurs when E_1 has also occurred; that is, n_{12}/n_1 is the conditional relative frequency of E_2, given that E_1 has occurred. This relative frequency n_{12}/n_1 is then equivalent to $(n_{12}/n)/(n_1/n)$. Using the frequency interpretation of probability for large n, n_{12}/n is approximately $P\{E_1 \cap E_2\}$, and n_1/n is approximately $P\{E_1\}$, so that the conditional relative frequency of E_2, given E_1, is approximately $P\{E_1 \cap E_2\}/P\{E_1\}$.

In essence, if one is interested in conditional probability, he or she is working with a reduced sample space, that is, from Ω to E_1, modifying other events accordingly. Also note that conditional probability has the four properties described in Sec. 3.4; that is,

1. $0 \le P\{E_2|E_1\} \le 1$.
2. If E_2 is an event that cannot occur, then $P\{E_2|E_1\} = 0$.
3. If the event E_2 is the entire sample space Ω, then $P\{\Omega|E_1\} = 1$.
4. If E_2 and E_3 are disjoint events in Ω, then

$$P\{(E_2 \cup E_3)|E_1\} = P\{E_2|E_1\} + P\{E_3|E_1\}.$$

In a similar manner, the conditional probability of the occurrence of the event E_1, given that the event E_2 has occurred, can be defined. If $P\{E_2\} > 0$, then

$$P\{E_1|E_2\} = \frac{P\{E_1 \cap E_2\}}{P\{E_2\}}.$$

The concept of conditional probability was introduced so that advantage could be taken of information about the occurrence or nonoccurrence of events. It is conceivable that information about the occurrence of the event E_1 yields no information about the occurrence or nonoccurrence of the event E_2. If $P\{E_2|E_1\} = P\{E_2\}$, or $P\{E_1|E_2\} = P\{E_1\}$, then E_1 and E_2 are said to be independent events. It then follows that if E_1 and E_2 are independent and $P\{E_1\} > 0$, then $P\{E_2|E_1\} = P\{E_1 \cap E_2\}/P\{E_1\} = P\{E_2\}$, so that $P\{E_1 \cap E_2\} = P\{E_1\}\, P\{E_2\}$. This can be taken as an alternative definition of independence of the events E_1 and E_2. It is usually difficult to show that events are independent by using the definitions of independence. Instead, it is generally simpler to use the information available about the experiment to postulate whether events are independent. This is usually based upon physical considerations. For example, if the demand for a product during a month is "known" *not* to affect the demand in subsequent months, then the events E_1 and E_2 defined previously can be said to be independent, in which case

$$\begin{aligned}
P\{E_2|E_1\} &= \frac{P\{E_1 \cap E_2\}}{P\{E_1\}} \\
&= \frac{P\{\omega = (10, 0),\ \omega = (10,1)\}}{P\{\omega = (10, 0),\ \omega = (10, 1),\ \ldots,\ \omega = (10, 99)\}}, \\
&= \frac{P\{E_1\}P\{E_2\}}{P\{E_1\}} = P\{E_2\}, \\
&= P\{\omega = (0, 0),\ \omega = (1, 0),\ \ldots,\ \omega = (99, 0),\ \omega = (0, 1), \\
&\qquad \omega = (1, 1),\ \ldots,\ \omega = (99, 1)\}.
\end{aligned}$$

The definition of independence can be extended to any number of events. E_1, $E_2, \ldots, E_n$ are said to be independent events if for *every* subset of these events $E_1^*, E_2^*, \ldots, E_k^*$,

$$P\{E_1^* \cap E_2^* \cap \cdots \cap E_k^*\} = P\{E_1^*\}P\{E_2^*\} \cdots P\{E_k^*\}.$$

Intuitively, this implies that knowledge of the occurrence of any of these events has no effect on the probability of occurrence of any other event.

3.6 Discrete Probability Distributions

It was pointed out that one is usually concerned with random variables and their associated probability distributions, and discrete random variables are those which take on a finite or countably infinite set of values. Furthermore, the CDF for a random variable is given by

$$F_X(b) = P\{\omega|X(\omega) \le b\}.$$

For a discrete random variable X, the event $\{\omega|X(\omega) \le b\}$ can be expressed as the union of disjoint sets; that is,

$$\{\omega|X(\omega) \leq b\} = \{\omega|X(\omega) = x_1\} \cup \{\omega|X(\omega) = x_2\} \cup \cdots \cup \{\omega|X(\omega) = x_{[b]}\},$$

where $x_{[b]}$ denotes the largest integer value of the x's less than or equal to b. It then follows that for the discrete random variable X, the CDF can be expressed as

$$\begin{aligned} F_X(b) &= P\{\omega|X(\omega) = x_1\} + P\{\omega|X(\omega) = x_2\} + \cdots + P\{\omega|X(\omega) = x_{[b]}\} \\ &= P\{X = x_1\} + P\{X = x_2\} + \cdots + P\{X = x_{[b]}\}. \end{aligned}$$

This last expression can also be expressed as

$$F_X(b) = \sum_{\text{all } k \leq b} P\{X = k\},$$

where k is an index that ranges over all the possible x values which the random variable X can take on.

Let $P_X(k)$ denote the probabilities $P\{X = k\}$, so that

$$F_X(b) = \sum_{\text{all } k \leq b} P_X(k).$$

The $P_X(k)$ are called the probability distribution of the discrete random variable X. When no ambiguity exists, $P_X(k)$ may be denoted by $P(k)$.

As an example, consider the discrete random variable that represents the demand for a product in a given month. Let $N = 99$. If it is assumed that $P_X(k) = P\{X = k\} = \frac{1}{100}$ for all $k = 0, 1, \ldots, 99$, then the CDF for this discrete random variable is given in Fig. 3.4. The probability distribution of this discrete random variable is shown in Fig. 3.5. Of course, the heights of the vertical lines in Fig. 3.5 are all equal because $P_X(0) = P_X(1) = P_X(2) = \cdots = P_X(99)$ in this case. For other random variables X, the $P_X(k)$ need not be equal, and hence the vertical lines will not be constant. In fact, all that is required for the $P_X(k)$ to form a probability distribution is that $P_X(k)$ be nonnegative and

$$\sum_{\text{all } k} P_X(k) = 1.$$

There are several important discrete probability distributions used in opera-

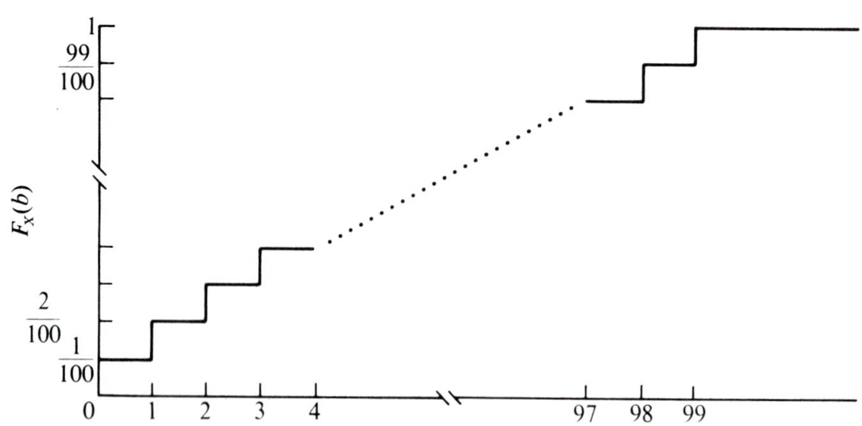

Figure 3.4 CDF of a discrete random variable.

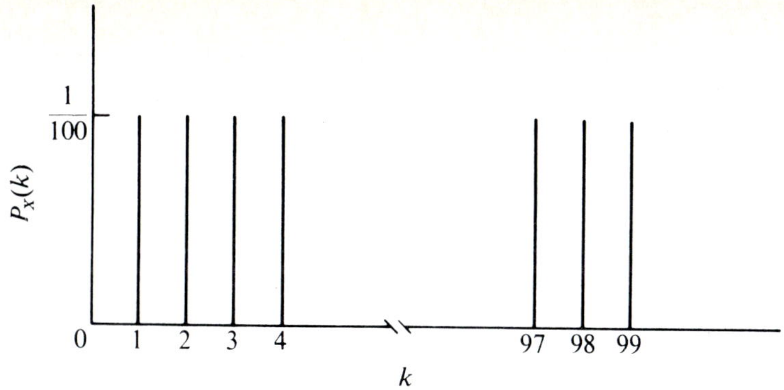

Figure 3.5 Probability distribution of a discrete random variable.

tions research work. The remainder of this section is devoted to a study of these distributions.

Binomial Distribution

A random variable X is said to have a binomial distribution if its probability distribution can be written as

$$P\{X = k\} = P_X(k) = \frac{n!}{k!(n-k)!}p^k(1-p)^{n-k},$$

where p is a constant lying between zero and 1, n is any positive integer, and k is also an integer such that $0 \leq k \leq n$. It is evident that $P_X(k)$ is always nonnegative, and it is easily proven that

$$\sum_{k=0}^{n} P_X(k) = 1.$$

Note that this distribution is a function of the two parameters n and p. The probability distribution of this random variable is shown in Fig. 3.6. An interesting interpretation of the binomial distribution is obtained when $n = 1$:

$$P\{X = 0\} = P_X(0) = 1 - p,$$

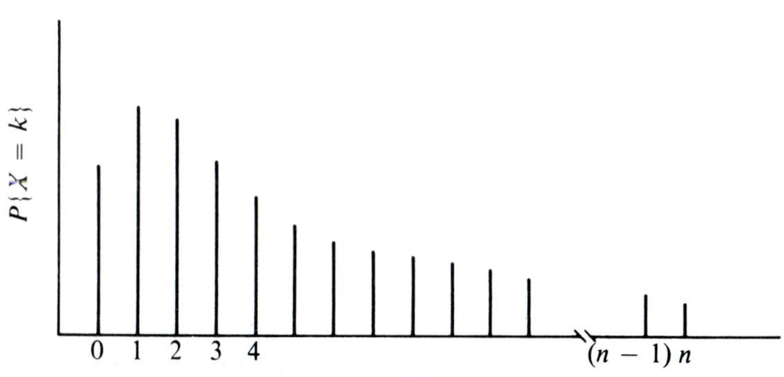

Figure 3.6 Binomial distribution with parameters n and p.

and

$$P\{X = 1\} = P_X(1) = p.$$

Such a random variable is said to have a Bernoulli distribution. Thus, if a random variable takes on two values, say, 0 or 1, with probability $1 - p$ or p, respectively, a Bernoulli random variable is obtained. The upturned face of a flipped coin is such an example: If a head is denoted by assigning it the number 0 and a tail by assigning it a 1, and if the coin is "fair" (the probability that a head will appear is $\frac{1}{2}$), the upturned face is a Bernoulli random variable with parameter $p = \frac{1}{2}$. Another example of a Bernoulli random variable is the quality of an item. If a defective item is denoted by 1 and a nondefective item by 0, if p represents the probability of an item being defective, and if $1 - p$ represents the probability of an item being nondefective, then the "quality" of an item (defective or nondefective) is a Bernoulli random variable.

If $X_1, X_2, \ldots, X_n$ are independent[1] Bernoulli random variables, each with parameter p, then it can be shown that the random variable

$$X = X_1 + X_2 + \cdots + X_n$$

is a binomial random variable with parameters n and p. Thus, if a fair coin is flipped 10 times, with the random variable X denoting the total number of tails (which is equivalent to $X_1 + X_2 + \cdots + X_{10}$), then X has a binomial distribution with parameters 10 and $\frac{1}{2}$, that is,

$$P\{X = k\} = \frac{10!}{k!(10 - k)!} \left(\frac{1}{2}\right)^k \left(\frac{1}{2}\right)^{10-k}.$$

Similarly, if the quality characteristics (defective or nondefective) of 50 items are independent Bernoulli random variables with parameter p, the total number of defective items in the 50 sampled, that is, $X = X_1 + X_2 + \cdots + X_{50}$, has a binomial distribution with parameters 50 and p, so that

$$P\{X = k\} = \frac{50!}{k!(50 - k)!} p^k(1 - p)^{50-k}.$$

Poisson Distribution

A random variable X is said to have a Poisson distribution if its probability distribution can be written as

$$P\{X = k\} = P_X(k) = \frac{\lambda^k e^{-\lambda}}{k!},$$

where λ is a positive constant (the parameter of this distribution), and k is any nonnegative integer. It is evident that $P_X(k)$ is nonnegative, and it is easily shown that

$$\sum_{k=0}^{\infty} \frac{\lambda^k e^{-\lambda}}{k!} = 1.$$

An example of a probability distribution of a Poisson random variable is shown in Fig. 3.7. (The CDF of the Poisson random variable is tabulated in Table A5.4 of Appendix 5.)

[1] The concept of independent random variables is introduced in Sec. 3.13. For the present purpose, random variables can be considered independent if their outcomes do not affect the outcomes of the other random variables.

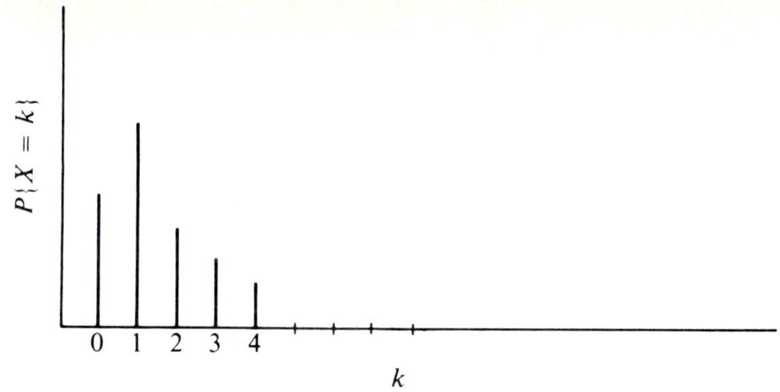

Figure 3.7 Poisson distribution.

The Poisson distribution is often used in operations research. Heuristically speaking, this distribution is appropriate in many situations where an "event" occurs over a period of time, like the arrival of a customer; when it is as likely that this "event" will occur in one interval as in any other; and also when the occurrence of an event has no effect on whether or not another occurs. Then the number of customer arrivals in a fixed time is often assumed to have a Poisson distribution. Similarly, the demand for a given product is also often assumed to have this distribution.

Geometric Distribution

A random variable X is said to have a geometric distribution if its probability distribution can be written as

$$P\{X = k\} = P_X(k) = p(1 - p)^{k-1},$$

where the parameter p is a constant lying between 0 and 1, and k takes on the values 1, 2, 3, ... It is clear that $P_X(k)$ is nonnegative, and it is easy to show that

$$\sum_{k=1}^{\infty} p(1 - p)^{k-1} = 1.$$

The geometric distribution is useful in the following situation. Suppose an experiment is performed that leads to a sequence of independent[1] Bernoulli random variables, each with parameter p; that is, $P\{X_i = 1\} = p$ and $P\{X_i = 0\} = 1 - p$, for all i. The random variable X, which is the number of trials occurring until the first Bernoulli random variable takes on the value 1, has a geometric distribution with parameter p.

3.7 Continuous Probability Distributions

Section 3.3 defined continuous random variables as those random variables that take on a continuum of values. The CDF for a continuous random variable $F_X(b)$ can

[1] The concept of independent random variables is introduced in Sec. 3.13. For now, random variables can be considered independent if their outcomes do not affect the outcomes of the other random variables.

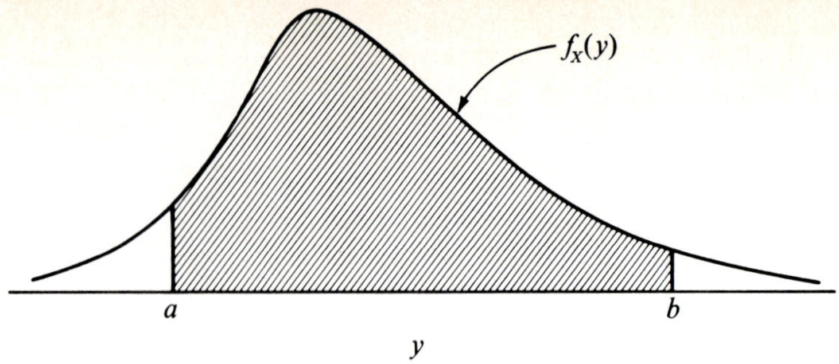

Figure 3.8 Density function of a random variable.

usually be written as

$$F_X(b) = P\{X(\omega) \leq b\} = \int_{-\infty}^{b} f_X(y)\, dy,$$

where $f_X(y)$ is known as the density function of the random variable X. From a notational standpoint, the subscript X is used to indicate the random variable that is under consideration. When there is no ambiguity, this subscript may be deleted, and $f_X(y)$ will be denoted by $f(y)$. It is evident that the CDF can be obtained if the density function is known. Furthermore, a knowledge of the density function enables one to calculate all sorts of probabilities, for example

$$P\{a < X \leq b\} = F(b) - F(a) = \int_{a}^{b} f_X(y)\, dy.$$

Note that strictly speaking the symbol $P\{a < X \leq b\}$ relates to the probability that the outcome ω of the experiment belongs to a particular event in the sample space, namely, that event such that $X(\omega)$ is between a and b whenever ω belongs to the event. However, the reference to the event will be suppressed, and the symbol P will be used to refer to the probability that X falls between a and b. It becomes evident from the previous expression for $P\{a < X \leq b\}$ that this probability can be evaluated by obtaining the area under the density function between a and b, as illustrated by the shaded area under the density function shown in Fig. 3.8. Finally, if the density function is known, it will be said that the probability distribution of the random variable is determined.

Naturally, the density function can be obtained from the CDF by using the relation

$$\frac{dF_X(y)}{dy} = \frac{d}{dy}\int_{-\infty}^{y} f_X(t)\, dt = f_X(y).$$

For a given value c, $P\{X = c\}$ has not been defined in terms of the density function. However, because probability has been interpreted as an area under the density function, $P\{X = c\}$ will be taken to be zero for all values of c. Having $P\{X = c\} = 0$ does not mean that the appropriate event E in the sample space [E contains those ω such that $X(\omega) = c$] is an impossible event. Rather, the event E can occur, but it occurs with probability zero. Since X is a continuous random variable, it takes on a continuum of possible values, so that selecting correctly the actual outcome before experimentation would be rather startling. Nevertheless, some outcome is obtained, so that it is not unreasonable to assume that the preselected outcome has probability zero of occurring. It then follows from $P\{X = c\}$ being equal to zero for all values c that for continuous random variables, and any a and b,

$$P\{a \le X \le b\} = P\{a < X \le b\} = P\{a \le X < b\} = P\{a < X < b\}.$$

Of course, this is not true for discrete random variables.

In defining the CDF for continuous random variables, it was implied that $f_X(y)$ was defined for values of y from minus to plus infinity because

$$F_X(b) = \int_{-\infty}^{b} f_X(y)\,dy.$$

This causes no difficulty, even for random variables that cannot take on negative values (e.g., the arrival time of the first customer) or are restricted to other regions, because $f_X(y)$ can be defined to be zero over the inadmissible segment of the real line. In fact, the only requirements of a density function are that

1. $f_X(y)$ be nonnegative.

2. $\int_{-\infty}^{\infty} f_X(y)\,dy = 1.$

It has already been pointed out that $f_X(y)$ cannot be interpreted as $P\{X = y\}$ because this probability is always zero. However, $f_X(y)\,dy$ can be interpreted as the probability that the random variable X lies in the infinitesimal interval $(y, y + dy)$, so that, loosely speaking, $f_X(y)$ is a measure of the frequency with which the random variable will fall into a "small" interval near y.

There are several important continuous probability distributions that are used in operations research work. The remainder of this section is devoted to a study of these distributions.

The Exponential Distribution

A continuous random variable whose density is given by

$$f_X(y) = \begin{cases} \dfrac{1}{\theta} e^{-y/\theta}, & \text{for } y \ge 0 \\ 0, & \text{for } y < 0 \end{cases}$$

is known as an exponentially distributed random variable. The exponential distribution is a function of the single parameter θ, where θ is any positive constant. $f_X(y)$ is a density function because it is a nonnegative and integrates to 1; that is,

$$\int_{-\infty}^{\infty} f_X(y)\,dy = \int_0^{\infty} \frac{1}{\theta} e^{-y/\theta}\,dy = -e^{-y/\theta}\Big|_0^{\infty} = 1.$$

The exponential density function is shown in Fig. 3.9.

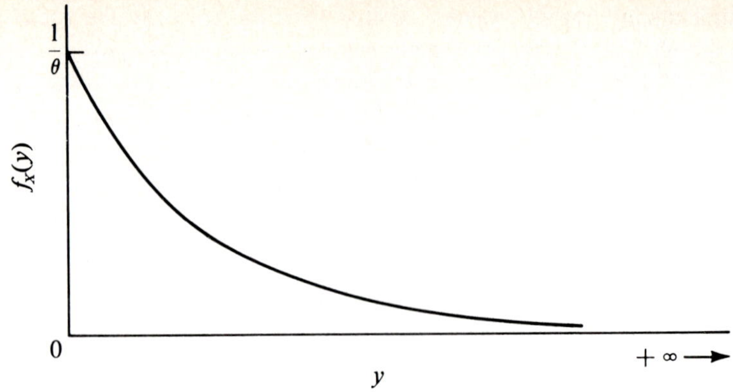

Figure 3.9 Density function of the exponential.

The CDF of an exponentially distributed random variable $F_X(b)$ is given by

$$F_X(b) = \int_{-\infty}^{b} f_X(y)\, dy$$

$$= \begin{cases} 0, & \text{for } b < 0 \\ \int_0^b \frac{1}{\theta} e^{-y/\theta}\, dy = 1 - e^{-b/\theta}, & \text{for } b \geq 0, \end{cases}$$

and is shown in Fig. 3.10.

The exponential distribution has had widespread use in operations research. The time between customer arrivals, the length of time of telephone conversations, and the life of electronic components are often assumed to have an exponential distribution. Such an assumption has the important implication that the random variable does not "age." For example, suppose that the life of a vacuum tube is assumed to have an exponential distribution. If the tube has lasted 1,000 hours, the probability of lasting an additional 50 hours is the same as the probability of lasting an additional 50 hours, given that the tube has lasted 2,000 hours. In other words, a brand-new tube is no "better" than one that has lasted 1,000 hours. This implication of the exponential is quite important and is often overlooked in practice.

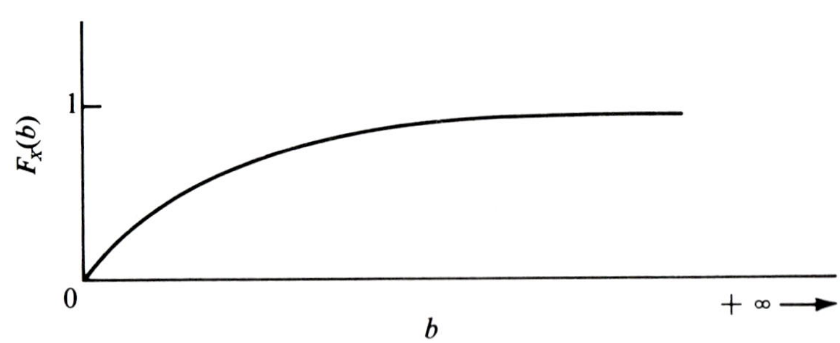

Figure 3.10 CDF of the exponential distribution.

The Gamma Distribution

A continuous random variable whose density is given by

$$f_X(y) = \begin{cases} \dfrac{1}{\Gamma(\alpha)\beta^{\alpha}} y^{(\alpha-1)} e^{-y/\beta}, & \text{for } y \geq 0 \\ 0, & \text{for } y < 0 \end{cases}$$

is known as a gamma-distributed random variable. This density is a function of the two parameters α and β, both of which are positive constants. $\Gamma(\alpha)$ is defined as

$$\Gamma(\alpha) = \int_0^{\infty} t^{\alpha-1} e^{-t}\, dt, \qquad \text{for all } \alpha > 0.$$

If α is an integer, then repeated integration by parts yields

$$\Gamma(\alpha) = (\alpha - 1)! = (\alpha - 1)(\alpha - 2)(\alpha - 3) \cdots 3 \cdot 2 \cdot 1.$$

A graph of a typical gamma density function is given in Fig. 3.11.

A random variable having a gamma density is useful in its own right as a mathematical representation of physical phenomena, or it may arise as follows: Suppose a customer's service time has an exponential distribution with parameter θ. The random variable T, the total time to service n (independent) customers, has a gamma distribution with parameters n and θ (replacing α and β, respectively); that is,

$$P\{T < t\} = \int_0^{t} \frac{1}{\Gamma(n)\theta^{n}} y^{(n-1)} e^{-y/\theta}\, dy.$$

Note that when $n = 1$ (or $\alpha = 1$) the gamma density becomes the density function of an exponential random variable. Thus, sums of independent, exponentially distributed random variables have a gamma distribution.

Another important distribution, the chi square, is related to the gamma distribution. If X is a random variable having a gamma distribution with parameters $\beta = 1$ and $\alpha = v/2$ (v is a positive integer), then a new random variable $Z = 2X$ is said to have a chi-square distribution with v degrees of freedom. The expression for the density function is given in Table 3.1. The percentage points of the chi-square distribution are given in Table A5.3 of Appendix 5. Percentage points of the distribution

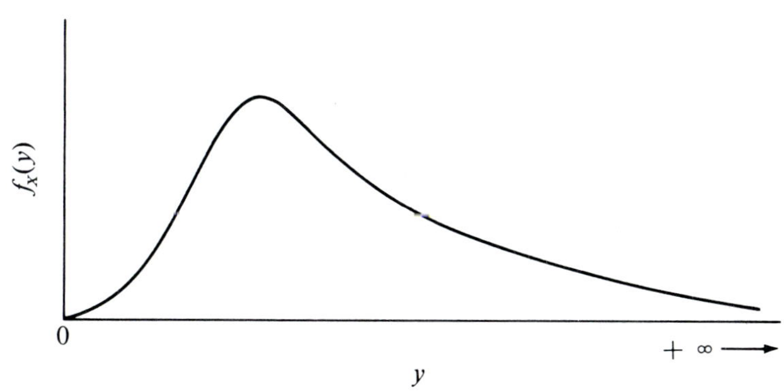

Figure 3.11 Gamma density function.

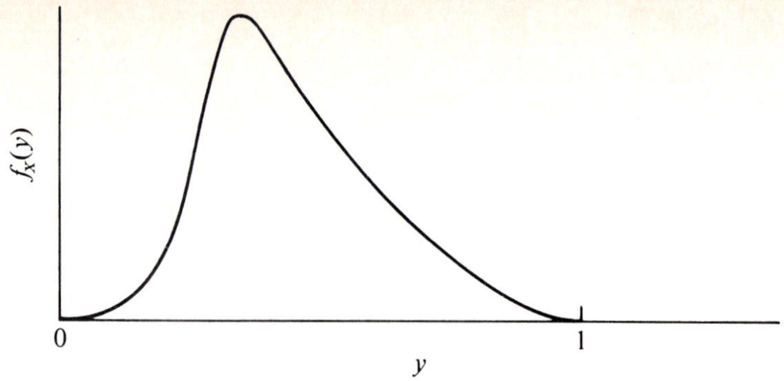

Figure 3.12 Beta density function.

of a random variable Z are the values z_α such that

$$P\{Z > z_\alpha\} = \alpha.$$

z_α is said to be the 100α percentage point of the distribution of the random variable Z.

The Beta Distribution

A continuous random variable whose density function is given by

$$f_X(y) = \begin{cases} \dfrac{\Gamma(\alpha + \beta)}{\Gamma(\alpha)\Gamma(\beta)} y^{(\alpha-1)}(1 - y)^{(\beta-1)}, & \text{for } 0 \le y \le 1 \\ 0, & \text{elsewhere} \end{cases}$$

is known as a beta-distributed random variable. This density is a function of the two parameters α and β, both of which are positive constants. A graph of a typical beta density function is given in Fig. 3.12.

Beta distributions form a useful class of distributions when a random variable is restricted to the unit interval. In particular, when $\alpha = \beta = 1$, the beta distribution is called the *uniform distribution* over the unit interval. Its density function is shown in Fig. 3.13, and it can be interpreted as having all the values between zero and one equally likely to occur. The CDF for this random variable is given by

$$F_X(b) = \begin{cases} 0, & \text{for } b < 0 \\ b, & \text{for } 0 \le b \le 1 \\ 1, & \text{for } b > 1. \end{cases}$$

If the density function is to be constant over some other interval, such as the interval $[c, d]$, a uniform distribution over this interval can also be obtained.[1] The density function is given by

$$f_X(y) = \begin{cases} \dfrac{1}{d - c}, & \text{for } c \le y \le d \\ 0, & \text{otherwise.} \end{cases}$$

[1] The beta distribution can also be generalized by defining the density function over some fixed interval, other than the unit interval.

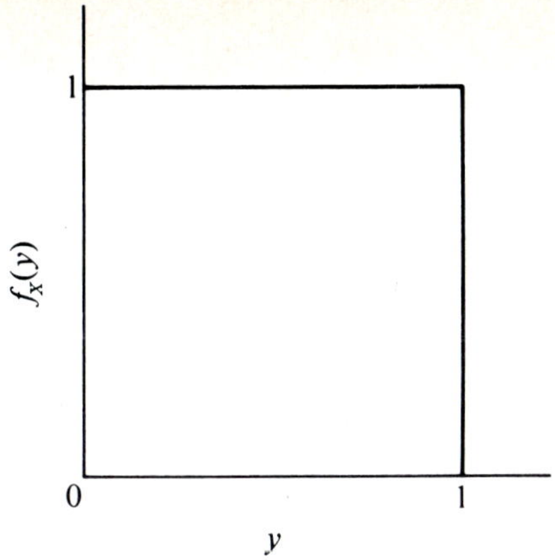

Figure 3.13 Uniform distribution over the unit interval.

Although such a random variable is said to have a uniform distribution over the interval $[c, d]$, it is no longer a special case of the beta distribution.

Another important distribution, Student's t, is related to the beta distribution. If X is a random variable having a beta distribution with parameters $\alpha = \frac{1}{2}$ and $\beta = v/2$ (v is a positive integer), then a new random variable $Z = \sqrt{vX/(1 - X)}$ is said to have a Student's t (or t) distribution with v degrees of freedom. The percentage points of the t distribution are given in Table A5.2 of Appendix 5.

A final distribution related to the beta distribution is the F distribution. If X is a random variable having a beta distribution with parameters $\alpha = v_1/2$ and $\beta = v_2/2$ (v_1 and v_2 are positive integers), then a new random variable $Z = v_2X/v_1(1 - X)$ is said to have an F distribution with v_1 and v_2 degrees of freedom. The percentage points of the F distribution are given in Table A5.5 of Appendix 5.

The Normal Distribution

One of the most important distributions in operations research is the normal distribution. A continuous random variable whose density function is given by

$$f_X(y) = \frac{1}{\sqrt{2\pi}\sigma} e^{-(y-\mu)^2/2\sigma^2}, \qquad \text{for } -\infty < y < \infty$$

is known as a normally distributed random variable. The density is a function of the two parameters μ and σ, where μ is any constant, and σ is positive. A graph of a typical normal density function is given in Fig. 3.14. This density function is a bell-shaped curve that is symmetric around μ. The CDF for a normally distributed random variable is given by

$$F_X(b) = \int_{-\infty}^{b} \frac{1}{\sqrt{2\pi}\sigma} e^{-(y-\mu)^2/2\alpha^2}\, dy.$$

By making the transformation $z = (y - \mu)/\sigma$, the CDF can be written as

$$F_X(b) = \int_{-\infty}^{(b-\mu)/\sigma} \frac{1}{\sqrt{2\pi}} e^{-z^2/2}\, dz.$$

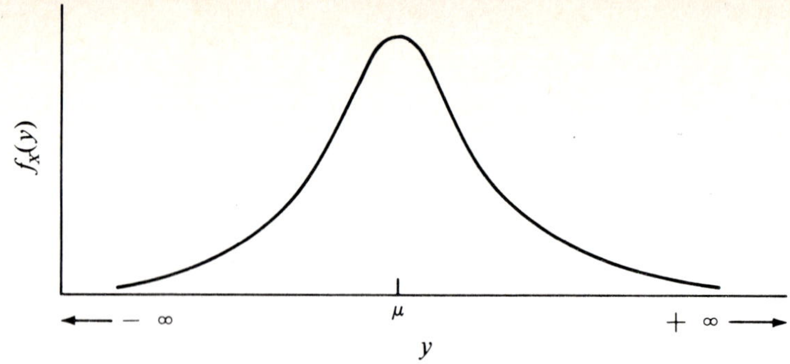

Figure 3.14 Normal density function.

Hence, although this function is not integrable, it is easily tabulated. Table A5.1 presented in Appendix 5 is a tabulation of

$$\alpha = \int_{K_\alpha}^{\infty} \frac{1}{\sqrt{2\pi}} e^{-z^2/2}\, dz$$

as a function of K_α. Hence, to find $F_X(b)$ (and any probability derived from it), Table A5.1 is entered with $K_\alpha = (b - \mu)/\sigma$, and

$$\alpha = \int_{K_\alpha}^{\infty} \frac{1}{\sqrt{2\pi}} e^{-z^2/2}\, dz$$

is read from it. $F_X(b)$ is then just $1 - \alpha$. Thus, if $P\{14 < X \le 18\} = F_X(18) - F_X(14)$ is desired, where X has a normal distribution with $\mu = 10$ and $\sigma = 4$, Table A5.1 is entered with $(18 - 10)/4 = 2$, and $1 - F_X(18) = 0.0228$ is obtained. The table is then entered with $(14 - 10)/4 = 1$, and $1 - F_X(14) = 0.1587$ is read. From these figures, $F_X(18) - F_X(14) = 0.1359$ is found. If K_α is negative, use can be made of the symmetry of the normal distribution because

$$F_X(b) = \int_{-\infty}^{(b-\mu)/\alpha} \frac{1}{\sqrt{2\pi}} e^{-z^2/2}\, dz = \int_{-(b-\mu)/\sigma}^{\infty} \frac{1}{\sqrt{2\pi}} e^{-z^2/2}\, dz.$$

In this case $-(b - \mu)/\sigma$ is positive, and $F_X(b) = \alpha$ is thereby read from the table by entering it with $-(b - \mu)/\sigma$. Thus, suppose it is desired to evaluate the expression

$$P\{2 < X \le 18\} = F_X(18) - F_X(2).$$

$F_X(18)$ has already been shown to be equal to $1 - 0.0228 = 0.9772$. To find $F_X(2)$ it is first noted that $(2 - 10)/4 = -2$ is negative. Hence, Table A5.1 is entered with $K_\alpha = +2$, and $F_X(2) = 0.0228$ is obtained. Thus,

$$F_X(18) - F_X(2) = 0.9772 - 0.0228 = 0.9544.$$

As indicated previously, the normal distribution is a very important one. In particular, it can be shown that if $X_1, X_2, \ldots, X_n$ are independent,[1] normally distributed random variables with parameters $(\mu_1, \sigma_1), (\mu_2, \sigma_2), \ldots, (\mu_n, \sigma_n)$, respectively, then $X = X_1 + X_2 + \cdots + X_n$ is also a normally distributed random

[1] The concept of independent random variables is introduced in Sec. 3.13. For now, random variables can be considered independent if their outcomes do not affect the outcomes of the other random variables.

variable with parameters

$$\sum_{i=1}^{n} \mu_i$$

and

$$\sqrt{\sum_{i=1}^{n} \sigma_i^2}.$$

In fact, even if $X_1, X_2, \ldots, X_n$ are not normal, then under very weak conditions

$$X = \sum_{i=1}^{n} X_i$$

tends to be normally distributed as n gets large. This is discussed further in Sec. 3.15.

Finally, if C is any constant and X is normal with parameters μ and σ, then the random variable CX is also normal with parameters $C\mu$ and $C\sigma$. Hence, it follows that if $X_1, X_2, \ldots, X_n$ are independent, normally distributed random variables, each with parameters μ and σ, the random variable

$$\bar{X} = \sum_{i=1}^{n} \frac{X_i}{n}$$

is also normally distributed with parameters μ and $\sigma/\sqrt{n}$.

3.8 Expectation

Although knowledge of the probability distribution of a random variable enables one to make all sorts of probability statements, a single value that may characterize the random variable and its probability distribution is often desirable. Such a quantity is the *expected value* of the random variable. One may speak of the expected value of the demand for a product or the expected value of the time of the first customer arrival. In the experiment where the arrival time of the first customer on 2 successive days was measured, the expected value of the average arrival time of the first customers on 2 successive days may be of interest.

Formally, the expected value of a random variable X is denoted by $E(X)$ and is given by

$$E(X) = \begin{cases} \displaystyle\sum_{\text{all } k} kP\{X = k\} = \sum_{\text{all } k} kP_X(k), & \text{if } X \text{ is a discrete random variable} \\ \displaystyle\int_{-\infty}^{\infty} yf_X(y)\,dy, & \text{if } X \text{ is a continuous random variable.} \end{cases}$$

For a discrete random variable it is seen that $E(X)$ is just the sum of the products of the possible values the random variable X takes on and their respective associated probabilities. In the example of the demand for a product, where $k = 0, 1, 2, \ldots, 98, 99$ and $P_X(k) = \frac{1}{100}$ for all k, the expected value of the demand is

$$E(X) = \sum_{k=0}^{99} kP_X(k) = \sum_{k=0}^{99} k\,\frac{1}{100} = 49.5.$$

Note that $E(X)$ need not be a value that the random variable can take on.

If X is a binomial random variable with parameters n and p, the expected value of X is given by

$$E(X) = \sum_{k=0}^{n} k \frac{n!}{k!(n-k)!} p^k(1-p)^{n-k}$$

and can be shown to equal np.

If the random variable X has a Poisson distribution with parameter λ,

$$E(X) = \sum_{k=0}^{\infty} k \frac{\lambda^k e^{-\lambda}}{k!}$$

and can be shown to equal λ.

Finally, if the random variable X has a geometric distribution with parameter p,

$$E(X) = \sum_{k=1}^{\infty} kp(1-p)^{k-1}$$

and can be shown to equal $1/p$.

For continuous random variables, the expected value can also be obtained easily. If X has an exponential distribution with parameter θ, the expected value is given by

$$E(X) = \int_{-\infty}^{\infty} y f_X(y)\, dy = \int_0^{\infty} y \frac{1}{\theta} e^{-y/\theta}\, dy.$$

This integral is easily evaluated to be

$$E(X) = \theta.$$

If the random variable X has a gamma distribution with parameter α and β, the expected value of X is given by

$$\int_{-\infty}^{\infty} y f_X(y)\, dy = \int_0^{\infty} y \frac{1}{\Gamma(\alpha)\beta^\alpha} y^{(\alpha-1)} e^{-y/\beta}\, dy = \alpha\beta.$$

If the random variable X has a beta distribution with parameters α and β, the expected value of X is given by

$$\int_{-\infty}^{\infty} y f_X(y)\, dy = \int_0^{1} y \frac{\Gamma(\alpha+\beta)}{\Gamma(\alpha)\Gamma(\beta)} y^{(\alpha-1)}(1-y)^{(\beta-1)}\, dy = \frac{\alpha}{\alpha+\beta}.$$

Finally, if the random variable X has a normal distribution with parameters μ and σ, the expected value of X is given by

$$\int_{-\infty}^{\infty} y f_X(y)\, dy = \int_{-\infty}^{\infty} y \frac{1}{\sqrt{2\pi}\sigma} e^{-(y-\mu)^2/2\alpha^2}\, dy = \mu.$$

The expectation of a random variable is quite useful in that it not only provides some characterization of the distribution, but it also has meaning in terms of the average of a sample. In particular, if a random variable is observed again and again and the arithmetic mean $\overline{X}$ is computed, then $\overline{X}$ tends to the expectation of the random variable X as the number of trials becomes large. A precise statement of this property is given in Sec. 3.14. Thus, if the demand for a product takes on the values $k = 0, 1, 2, \ldots, 98, 99$, each with $P_X(k) = \frac{1}{100}$ for all k, and if demands of $x_1, x_2, \ldots, x_n$ are observed on successive days, then the average of these values, $(x_1 + x_2 + \cdots + x_n)/n$, should be close to $E(X) = 49.5$ if n is sufficiently large.

It is not necessary to confine the discussion of expectation to discussion of the expectation of a random variable X. If Z is some function of X, say, $Z = g(X)$, then $g(X)$ is also a random variable. The expectation of $g(X)$ can be defined as

$$E[g(X)] = \begin{cases} \sum_{\text{all } k} g(k)P\{X = k\} = \sum_{\text{all } k} g(k)P_X(k), & \text{if } X \text{ is a discrete random variable} \\ \int_{-\infty}^{\infty} g(y)f_X(y)\,dy, & \text{if } X \text{ is a continuous random variable.} \end{cases}$$

An interesting theorem, known as the "theorem of the unconscious statistician"[1] states that if X is a continuous random variable having density $f_X(y)$ and $Z = g(X)$ is a function of X having density $h_Z(y)$, then

$$E(Z) = \int_{-\infty}^{\infty} yh_Z(y)\,dy = \int_{-\infty}^{\infty} g(y)f_X(y)\,dy.$$

Thus, the expectation of Z can be found by using its definition in terms of the density of Z or, alternatively, by using its definition as the expectation of a function of X with respect to the density function of X. The identical theorem is true for discrete random variables.

3.9 Moments

If the function g described in the previous section is given by

$$Z = g(X) = X^j,$$

where j is a positive integer, then the expectation of X^j is called the jth moment about the origin of the random variable X and is given by

$$E(X^j) = \begin{cases} \sum_{\text{all } k} k^j P_X(k), & \text{if } X \text{ is a discrete random variable} \\ \int_{-\infty}^{\infty} y^j f_X(y)\,dy, & \text{if } X \text{ is a continuous random variable.} \end{cases}$$

Note that when $j = 1$ the first moment coincides with the expectation of X. This is usually denoted by the symbol μ and is often called the mean or average of the distribution.

Using the theorem of the unconscious statistician, the expectation of $Z = g(X) = CX$ can easily be found, where C is a constant. If X is a continuous random variable, then

$$E(CX) = \int_{-\infty}^{\infty} Cyf_X(y)\,dy = C\int_{-\infty}^{\infty} yf_X(y)\,dy = CE(X).$$

[1] The name for this theorem is motivated by the fact that a statistician often uses its conclusions without consciously worrying about whether the theorem is true.

Thus, the expectation of a constant times a random variable is just the constant times the expectation of the random variable. This is also true for discrete random variables.

If the function g described in the previous section is given by $Z = g(X) = (X - E(X))^j = (X - \mu)^j$, where j is a positive integer, then the expectation of $(X - \mu)^j$ is called the jth moment about the mean of the random variable X and is given by

$$E(X - E(X))^j = E(X - \mu)^j = \begin{cases} \sum_{\text{all } k} (k - \mu)^j P_X(k), & \text{if } X \text{ is a discrete random variable} \\ \int_{-\infty}^{\infty} (y - \mu)^j f_X(y)\, dy, & \text{if } X \text{ is a continuous random variable.} \end{cases}$$

Note that if $j = 1$, then $E(X - \mu) = 0$. If $j = 2$, then $E(X - \mu)^2$ is called the variance of the random variable X and is often denoted by σ^2. The square root of the variance σ is called the standard deviation of the random variable X. It is easily shown, in terms of definitions, that $\sigma^2 = E(X - \mu)^2 = E(X^2) - \mu^2$; that is, the variance can be written as the second moment about the origin minus the square of the mean.

It has already been shown that if $Z = g(X) = CX$, then $E(CX) = CE(X) = C\mu$, where C is any constant and μ is $E(X)$. The variance of the random variable $Z = g(X) = CX$ is also easily obtained. By definition, if X is a continuous random variable, the variance of Z is given by

$$E(Z - E(Z))^2 = E(CX - CE(X))^2 = \int_{-\infty}^{\infty} (Cy - C\mu)^2 f_X(y)\, dy$$

$$= C^2 \int_{-\infty}^{\infty} (y - \mu)^2 f_X(y)\, dy = C^2\sigma^2.$$

Thus, the variance of a constant times a random variable is just the square of the constant times the variance of the random variable. This is also true for discrete random variables. Finally, the variance of a constant is easily seen to be zero.

It has already been shown that if the demand for a product takes on the values $0, 1, 2, \ldots, 99$, each with probability $\frac{1}{100}$, then $E(X) = \mu = 49.5$. Similarly,

$$\sigma^2 = \sum_{k=0}^{99} (k - \mu)^2 P_X(k) = \sum_{k=0}^{99} k^2 P_X(k) - \mu^2$$

$$= \sum_{k=0}^{99} \frac{k^2}{100} - (49.5)^2 = 833.25.$$

Table 3.1 gives the means and variances of the random variables that are often useful in operations research. Note that for some random variables a single moment, the mean, provides a complete characterization of the distribution, for example, the Poisson random variable. For some random variables the mean and variance provide a complete characterization of the distribution, for example, the normal. In fact, if all the moments of a probability distribution are known, this is usually equivalent to specifying the entire distribution.

It was seen that the mean and variance may be sufficient to completely characterize a distribution, for example, the normal. However, what can be said, in general, about a random variable whose mean μ and variance σ^2 are known, but nothing

Table 3.1 **Table of Common Distributions**

Distribution of Random Variable X	Form	Parameters	Expected Value	Variance	Range of Random Variable
Binomial	$P_X(k) = \frac{n!}{k!(n-k)!} p^k(1-p)^{n-k}$	n, p	np	$np(1-p)$	$0, 1, 2, \ldots, n$
Poisson	$P_X(k) = \frac{\lambda^k e^{-\lambda}}{k!}$	λ	λ	λ	$0, 1, 2, \ldots.$
Geometric	$P_X(k) = p(1-p)^{k-1}$	p	$\frac{1}{p}$	$\frac{1-p}{p^2}$	$1, 2, \ldots.$
Exponential	$f_X(y) = \frac{1}{\theta} e^{-y/\theta}$	θ	θ	θ^2	$(0, \infty)$
Gamma	$f_X(y) = \frac{1}{\Gamma(\alpha)\beta^\alpha} y^{(\alpha-1)} e^{-y/\beta}$	α, β	$\alpha\beta$	$\alpha\beta^2$	$(0, \infty)$
Beta	$f_X(y) = \frac{\Gamma(\alpha+\beta)}{\Gamma(\alpha)\Gamma(\beta)} y^{(\alpha-1)}(1-y)^{(\beta-1)}$	α, β	$\frac{\alpha}{\alpha+\beta}$	$\frac{\alpha\beta}{(\alpha+\beta)^2(\alpha+\beta+1)}$	$(0, 1)$
Normal	$f_X(y) = \frac{1}{\sqrt{2\pi}\sigma} e^{-(y-\mu)^2/2\sigma^2}$	μ, σ	μ	σ^2	$(-\infty, \infty)$
Student's t	$f_X(y) = \frac{1}{\sqrt{\pi v}} \frac{\Gamma([v+1]/2)}{\Gamma(v/2)} (1 + y^2/v)^{-(v+1)/2}$	v	0(for $v > 1$)	$v/(v-2)$(for $v > 2$)	$(-\infty, \infty)$
Chi square	$f_X(y) = \frac{1}{2^{v/2}\Gamma(v/2)} y^{(v-2)/2} e^{-y/2}$	v	v	$2v$	$(0, \infty)$
F	$f_X(y) = \frac{\Gamma\left(\frac{v_1+v_2}{2}\right) v_1^{v_1/2} v_2^{v_2/2}}{\Gamma\left(\frac{v_1}{2}\right)\Gamma\left(\frac{v_2}{2}\right)} \frac{(y)^{(v_1/2)-1}}{(v_2+v_1 y)^{(v_1+v_2)/2}}$	v_1, v_2	$\frac{v_2}{v_2-2}$ for $v_2 > 2$.	$\frac{v_2^2(2v_2+2v_1-4)}{v_1(v_2-2)^2(v_2-4)}$ for $v_2 > 4$.	$(0, \infty)$

else about the form of the distribution is specified? This can be expressed in terms of *Chebyshev's inequality*, which states that for any positive number C,

$$P\{\mu - C\sigma \leq X \leq \mu + C\sigma\} > 1 - \frac{1}{C^2},$$

where X is any random variable having mean μ and variance σ^2. For example, if $C = 3$, it follows that $P\{\mu - 3\sigma \leq X \leq \mu + 3\sigma\} > 1 - \frac{1}{9} = 0.8889$. However, if X is known to have a normal distribution, then $P\{\mu - 3\sigma \leq X \leq \mu + 3\sigma\} = 0.9973$. Note that the Chebyshev inequality only gives a lower bound on the probability (usually a very conservative one), so there is no contradiction here.

3.10 Bivariate Probability Distribution

Thus far the discussion has been concerned with the probability distribution of a single random variable, for example, the demand for a product during the first month or the demand for a product during the second month. In an experiment that measures the demand during the first 2 months, it may well be important to look at the probability distribution of the vector random variable (X_1, X_2), the demand during the first month, and the demand during the second month, respectively.

Define the symbol

$$E_{b_1,b_2}^{X_1,X_2} = \{\omega | X_1(\omega) \leq b_1, X_2(\omega) \leq b_2\},$$

or equivalently,

$$E_{b_1,b_2}^{X_1,X_2} = \{X_1 \leq b_1, X_2 \leq b_2\},$$

as the set of outcomes ω in the sample space forming the event $E_{b_1,b_2}^{X_1,X_2}$, such that the random variable X_1 takes on values less than or equal to b_1, and X_2 takes on values less than or equal to b_2. Then $P\{E_{b_1,b_2}^{X_1,X_2}\}$ denotes the probability of this event. In the above example of the demand for a product during the first 2 months, suppose that the sample space Ω consists of the set of all possible points ω, where ω represents a pair of nonnegative integer values (x_1, x_2). Assume that x_1 and x_2 are bounded by 99. Thus there are $(100)^2\omega$ points in Ω. Suppose further that each point ω has associated with it a probability equal to $1/(100)^2$, except for the points $\omega = (0, 0)$ and $\omega = (99, 99)$. The probability associated with the event $\{0, 0\}$ will be $1.5/(100)^2$, that is, $P\{0,0\} = 1.5/(100)^2$, and the probability associated with the event $\{99,99\}$ will be $0.5/(100)^2$; that is, $P\{99,99\} = 0.5/(100)^2$. Thus, if there is interest in the "bivariate" random variable (X_1, X_2), the demand during the first and second months, respectively, then the event

$$\{X_1 \leq 1, X_2 \leq 3\}$$

is the set

$$E_{1,3}^{X_1,X_2} = \{(0, 0), (0, 1), (0, 2), (0, 3), (1, 0), (1, 1), (1, 2), (1, 3)\}.$$

Furthermore,

$$\begin{aligned} P\{E_{1,3}^{X_1,X_2}\} &= \frac{1.5}{(100)^2} + \frac{1}{(100)^2} + \frac{1}{(100)^2} + \frac{1}{(100)^2} + \frac{1}{(100)^2} \\ &\quad + \frac{1}{(100)^2} + \frac{1}{(100)^2} + \frac{1}{(100)^2} \\ &= \frac{8.5}{(100)^2}, \end{aligned}$$

so that

$$P\{X_1 \leq 1, X_2 \leq 3\} = P\{E_{1,3}^{X_1,X_2}\} = \frac{8.5}{(100)^2}.$$

A similar calculation can be made for any value of b_1 and b_2.

For any given bivariate random variable (X_1, X_2), $P\{X_1 \leq b_1, X_2 \leq b_2\}$ is denoted by $F_{X_1X_2}(b_1, b_2)$ and is called the joint cumulative distribution function (CDF) of the bivariate random variable (X_1, X_2) and is defined for all real values of b_1 and b_2. Where there is no ambiguity the joint CDF may be denoted by $F(b_1, b_2)$. Thus, attached to every bivariate random variable is a joint CDF. This is not an arbitrary function but is induced by the probabilities associated with events defined over the sample space Ω such that $\{\omega | X_1(\omega) \leq b_1, X_2(\omega) \leq b_2\}$.

The joint CDF of a random variable is a numerically valued function defined for all b_1, b_2, such that $-\infty \leq b_1, b_2 \leq \infty$, having the following properties:

1. $F_{X_1X_2}(b_1, \infty) = P\{X_1 \le b_1,\ X_2 \le \infty\} = P\{X_1 \le b_1\} = F_{X_1}(b_1)$, where $F_{X_1}(b_1)$ is just the CDF of the univariate random variable X_1.
2. $F_{X_1X_2}(\infty, b_2) = P\{X_1 \le \infty,\ X_2 \le b_2\} = P\{X_2 \le b_2\} = F_{X_2}(b_2)$, where $F_{X_2}(b_2)$ is just the CDF of the univariate random variable X_2.
3. $F_{X_1X_2}(b_1, -\infty) = P\{X_1 \le b_1, X_2 \le -\infty\} = 0$,
   $F_{X_1X_2}(-\infty, b_2) = P\{X_1 \le -\infty, X_2 \le b_2\} = 0$.
4. $F_{X_1X_2}(b_1 + \Delta_1,\ b_2 + \Delta_2) - F_{X_1X_2}(b_1 + \Delta_1,\ b_2) - F_{X_1X_2}(b_1,\ b_2 + \Delta_2) + F_{X_1X_2}(b_1, b_2) \ge 0$, for every $\Delta_1, \Delta_2 \ge 0$, and b_1, b_2.

Using the definition of the event $E_{b_1,b_2}^{X_1,X_2}$, events of the form

$$\{a_1 < X_1 \le b_1, a_2 < X_2 \le b_2\}$$

can be described as the set of outcomes ω in the sample space such that the bivariate random variable (X_1, X_2) takes on values such that X_1 is greater than a_1 but does not exceed b_1 *and* X_2 is greater than a_2 but does not exceed b_2. $P\{a_1 < X_1 \le b_1, a_2 < X_2 \le b_2\}$ can easily be seen to be

$$F_{X_1X_2}(b_1, b_2) - F_{X_1X_2}(b_1, a_2) - F_{X_1X_2}(a_1, b_2) + F_{X_1X_2}(a_1, a_2).$$

It was noted that single random variables are generally characterized as discrete or continuous random variables. A bivariate random variable can be characterized in a similar manner. A bivariate random variable (X_1, X_2) is called a *discrete bivariate random variable* if both X_1 and X_2 are discrete random variables. Similarly, a bivariate random variable (X_1, X_2) is called a *continuous bivariate random variable* if both X_1 and X_2 are continuous random variables. Of course, bivariate random variables that are neither discrete nor continuous can exist, but these will not be important in this book.

The joint CDF for a discrete random variable $F_{X_1X_2}(b_1, b_2)$ is given by

$$\begin{aligned} F_{X_1X_2}(b_1, b_2) &= P\{\omega | X_1(\omega) \le b_1, X_2(\omega) \le b_2\} \\ &= \sum_{\text{all } k \le b_1} \sum_{\text{all } l \le b_2} P\{\omega | X_1(\omega) = k, X_2(\omega) = l\} \\ &= \sum_{\text{all } k \le b_1} \sum_{\text{all } l \le b_2} P_{X_1X_2}(k, l), \end{aligned}$$

where $\{\omega | X_1(\omega) = k, X_2(\omega) = l\}$ is the set of outcomes ω in the sample space, such that the random variable X_1 takes on the value k, and the variable X_2 takes on the value l; and $P\{\omega | X_1(\omega) = k, X_2(\omega) = l\} = P_{X_1X_2}(k, l)$ denotes the probability of this event. The $P_{X_1X_2}(k, l)$ are called the joint probability distribution of the discrete bivariate random variable (X_1, X_2). Thus, in the previous example, $P_{X_1X_2}(k, l) = 1/(100)^2$ for all k, l that are integers between 0 and 99, except for $P_{X_1X_2}(0, 0) = 1.5/(100)^2$ and $P_{X_1X_2}(99, 99) = 0.5/(100)^2$.

For a continuous random variable, the joint CDF $F_{X_1X_2}(b_1, b_2)$ can usually be written as

$$F_{X_1X_2}(b_1, b_2) = P\{\omega | X_1(\omega) \le b_1, X_2(\omega) \le b_2\} = \int_{-\infty}^{b_1} \int_{-\infty}^{b_2} f_{X_1X_2}(s, t)\, ds\, dt,$$

where $f_{X_1X_2}(s, t)$ is known as the joint density function of the bivariate random variable (X_1, X_2). A knowledge of the joint density function enables one to calculate all sorts of probabilities, for example,

$$P\{a_1 < X_1 \leq b_1, a_2 < X_2 \leq b_2\} = \int_{a_1}^{b_1}\int_{a_2}^{b_2} f_{X_1X_2}(s, t)\, ds\, dt.$$

Finally, if the density function is known, it is said that the probability distribution of the random variable is determined.

The joint density function can be viewed as a surface in three dimensions, where the volume under this surface over regions in the s, t plane correspond to probabilities. Naturally, the density function can be obtained from the CDF by using the relation

$$\frac{\partial^2 F_{X_1X_2}(s, t)}{\partial s\, \partial t} = \frac{\partial^2}{\partial s\, \partial t}\int_{-\infty}^{s}\int_{-\infty}^{t} f_{X_1X_2}(u, v)\, du\, dv = f_{X_1X_2}(s, t).$$

In defining the joint CDF for a bivariate random variable, it was implied that $f_{X_1X_2}(s, t)$ was defined over the entire plane because

$$F_{X_1X_2}(b_1, b_2) = \int_{-\infty}^{b_1}\int_{-\infty}^{b_2} f_{X_1X_2}(s, t)\, ds\, dt$$

(which is analogous to what was done for a univariate random variable). This causes no difficulty, even for bivriate random variables having one or more components that cannot take on negative values or are restricted to other regions. In this case, $f_{X_1X_2}(s, t)$ can be defined to be zero over the inadmissible part of the plane. In fact, the only requirements for a function to be a bivariate density function are that

1. $f_{X_1X_2}(s, t)$ be nonnegative, and
2. $\int_{-\infty}^{\infty}\int_{-\infty}^{\infty} f_{X_1X_2}(s, t)\, ds\, dt = 1.$

3.11 Marginal and Conditional Probability Distributions

In Sec. 3.10 the discussion was concerned with the joint probability distribution of a bivariate random variable (X_1, X_2). However, there may also be interest in the probability distribution of the random variables X_1 and X_2 considered separately. It was shown that if $F_{X_1X_2}(b_1, b_2)$ represents the joint CDF of (X_1, X_2), then $F_{X_1}(b_1) = F_{X_1X_2}(b_1, \infty) = P\{X_1 \leq b_1, X_2 \leq \infty\} = P\{X_1 \leq b_1\}$ is the CDF for the univariate random variable X_1, and $F_{X_2}(b_2) = F_{X_1X_2}(\infty, b_2) = P\{X_1 \leq \infty, X_2 \leq b_2\} = P\{X_2 \leq b_2\}$ is the CDF for the univariate random variable X_2.

If the bivariate random variable (X_1, X_2) is discrete, it was noted that the

$$P_{X_1X_2}(k, l) = P\{X_1 = k, X_2 = l\}$$

describe its joint probability distribution. The probability distribution of X_1 individually, $P_{X_1}(k)$, now called the marginal probability distribution of the discrete random variable X_1, can be obtained from the $P_{X_1X_2}(k, l)$. In particular,

$$F_{X_1}(b_1) = F_{X_1X_2}(b_1, \infty) = \sum_{\text{all } k \leq b_1}\sum_{\text{all } l} P_{X_1X_2}(k, l) = \sum_{\text{all } k \leq b_1} P_{X_1}(k),$$

so that

$$P_{X_1}(k) = P\{X_1 = k\} = \sum_{\text{all } l} P_{X_1X_2}(k, l).$$

Similarly, the marginal probability distribution of the discrete random variable X_2 is given by

$$P_{X_2}(l) = P\{X_2 = l\} = \sum_{\text{all } k} P_{X_1X_2}(k, l).$$

If the experiment (described previously) which measures the demand for a product during the first 2 months is considered, it is seen that the marginal distribution of X_1 is given by

$$P_{X_1}(0) = \sum_{\text{all } l} P_{X_1X_2}(0, l)$$

$$= P_{X_1X_2}(0, 0) + P_{X_1X_2}(0, 1) + \cdots + P_{X_1X_2}(0, 99)$$

$$= \frac{1.5}{(100)^2} + \frac{1}{(100)^2} + \cdots + \frac{1}{(100)^2} = \frac{100.5}{(100)^2},$$

$$P_{X_1}(1) = P_{X_1}(2) = \cdots P_{X_1}(98) = \sum_{\text{all } l} P_{X_1X_2}(k, l),\ k = 1, 2, \ldots, 98$$

$$= \frac{100}{(100)^2},$$

$$P_{X_1}(99) = \sum_{\text{all } l} P_{X_1X_2}(99, l)$$

$$= P_{X_1X_2}(99, 0) + P_{X_1X_2}(99, 1) + \cdots + P_{X_1X_2}(99, 99)$$

$$= \frac{1}{(100)^2} + \frac{1}{(100)^2} + \cdots + \frac{0.5}{(100)^2} = \frac{99.5}{(100)^2}.$$

Note that this is indeed a probability distribution in that

$$P_{X_1}(0) + P_{X_1}(1) + \cdots + P_{X_1}(99) = \frac{100.5}{(100)^2} + \frac{100}{(100)^2} + \cdots + \frac{99.5}{(100)^2} = 1.$$

Similarly, the marginal distribution of X_2 is given by

$$P_{X_2}(0) = \sum_{\text{all } k} P_{X_1X_2}(k, 0)$$

$$= P_{X_1X_2}(0, 0) + P_{X_1X_2}(1, 0) + \cdots + P_{X_1X_2}(99, 0)$$

$$= \frac{1.5}{(100)^2} + \frac{1}{(100)^2} + \cdots + \frac{1}{(100)^2} = \frac{100.5}{(100)^2},$$

$$P_{X_2}(1) = P_{X_2}(2) = \cdots = P_{X_2}(98) = \sum_{\text{all } k} P_{X_1X_2}(k, l),\ l = 1, 2, \ldots, 98$$

$$= \frac{100}{(100)^2},$$

$$P_{X_2}(99) = \sum_{\text{all } k} P_{X_1X_2}(k, 99)$$

$$= P_{X_1X_2}(0, 99) + P_{X_1X_2}(1, 99) + \cdots + P_{X_1X_2}(99, 99)$$

$$= \frac{1}{(100)^2} + \frac{1}{(100)^2} + \cdots + \frac{0.5}{(100)^2} = \frac{99.5}{(100)^2}.$$

If the bivariate random variable (X_1, X_2) is continuous, then $f_{X_1X_2}(s, t)$ represents the joint density. The density function of X_1 individually, $f_{X_1}(s)$, now called the marginal density function of the continuous random variable X_1, can be obtained from the $f_{X_1X_2}(s, t)$. In particular,

$$F_{X_1}(b_1) = F_{X_1X_2}(b_1, \infty) = \int_{-\infty}^{b_1}\int_{-\infty}^{\infty} f_{X_1X_2}(s, t)\, dt\, ds = \int_{-\infty}^{b_1} f_{X_1}(s)\, ds,$$

so that

$$f_{X_1}(s) = \int_{-\infty}^{\infty} f_{X_1X_2}(s, t)\, dt.$$

Similarly, the marginal density function of the continuous random variable X_2 is given by

$$f_{X_2}(t) = \int_{-\infty}^{\infty} f_{X_1X_2}(s, t)\, ds.$$

As indicated in Section 3.5, experiments are often performed where some results are obtained early in time and further results later in time. For example, in the previously described experiment that measures the demand for a product during the first two months, the demand for the product during the first month is observed at the end of the first month. This information can be utilized in making probability statements about the demand during the second month. In particular, if the bivariate random variable (X_1, X_2) is discrete, the conditional probability distribution of X_2, given X_1, can be defined as

$$P_{X_2|X_1=k}(l) = P\{X_2 = l | X_1 = k\} = \frac{P_{X_1X_2}(k, l)}{P_{X_1}(k)}, \qquad \text{if } P_{X_1}(k) > 0,$$

and the conditional probability distribution of X_1, given X_2, as

$$P_{X_1|X_2=l}(k) = P\{X_1 = k | X_2 = l\} = \frac{P_{X_1X_2}(k, l)}{P_{X_2}(l)}, \text{ if } P_{X_2}(l) > 0.$$

Note that for a given $X_2 = l$, $P_{X_1|X_2=l}(k)$ satisfies all the conditions for a probability distribution for a discrete random variable. $P_{X_1|X_2=l}(k)$ is nonnegative, and furthermore,

$$\sum_{\text{all } k} P_{X_1|X_2=l}(k) = \sum_{\text{all } k} \frac{P_{X_1X_2}(k, l)}{P_{X_2}(l)} = \frac{P_{X_2}(l)}{P_{X_2}(l)} = 1.$$

Again, returning to the demand for a product during the first 2 months, if it were known that there was no demand during the first month, then

$$P_{X_2|X_1=0}(l) = P\{X_2 = l | X_1 = 0\} = \frac{P_{X_1X_2}(0, l)}{P_{X_1}(0)} = \frac{P_{X_1X_2}(0, l)}{100.5/(100)^2}.$$

Hence

$$P_{X_2|X_1=0}(0) = \frac{P_{X_1X_2}(0, 0)}{(100.5)/(100)^2} = \frac{1.5}{100.5},$$

and

$$P_{X_2|X_1=0}(l) = \frac{1}{100.5} \qquad l = 1, 2, \ldots, 99.$$

If the bivariate random variable (X_1, X_2) is continuous with joint density function $f_{X_1X_2}(s, t)$, and the marginal density function of X_1 is given by $f_{X_1}(s)$, then the conditional density function of X_2, given $X_1 = s$, is defined as

$$f_{X_2|X_1=s}(t) = \frac{f_{X_1X_2}(s, t)}{f_{X_1}(s)}, \qquad \text{if } f_{X_1}(s) > 0.$$

Similarly, if the marginal density function of X_2 is given by $f_{X_2}(t)$, then the conditional density function of X_1, given $X_2 = t$, is defined as

$$f_{X_1|X_2=t}(s) = \frac{f_{X_1X_2}(s, t)}{f_{X2}(t)}, \qquad \text{if } f_{X_2}(t) > 0.$$

Note that, given $X_1 = s$ and $X_2 = t$, the conditional density functions, $f_{X_2|X_1=s}(t)$ and $f_{X_1|X_2=t}(s)$, respectively, satisfy all the conditions for a density function. They are nonnegative, and furthermore,

$$\int_{-\infty}^{\infty} f_{X_2|X_1=s}(t)\,dt = \int_{-\infty}^{\infty} \frac{f_{X_1X_2}(s, t)\,dt}{f_{X_1}(s)}$$

$$= \frac{1}{f_{X_1}(s)} \int_{-\infty}^{\infty} f_{X_1X_2}(s, t)\,dt = \frac{f_{X_1}(s)}{f_{X_1}(s)} = 1,$$

and
$$\int_{-\infty}^{\infty} f_{X_1|X_2=t}(s)\,ds = \int_{-\infty}^{\infty} \frac{f_{X_1X_2}(s, t)\,ds}{f_{X_2}(t)}$$

$$= \frac{1}{f_{X_2}(t)} \int_{-\infty}^{\infty} f_{X_1X_2}(s, t)\,ds = \frac{f_{X_2}(t)}{f_{X_2}(t)} = 1.$$

As an example of the use of these concepts for a continuous bivariate random variable, consider an experiment that measures the time of the first arrivals at a store on each of two successive days. Suppose that the joint density function for the random variable (X_1, X_2), which represents the arrival time on the first and second days, respectively, is given by

$$f_{X_1X_2}(s, t) = \begin{cases} \dfrac{1}{\theta^2} e^{-(s+t)/\theta}, & \text{for } s, t \geq 0 \\ 0, & \text{otherwise.} \end{cases}$$

The marginal density function of X_1 is given by

$$f_{X_1}(s) = \begin{cases} \displaystyle\int_0^{\infty} \frac{1}{\theta^2} e^{-(s+t)/\theta}\,dt = \frac{1}{\theta} e^{-s/\theta}, & \text{for } s \geq 0 \\ 0, & \text{otherwise,} \end{cases}$$

and the marginal density function of X_2 is given by

$$f_{X_2}(t) = \begin{cases} \displaystyle\int_0^{\infty} \frac{1}{\theta^2} e^{-(s+t)/\theta}\,ds = \frac{1}{\theta} e^{-t/\theta}, & \text{for } t \geq 0 \\ 0, & \text{otherwise.} \end{cases}$$

If it is announced that the arrival time of the first customer on the first day occurred at time s, the conditional density of X_2, given $X_1 = s$, is given by

$$f_{X_2|X_1=s}(t) = \frac{f_{X_1X_2}(s, t)}{f_{X_1}(s)} = \frac{(1/\theta^2)e^{-(s+t)/\theta}}{(1/\theta)e^{-s/\theta}} = \frac{1}{\theta}e^{-t/\theta}.$$

It is interesting to note at this point that the conditional density of X_2, given $X_1 = s$, is independent of s and, furthermore, is the same as the marginal density of X_2.

3.12 Expectations for Bivariate Distributions

Section 3.8 defined the expectation of a function of a univariate random variable. The expectation of a function of a bivariate random variable (X_1, X_2) may be defined in a similar manner. Let $g(X_1, X_2)$ be a function of the bivariate random variable (X_1, X_2). Let

$$P_{X_1X_2}(k, l) = P\{X_1 = k, X_2 = l\}$$

denote the joint probability distribution if (X_1, X_2) is a discrete random variable, and let $f_{X_1X_2}(s, t)$ denote the joint density function if (X_1, X_2) is a continuous random variable. The expectation of $g(X_1, X_2)$ is now defined as

$$E[g(X_1, X_2)] = \begin{cases} \sum_{\text{all } k,l} g(k, l)P_{X_1X_2}(k, l), & \text{if } X_1, X_2 \text{ is a discrete random variable} \\ \int_{-\infty}^{\infty}\int_{-\infty}^{\infty} g(s, t)f_{X_1X_2}(s, t)\, ds\, dt, & \text{if } X_1, X_2 \text{ is a continuous random variable.} \end{cases}$$

An alternative definition can be obtained by recognizing that $Z = g(X_1, X_2)$ is itself a univariate random variable and hence has a density function if Z is continuous and a probability distribution if Z is discrete. The expectation of Z for these cases has already been defined in Sec. 3.8. Of particular interest here is the extension of the theorem of the unconscious statistician, which states that if (X_1, X_2) is a continuous random variable and if Z has a density function $h_Z(y)$, then

$$E(Z) = \int_{-\infty}^{\infty} yh_Z(y)\, dy = \int_{-\infty}^{\infty}\int_{-\infty}^{\infty} g(s, t)f_{X_1X_2}(s, t)\, ds\, dt.$$

Thus, the expectation of Z can be found by using its definition in terms of the density of the univariate random variable Z or, alternatively, by use of its definition as the expectation of a function of the bivariate random variable (X_1, X_2) with respect to its joint density function. The identical theorem is true for a discrete bivariate random variable, and, of course, both results are easily extended to n variate random variables.

There are several important functions g that should be considered. All the results will be stated for continuous random variables, but equivalent results also hold for discrete random variables.

If $g(X_1, X_2) = X_1$, it is easily seen that

$$E(X_1) = \int_{-\infty}^{\infty}\int_{-\infty}^{\infty} sf_{X_1X_2}(s, t)\, ds\, dt = \int_{-\infty}^{\infty} sf_{X_1}(s)\, ds.$$

Note that this is just the expectation of the univariate random variable X_1 with respect to its marginal density.

In a similar manner, if $g(X_1, X_2) = [X_1 - E(X_1)]^2$, then

$$E[X_1 - E(X_1)]^2 = \int_{-\infty}^{\infty}\int_{-\infty}^{\infty} [s - E(X_1)]^2 f_{X_1X_2}(s, t)\, ds\, dt$$

$$= \int_{-\infty}^{\infty} [s - E(X_1)]^2 f_{X_1}(s)\, ds,$$

which is just the variance of the univariate random variable X_1 with respect to its marginal density.

If $g(X_1, X_2) = [X_1 - E(X_1)][X_2 - E(X_2)]$, then $E[g(X_1, X_2)]$ is called the covariance of the random variable (X_1, X_2); that is,

$$E[X_1 - E(X_1)][X_2 - E(X_2)] = \int_{-\infty}^{\infty}\int_{-\infty}^{\infty} [s - E(X_1)][t - E(X_2)] f_{X_1X_2}(s, t)\, ds\, dt.$$

An easy computational formula is provided by the identity

$$E[X_1 - E(X_1)][X_2 - E(X_2)] = E(X_1X_2) - E(X_1)E(X_2).$$

The correlation coefficient between X_1 and X_2 is defined to be

$$\rho = \frac{E[X_1 - E(X_1)][X_2 - E(X_2)]}{\sqrt{E[X_1 - E(X_1)]^2 E[X_2 - E(X_2)]^2}}.$$

It is easily shown that $-1 \le \rho \le +1$.

The final results pertain to a linear combination of random variables. Let $g(X_1, X_2) = C_1X_1 + C_2X_2$, where C_1 and C_2 are constants. Then

$$E[g(X_1, X_2)] = \int_{-\infty}^{\infty}\int_{-\infty}^{\infty} (C_1 s + C_2 t) f_{X_1X_2}(s, t)\, ds\, dt,$$

$$= C_1 \int_{-\infty}^{\infty} s f_{X_1}(s)\, ds + C_2 \int_{-\infty}^{\infty} t f_{X_2}(t)\, dt,$$

$$= C_1E(X_1) + C_2E(X_2).$$

Thus, the expectation of a linear combination of univariate random variables is just the sum of the respective coefficient times the expectations of the random variables. It easily follows that

$$E[C_1X_1 + C_2X_2 + \cdots + C_nX_n] = C_1E(X_1) + C_2E(X_2) + \cdots + C_nE(X_n).$$

If $$g(X_1, X_2) = [C_1X_1 + C_2X_2 - \{C_1E(X_1) + C_2E(X_2)\}]^2,$$

then
$$\begin{aligned} E[g(X_1, X_2)] &= \text{variance } (C_1X_1 + C_2X_2) \\ &= C_1^2E[X_1 - E(X_1)]^2 + C_2^2E[X_2 - E(X_2)]^2 \\ &\quad + 2C_1C_2E[X_1 - E(X_1)][X_2 - E(X_2)], \\ &= C_1^2 \text{ variance } (X_1) + C_2^2 \text{ variance } (X_2) \\ &\quad + 2C_1C_2 \text{ covariance } (X_1X_2). \end{aligned}$$

For n univariate random variables, the variance of a linear combination $C_1X_1 + C_2X_2 + \cdots + C_nX_n$ is given by

$$\sum_{i=1}^{n} C_i^2 \text{ variance } (X_i) + 2 \sum_{j=2}^{n} \sum_{i=1}^{j-1} C_iC_j \text{ covariance } (X_iX_j).$$

3.13 Independent Random Variables and Random Samples

The concept of independent events has already been defined; that is, E_1 and E_2 are independent events if, and only if,

$$P\{E_1 \cap E_2\} = P\{E_1\}P\{E_2\}.$$

From this definition the very important concept of independent random variables can be introduced. For a bivariate random variable (X_1, X_2) and constants b_1 and b_2, denote by E_1 the event containing those ω such that $X_1(\omega) \le b_1$, $X_2(\omega)$ is anything; that is,

$$E_1 = \{\omega | X_1(\omega) \le b_1, X_2(\omega) \le \infty\}.$$

Similarly, denote by E_2 the event containing those ω such that $X_1(\omega)$ is anything and $X_2(\omega) \le b_2$; that is,

$$E_2 = \{\omega | X_1(\omega) \le \infty, X_2(\omega) \le b_2\}.$$

Furthermore, the event $E_1 \cap E_2$ is given by

$$E_1 \cap E_2 = \{\omega | X_1(\omega) \le b_1, X_2(\omega) \le b_2\}.$$

The random variables X_1 and X_2 are said to be independent if events of the form given by E_1 and E_2 are independent events for all b_1 and b_2. Using the definition of independent events, then, the random variables X_1 and X_2 are called independent random variables if

$$P\{X_1 \le b_1, X_2 \le b_2\} = P\{X_1 \le b_1\}P\{X_2 \le b_2\}$$

for all b_1 and b_2. Therefore X_1 and X_2 are independent if

$$F_{X_1X_2}(b_1, b_2) = P\{X_1 \le b_1, X_2 \le b_2\} = P\{X_1 \le b_1\}P\{X_2 \le b_2\}$$

$$= F_{X_1}(b_1)F_{X_2}(b_2).$$

Thus, the independence of the random variables X_1 and X_2 implies that the joint CDF factors into the product of the CDF's of the individual random variables. Furthermore, it is easily shown that if (X_1, X_2) is a discrete bivariate random variable, then X_1 and X_2 are independent random variables if, and only if, $P_{X_1X_2}(k, l) = P_{X_1}(k)P_{X_2}(l)$; in other words, $P\{X_1 = k, X_2 = l\} = P\{X_1 = k\}P\{X_2 = l\}$, for all k and l. Similarly, if (X_1, X_2) is a continuous bivariate random variable, then X_1 and X_2 are independent random variables if, and only if,

$$f_{X_1X_2}(s, t) = f_{X_1}(s)f_{X_2}(t),$$

for all s and t. Thus, if X_1, X_2 are to be independent random variables, the joint density (or probability) function must factor into the product of the marginal density functions of the random variables. Using this result, it is easily seen that if X_1, X_2 are independent random variables, then the covariance of X_1, X_2 must be zero. Hence, the results on the variance of linear combinations of random variables given in Sec. 3.12 can be simplified when the random variables are independent; that is,

$$\text{Variance}\left(\sum_{i=1}^{n} C_i X_i\right) = \sum_{i=1}^{n} C_i^2 \text{ variance } (X_i)$$

when the X_i are independent.

Another interesting property of independent random variables can be deduced from the factorization property. If (X_1, X_2) is a discrete bivariate random variable, then X_1 and X_2 are independent if, and only if,

$$P_{X_1|X_2=l}(k) = P_{X_1}(k), \qquad \text{for all } k \text{ and } l.$$

Similarly, if (X_1, X_2) is a continuous bivariate random variable, then X_1 and X_2 are independent if, and only if,

$$f_{X_1|X_2=t}(s) = f_{X_1}(s), \qquad \text{for all } s \text{ and } t.$$

In other words, if X_1 and X_2 are independent, a knowledge of the outcome of one, say, X_2, gives no information about the probability distribution of the other, say, X_1. It was noted in the example in Sec. 3.11 on the time of first arrivals that the conditional density of the arrival time of the first customer on the second day, given that the first customer on the first day arrived at time s, was equal to the marginal density of the arrival time of the first customer on the second day. Hence, X_1 and X_2 were independent random variables. In the example of the demand for a product during 2 consecutive months it was seen that

$$P_{X_2|X_1=0}(0) = \frac{1.5}{100.5} \neq P_{X_2}(0) = \frac{100.5}{(100)^2}.$$

Hence, the demands during each month were dependent (not independent) random variables.

The definition of independent random variables generally does not lend itself to determine whether or not random variables are independent in a probabilistic sense by looking at their outcomes. Instead, by analyzing the physical situation the experimenter usually is able to make a judgment about whether the random variables are independent by ascertaining if the outcome of one will affect the probability distribution of the other.

The definition of independent random variables is easily extended to three or more random variables. For example, if the joint CDF of the n-dimensional random variable $(X_1, X_2, \ldots, X_n)$ is given by $F_{X_1X_2\ldots X_n}(b_1, b_2, \ldots, b_n)$ and $F_{X_1}(b_1)$, $F_{X_2}(b_2), \ldots, F_{X_n}(b_n)$ represents the CDF's of the univariate random variables X_1, $X_2, \ldots, X_n$, respectively, then $X_1, X_2, \ldots, X_n$ are independent random variables if, and only if,

$$F_{X_1X_2\ldots X_n}(b_1, b_2, \ldots, b_n) = F_{X_1}(b_1)F_{X_2}(b_2)\cdots F_{X_n}(b_n), \quad \text{for all } b_1, b_2, \ldots, b_n.$$

Having defined the concept of independent random variables, we can now introduce the term *random sample*. A random sample simply means a sequence of independent and identically distributed random variables. Thus $X_1, X_2, \ldots, X_n$ constitute a random sample of size n if the X_i are independent and identically distributed random variables. For example, in Sec. 3.6 it was pointed out that if $X_1, X_2, \ldots, X_n$ are independent Bernoulli random variables, each with parameter p (that is, if the X's are a random sample), then the random variable

$$X = \sum_{i=1}^{n} X_i$$

has a binomial distribution with parameters n and p.

3.14 Law of Large Numbers

Section 3.8 pointed out that the mean of a random sample tends to converge to the expectation of the random variables as the sample size increases. In particular, suppose the random variable X, the demand for a product, may take on one of the possible values $k = 0, 1, 2, \ldots, 98, 99$, each with $P_X(k) = \frac{1}{100}$ for all k. Then $E(X)$ is easily seen to be 49.5. If a random sample of size n is taken, that is, the demands are observed for n days, with each day's demand being independent and identically distributed random variables, it was noted that the random variable $\overline{X}$ should take on a value close to 49.5 if n is large. This result can be stated precisely as the *law of large numbers*.

Law of Large Numbers

Let the random variables $X_1, X_2, \ldots, X_n$ be independent, identically distributed random variables (a random sample of size n), each having mean μ. Consider the random variable that is the sample mean $\overline{X}$:

$$\overline{X} = \frac{X_1 + X_2 + \cdots + X_n}{n}.$$

Then for any constant $\varepsilon > 0$,

$$\lim_{n \to \infty} P\{|\overline{X} - \mu| > \varepsilon\} = 0.$$

The interpretation of the law of large numbers is that as the sample size increases, the probability is "close" to 1 that $\overline{X}$ is "close" to μ. Assuming that the variance of each X_i is $\sigma^2 < \infty$, this result is easily proved by using Chebyshev's inequality (stated in Sec. 3.9). Since each X_i has mean μ and variance σ^2, $\overline{X}$ also has mean μ, but its variance is σ^2/n. Hence, applying Chebyshev's inequality to the random variable $\overline{X}$, it is evident that

$$P\left\{\mu - \frac{C\sigma}{\sqrt{n}} \le \overline{X} \le \mu + \frac{C\sigma}{\sqrt{n}}\right\} > 1 - \frac{1}{C^2}.$$

This is equivalent to

$$P\left\{|\overline{X} - \mu| > \frac{C\sigma}{\sqrt{n}}\right\} < \frac{1}{C^2}.$$

Let $C\sigma/\sqrt{n} = \varepsilon$, so that $C = \varepsilon\sqrt{n}/\sigma$. Thus,

$$P\{|\overline{X} - \mu| > \varepsilon\} < \frac{\sigma^2}{\varepsilon^2 n},$$

so that

$$\lim_{n\to\infty} P\{|\overline{X} - \mu| > \varepsilon\} = 0,$$

as was to be proved.

3.15 Central Limit Theorem

Section 3.7 pointed out that sums of independent normally distributed random variables are themselves normally distributed, and that even if the random variables are *not* normally distributed, the distribution of their sum still tends toward normality. This latter statement can be made precise by means of the *central limit theorem*.

Central Limit Theorem

Let the random variables $X_1, X_2, \ldots, X_n$ be independent with means $\mu_1, \mu_2, \ldots, \mu_n$, respectively, and variance $\sigma_1^2, \sigma_2^2, \ldots, \sigma_n^2$, respectively. Consider the random variable Z_n,

$$Z_n = \frac{\sum_{i=1}^n X_i - \sum_{i=1}^n \mu_i}{\sqrt{\sum_{i=1}^n \sigma_i^2}}.$$

Then, under certain regularity conditions, Z_n is approximately normally distributed with zero mean and unit variance in the sense that

$$\lim_{n\to\infty} P\{Z_n \le b\} = \int_{-\infty}^{b} \frac{1}{\sqrt{2\pi}} e^{-y^2/2}\, dy.$$

Note that if the X_i form a random sample, with each X_i having mean μ and variance σ^2, then $Z_n = (\overline{X} - \mu)\sqrt{n}/\sigma$.[1] Hence, sample means from random samples tend toward normality in the sense just described by the central limit theorem even if the X_i are not normally distributed.

It is difficult to give sample sizes beyond which the central limit theorem applies, and approximate normality can be assumed for sample means. This, of course, does depend upon the form of the underlying distribution. From a practical point of view, moderate sample sizes, like 10, are often sufficient.

3.16 Functions of Random Variables

Section 3.8 introduced the theorem of the unconscious statistician and pointed out that if a function $Z = g(X)$ of a continuous random variable is considered, its expectation can be taken with respect to the density function $f_X(y)$ of X or the density

[1] Under these conditions, the central limit theorem actually holds without assuming any other regularity conditions.

function $h_Z(y)$ of Z. In discussing this choice, it was implied that the density function of Z was known. In general, then, given the cumulative distribution function $F_X(b)$ of a random variable X, there may be interest in obtaining the cumulative distribution function $H_Z(b)$ of a random variable $Z = g(X)$. Of course, it is always possible to go back to the sample space and determine $H_Z(b)$ directly from probabilities associated with the sample space. However, alternate methods for doing this are desirable.

If X is a discrete random variable, the values k that the random variable X takes on and the associated $P_X(k)$ are known. If $Z = g(X)$ is also discrete, denote by m the values that Z takes on. The probabilities $Q_Z(m) = P\{Z = m\}$ for all m are required. The general procedure is to enumerate for each m all the values of k such that

$$g(k) = m.$$

$Q_Z(m)$ is then determined as

$$Q_Z(m) = \sum_{\substack{\text{all } k \\ \text{such that} \\ g(k)=m}} P_X(k).$$

To illustrate, consider again the example involving the demand for a product in a single month. Let this random variable be noted by X, and let $k = 0, 1, \ldots, 99$ with $P_X(k) = \frac{1}{100}$, for all k. Consider a new random variable Z that takes on the value of 0 if there is no demand and 1 if there is *any* demand. This random variable may be useful for determining whether any is needed for shipping. The probabilities

$$Q_Z(0) \quad \text{and} \quad Q_Z(1)$$

are required. If $m = 0$, the only value of k such that $g(k) = 0$ is $k = 0$. Hence,

$$Q_Z(0) = \sum_{\substack{\text{all } k \\ \text{such that} \\ g(k)=0}} P_X(k) = P_X(0) = \tfrac{1}{100}.$$

If $m = 1$, the values of k such that $g(k) = 1$ are $k = 1, 2, 3, \ldots, 98, 99$. Hence,

$$Q_Z(1) = \sum_{\substack{\text{all } k \\ \text{such that} \\ g(k)=1}} P_X(k)$$

$$= P_X(1) + P_X(2) + P_X(3) + \cdots + P_X(98) + P_X(99) = \tfrac{99}{100}.$$

If X is a continuous random variable, then both the CDF $F_X(b)$ and the density function $f_X(y)$ may be assumed to be known. If $Z = g(X)$ is also a continuous random variable, either the CDF $H_Z(b)$ or the density function $h_Z(y)$ is sought. To find $H_Z(b)$, note that

$$H_Z(b) = P\{Z \le b\} = P\{g(X) \le b\} = P\{A\},$$

where A consists of all points such that $g(X) \le b$. Thus $P\{A\}$ can be determined from the density function or CDF of the random variable X. For example, suppose that the CDF for the time of the first arrival in a store is given by

$$F_X(b) = \begin{cases} 1 - e^{-b/\theta}, & \text{for } b \ge 0 \\ 0, & \text{for } b < 0, \end{cases}$$

where $\theta > 0$. Suppose further that the random variable $Z = g(X) = X + 1$, which represents an hour after the first customer arrives, is of interest, and the CDF of Z,

$H_Z(b)$, is desired. To find this CDF note that

$$H_Z(b) = P\{Z \le b\} = P\{X + 1 \le b\} = P\{X \le b - 1\}$$

$$= \begin{cases} 1 - e^{-(b-1)/\theta}, & \text{for } b \ge 1 \\ 0, & \text{for } b < 1. \end{cases}$$

Furthermore, the density can be obtained by differentiating the CDF; that is,

$$h_Z(y) = \begin{cases} \dfrac{1}{\theta} e^{-(y-1)/\theta}, & \text{for } y \ge 1 \\ 0, & \text{for } y < 1. \end{cases}$$

Another technique can be used to find the density function directly if $g(X)$ is monotone and differentiable; it can be shown that

$$h_Z(y) = f_X(s) \left| \frac{ds}{dy} \right|,$$

where s is expressed in terms of y. In the example, $Z = g(X) = X + 1$, so that y, the value the random variable Z takes on, can be expressed in terms of s, the value the random variable X takes on; that is, $y = g(s) = s + 1$. Thus,

$$s = y - 1,\ f_X(s) = \frac{1}{\theta} e^{-s/\theta} = \frac{1}{\theta} e^{-(y-1)/\theta}, \quad \text{and} \quad \frac{ds}{dy} = 1.$$

Hence
$$h_Z(y) = \frac{1}{\theta} e^{-(y-1)/\theta} |1| = \frac{1}{\theta} e^{-(y-1)/\theta},$$

which is the result previously obtained.

All the discussion in this section concerned functions of a single random variable. If (X_1, X_2) is a bivariate random variable, there may be interest in the probability distribution of such functions as $X_1 + X_2$, X_1X_2, X_1/X_2, and so on. If (X_1, X_2) is discrete, the technique for single random variables is easily extended. A detailed discussion of the techniques available for continuous bivariate random variables is beyond the scope of this text; however, a few notions related to independent random variables will be discussed.

If (X_1, X_2) is a continuous bivariate random variable, and X_1 and X_2 are independent, then its joint density is given by

$$f_{X_1X_2}(s, t) = f_{X_1}(s) f_{X_2}(t).$$

Consider the function

$$Z = g(X_1, X_2) = X_1 + X_2.$$

The CDF for Z can be expressed as $H_Z(b) = P\{Z \le b\} = P\{X_1 + X_2 \le b\}$. This can be evaluated by integrating the bivariate density over the region such that $s + t \le b$; that is

$$H_Z(b) = \iint_{s+t \le b} f_{X_1}(s) f_{X_2}(t)\, ds\, dt$$

$$= \int_{-\infty}^{\infty} \int_{-\infty}^{b-t} f_{X_1}(s) f_{X_2}(t)\, ds\, dt.$$

Differentiating with respect to b yields the density function

$$h_Z(y) = \int_{-\infty}^{\infty} f_{X_2}(t) f_{X_1}(y - t)\, dt.$$

This can be written alternately as

$$h_Z(y) = \int_{-\infty}^{\infty} f_{X_1}(s) f_{X_2}(y - s)\, ds.$$

Note that the integrand may be zero over part of the range of the variable, as shown in the following example.

Suppose that the times of the first arrival on two successive days, X_1 and X_2, are independent, identically distributed random variables having density

$$f_{X_1}(s) = \begin{cases} \dfrac{1}{\theta} e^{-s/\theta}, & \text{for } s \geq 0 \\ 0, & \text{otherwise.} \end{cases}$$

$$f_{X_2}(t) = \begin{cases} \dfrac{1}{\theta} e^{-t/\theta}, & \text{for } t \geq 0 \\ 0, & \text{otherwise.} \end{cases}$$

To find the density of $Z = X_1 + X_2$, note that

$$f_{X_1}(s) = \begin{cases} \dfrac{1}{\theta} e^{-s/\theta}, & \text{for } s \geq 0 \\ 0, & \text{for } s < 0, \end{cases}$$

and

$$f_{X_2}(y - s) = \begin{cases} \dfrac{1}{\theta} e^{-(y-s)/\theta}, & \text{if } y - s \geq 0 \text{ so that } s \leq y \\ 0, & \text{if } y - s < 0 \text{ so that } s > y. \end{cases}$$

Hence
$$f_{X_1}(s) f_{X_2}(y - s) = \begin{cases} \dfrac{1}{\theta} e^{-s/\theta} \dfrac{1}{\theta} e^{-(y-s)/\theta} = \dfrac{1}{\theta^2} e^{-y/\theta}, & \text{if } 0 \leq s \leq y \\ 0, & \text{otherwise.} \end{cases}$$

Hence
$$h_Z(y) = \int_{-\infty}^{\infty} f_{X_1}(s) f_{X_2}(y - s)\, ds = \int_0^y \frac{1}{\theta^2} e^{-y/\theta}\, ds$$

$$= \frac{y}{\theta^2} e^{-y/\theta}.$$

Note that this is just a gamma distribution, with parameters $\alpha = 2$ and $\beta = \theta$. Hence, as indicated in Sec. 3.7, the sum of two independent, exponentially distributed random variables has a gamma distribution. This example illustrates how to find the density function for finite sums of independent random variables. Combining this result with those for univariate random variables leads to easily finding the density function of linear combinations of independent random variables.

A final result on the distribution of functions of random variables concerns functions of normally distributed random variables. The chi-square and the t and F distributions, introduced in Sec. 3.7, can be generated from functions of normally distributed random variables. These distributions are particularly useful in the study of statistics. In particular, let $X_1, X_2, \ldots, X_v$ be independent, normally distributed random variables having zero mean and unit variance. The random variable

$$\chi^2 = X_1^2 + X_2^2 + \cdots + X_v^2$$

can be shown to have a chi-square distribution with v degrees of freedom. A random variable having a t distribution may be generated as follows. Let X be a normally distributed random variable having zero mean and unit variance and χ^2 be a chi-square random variable (independent of X) with v degrees of freedom. The random variable

$$t = \frac{\sqrt{v}X}{\sqrt{\chi^2}}$$

can be shown to have a t distribution with v degrees of freedom. Finally, a random variable having an F distribution can be generated from a function of two independent chi-square random variables. Let χ_1^2 and χ_2^2 be independent chi-square random variables, with v_1 and v_2 degrees of freedom, respectively. The random variable

$$F = \frac{\chi_1^2/v_1}{\chi_2^2/v_2}$$

can be shown to have an F distribution with v_1 and v_2 degrees of freedom.

SELECTED REFERENCES

1. DeGroot, M. H.: *Probability and Statistics*, 2d ed., Addison-Wesley, Reading, Mass., 1986.
2. Feller, W.: *An Introduction to Probability Theory and Its Applications*, vol. 1, 2d ed., Wiley, New York, 1957.
3. Hoel, D., S. Port, and C. Stone: *Introduction to Probability Theory*, Houghton-Mifflin, Boston, Mass., 1971.
4. Meyer, P. L.: *Introductory Probability and Statistical Applications*, Addison-Wesley, Reading, Mass., 1965.
5. Parzen, E.: *Modern Probability Theory and Its Applications*, Wiley, New York, 1960.
6. Ross, S.: *A First Course in Probability*, 3d ed., Macmillan, New York, 1988.

PROBLEMS

1.* A cube has its six sides colored red, white, blue, green, yellow, and violet. It is assumed that these six sides are equally likely to show when the cube is tossed. The cube is tossed once.

(*a*) Describe the sample space.

(*b*) Consider the random variable that assigns the number 0 to red and white, the number 1 to green and blue, and the number 2 to yellow and violet. What is the distribution of this random variable?

(*c*) Let $Y = (X + 1)^2$, where X is the random variable in part (*b*). Find $E(Y)$.

2. Suppose the sample space Ω consists of the four points

$$\omega_1, \omega_2, \omega_3, \omega_4,$$

and the associated probabilities over the events are given by

$$P\{\omega_1\} = \tfrac{1}{3}, P\{\omega_2\} = \tfrac{1}{5}, P\{\omega_3\} = \tfrac{3}{10}, P\{\omega_4\} = \tfrac{1}{6}.$$

Define the random variable X_1 by

$$X_1(\omega_1) = 1,$$

$$X_1(\omega_2) = 1,$$

$$X_1(\omega_3) = 4,$$

$$X_1(\omega_4) = 5,$$

and the random variable X_2 by

$$X_2(\omega_1) = 1,$$

$$X_2(\omega_2) = 1,$$

$$X_2(\omega_3) = 1,$$

$$X_2(\omega_4) = 5.$$

(*a*) Find the probability distribution of X_1, that is, $P_{X_1}(i)$.
(*b*) Find $E(X_1)$.
(*c*) Find the probability distribution of the random variable $X_1 + X_2$, that is, $P_{X_1+X_2}(i)$.
(*d*) Find $E(X_1 + X_2)$ and $E(X_2)$.
(*e*) Find $F_{X_1X_2}(b_1, b_2)$.
(*f*) Compute the correlation coefficient between X_1 and X_2.
(*g*) Compute $E[2X_1 - 3X_2]$.

3. During the course of a day a machine turns out two items, one in the morning and one in the afternoon. The quality of each item is measured as good (G), mediocre (M), or bad (B). The long-run fraction of good items the machine produces is $\frac{1}{2}$, the fraction of mediocre items is $\frac{1}{3}$, and the fraction of bad items is $\frac{1}{6}$.
(*a*) In a column, write the sample space for the experiment that consists of observing the day's production.
(*b*) Assume a good item returns a profit of \$2, a mediocre item a profit of \$1, and a bad item yields nothing. Let X be the random variable describing the total profit for the day. In a column adjacent to the column in part (*a*), write the value of this random variable corresponding to each point in the sample space.
(*c*) Assuming that the qualities of the morning and afternoon items are independent, in a third column associate with every point in the sample space a probability for that point.
(*d*) Write the set of all possible outcomes for the random variable X. Give the probability distribution function for the random variable.
(*e*) What is the expected value of the day's profit?

4.* The random variable X has density function f given by

$$f_X(y) = \begin{cases} \theta, & \text{for } 0 \le y \le \theta \\ K, & \text{for } \theta < y \le 1 \\ 0, & \text{elsewhere.} \end{cases}$$

(*a*) Determine K in terms of θ.
(*b*) Find $F_X(b)$, the CDF of X.
(*c*) Find $E(X)$.
(*d*) Suppose $\theta = \frac{1}{3}$. Is $P\{X - \frac{1}{3} < a\} = P\{-(X - \frac{1}{3} < a\}$?

5. Let X be a discrete random variable, with probability distribution

$$P\{X = x_1\} = \tfrac{1}{4}$$

and $$P\{X = x_2\} = \tfrac{3}{4}.$$

(*a*) Determine x_1 and x_2, so that

$$E(X) = 0 \text{ and variance } (X) = 10.$$

(*b*) Sketch the CDF of X.

6. The life X, in hours, of a certain kind of radio tube has a probability density function given by

$$f_X(y) = \begin{cases} \dfrac{100}{y^2}, & \text{for } y \geq 100 \\ 0, & \text{for } y < 100. \end{cases}$$

(*a*) What is the probability that a tube will survive 250 hours of operation?
(*b*) Find the expected value of the random variable.

7.* The random variable X can take on only the values $0, \pm 1, \pm 2$, and

$$P\{-1 < X < 2\} = 0.4, \qquad P\{X = 0\} = 0.3,$$

$$P\{|X| \leq 1\} = 0.6, \qquad P\{X \geq 2\} = P\{X = 1 \text{ or } -1\}.$$

(*a*) Find the probability distribution of X.
(*b*) Graph the CDF of X.
(*c*) Compute $E(X)$.

8. Let X be a random variable with density

$$f_X(y) = \begin{cases} K(1 - y^2), & \text{for } -1 < y < 1 \\ 0, & \text{otherwise.} \end{cases}$$

(*a*) What value of K will make $f_X(y)$ a true density?
(*b*) What is the CDF of X?
(*c*) Find $E(2X - 1)$.
(*d*) Find the variance (X).
(*e*) Find the approximate value of $P\{\overline{X} > 0\}$, where $\overline{X}$ is the sample mean from a random sample of size $n = 100$ from the above distribution. (*Hint*: Note that n is "large.")

9. The distribution of X, the life of a transistor, in hours, is approximated by a triangular distribution as follows:

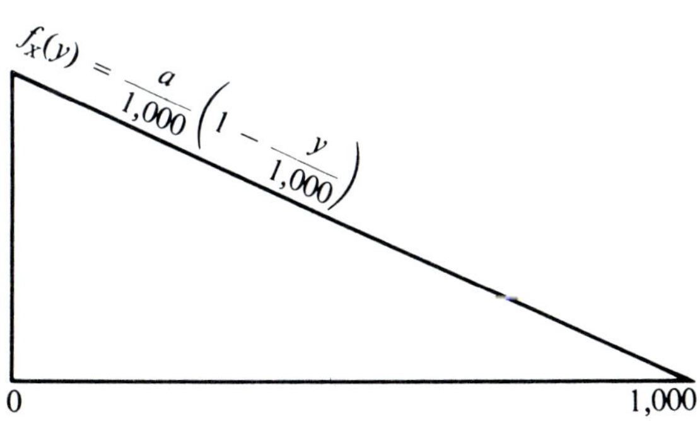

(*a*) What is the value of a?
(*b*) Find the expected value of the life of transistors.
(*c*) Find the CDF, $F_X(b)$, for this density. Note that this must be defined for all b between plus and minus infinity.

(*d*) If X represents the random variable, the life of a transistor, let $Z = 3X$ be a new random variable. Using the results of (*c*), find the CDF of Z.

10.* The number of orders per week, X, for radios can be assumed to have a Poisson distribution with parameter $\lambda = 25$.

(*a*) Find $P\{X \geq 25\}$ and $P\{X = 20\}$.

(*b*) If the number of radios in the inventory is 35, what is the probability of a shortage occurring in a week?

11. Consider the following game. Player A flips a fair coin until a head appears. She pays player B 2^n dollars, where n is the number of tosses required until a head appears. For example, if a head appears on the first trial, player A pays player B \$2. If the game results in 4 tails followed by a head, player A pays player B $2^5 = \$32$. Therefore the payoff to player B is a random variable that takes on the values 2^n for $n = 1, 2, \ldots$ and whose probability distribution is given by $(\frac{1}{2})^n$ for $n = 1, 2, \ldots$, that is, if X denotes the payoff to player B,

$$P(X = 2^n) = (\tfrac{1}{2})^n \qquad \text{for } n = 1, 2, \ldots .$$

The usual definition of a fair game between two players is for each player to have equal expectation for the amount to be won.

(*a*) How much should player B pay to player A so that this game will be fair?

(*b*) What is the variance of X?

(*c*) What is the probability of player B winning no more than \$8 in one play of the game?

12. The demand D for a product in a week is a random variable taking on the values of $-1, 0, 1$ with probabilities $\frac{1}{8}$, $\frac{5}{8}$, and $C/8$, respectively. A demand of -1 implies that an item is returned.

(*a*) Find C, $E(D)$, and variance D.

(*b*) Find $E(e^{D^2})$.

(*c*) Sketch the CDF of the random variable D, labeling all the necessary values.

13.* In a certain chemical process three bottles of a standard fluid are emptied into a large container. A study of the individual bottles shows that the mean value of the contents is 15 ounces and the standard deviation is 0.08 ounces. If three bottles form a random sample,

(*a*) Find the expected value and the standard deviation of the volume of liquid emptied into the large container.

(*b*) If the content of the individual bottles is normally distributed, what is the probability that the volume of liquid emptied into the larger container will be in excess of 45.2 ounces?

14. Consider the density function of a random variable X defined by

$$f_X(y) = \begin{cases} 0, & \text{for } \quad y < 0 \\ 6y(1 - y), & \text{for } 0 \leq y \leq 1 \\ 0, & \text{for } \quad 1 < y. \end{cases}$$

(*a*) Find the CDF corresponding to this density function. (Be sure you describe it completely.)

(*b*) Calculate the mean and variance.

(*c*) What is the probability that a random variable having this density will exceed 0.5?

(*d*) Consider the experiment where six independent random variables are observed, each random variable having the density function given above. What is the expected value of the sample mean of these observations?

(*e*) What is the variance of the sample mean described in part (*d*)?

15. A transistor radio operates on two $1\frac{1}{2}$-volt batteries, so that nominally it operates on 3 volts. Suppose the actual voltage of a single new battery is normally distributed with mean

$1\frac{1}{2}$ volts and variance 0.0625. The radio will not operate "properly" at the outset if the voltage falls outside the range $2\frac{3}{4}$ to $3\frac{1}{4}$ volts.

(a) What is the probability that the radio will not operate "properly?"

(b) Suppose that the assumption of normality is not valid. Give a bound on the probability that the radio will not operate "properly."

16.* The life of electric light bulbs is known to be a normally distributed random variable with unknown mean μ and standard deviation 200 hours. The value of a lot of 1,000 bulbs is (1,000)(1/5,000) μ dollars. A random sample of n bulbs is to be drawn by a prospective buyer, and 1,000(1/5,000) $\overline{X}$ dollars paid to the manufacturer. How large should n be so that the probability is 0.90 that the buyer does not overpay or underpay the manufacturer by more than \$15?

17. A joint random variable (X_1, X_2) is said to have a bivariate normal distribution if its joint density is given by

$$f_{X_1X_2}(s, t) = \frac{1}{2\pi\sigma_{X_1}\sigma_{X_2}\sqrt{1-\rho^2}}$$

$$\exp\left\{-\frac{1}{2(1-\rho^2)}\left[\left(\frac{s-\mu_{X_1}}{\sigma_{X_1}}\right)^2 - 2\rho\frac{(s-\mu_{X_1})(t-\mu_{X_2})}{\sigma_{X_1}\sigma_{X_2}} + \left(\frac{t-\mu_{X_2}}{\sigma_{X_2}}\right)^2\right]\right\}$$

for $-\infty < s < \infty$ and $-\infty < t < \infty$.

(a) Show that $E(X_1) = \mu_{X_1}$ and $E(X_2) = \mu_{X_2}$.

(b) Show that variance $(X_1) = \sigma_{X_1}^2$, variance $(X_2) = \sigma_{X_2}^2$, and the correlation coefficient is ρ.

(c) Show that marginal distributions of X_1 and X_2 are normal.

(d) Show that the conditional distribution of X_1, given $X_2 = x_2$, is normal with mean

$$\mu_{X_1} + \rho\frac{\sigma_{X_1}}{\sigma_{X_2}}(x_2 - \mu_{X_2})$$

and variance $\sigma_{X_1}^2(1-\rho^2)$.

18. The joint demand for a product over 2 months is a continuous random variable (X_1, X_2) having a joint density given by

$$f_{X_1,X_2}(s, t) = \begin{cases} c, & \text{if } 100 \le s \le 150, \text{ and } 50 < t \le 100 \\ 0, & \text{otherwise.} \end{cases}$$

(a) Find c.

(b) Find $F_{X_1X_2}(b_1, b_2)$, $F_{X_1}(b_1)$, and $F_{X_2}(b_2)$.

(c) Find $f_{X_2|X_1=s}(t)$.

19.* Two machines produce a certain item. The capacity per day of machine 1 is 1 unit and that of machine 2 is 2 units. Let (X_1, X_2) be the discrete random variable that measures the actual production on each machine per day. Each entry in the table below represents the joint probability, for example, $P_{X_1X_2}(0, 0) = \frac{1}{8}$.

X_2 \ X_1	0	1
0	$\frac{1}{8}$	0
1	$\frac{1}{4}$	$\frac{1}{8}$
2	$\frac{1}{8}$	$\frac{3}{8}$

(*a*) Find the marginal distributions of X_1 and X_2.
(*b*) Find the conditional distribution of X_1, given $X_2 = 1$.
(*c*) Are X_1 and X_2 independent random variables?
(*d*) Find $E(X_1)$, $E(X_2)$, variance (X_1), and variance (X_2).
(*e*) Find the probability distribution of $(X_1 + X_2)$.

20. Suppose that $E_1, E_2, \ldots, E_m$ are mutually exclusive events such that $E_1 \cup E_2 \cup \cdots \cup E_m = \Omega$; that is, exactly one of the E events will occur. Denote by F any event in the sample space. Note that

$$F = FE_1 \cup FE_2 \cup \cdots \cup FE_m$$

and that FE_i, $i = 1, 2, \ldots, m$ are also mutually exclusive.

(*a*) Show that $P\{F\} = \sum_{i=1}^{m} P\{FE_i\} = \sum_{i=1}^{m} P\{F|E_i\}P\{E_i\}$.

(*b*) Show that $P\{E_i|F\} = \dfrac{P\{F|E_i\}P\{E_i\}}{\sum_{i=1}^{m} P\{F|E_i\}P\{E_i\}}$.

(This result is called Bayes' formula and is useful when it is known that the event F has occurred and there is interest in determining which one of the E_i also occurred.)

4

Statistical Inference

4.1 Introduction

In the previous chapter, the basic notions of probability theory were introduced. Little mention was made of decision making since the discussion was concerned with describing natural phenomena in terms of a probabilistic model. Random variables and their associated probability distributions were always assumed to be completely specified. Even in this situation, and in the absence of sample data, decisions may be required.

A more common occurrence involves decision making when sample data are available; this is the concern of statistics. Statistics is a science which deals with making decisions from observed data in the face of uncertainty. This chapter presents a framework for making decisions when sample data are both available and unavailable.

4.2 Decision-Making Framework

Consider the following inventory problem. A fleet of a new model airplane is to be purchased by a major airline carrier, and the number of spare engines accompanying the order must be determined. The airline carrier must order these spare engines in batches of 5 and can only choose among 15, 20, or 25 spares. The supplier of these engines has two plants, and the airline carrier must make its decision prior to knowing which plant will be used. From past experience it is known that the number of spare engines required when production takes place at plant A is approximated by a Poisson distribution with parameter $\lambda = 21$, whereas the number of spare engines required when production takes place at plant B is approximated by a Poisson distribution with parameter $\lambda = 24$. The cost of a spare engine purchased now is \$250,000, whereas the cost of purchasing a spare engine at a later date will be \$500,000. Holding costs and interest charges are to be neglected. Spares must always be supplied if they are demanded, and unused engines will be scrapped when the airplanes become obsolete.

Before seeking a solution to this problem, it is worthwhile to formulate a general framework for decision making. The decision-maker must choose an action a from a set A of possible actions. In the inventory example, the set A consists of three points, a_1, a_2, a_3, corresponding to ordering 15, 20, or 25 spare engines, respectively. In taking an action, the decision-maker must be aware of its consequences, which will be a function of the "state of nature." A "state of nature," θ, is a representation of a possible situation to which the action will apply. Generally, the "state of nature" is a representation of the probabilistic model of the physical phenomenon being studied and is characterized by the parameter of a family of probability distributions. The set of possible values that θ can assume will be denoted by Θ. In the inventory example Θ consists of two points, θ_1 and θ_2, which can be $\lambda = 21$ and $\lambda = 24$, the parameters of two possible Poisson distributions. In order to measure the consequences of a decision-maker's action, it will be assumed that there exists a *loss function* $l(a, \theta)$ which reflects the loss from taking action a when the state of nature is θ, and which is defined for each combination of a and θ. If the problem is formulated in terms of gains, a gain can be termed as a negative loss. The loss function is generally measured in monetary terms, although other "utility" functions can be used. Note that $l(a, \theta)$ is assumed to be a function only of a and θ. The interpretation of the loss function requires further explanation. In the inventory example, the "cost" incurred when, say, action a_2 is taken and the state of nature is θ_1 (when 15 spares are ordered and $\lambda = 21$) depends upon the number of spare engines actually required. The number of spare engines actually required is a random variable whose probability distribution is a function of $\lambda = 21$. In this situation, the loss is measured as the *expected* "cost" and is a function of a_2 and θ_1. Thus, in general, when "costs" associated with a and θ are random variables, the loss function is measured as the *expected* "cost."

For the inventory example, Table 4.1, the loss table for the inventory problem, can now be computed.

An entry in the table, for example, the entry corresponding to taking action a_2 when the true state of nature is θ_1, is computed as follows. Let the random variable Z represent the total cost of spare engines purchased, which is a function of the

Table 4.1 **Loss Table for the Inventory Problem**

Action \ State of Nature	$\theta_1 : \lambda = 21$	$\theta_2 : \lambda = 24$
a_1: order 15	6.8265×10^6	8.265×10^6
a_2: order 20	6.178×10^6	7.270×10^6
a_3: order 25	6.514×10^6	7.002×10^6

demand. Thus, when action a_2 is taken, and the true state of nature is $\theta_1 = 21$ ($\theta_1 = 21$ is written in place of θ_1: $\lambda = 21$ for economy of notation), then

$$Z(D) = \begin{cases} (250{,}000)(20) + (500{,}000)(D - 20), & \text{for } D > 20, \\ (250{,}000)(20), & \text{for } D \le 20, \end{cases}$$

where D, the number of spare engines required, is a random variable having a Poisson distribution with parameter $\lambda = 21$.

As mentioned above, the expected total cost of spare engines purchased is required for completion of the loss table. By applying the material in Secs. 3.6.2 and 3.8, the expected value of the total cost of spare engines purchased, $E(Z)$, can be calculated as

$$\begin{aligned} E(Z) &= \sum_{k=0}^{\infty} Z(k)P_D(k) = \sum_{k=0}^{20} Z(k)P_D(k) + \sum_{k=21}^{\infty} Z(k)P_D(k) \\ &= \sum_{k=0}^{20} 5{,}000{,}000P_D(k) + \sum_{k=21}^{\infty} (5{,}000{,}000 + 500{,}000k - 10{,}000{,}000)P_D(k) \\ &= 5{,}000{,}000P\{D \le 20\} - 5{,}000{,}000P\{D > 20\} + 500{,}000 \sum_{k=21}^{\infty} kP_D(k) \\ &= -5{,}000{,}000 + 10{,}000{,}000P\{D \le 20\} + 500{,}000 \sum_{k=21}^{\infty} \frac{k(21)^k e^{-21}}{k!}, \end{aligned}$$

where $P\{D \le 20\}$ is the probability that a Poisson random variable with parameter $\lambda = 21$ is less than or equal to 20. By referring to Table A5.4 of Appendix 5, it is seen that

$$P\{D \le 20\} = 0.471$$

and that

$$\begin{aligned} \sum_{k=21}^{\infty} \frac{k(21)^k e^{-21}}{k!} &= 21 \sum_{k=21}^{\infty} \frac{(21)^{k-1} e^{-21}}{(k-1)!} \\ &= 21 \sum_{j=20}^{\infty} \frac{(21)^j e^{-21}}{j!} \\ &= 21(1 - 0.384) = 12.936. \end{aligned}$$

Hence, $E(Z) = -5{,}000{,}000 + 4{,}710{,}000 + 6{,}468{,}000 = 6{,}178{,}000$, and the entry is obtained. Similar calculations are made for the other five entries in the table. Although the calculations required are tedious, this table may now be assumed to be given; that is, the loss function $l(a, \theta)$ is well-defined for all θ and a.

From the foundations of utility theory, the regret function is an alternative to the loss function in measuring the consequences of taking an action a when the true state of nature is θ. The *regret function* $r(a, \theta)$ is defined to be

$$r(a, \theta) = l(a, \theta) - \min_{a \in A} l(a, \theta),$$

that is, the regret for all a and θ is obtained by subtracting the minimum loss for this θ from the loss for this particular a and θ. The loss function and the regret function generally are not equivalent. The philosophy for using regret instead of loss can be summarized as follows. If the true state of nature is known, and the decision-maker

Table 4.2 **Regret Table for the Inventory Problem**

Action \ State of Nature	$\theta_1: \lambda = 21$	$\theta_2: \lambda = 24$
a_1: order 15	0.6485×10^6	1.263×10^6
a_2: order 20	0	0.268×10^6
a_3: order 25	0.336×10^6	0

takes the best action, she still incurs a loss, although through no fault of her own since she took the best possible action. Therefore, instead of penalizing her by this quantity, it is subtracted from each loss incurred by actions taken under this state of nature. The difference $r(a, \theta)$ is called the regret and represents the loss which could have been avoided had the best action for that state of nature been taken. Thus, a regret table always has a zero entry as one of its entries for a given state of nature. In the inventory example, the regret table is given in Table 4.2.

If the true state of nature is $\theta_1 = 21$, the best action to take is a_2, order 20 items. If action a_3 is taken, and 25 are ordered, a cost of 0.336×10^6 is "regretted."

In some formulations, losses are stated in terms of regret immediately. If the best action is taken for a given state of nature, the loss is often assumed to be zero for this combination. In fact, in the next section such situations are studied extensively.

If the true state of nature were known, it would be simple to choose the correct action, that is, that action which has minimum loss or regret. Unfortunately, the true state of nature is not generally known, and choosing a "correct action" is not simple. In the inventory example, if $\theta = 21$, the best action is to purchase 20 spare engines, whereas, if $\theta = 24$, the best action is to purchase 25 spare engines.

4.3 Tests of Hypothesis

A complete general decision analysis framework is presented in Chapter 14. This section is concerned with a special framework for decision making when there are only two actions, a_1 and a_2, open to the decision-maker, and decisions are to be based upon the outcome of a random sample. The decisions will still be related to making statements about the states of nature. The objective will be to partition Θ, the set of all possible values of the states of nature, into two mutually exclusive sets, Ω_1 and Ω_2, having the following property. If θ is in Ω_1, the action a_1 is the preferred action, whereas, if θ is in Ω_2, the action a_2 is the preferred action. The loss incurred when action a is taken and the state of nature is θ is denoted by $l(a, \theta)$, and it is assumed that

$$l(,\mathrm{a}_1, \theta) = 0, \qquad \text{if } \theta \text{ is in } \Omega_1,$$
$$l(a_2, \theta) = 0, \qquad \text{if } \theta \text{ is in } \Omega_2,$$

that is, the loss function is a regret function. A decision function, d, is a function which assigns either action a_1 or a_2 for each possible sample vector $(x_1, x_2, \ldots, x_n)$.

Since this type of problem has been studied extensively in the statistical literature, its own terminology has evolved. Statements that θ is contained in a subset of

Θ are said to be hypotheses about nature. When the subset consists of a single point, an assertion about the state is called a *simple hypothesis*. A statement about a subset containing more than one simple hypothesis is called a *composite hypothesis*.

A composite hypothesis is said to be true if one of the simple hypotheses that constitute the composite hypothesis is the true state of nature. The set Ω_1 is associated with the hypothesis H_1: θ is in Ω_1. The set Ω_2 is associated with the alternative hypothesis H_2: θ is in Ω_2. The action a_1 is called accepting the hypothesis H_1, and the action a_2 is called rejecting the hypothesis H_1 (which implies accepting the alternative hypothesis H_2). The decision function d leads to accepting or rejecting the hypothesis H_1 and is called a test of the hypothesis. Actually, the decision function partitions the sample space into two disjoint parts, those points which lead to accepting H_1, and those points which lead to rejecting H_1 (and consequently accepting H_2). The set of points in the sample space which leads to rejection of H_1 is called a "critical region" of the test. Thus, if a sample point falls into the critical region, action a_2 is taken and the hypothesis, H_1: θ is in Ω_1, is rejected. If the sample point falls outside the critical region, action a_1 is taken and the hypothesis, H_1: θ is in Ω_1, is accepted. Therefore, choosing a critical region is equivalent to choosing a decision function or test, and conversely.

It is often possible to show that "optimal" tests depend on the sample only through a function T of the random sample. Such a function of the random sample is called a *test statistic*. The decision function is then of the form $d(T)$. Again, the decision function partitions the range of values of T into two disjoint parts, those points which lead to accepting H_1, and those points which lead to rejecting H_1. Those points which lead to rejecting H_1 are called the *critical region* of the test. Henceforth, the critical region of a test will be labeled C, and will denote the set of sample points, or the set of values of the test statistic T, which lead to rejection of H_1.

Type I and Type II Errors

As noted previously, there is no loss incurred when action a_1 is taken and θ is in Ω_1, or when action a_2 is taken and θ is in Ω_2. This is then equivalent to incurring no loss when the hypothesis H_1 is accepted when the true state of nature is in Ω_1, or when the hypothesis H_1 is rejected when the true state of nature is in Ω_2. When the hypothesis H_1 is rejected (action a_2 is taken) and the true state of nature is in Ω_1, the loss $l(a, \theta)$ will be denoted by $\epsilon_1(\theta)$, and it is called the *type I error*. When the hypothesis H_1 is accepted (action a_1 is taken) and the true state of nature is in Ω_2, the loss $l(a, \theta)$ will be denoted by $\epsilon_2(\theta)$, and it is called the *type II error*. Thus, the loss function can be written as

$$l(a_1, \theta) = \begin{cases} 0, & \text{if } \theta \text{ is in } \Omega_1 \\ \epsilon_2(\theta), & \text{if } \theta \text{ is in } \Omega_2 \end{cases}$$

and

$$l(a_2, \theta) = \begin{cases} \epsilon_1(\theta), & \text{if } \theta \text{ is in } \Omega_1 \\ 0, & \text{if } \theta \text{ is in } \Omega_2. \end{cases}$$

For a given decision rule, or critical region C, the risk function $R(C, \theta)$ can be evaluated as follows.

$$R(C, \theta) = \begin{cases} \epsilon_1(\theta)P(C|\theta), & \text{if } \theta \text{ is in } \Omega_1 \\ \epsilon_2(\theta)P(\overline{C}|\theta), & \text{if } \theta \text{ is in } \Omega_2, \end{cases}$$

Table 4.3 **Special Loss Function for Hypothesis Testing**

Action \ State of Nature	H_1 True	H_2 True
a_1: Accept H_1	0	1
a_2: Reject H_1 (Accept H_2)	1	0

where $P(C|\theta)$ is the probability that the sample point or the test statistic, whichever is appropriate, falls into the critical region C when θ is in Ω_1, and $P(\bar{C}|\theta)$ is the probability that the sample point or the test statistic, whichever is appropriate, falls outside the critical region C when θ is in Ω_2.

A good decision rule is sought which will minimize the risk for every value of θ in Θ. Unfortunately, this is not generally possible, so that additional criteria or measures must be introduced. Other problems may also arise in using the risk as the appropriate measure. In many of the problems that the decision-maker faces in the context of testing hypotheses, the loss function is not known (or not known with sufficient accuracy) to merit its use. When the explicit form of the loss (or regret) function is not known, an acceptable loss function may be given by entries in Table 4.3.

For this loss function and a critical region C, the risk function $R(C, \theta)$ is now given by

$$R(C, \theta) = \begin{cases} P(C|\theta), & \text{if } \theta \text{ is in } \Omega_1 \\ P(\bar{C}|\theta), & \text{if } \theta \text{ is in } \Omega_2 \end{cases}.$$

In this case, a reasonable decision rule will be one which, in some sense, minimizes $P(C|\theta)$ and $P(\bar{C}|\theta)$. The usual method is to choose a value α, and then find the decision procedure (if one exists) which minimizes $P(\bar{C}|\theta)$ for all θ in Ω_2 subject to $P(C|\theta) \leq \alpha$. Note that $P(C|\theta)$ is just the probability of rejecting the hypothesis when it is in fact true and is called the probability of the type I error. This probability is also known as the *level of significance* and is often denoted by α. $P(\bar{C}|\theta)$ is the probability of accepting the hypothesis when it is in fact false and is called the probability of the type II error. It is often denoted by β [or $\beta(\theta)$].

As an example of the material presented on tests of hypotheses, return to the inventory problem. The hypothesis to be tested is H_1: the true state of nature is $\theta = 21$, and the alternative to the hypothesis is H_2: the true state of nature is $\theta = 24$.[1] Thus, Ω_1 consists of the single point, $\theta = 21$, and Ω_2 consists of the single point, $\theta = 24$. Assume that the action a_1 (order 15 spares) is no longer permitted. Hence, there are only two remaining actions, a_2 and a_3, that is, order 20 and 25 spares, respectively. Relabel a_2 by a_1' and a_3 by a_2'. Then a_1' denotes the action that accepts H_1, the hypothesis that $\theta = 21$, which implies the ordering of 20 spare engines; a_2' denotes the action which accepts H_2, the alternative to the hypothesis, which implies the ordering of 25 spare engines. The losses are given in Table 4.4.

Thus, $$\epsilon_2(\theta) = 0.268 \times 10^6,$$

[1]The hypothesis and alternative can be interchanged without changing the problem.

Table 4.4 **Loss Table for a Test of Hypothesis for the Inventory Problem**

Action \ State of Nature	H_1 true $\theta = 21$	H_2 true $\theta = 24$
a_1': accept H_1	0	0.268×10^6
a_2': reject H_1 (accept H_2)	0.336×10^6	0

and
$$\epsilon_1(\theta) = 0.336 \times 10^6.$$

Suppose that the data on a previous model airplane are available, and the decision-maker uses the decision rule to accept the hypothesis if the number of spare engines demanded does not exceed 23. The critical region C is given by the set of points $\{24, 25, 26, \ldots\}$. Hence,

$$P(C|\theta = 21) = \sum_{k=24}^{\infty} P_X(k),$$

where X is a Poisson random variable with parameter $\theta = \lambda = 21$. From A5.4 of Appendix 5, $P(C|\theta = 21)$ is seen to equal 0.284.

$$P(\overline{C}|\theta = 24) = \sum_{k=0}^{23} P_X(k),$$

where again X is a Poisson random variable, now with parameter $\theta = \lambda = 24$. Thus, $P(\overline{C}|\theta) = 24) = 0.473$.

The risk is given by

$$R(C, \theta) = \begin{cases} (0.336 \times 10^6)(0.284) = 0.0954 \times 10^6, & \text{if } \theta = 21 \\ (0.268 \times 10^6)(0.473) = 0.127 \times 10^6, & \text{if } \theta = 24. \end{cases}$$

If the loss function is not known, or if the special loss function for hypothesis testing (given in Table 4.4) is appropriate, and this decision procedure is used, the probability of type I error is $\alpha = 0.284$, and the probability of type II error is $\beta = 0.473$. If this type I error is satisfactory, the optimal decision procedure would be that rule which had a value of α of 0.284 or less, and which minimized β. It can be shown that the decision rule presented is this optimal decision function.

By choosing a critical region which minimizes $P(\overline{C}|\theta)$ for all θ in Ω_2 subject to $P(C|\theta) \leq \alpha$ as a criterion for optimality, optimal decision rules can often be derived. Once the form of these rules is known, the probability of accepting the hypothesis H_1 can be plotted against the true state of nature. Such a plot is known as an *operating characteristic* (OC) curve. This OC curve is a function of the sample size as well as the decision functions. The decision-maker usually chooses an OC curve which reflects his attitudes toward the allowable probability of type I and type II errors, and then he chooses his sample size to guarantee that his errors will not exceed these allowances.

As another example, suppose that the time to failure of an electronic device can be approximated by a normally distributed random variable having known standard deviation of 200 hours. The manufacturer of this device claims that the mean time to failure is at least 5,000 hours. A purchaser will sample these devices, and decide whether or not the claim is warranted. The hypothesis to be tested is that the mean time to failure, μ, is at least 5,000 hours, that is

$$H_1: \mu \geq 5{,}000.$$

The alternative hypothesis is that the mean time to failure, μ, is less than 5,000 hours, that is,

$$H_2: \mu < 5{,}000.$$

The test statistic chosen is the sample mean, $\overline{X}$, based upon a random sample of 25 units. When $\mu = 5{,}000$, the consumer is willing to accept the hypothesis H_1 with probability 0.95; that is, the maximum probability of a type I error is 0.05, or equivalently, the level of significance is 0.05. It is evident that the hypothesis H_1 is to be rejected if $\overline{X}$ is too small, say, if $\overline{X} < c$. Hence, the critical region, $(-\infty, c)$, is determined by finding the value of c such that

$$P(\overline{X} < c|\mu = 5{,}000) = 0.05.$$

Under these conditions $\overline{X}$ has a normal distribution with mean equal to 5,000 and standard deviation equal to 40 (200/5). Hence,

$$P(\overline{X} < c|\mu = 5{,}000) = 0.05$$

is equivalent to

$$P\left(Y < \frac{c - 5{,}000}{40}\right) = 0.05,$$

where Y is a normally distributed random variable having mean equal to 0 and standard deviation equal to 1.

Hence, from Table A.1 in Appendix 5

$$\frac{c - 5{,}000}{40} = -1.645,$$

or equivalently,

$$c = 5{,}000 - 65.80 = 4{,}934.$$

Thus, the hypothesis H_1 is accepted if the sample mean is at least 4,934, and rejected otherwise.

Suppose that the probability of a type II error is desired when the true mean time to failure is 4,900 hours. This can be obtained from the expression

$$P(\overline{X} \geq 4{,}934|\mu = 4{,}900).$$

This is equivalent to

$$P\left(Y \geq \frac{4{,}934 - 4{,}900}{40}\right) = P(Y \geq 0.85).$$

The probability that Y is at least equal to 4,934 is just the probability that a normally distributed random variable having mean equal to 0 and standard deviation equal to 1 is at least 0.85. This is determined from Table A.1 in Appendix 5, and found to be equal to 0.198. Of course, if this value is unacceptable as a type II error (too large), it can be reduced by using a large sample size. Indeed, if two points on the OC curve are specified, the required sample size is uniquely determined.

Finally, it is important to consider what "accepting the hypothesis H_1" means. Accepting H_1 does *not* mean that H_1 is true. Instead, it means that the observed data is consistent with H_1 being true. However, if more data had been observed, the resultant data might become inconsistent with H_1, and lead to a rejection of H_1. For example, an $\overline{X}$ equal to 4,950 is consistent with a true mean time to failure of 5,000 based upon a sample of size 25. However, such an $\overline{X}$ is inconsistent with a true mean time to failure of 5,000 based upon a sample of 250, and would have led to rejection of the hypothesis (assuming the same level of significance).

4.4 Point Estimation

In many decision-making problems, the decision-maker is willing to assume that the observations obtained come from a family of distributions completely specified except for the parameters. For example, the decision-maker may be willing to assume that the time between arrivals of customers to a store follows an exponential distribution. This implies that the density function of the random variable, time, is given by

$$f_X(y) = \begin{cases} \dfrac{1}{\theta} e^{-y/\theta}, & \text{if } y \geq 0 \\ 0, & \text{if } y < 0, \end{cases}$$

where θ is the unknown parameter of this distribution. The general problem to be considered here is how a decision-maker should estimate such a parameter θ on the basis of observing a random sample, $X_1, X_2, \ldots, X_n$. The decision function he or she chooses associates with each sample point an action, the action being an estimate, $\hat{\theta}$, of the parameter θ. Thus,

$$\hat{\theta} = d(x_1, x_2, \ldots, x_n)$$

and is called an *estimator* of θ. The function $\hat{\theta}$ is an estimator, and the value the function takes on (after the experiment) is called an *estimate* of θ. A reasonable loss function used in problems concerned with point estimation is given by

$$l(\hat{\theta}, \theta) = K(\theta)(\hat{\theta} - \theta)^2,$$

where $K(\theta) > 0$ for all values of $\hat{\theta}$. This quadratic loss function has the desirable property that the loss increases quite rapidly as $|\hat{\theta} - \theta|$ increases. Furthermore, many smooth functions which are 0 at $\hat{\theta} = \theta$ can be reasonably approximated by this quadratic function for $\hat{\theta}$ near θ. A possible disadvantage of using such a loss function is the determination of $K(\theta)$. Many real-world problems are solved by assuming that $K(\theta) = 1$, in which case the quadratic loss function is called the squared error loss function. Note that the loss and regret are the same when the quadratic loss function is used.

In order to compare estimators, it is necessary to compute their risks. If $\hat{\theta}_1 = d(x_1, x_2, \ldots, x_n)$ is an estimator, the risk is given by

$$R(d, \theta) = E[l(\hat{\theta}, \theta)].$$

A convenient way of comparing two estimators, $\hat{\theta}_1$, and $\hat{\theta}_2$, is to examine the ratio of their risks,

$$\frac{\mathrm{R}(d_1, \theta)}{R(d_2, \theta)},$$

for the various values of θ. This ratio is known as the *relative efficiency* of $\hat{\theta}_1$ with respect to $\hat{\theta}_2$. Thus, if squared error is used (as assumed in the remaining sections of this chapter), the relative efficiency is given by

$$\frac{R(d_1, \theta)}{R(d_2, \theta)} = \frac{E(\hat{\theta}_1 - \theta)^2}{E(\hat{\theta}_2 - \theta)^2},$$

where $E[\hat{\theta}_i - \theta]^2$ is called the mean square error for the estimator $\hat{\theta}_i$. If the relative efficiency is greater than 1 for all θ, it is said that $\hat{\theta}_2$ is a "better" estimator than $\hat{\theta}_1$. Similarly, if the relative efficiency is smaller than 1 for all θ, then $\hat{\theta}_1$ is a better estimator than $\hat{\theta}_2$. If an estimator, $\hat{\theta}_0$, exists that has risk no greater than that associated with any other estimator for every value of θ, then $\hat{\theta}_0$ is said to be an optimal estimator. Unfortunately, optimal estimators do not exist for many problems.

Continuing with the example of the exponential density function for the time between arrivals, suppose that a random sample is taken. Let $\hat{\theta} = \overline{X}$ be the estimator of θ. The mean square error is given by

$$R(\hat{\theta}, \theta) = E[\overline{X} - \theta]^2.$$

Before this expression is evaluated, it is worthwhile determining $E(X)$. This expectation is given by

$$E(X) = \int_0^\infty y \frac{1}{\theta} e^{-y/\theta}\, dy = \theta.$$

Hence, $E(\overline{X})$ also equals θ, so that the mean square error coincides with the variance of $\overline{X}$. Furthermore, the variance of $\overline{X}$ is just the variance of X divided by the sample size. It then follows that

$$R(\hat{\theta}, \theta) = \frac{\text{variance } (X)}{n}.$$

The variance of X is easily evaluated to be θ^2 so that

$$R(\hat{\theta}, \theta) = \frac{\theta^2}{n}.$$

Unbiased Estimators

In addition to looking at the risk of an estimator, a decision-maker seeks other properties that may be desirable. One such property is unbiasedness. An estimator $\hat{\theta}$ is said to be unbiased if

$$E(\hat{\theta}) = \theta \qquad \text{for all } \theta.$$

If an estimator is unbiased, the mean square error is equivalent to the variance so that

$$R(\hat{\theta}, \theta) = \text{variance } (\hat{\theta}).$$

If $\hat{\theta}$ is not unbiased, the difference between $E(\hat{\theta})$ and θ is called the bias, that is,

$$\text{bias} = E(\hat{\theta}) - \theta.$$

Such an estimator is called a *biased estimator*. In the above example involving the time between customer arrivals, $\overline{X}$ is an unbiased estimator of θ. Furthermore, the risk, θ^2/n, is the variance of $\overline{X}$. There are many unbiased estimators, for example, the first observation X_1 is one. Note that $\overline{X}$ is a better estimator than X_1 for all n in the sense of relative efficiency in that

$$\frac{R(\bar{X}, \theta)}{R(X_1, \theta)} = \frac{\theta^2/n}{\theta^2} = \frac{1}{n}$$

is always less than 1.

Although an unbiased estimator is intuitively appealing, it should be noted that a biased estimator can sometimes be better in the sense of having a lower mean square error.

Consistent Estimators

Intuitively, another desirable property for estimators to possess is that the risk becomes small as the sample size increases, or, equivalently, as the sample size n gets large, $\hat{\theta}_n$ (the estimator based upon a sample of size n) should tend to θ. A precise statement is as follows: let $\hat{\theta}_1, \hat{\theta}_2, \ldots, \hat{\theta}_n, \ldots$ be a sequence of estimators of θ. This sequence of estimators is said to be consistent if for every $\epsilon > 0$

$$\lim_{n \to \infty} P\{\theta - \epsilon < \hat{\theta}_n < \theta + \epsilon\} = 1$$

for each value of θ. In the current example, it is easily shown that $\bar{X}$ is a consistent estimator of θ.

Efficient Unbiased Estimators

A useful lower bound on the variance of an unbiased estimator of a parameter θ is known as the Cramer-Rao inequality. Let $\hat{\theta}$ be an unbiased estimator based upon a random sample of size n. The Cramer-Rao inequality states (assuming some regularity conditions hold) that

$$\text{variance } (\hat{\theta}) \geq \frac{1}{nE\left\{\left[\frac{\partial}{\partial\theta} \log f_X(X, \theta)\right]^2\right\}},$$

where $f_X(X, \theta)$ is the density function (probability distribution if X is discrete) of the random variable X. Thus, if an unbiased estimator can be found which achieves this lower bound, it must be the "optimal" estimator among the class of all unbiased estimators.

To illustrate, consider again the exponential distribution so that

$$\log f_X(X, \theta) = \log\left[\frac{1}{\theta} e^{-X/\theta}\right] = -\log\theta - \frac{X}{\theta}.$$

Furthermore,

$$\frac{\partial}{\partial\theta}\left[-\log\theta - \frac{X}{\theta}\right] = -\frac{1}{\theta} + \frac{X}{\theta^2},$$

and its square is given by

$$\frac{1}{\theta^2} - \frac{2X}{\theta^3} + \frac{X^2}{\theta^4}.$$

Hence,

$$E\left\{\left[\frac{\partial}{\partial\theta} \log f_X(X, \theta)\right]^2\right\} = \frac{1}{\theta^2} - \frac{2E(X)}{\theta^3} + \frac{E(X^2)}{\theta^4} = \frac{1}{\theta^2}$$

so that

$$\text{variance}\ (\hat{\theta}) \geq \frac{\theta^2}{n}.$$

Since it has been shown that the variance of $\overline{X}$ equals θ^2/n, it follows that $\overline{X}$ must be the "optimal" estimator among the class of all unbiased estimators.

However, it often happens that an unbiased estimator cannot be exhibited which achieves the Cramer-Rao lower bound.

Estimators Determined by the Method of Moments

In the previous sections, properties of estimators were discussed, but methods for obtaining estimators were not presented. The oldest, and perhaps simplest method is the *method of moments*. Suppose that there are k parameters to be estimated for a density function, $f_X(y, \theta_1, \theta_2, \ldots, \theta_k)$. Let $\mu'_1, \mu'_2, \ldots, \mu'_k$ denote the first, second, . . . , kth moment about the origin of the random variable X, respectively, that is

$$\mu'_i = \int_{-\infty}^{\infty} y^i f_X(y, \theta_1, \theta_2, \ldots, \theta_k)\, dy.$$

These moments will, in general, be a function of the k parameters, $\theta_1, \theta_2, \ldots, \theta_k$. Let $X_1, X_2, \ldots, X_n$ be a random sample of size n. Denote by $m'_1, m'_2, \ldots, m'_k$ the first, second, . . . , kth sample moment, that is,

$$m'_i = \sum_{j=1}^{n} \frac{X_j^i}{n}.$$

Let $\hat{\theta}_1, \hat{\theta}_2, \ldots, \hat{\theta}_k$ be the solution to the following k simultaneous equations in k unknowns,

$$\mu'_i = m'_i, \qquad \text{for } i = 1, 2, \ldots, k.$$

These solutions represent the method of moments estimators. For the exponential distribution, there is only one parameter θ and $\mu'_1 = \theta$. Furthermore, $m'_1 = \overline{X}$ so that $\overline{X}$ is the method of moments estimator of θ.

A more complicated example concerns the normal distribution. The two parameters are μ and σ^2, the mean and variance, respectively. Furthermore,

$$\mu'_1 = \mu,$$

$$\mu'_2 = \sigma^2 + \mu^2.$$

Also,

$$m'_1 = \overline{X},$$

$$m'_2 = \frac{\Sigma X_i^2}{n}.$$

Hence, the two equations are

$$\overline{X} = \mu,$$

$$\frac{\Sigma X_i^2}{n} = \sigma^2 + \mu^2.$$

Solving these equations simultaneously leads to

$$\hat{\theta}_1 = \overline{X},$$

$$\hat{\theta}_2 = \frac{\Sigma X_i^2}{n} - \overline{X}^2.$$

Under quite general conditions, the method of moments can be shown to lead to estimates which are consistent. Furthermore, their distributions tend to be normal as the sample size gets large.

Estimators Determined by the Method of Maximum Likelihood

The reliability at time t of a piece of complex electronic equipment is defined as the probability that it will perform successfully until at least time t. Suppose that 20 pieces are tested, and all that is measured is whether or not the equipment failed before time t. If the equipment performs successfully, record a 1. If it fails, record a zero. This can be represented in terms of a mathematical model as drawing a random sample of size 20 from a Bernoulli distribution with parameter p (Sec. 3.6.1), that is X_1, X_2, . . . , X_{20} are random variables with

$$\left.\begin{aligned} P\{X_i = 1\} &= P_{X_i}(1) = p \\ P\{X_i = 0\} &= P_{X_i}(0) = 1 - p \end{aligned}\right\} \qquad \text{for } i = 1, 2, \ldots, 20.$$

An equivalent way of describing this probability distribution is given by the expression

$$P\{X_i = x_i\} = p^{x_i}(1 - p)^{1-x_i}, \qquad x_i = 0, 1, \qquad \text{and } i = 1, 2, \ldots, 20.$$

The reliability at time t is then

$$P_X(1) = p.$$

In general, let $X_1, X_2, \ldots, X_n$ be a random sample from a discrete probability distribution $P_X(k)$. Define the *likelihood function*, L, by

$$L = P\{X_1 = x_1\}P\{X_2 = x_2\} \cdots P\{X_n = x_n\},$$

where x_i represents the actual outcome of the random variable X_i. Therefore, L is just the probability of obtaining the observed sample and is a function of the parameters of the probability distribution. Thus, for the reliability example,

$$\begin{aligned} L &= [p^{x_1}(1 - p)^{1-x_1}][p^{x_2}(1 - p)^{1-x_2}] \cdots [p^{x_n}(1 - p)^{1-x_n}] \\ &= p^{\Sigma x_i}(1 - p)^{n-\Sigma x_i}. \end{aligned}$$

The principle of maximum likelihood says to find that value of the parameter θ which maximizes the likelihood function, and call this value the maximum likelihood estimate of θ. Thus, the likelihood function can be viewed as only a function of the parameter θ (the x_i are now fixed constants), and the value of θ is sought which maximizes the probability of obtaining this particular sample.

Often it is easier to maximize the logarithm of the likelihood function rather than the likelihood function itself. This is permissible since the logarithm is a monotonic function, so these two maxima will occur at the same place. The logarithm of the likelihood function will be denoted by $\mathcal{L}$. Thus, in the reliability example,

$$\mathcal{L} = \Sigma x_i \log p + (n - \Sigma x_i) \log (1 - p).$$

To maximize this function, it is necessary to solve the following equation for p:

$$\frac{d\mathcal{L}}{dp} = \frac{\Sigma x_i}{p} - \frac{(n - \Sigma x_i)}{(1 - p)} = 0.$$

The solution is $\hat{p} = \Sigma x_i/n = \bar{x}$. Hence, the maximum likelihood estimator is the total number of pieces of equipment which do not fail before time t divided by the number tested.

In general then, to find maximum likelihood estimators, it is necessary to find that value of θ which satisfies the equation

$$\frac{dL}{d\theta} = 0$$

or, equivalently,

$$\frac{d\mathcal{L}}{d\theta} = 0.$$

Of course, it is necessary to show that this leads to a maximum and not a minimum or point of inflection (see Appendix 2).

For the case of a continuous random variable, maximizing the probability of obtaining the observed sample leads to difficulties since such a probability is zero. However, if the "probability of obtaining the observed sample" is interpreted as the probability of each random variable falling into its interval $(x_i, x_i + dx_i)$, with the length of the interval suppressed, then the interpretation given previously is applicable. Hence, the likelihood function, L, is defined as

$$L = f_{X_1}(x_1, \theta) f_{X_2}(x_2, \theta) \ldots f_{X_n}(x_n, \theta),$$

where $f_{X_i}(x_i, \theta)$ is the value of the density function of the random variable X_i at the sample point x_i, given that the value of the parameter is θ. The principle of maximum likelihood says to find that value of the parameter θ which maximizes the likelihood function. Call this value the maximum likelihood estimate of θ. Returning to the exponential distribution,

$$L = \frac{1}{\theta} e^{-x_1/\theta} \frac{1}{\theta} e^{-x_2/\theta} \ldots \frac{1}{\theta} e^{-x_n/\theta}$$

$$= \frac{1}{\theta^n} e^{-\Sigma x_i/\theta}.$$

Again, it is easier to work with $\mathcal{L} = \log L$. Hence,

$$\mathcal{L} = -n \log \theta - \frac{\Sigma x_i}{\theta}.$$

In order to find the value of θ which maximizes $\mathcal{L}$, it is necessary to solve the equation

$$\frac{d\mathcal{L}}{d\theta} = -\frac{n}{\theta} + \frac{\Sigma x_i}{\theta^2} = 0.$$

$\hat{\theta} = \Sigma x_i/n$ is the value of θ which maximizes $\mathcal{L}$, and hence is the maximum likelihood *estimate* of θ. The maximum likelihood *estimator* is then

$$\hat{\theta} = \frac{\Sigma X_i}{n},$$

which coincides with the method of moments estimator.

If the likelihood function contains k parameters, that is

$$L = f_{X_1}(x_1, \theta_1, \theta_2, \ldots, \theta_k) f_{X_2}(x_2, \theta_1, \theta_2, \ldots, \theta_k) \ldots f_{X_n}(x_n, \theta_1, \theta_2, \ldots, \theta_k),$$

then the maximum likelihood estimators of $\theta_1, \theta_2, \ldots, \theta_k$ are the values $\hat{\theta}_1, \hat{\theta}_2, \ldots, \hat{\theta}_k$ which maximize L. These are often obtained as the simultaneous solution to the equations

$$\frac{\partial L}{\partial \theta_1} = \frac{\partial L}{\partial \theta_2} = \cdots = \frac{\partial L}{\partial \theta_k} = 0.$$

(See Appendix 2 on maximizing functions of several variables.)

Maximum likelihood estimators are the most commonly used estimators in practice. They are often easy to obtain, and have many other desirable properties (in addition to maximizing the likelihood function). In particular, maximum likelihood estimators, though often biased, are consistent. They are also asymptotically efficient in that $\sqrt{n}\hat{\theta}$ has a variance which achieves the Cramer–Rao lower bound, that is,

$$\text{variance } (\sqrt{n}\hat{\theta}) \rightarrow \frac{1}{E\left\{\left[\dfrac{\partial}{\partial \theta} \log f_X(X, \theta)\right]^2\right\}},$$

as n gets large. Furthermore, the random variable $\sqrt{n}(\hat{\theta} - \theta)$ approaches the normal distribution with zero mean and the variance given above. Finally, maximum likelihood estimators possess the property of invariance. That is, if $\hat{\theta}$ is the maximum likelihood estimator of θ, and $h(\theta)$ is a function of θ, with a single inverse, then the maximum likelihood estimator of $h(\theta)$ is $h(\hat{\theta})$.

Bayes' Estimators

In certain cases, the decision-maker has some advance information about the states of nature, say, described in terms of a probability distribution, called a *prior* distribution. If the states of nature are continuous, this probability density function of θ is denoted by $P_\theta(y)$. For example, if each value of θ is equally likely between 0 and 1, then the prior distribution is just the uniform distribution between 0 and 1. Data are then observed, and a function X of the random sample $X_1, X_2, \ldots, X_n$ is obtained. It is intuitive that the observed data, whose distribution depends upon θ should be used to update the prior distribution. If θ is a continuous random variable, and if X is, say, discrete, denote by $Q_{X|\theta=y}(x)$ the probability that the function X is equal to x, given that θ has value y. The updated prior distribution of θ, $h_{\theta|X=x}(y)$, is just the conditional distribution of θ, given $X = x$, and is called the *posterior distribution* of θ. The posterior distribution is obtained from Bayes' theorem, and it can be expressed by

$$h_{\theta|X=x}(y) = \frac{Q_{X|\theta=y}(x)\, P_\theta(y)}{\int Q_{X|\theta=y}(x)\, P_\theta(y)\, dy}.$$

Given a mean square error loss function, that is, $(\hat{\theta} - \theta)^2$, the posterior expected loss is $E[\hat{\theta} - \theta]^2$, where the expectation is taken with respect to the posterior distribution of θ. However, $E[\hat{\theta} - \theta]^2 = E[\theta - \hat{\theta}]^2$ is the second moment of the "random variable" θ and is known to be a minimum if it is taken about $E(\theta)$. Hence, the expected value of θ (evaluated with respect to the posterior distribution of θ) is called the Bayes' action, that is,

$$\hat{\theta} = E(\theta) = \int_{-\infty}^{\infty} y h_{\theta|X=x}(y)\, dy.$$

As an example, return to the reliability problem. Assume that there exists a prior distribution of p which is uniform over the interval [0, 1]. The posterior distribution of p, given k success, can be computed from the expressions given above, which yields

$$h_{p|X=k}(y) = \binom{n}{k} (n + 1)y^k(1 - y)^{n-k}.$$

Therefore, when $n = 20$, the Bayes' action (Bayes' estimator of p) is given by

$$E(p) = \int_0^1 y\binom{20}{k} (21)y^k(1 - y)^{20-k}\, dy = \frac{k + 1}{22}.$$

Thus, the Bayes' estimator of p does not coincide with the maximum likelihood estimator of p. (Sec. 4.45 showed the latter estimate is given by $k/20$.)

Under quite general conditions, Bayes' estimators corresponding to any given prior distribution can be shown to be consistent, asymptotically efficient, and asymptotically normally distributed. In fact, Bayes' estimators and maximum likelihood estimators differ only slightly when n is large.

4.5 Interval Estimation

The preceding section indicated that certain kinds of point estimates have many desirable properties. Unfortunately, they also have the undesirable property of usually being incorrect! For example, if the estimator $\hat{\theta}$ is a continuous random variable, then the probability is zero that it equals the parameter θ being estimated. Hence, an estimate of a parameter that is accompanied by some interval around this estimate is desirable, together with some measure of assurance that the interval includes the true parameter. This kind of approach is described below.

Confidence Interval Estimation

A technique for finding intervals as described above is available. Such intervals are known as *confidence intervals*, and the measure of assurance is given by a *confidence coefficient*. As an illustration of how to obtain them, suppose that the daily demand for a given product can be approximated by a normally distributed random variable having unknown mean but known standard deviation, say, 20. Suppose that a random sample of 25 days is to be taken, and an interval estimate of the mean demand μ is desired. Let the random variable $\overline{X}$ be the sample mean. Hence, $\overline{X}$ is also normally distributed with mean μ, and its standard deviation is $20/5 = 4$. Thus,

$$P\left\{-K_{\alpha/2} \leq \frac{\overline{X} - \mu}{4} \leq K_{\alpha/2}\right\} = 1 - \alpha$$

so that $$P\{\overline{X} - 4K_{\alpha/2} \leq u \leq \overline{X} + 4K_{\alpha/2}\} = 1 - \alpha,$$

where $K_{\alpha/2}$ is the 100 $\alpha/2$ percentage point of the normal distribution. $\overline{X} - 4K_{\alpha/2}$ is called the lower confidence limit, $\overline{X} + 4K_{\alpha/2}$ is called the upper confidence limit, and $(\overline{X} - 4K_{\alpha/2}, \overline{X} + 4K_{\alpha/2})$ is known as the confidence interval. $(1 - \alpha)$ is called the confidence coefficient. For example, if, after the experiment, $\overline{X}$ takes on the value 93 and α is chosen to be $0.05(K_{\alpha/2} = 1.96)$, then the lower confidence limit is given by $93 - 7.84 = 85.16$ and the upper confidence limit is given by $93 + 7.84 = 100.84$. The confidence coefficient is 0.95. The decision-maker can then make the following announcement: *μ lies in the interval (85.16, 100.84), and this statement is made with confidence 0.95*. The meaning of this assertion must be examined carefully. If the decision-maker is going to make such statements about 100 (independent) products using the above method (correcting for the proper standard deviations), and if she later became aware of the correct means, she would have found that, on the average, 95 of the 100 intervals would have included the true values of μ. Since μ, in this context, is not to be interpreted as a random variable, the statement,

$$P\{85.16 \leq \mu \leq 100.84\} = 0.95,$$

has no meaning. (This probability is either one or zero, depending upon whether the mean is in *this* interval or not.) The assertion of "confidence 0.95" means that *before* the sample was drawn, the probability was 0.95 that the interval to be constructed would include the true mean.

The general expression for finding a confidence interval for μ with confidence coefficient $1 - \alpha$ when σ is known is given by

$$\left(\overline{X} - \frac{K_{\alpha/2}\sigma}{\sqrt{n}}, \overline{X} + \frac{K_{\alpha/2}\sigma}{\sqrt{n}}\right).$$

A useful method for constructing confidence intervals for any single parameter θ is as follows. Let $\hat{\theta}$ be an estimator of θ whose distribution depends on θ. Find a function of $\hat{\theta}$ and θ, say, $Z(\hat{\theta}, \theta)$, which has a distribution independent of θ and any other (nuisance) parameters present. Then a probability statement of the form

$$P\{c \leq Z \leq d\} = 1 - \alpha$$

can be inverted to obtain an equivalent inequality of the form

$$P\{g(Z) \leq \theta \leq h(Z)\} = 1 - \alpha.$$

The random variables $g(Z)$ and $h(Z)$ are the lower and upper confidence limits, respectively. These limits are random variables; sometimes they include θ, and sometimes they exclude θ. The probability is $(1 - \alpha)$ that they include θ.

As an example, a confidence interval can be obtained for the mean, μ, of a normal distribution when the variance, σ^2, is unknown. Let $\overline{X}$ be the random variable, the sample mean, obtained from a random sample of size n. It can be shown that

$$Z = \frac{(\overline{X} - \mu)\sqrt{n}}{s}$$

has a t distribution (see the end of Section 3.7.3) with $(n - 1)$ degrees of freedom, where

$$s^2 = \sum_{i=1}^{n} \frac{(X_i - \overline{X})^2}{n - 1}.$$

Thus, $\overline{X}$ is an estimator of μ whose distribution depends on μ. However, Z has a distribution independent of μ, so that

$$P\left\{-t_{\alpha/2;n-1} \le \frac{(\overline{X} - \mu)\sqrt{n}}{s} \le t_{\alpha/2;n-1}\right\} = 1 - \alpha,$$

where $t_{\alpha/2;n-1}$ is the 100 $\alpha/2$ percentage point of the t distribution with $(n - 1)$ degrees of freedom, as given in Table A5.2. Inverting this inequality leads to

$$P\left\{\overline{X} - t_{\alpha/2;n-1}\frac{s}{\sqrt{n}} \le u \le \overline{X} + t_{\alpha/2;n-1}\frac{s}{\sqrt{n}}\right\} = 1 - \alpha,$$

so that the random variables

$$g(Z) = \overline{X} - t_{\alpha/2;n-1}\frac{s}{\sqrt{n}}$$

and

$$h(Z) = \overline{X} + t_{\alpha/2;n-1}\frac{s}{\sqrt{n}}$$

are the lower and upper confidence limits, respectively.

This method is often applicable to real-world problems. However, there are certain cases where it is not possible to find functions of the desired form which have distributions independent of the parameters. In these cases, other methods must be applied.

Posterior Distributions

An alternative to constructing a confidence interval can be obtained by using a Bayesian framework. If some information about the parameter θ is available, and if the decision-maker is willing to translate this information into a prior distribution for θ, then the posterior distribution given the sample data can be used to make probability statements about the parameter. In fact, probability statements relative to the posterior distribution of θ can be used as an alternative to the confidence statements. Of course, such probability statements depend upon the prior distribution chosen. It is interesting to note that when the mean μ of a normal distribution with known standard deviation is assumed to have a uniform prior weighting, the confidence interval with confidence coefficient $(1 - \alpha)$ coincides with the $(1 - \alpha)$ probability interval for μ relative to the posterior distribution. For other prior distributions, this is not necessarily true, nor is it generally true for other parameters using uniform prior distributions.

4.6 Linear Regression

Statistical problems often are concerned with data where there exists a relationship between two variables. This section highlights the results when the relationship is linear. For example, suppose that a publisher of textbooks is concerned about the initial production run for his books. He sells books both through bookstores and through direct-mail orders. This latter method comes about after an extensive advertising campaign through publishing media and direct mail. The advertising campaign is conducted prior to the publication of the book. The sales manager has noted that there is a rather interesting linear relationship that exists between the number of mail orders and the number sold through bookstores during the first year. He suggested that this relationship be exploited for use in setting the initial production run for his subsequent books.

Thus, if the number of mail-order sales for a book is denoted by X, and the number of bookstore sales by Y, then the random variables X and Y exhibit ''a degree of association.'' There is no functional relationship between these two random variables, that is, given the number of mail-order sales, one does not expect to determine *exactly* the number of bookstore sales. Note that for any given number of mail-order sales there is a range of possible bookstore sales, and vice versa. This variation may be partially due to measurement errors (incorrect counts, for example), but it can be attributed primarily to individual variation among published books. Thus, no unique functional relationship between mail-order sales and bookstore sales can be expected. However, it is anticipated that bookstore sales, for a given observed mail-order sales, increase as mail-order sales increase. If this is so, what then is meant by the statement, ''The sales manager has noted that there is a rather interesting linear relationship that exists between the number of mail orders and the number sold through bookstores during the first year''? Such a statement implies that the *expected value* of the number of bookstore sales is linear with respect to the number of mail-order sales, that is,

$$E[Y|X = x] = \alpha + \beta x.$$

Thus, if the number of mail-order sales is x for many different books, the average number of corresponding bookstore sales would tend to be approximately $\alpha + \beta x$.

Other examples of this ''degree of association model'' can easily be found. An educator may be interested in the relationship between a student's performance on the college entrance examination and his subsequent performance in college. An engineer may be interested in the relationship between tensile strength and hardness of a material. An economist may wish to predict a measure of inflation as a function of the cost of living index, and so on.

The ''degree of association model'' is not the only model of interest. In some cases, there exists a functional relationship between two variables, possibly linearly linked. For example, Ohm's law, which is extremely useful in the design and analysis of electrical circuits, is of this form. The algebraic expression of Ohm's law is given by

$$I = \frac{V}{R},$$

where I is the current measured in amperes, R is the resistance in ohms, and V is the potential difference in volts. If an experiment were run, holding R fixed (perhaps unknown), and plotting I against V, these points *should* fall on a straight line through the origin with slope I/R. However, the points will not lie exactly on a straight line due to errors of measurement. Generally, one of the two variables is assumed to be known without error, and this variable will be denoted by x. (In practice, the error in x may not be exactly zero, but will be assumed to be negligible compared to the error in Y.) Other examples occur in calibration problems in engineering, and bioassay in medicine. In the calibration of an instrument, the instrument is usually calibrated to some standard (which is known exactly) over a wide range of values; the resultant plot (the calibration curve) is often linear. An example of a bioassay is an assay for immunoglobulins in human sera. Human serum contains three known antibodies (gamma globulins): γ_A, γ_G, γ_M. These are often routinely assayed by the method of radial immunodiffusion. A fixed amount of human serum is placed on a special plate, and a ring forms. The diameter of the ring is a function of the concentration of the immunoglobulin in the serum. Standard dilutions of immunoglobulin can be placed on these special plates, and pairs of points plotted. The ring-size diameter plotted against the logarithm of the concentration of the standard is a linear function.

Calling these plots linear functions does not imply that each point falls on the line. Again, due to measurement errors in Y, all that can be said is that

$$E[Y|x] = \alpha + \beta x,$$

that is, the expected value of Y, given x, is linear. Such a statement is equivalent to saying that a linear regression exists. Note that both the "degree of association" and "exact functional relationship" models lead to the same linear regression, and their subsequent treatment is almost identical. hence, the bookstore example will be explored further to illustrate how to treat both kinds of models.

Method of Least Squares

Suppose that bookstore sales and mail-order sales are given for 15 books. These data appear in Table 4.5, and the resulting plot is given in Fig. 4.1.

It is evident that the points in Fig. 4.1 do not lie on a straight line. Hence, it is not clear where the line should be drawn to show the linear relationship. Suppose that an arbitrary line, given by the expression $\tilde{y} = a + bx$, is drawn through the data. A measure of how well this line fits the data can be obtained by computing the sum of squares of the vertical deviations of the actual points from the fitted line. Thus, let y_i represent the bookstore sales of the ith book, and x_i the corresponding mail-order sales. Denote by $\tilde{y}_i$ the point on the fitted line corresponding to the mail-order sales of x_i. The proposed measure of fit is then given by

$$Q = (y_1 - \tilde{y}_1)^2 + (y_2 - \tilde{y}_2)^2 + \cdots + (y_{15} - \tilde{y}_{15})^2 = \sum_{i=1}^{15} (y_i - \tilde{y}_i)^2.$$

The usual method for identifying the "best" fitted line is the *method of least squares*. This method chooses that line, $a + bx$, which makes Q a minimum. Thus, a and b are obtained simply by setting the partial derivatives of Q with respect to a and b equal to zero, and solving the resultant equations. This yields the desired solution,

$$b = \frac{\sum_{i=1}^{n} (x_i - \bar{x})(y_i - \bar{y})}{\sum_{i=1}^{n} (x_i - \bar{x})^2} = \frac{\sum_{i=1}^{n} (x_i - \bar{x})y_i}{\sum_{i=1}^{n} (x_i - \bar{x})^2},$$

and

$$a = \bar{y} - b\bar{x},$$

where

$$\bar{x} = \sum_{i=1}^{n} \frac{x_i}{n}$$

and

$$\bar{y} = \sum_{i=1}^{n} \frac{y_i}{n}.$$

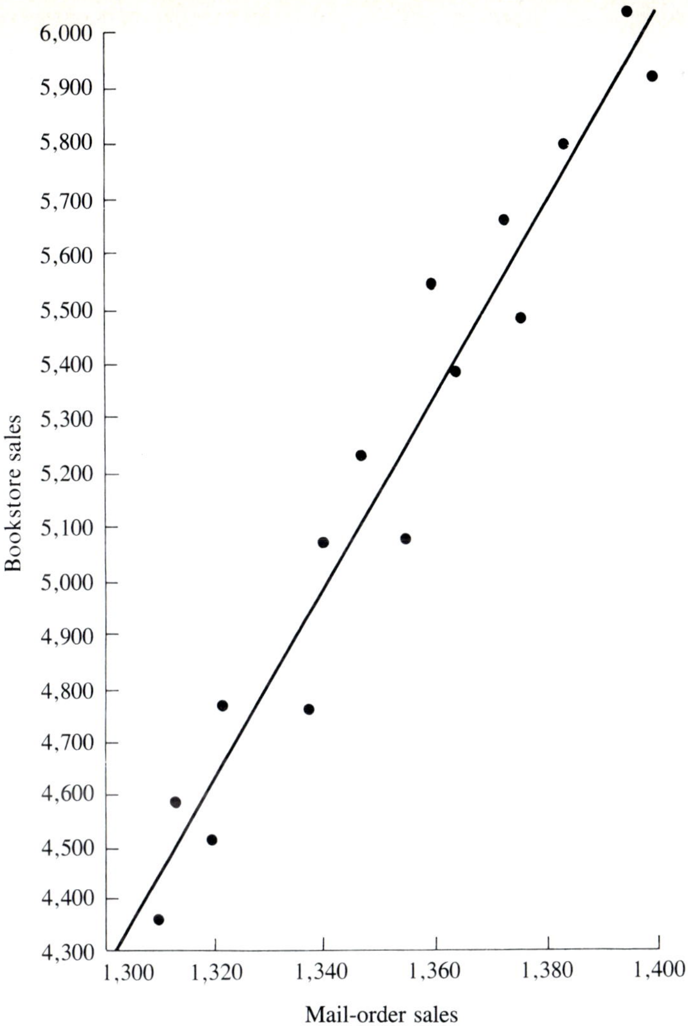

Figure 4.1 Plot of mail-order sales versus bookstore sales.

For the book example, $\bar{x} = 1{,}353.1$, $\bar{y} = 5{,}219.3$,

$$\sum_{i=1}^{15} (x_i - \bar{x})^2 = 11{,}966,$$

and $a = -19{,}041.9$, $b = 17.930$. Hence, the least-squares estimate of the bookstore sales $\tilde{y}$, when the mail-order sales is x, is given by

$$\tilde{y} = -19{,}041.9 + 17.930x,$$

and this line is drawn in Fig. 4.1.

Table 4.5 **Table of Mail-Order and Bookstore Sales**

Mail-Order Sales	Bookstore Sales
1,310	4,360
1,313	4,590
1,320	4,520
1,322	4,770
1,338	4,760
1,340	5,070
1,347	5,230
1,355	5,080
1,360	5,550
1,364	5,390
1,373	5,670
1,376	5,490
1,384	5,810
1,395	6,060
1,400	5,940

Properties of Least-Squares Estimates

Although the method of least squares is intuitively appealing, it appears to be based on a rather arbitrary criterion. In fact, one can think of many other competitive measures, such as the "sum of the absolute deviations," "sum of the perpendicular deviations," and so on. However, there are three properties of least squares that make it a particularly appealing measure. These are stated below.

Gauss-Markov property: Suppose that the underlying model of the relationship between the two variables is given by

$$E(Y_i|x_i) = \alpha + \beta x_i,$$

where the x_i are viewed as constants, and the Y_i are random variables having mean $E(Y_i|x_i)$ and variance σ^2 which is independent of i. A random sample of n pairs, $(x_1, Y_1), (x_2, Y_2), \ldots, (x_n, Y_n)$ is taken. The line, $\alpha + \beta x$, is to be estimated by a line $\tilde{y} = a + bx$, where a is to be an estimate of α, and b is to be an estimate of β. The Gauss-Markov theorem states that a and b, the least-squares estimate of α and β, respectively, are the minimum variance unbiased estimates of these parameters among the class of all linear estimates, that is, estimates of the form

$$\sum_{i=1}^{n} c_i y_i.$$

Note that since the x_i are assumed to be fixed constants, the least-squares estimates are indeed linear functions of the y_i.

Maximum likelihood property: If the above assumptions for Y_i still hold, that is, the Y_i are random variables having mean $E(Y_i|x_i) = \alpha + \beta x_i$ and variance σ^2 independent of i, and if, in addition, the Y_i are normally distributed, then the least-squares estimates for α and β (based upon a random sample of size n) coincide with the maximum likelihood estimates. Furthermore, the probability distributions of these estimates are normally distributed.

Extention to nonlinear regression: The third property is that, even if the underlying relationship between the two variables is not linear, the method of least squares generalizes easily to the problem of estimating the nonlinear relationship.

For the remainder of this section on linear regression, it will be assumed that

(i) A random sample of n pairs $(x_1, Y_1), (x_2, Y_2), \ldots, (x_n, Y_n)$ is to be taken.
(ii) The Y_i are normally distributed with mean, $\alpha + \beta x_i$, and variance σ^2 (independent of i).

Assumption (ii) is worth exploring. The procedures to be presented are "robust" in the sense that the assumption of normality is not crucial for obtaining good estimates. However, the assumption of constant variance is quite important.

Under assumptions (i) and (ii) the least-squares estimate of the intercept α and the slope β, given by a and b, respectively,

1. coincide with the maximum likelihood estimates,
2. are linear combinations of the y's (x's are assumed to be fixed constants),
3. are unbiased, i.e., $E(a) = \alpha$ and $E(b) = \beta$,
4. are the minimum variance unbiased estimates among the class of all linear estimates with

$$\text{variance}(a) = \sigma^2 \left[\frac{1}{n} + \frac{\bar{x}^2}{\sum_{i=1}^{n} (x_i - \bar{x})^2} \right],$$

$$\text{variance}(b) = \frac{\sigma^2}{\sum_{i=1}^{n} (x_i - \bar{x})^2},$$

5. are normally distributed with parameters as given in (3) and (4).

Finally, an unbiased estimate of σ^2 is given by $s^2_{y|x}$, where

$$s^2_{y|x} = \sum_{i=1}^{n} \frac{(y_i - \tilde{y}_i)^2}{n - 2}.$$

Furthermore, $(n - 2)s^2_{y|x}/\sigma^2$ has a chi-square distribution with $(n - 2)$ degrees of freedom (see Chapter 3) and is independent of the random variables, a, b, and $\tilde{y}$. This result can be used to help derive tests of hypothesis about α, β, and σ^2.

Confidence Interval Estimation of $E(Y|x = x_*)$

A very important reason for obtaining the linear relationship between two variables is to use the line for future decision making. From the regression line, it is possible to estimate $E(Y|x)$ by a point estimate and a confidence interval. For example, the book publisher might want to use this approach to estimate the expected number of bookstore sales corresponding to a mail-order sale of, say, 1,400 by both a point estimate and confidence interval estimate. He may be interested in this for sales projections. A point estimate corresponding to $x = x_*$ is given by

$$\tilde{y}_* = a + bx_*.$$

The endpoints of a $(100)(1 - \alpha)$ percent confidence interval are given by

$$a + bx_* - t_{\alpha/2;n-2}s_{y|x}\sqrt{\left[\frac{1}{n} + \frac{(x_* - \bar{x})^2}{\Sigma(x_i - \bar{x})^2}\right]},$$

and

$$a + bx_* + t_{\alpha/2;n-2}s_{y|x}\sqrt{\left[\frac{1}{n} + \frac{(x_* - \bar{x})^2}{\Sigma(x_i - \bar{x})^2}\right]},$$

where $s_{y|x}^2$ is the estimate of σ^2, and $t_{\alpha/2;n-2}$ is the $100\alpha/2$ percentage point of the t distribution with $n - 2$ degrees of freedom (see Table A5.2 of Appendix 5). It should be noted that the interval is most narrow where $x_* = \bar{x}$, and becomes wider as x_* departs from the mean.

In the book example, $s_{y|x}^2$ is computed from the data in Table 4.5 to be 17,030. If a 95 percent confidence interval is required, Table A5.2 gives $t_{.025;13} = 2.160$. The results derived in the preceding section yield 6,060 as the point estimate of $E(Y|1{,}400)$. Hence, the lower confidence limit corresponding to a mail order of 1,400 is 5,918, and the upper confidence limit is 6,202. The fact that the confidence interval was obtained at a data point ($x = 1{,}400$) is purely coincidental.

Predictions

The confidence interval statement for the expected number of bookstore sales corresponding to a mail order of 1,400 may be useful for budgeting purposes, but it is not too useful for making decisions about the *actual* number of books to be printed. Instead of obtaining bounds on the *expected* number of bookstore sales, this kind of decision requires bounds on what the *actual* bookstore sales will be, that is, a *prediction interval* on the value that the random variable (bookstore sales) can take on. The two endpoints of such an interval are given *by* the expressions

$$a + bx_+ - t_{\alpha/2;n-2}s_{y|x}\sqrt{\left[1 + \frac{1}{n} + \frac{(x_+ - \bar{x})^2}{\sum_{i=1}^{n}(x_i - \bar{x})^2}\right]},$$

$$a + bx_+ + t_{\alpha/2;n-2}s_{y|x}\sqrt{\left[1 + \frac{1}{n} + \frac{(x_+ - \bar{x})^2}{\sum_{i=1}^{n}(x_i - \bar{x})^2}\right]}.$$

For a given value x_+, the probability is $1 - \alpha$ that the value of the future Y_+ associated with the x_+ will fall in this interval. Thus, if x_+ is 1,400, then the corresponding 95 percent prediction interval for the number of bookstore sales is given by 6,060 ± 316, which is naturally wider than the confidence interval for the expected number of bookstore sales.

Whereas the book publisher can find an interval which will contain bookstore sales corresponding to a particular mail-order sale with probability $1 - \alpha$, she is unable to use this type of result over and over and still maintain a measure for making correct statements. The reason is that these statements would all be based upon the same statistical data, so that they would not be statistically independent. If the statements are independent, and k future bookstore sales are to be predicted with each statement being made with probability $1 - \alpha$, then the probability is $(1 - \alpha)^k$ that *all k* predictions of future bookstore sales are correct. However, if k is large or possibly unknown, even this technique based upon the (incorrect) assumption of independence would be useless. A solution to this problem can be obtained by using *simultaneous tolerance intervals*. Using this technique, the book publisher can take the mail-order sales of any book, find an interval (based on the previously determined fitted line) that will contain the actual bookstore sales with probability at least $1 - \alpha$, and can repeat this for any number of books having the same or different mail-order sales. Furthermore, the probability is P that all of these predictions are correct. An alternative interpretation is as follows. If every book publisher followed this procedure, each using her own fitted line, then 100 P percent of the book publishers (on the average) would find that at least $100(1 - \alpha)$ percent of the bookstore sales of their books would fall into the predicted intervals. The expression for the endpoints of each such interval is given by

$$a + bx_+ - c^{**}s_{y|x}\sqrt{\left[\frac{1}{n} + \frac{(x_+ - \bar{x})^2}{\sum_{i=1}^{n}(x_i - \bar{x})^2}\right]}$$

and

$$a + bx_+ + c^{**}s_{y|k}\sqrt{\left[\frac{1}{n} + \frac{(x_+ - \bar{x})^2}{\sum_{i=1}^{n}(x_i - \bar{x})^2}\right]},$$

where c^{**} is given in Table 4.6. c^{**} is clearly a function of n, P, and α.

Thus, the book publisher can state that the bookstore sales corresponding to a known mail-order sales will fall in the interval constructed using the expressions given above. Such statements can be made for as many books as the publisher desires. Furthermore, the probability is P that at least $100(1 - \alpha)$ percent of bookstore sales corresponding to mail-order sales will fall in these intervals. If P is chosen as 0.90 and $\alpha = 0.05$, the appropriate value of c^{**} is 11.625. Hence, the number of bookstore sales corresponding to mail-order sales of 1,400 books will fall in the interval 6,060 ±759. If another book had mail-order sales of 1,353, the bookstore sales would fall in the interval 5,258 ± 390, and so on. At least 95 percent of the bookstore sales will fall into their predicted intervals, and these statements are made with confidence 0.90.

Table 4.6 **Values† of** c^{**}

n	$\alpha = 0.50$	$\alpha = 0.25$	$\alpha = 0.10$	$\alpha = 0.05$	$\alpha = 0.01$	$\alpha = 0.001$
			$P = .90$			
4	7.471	10.160	13.069	14.953	18.663	23.003
6	5.380	7.453	9.698	11.150	14.014	17.363
8	5.037	7.082	9.292	10.722	13.543	16.837
10	4.983	7.093	9.366	10.836	13.733	17.118
12	5.023	7.221	9.586	11.112	14.121	17.634
14	5.101	7.394	9.857	11.447	14.577	18.232
16	5.197	7.586	10.150	11.803	15.057	18.856
18	5.300	7.786	10.449	12.165	15.542	19.484
20	5.408	7.987	10.747	12.526	16.023	20.104
			$P = .95$			
4	10.756	14.597	18.751	21.445	26.760	32.982
6	6.652	9.166	11.899	13.669	17.167	21.266
8	5.933	8.281	10.831	12.484	15.750	19.568
10	5.728	8.080	10.632	12.286	15.553	19.369
12	5.684	8.093	10.701	12.391	15.724	19.619
14	5.711	8.194	10.880	12.617	16.045	20.050
16	5.771	8.337	11.107	12.898	16.431	20.559
18	5.848	8.499	11.357	13.204	16.845	21.097
20	5.937	8.672	11.619	13.521	17.272	21.652
			$P = .99$			
4	24.466	33.019	42.398	48.620	60.500	74.642
6	10.444	14.285	18.483	21.215	26.606	32.920
8	8.290	11.453	14.918	17.166	21.652	26.860
10	7.567	10.539	13.796	15.911	20.097	24.997
12	7.258	10.182	13.383	15.479	19.579	24.403
14	7.127	10.063	13.267	15.355	19.485	24.316
16	7.079	10.055	13.306	15.410	19.582	24.467
18	7.074	10.111	13.404	15.552	19.794	24.746
20	7.108	10.198	13.566	15.745	20.065	25.122

† Reprinted by permission from "Simultaneous Tolerance Intervals in Regression," by Lieberman, G. J., and R. G. Miller: *Biometrika,* **50**, 1 and 2, 164 (1963).

4.7 Choice of Distribution Functions

One of the most important problems in statistics, and one of the most difficult, is the choice of the parametric form of a probability distribution. It is especially important because the assumption made about the form of the probability distribution is usually quite critical in determining the decision rules to be used. Such an assumption often must be made in the absence of any data. Sometimes it may be possible to use large sample properties of the decision rules such as obtained, for example, by invoking the central limit theorem. An alternative approach is to base the decisions upon procedures which do not use any information about the probability distribution (nonparametric procedures).

When sample data are available, decision-makers have both subjective and objective procedures available to aid them in their choice. In the category of subjective procedures are analyses made by plotting histograms and sample cumulative distribution functions. Objective procedures are obtained by using goodness-of-fit tests.

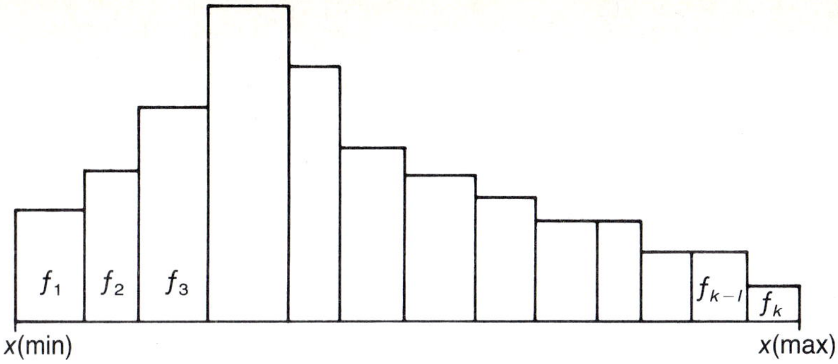

Figure 4.2 Histogram.

Histograms and Sample Cumulative Distribution Functions

Suppose that the unknown density function of a continuous random variable is under consideration. Empirical frequencies obtained from sample data can be related to density functions as follows. Suppose a random sample of size n is obtained. Let x_1, x_2, . . . , x_n denote the values that the random variable in this random sample takes on. Divide the abscissa between the largest and smallest x's into k equal intervals. Let f_1, f_2, . . . , f_k denote the proportion (relative frequency) of the values, x_1, x_2, . . . , x_n, that fall into the first, second, . . . , kth interval, respectively. Plot a bar graph such as that shown in Figure 4.2, where the height of the ith bar is determined so that its area is f_i. This plot is called a *histogram*, and the total area of the histogram is one.

As the sample size gets large and the intervals small, the histogram tends to be similar in shape to that of the density function of X. Hence, by looking at a histogram, one can assess visually whether it seems to possess the characteristics of a presumed density function. For example, the histogram may suggest whether or not the underlying density is symmetric, unimodal, and so on.

In Sec. 3.4, the sample cumulative distribution function $F_n(x)$ was defined, and such a plot appears in Fig. 3.3. $F_n(x)$ can be interpreted as the fraction of outcomes of the experiment less than or equal to x. It can be shown that as n (the number of repetitions of the experiment) gets large, the sample cumulative distribution function approaches the cumulative distribution function of the random variable X. Again then, by looking at a sample cumulative distribution function, one can assess visually whether it seems to possess the characteristics of a presumed distribution function. In this case, however, additional aid is available.

There exists graph paper known as normal probability paper. By properly choosing the scale of the ordinate, one can plot the cumulative distribution function of the normal distribution as a straight line. Normal probability paper is a graph paper possessing such a special grid for the ordinate. Hence, the cumulative distribution function of a normally distributed random variable will plot as a straight line on this paper, where the mean is the abscissa corresponding to the 0.50 ordinate, and the standard deviation is the distance between the abscissa of the 0.8413 point and the abscissa of the 0.50 point. Hence, if a plot of the *sample* cumulative distribution function appears to be nearly a straight line on normal probability paper, this would suggest that the probability distribution is indeed normal. Similar paper can be constructed for other distribution functions.

Both of the above techniques suffer from the fact that many different distributions may sometimes yield random samples that have an indistinguishable appearance when they are plotted on either a histogram or on probability paper. Thus, it often is not possible to identify objectively a particular probability distribution on the basis of a random sample. Therefore, objective procedures for doing this are of interest, and two such procedures are described in the next section.

Chi-Square Goodness of Fit and Kolmogorov-Smirnov

One objective procedure for verifying whether or not a probability distribution has a specified form is the chi-square goodness-of-fit test. This test compares a set of sample frequencies with a set of frequencies that would be expected on the basis of some hypothesis. If this comparison is favorable, the hypothesis is accepted; otherwise it is rejected. The test procedure is as follows: let $f_1, f_2, \ldots, f_k$ be the sample frequencies of k classes, and let $f_1^o, f_2^o, \ldots, f_k^o$ be the frequencies that would be expected if the hypothesis about the form of the distribution is true. If the hypothesis is true, then

$$\chi^2 = \sum_{i=1}^{k} \frac{(f_i - f_i^o)^2}{f_i^o}$$

will tend to have a chi-square distribution as k gets large. The number of degrees of freedom associated with this chi-square random variable will be $(k - 1)$ minus the number of parameters that must be estimated from the sample data. For example, if the hypothesis is that the distribution is normal with mean 7 and variance 6, the degrees of freedom are just $(k - 1)$. On the other hand, if the hypothesis is that the distribution is normal without specifying its mean and variance, then in order to obtain f_i^o, the mean and variance must be estimated from the data (usually by the method of maximum likelihood), and the degrees of freedom become $k - 3$.

The choice of k is rather arbitrary, but a good rule of thumb indicates that k should be chosen so that each $f_i^o \geq 5$. The test of the hypothesis is carried out by comparing the value that the random variable χ^2 takes on, with a selected critical value of the chi-square distribution (chosen from the table of the percentage points of the chi-square distribution given in Table A5.3 of Appendix 5). The hypothesis is rejected if χ^2 exceeds this critical value. If the critical value chosen corresponds to the α percentage point, then the probability of accepting the hypotheses when it is true is approximately $(1 - \alpha)$.

The Kolmogorov-Smirnov test is a competitor of the chi-square goodness of fit whenever the hypothesized form of the distribution is completely specified (including the values of all parameters). In fact, it is a more powerful test and so should be used in this situation. This test compares the sample cumulative distribution function $F_n(x)$ with the hypothesized cumulative distribution function $F(x)$. In particular, define the random variable D_n as

$$D_n = \max_{\text{all } x} |F_n(x) - F(x)|.$$

The distribution of D_n is independent of $F(x)$. It has been computed for various sample sizes and can be found in most elementary textbooks on mathematical statistics. If $D_n^*(\alpha)$ is the α percentage point of the distribution of D_n, and if the hypothesis that $F(x)$ is the appropriate cumulative distribution function is rejected when $D_n > D_n^*(\alpha)$, then the probability is $(1 - \alpha)$ of accepting the hypothesis when it is indeed true.

SELECTED REFERENCES

1. DeGroot, M. H.: *Probability and Statistics*, 2d ed., Addison-Wesley, Reading, Mass., 1986.
2. Freund, J. E., and R. E. Walpole: *Mathematical Statistics*, 4th ed., Prentice-Hall, Englewood Cliffs, N.J., 1987.
3. Hoel, D., S. Port, and C. Stone: *Introduction to Statistical Theory*, Houghton-Mifflin, Boston, Mass., 1971.
4. Hogg, R., and A. Craig: *Introduction to Mathematical Statistics*, 4th ed., Macmillan, New York, 1978.
5. Mood, A. M., and F. Graybill: *Introduction to the Theory of Statistics*, 3d ed., McGraw-Hill, New York, 1974.

PROBLEMS

1.* Suppose X_1 and X_2 form a random sample of size 2 from a distribution such that

$$E(X) = \mu$$

and

$$E(X - \mu)^2 = \sigma^2.$$

Two estimators of μ are

$$T_1 = \frac{X_1 + X_2}{2} = \overline{X},$$

and

$$T_2 = \frac{X_1 + 2X_2}{3}.$$

Which is a better estimate?

2. The probability density function of a certain random variable X is given by

$$f_X(y) = \begin{cases} 2y/\theta^2, & \text{for } 0 \le y \le \theta \\ 0, & \text{otherwise,} \end{cases}$$

where θ is a parameter of the density.

Consider the problem of estimating θ from a random sample $X_1, \ldots, X_n$ of n observations from the given density. Two methods for forming such estimates are the method of maximum likelihood and the method of moments. For this problem, the maximum likelihood estimator, denoted by T_1, is the largest of the values $X_1, X_2, \ldots, X_n$; the method of moments estimator, T_2, is three halves of the sample mean. Some information about these estimates is:

$$T_1 = \max\{X_1, X_2, \ldots, X_n\},$$

$$E(T_1) = \frac{2n}{2n + 1}\theta,$$

$$\text{variance } (T_1) = \frac{n}{(n + 1)(2n + 1)^2}\theta^2,$$

$$T_2 = \left(\frac{3}{2}\right)\overline{X} = \frac{3(X_1 + \cdots + X_n)}{2n},$$

$$E(T_2) = \theta,$$

$$\text{variance } (T_2) = \frac{\theta^2}{8n}.$$

Compute the mean square error for both estimates. Using mean square error as a criterion, which estimate is better if $n = 2$? If $n = 3$, which is better?

3. Consider the function

$$f_X(y) = \begin{cases} \dfrac{K}{\theta}\left(1 - \dfrac{1}{\theta}y\right), & 0 \le y \le \theta \\ 0, & y < 0 \text{ and } y > \theta. \end{cases}$$

(*a*) What value(s) of K and θ make $f_X(y)$ a bona-fide probability density function?
(*b*) Find $E(X)$.
Consider the class of all estimators of θ of the form $C\overline{X}$ (where C is a constant and $\overline{X}$ is computed from a sample of size n).
(*c*) What value of C makes $C\overline{X}$ an unbiased estimator of θ?
(*d*) Find the minimum mean square error estimator in the class $C\overline{X}$.
(*e*) Show that the minimum mean square error estimator approaches the unbiased estimator for large n.

4.* Suppose X_1 and X_2 are independent observations having a common distribution with mean $= \mu$ and variance $= 1$. Consider point estimates of μ of the following form:

$$\hat{\mu} = \lambda X_1 + (1 - \lambda)X_2,$$

where $0 \le \lambda \le 1$.
(*a*) For what values of λ is $\hat{\mu}$ unbiased?
(*b*) Noting that $\hat{\mu} - \mu = \lambda(X_1 - \mu) + (1 - \lambda)(X_2 - \mu)$, compute the expected squared error of $\hat{\mu}$. Is there a value of λ which yields a uniformly best estimate of this type?

5. A radio manufacturer wishes to test the reliability of radios. Let p be the probability that a radio functions properly when turned on, and let $(1 - p)$ be the probability that the radio fails when turned on. Suppose that radios are tested as follows: The radio is turned on and off repeatedly until failure occurs. Let N be the number of the trial at which the first failure occurs. (Each turn-on constitutes a trial.) Then N is a discrete random variable with the distribution

$$P\{N = n\} = (1 - p)\, p^{n-1}, \qquad \text{for } n = 1, 2, 3, \ldots .$$

Suppose a random sample of r radios is taken, and each radio is tested by the above procedure. Let N_i denote the outcome of testing the ith radio.
(*a*) Find the maximum likelihood estimate of p.
(*b*) The reliability $R(n_0)$ of a radio is defined to be the probability that the radio will not fail in the first n_0 trials.

$$R(n_0) = P\{N > n_0\} = 1 - P\{N \le n_0\} = 1 - \sum_{n=1}^{n_0} (1 - p)p^{n-1}.$$

Find the maximum likelihood estimate of $R(n_0)$.

6. A buyer of lots of material notes that the incoming quality P has density given by

$$f_P(y) = \theta y^{\theta - 1}, \qquad \text{for } 0 < y < 1, \text{ and for } 0 < \theta < \infty.$$

Suppose that a random sample of n lots is screened and $p_1, p_2, \ldots, p_n$ obtained.
(*a*) Find $\hat{\theta}$, the maximum likelihood estimate of θ.
(*b*) Find the asymptotic variance of $\sqrt{n}\hat{\theta}$.
(*c*) Using the asymptotic properties of $\hat{\theta}$, find a lower 95 percent confidence bound for θ.

7. Let $X_1, \ldots, X_n$ be a random sample of size n from a Poisson distribution with parameter λ, that is,

$$P_X(k) = \frac{\lambda^k e^{-\lambda}}{k!}.$$

Let λ be considered as a random variable having a prior density function

$$f_\lambda(y) = \begin{cases} e^{-y}, & \text{for } 0 < y < \infty \\ 0, & \text{otherwise.} \end{cases}$$

The loss function to be considered is

$$(\hat{\lambda} - \lambda)^2.$$

(*a*) Let $T = \Sigma X_i$, and noting that the sum of n independent Poisson random variables, each with parameter λ, is also distributed as a Poisson random variable but with parameter $n\lambda$, show that the posterior distribution of λ, given $T = t$, is given by

$$h_{\lambda|T=t}(y) = \frac{y^t e^{-(n+1)y}(n+1)^{t+1}}{\Gamma(\theta + 1)}.$$

(*b*) Show that the Bayes' estimator is given by

$$\frac{(T + 1)}{(n + 1)}.$$

(*c*) Find the maximum likelihood estimator.

8. Let $X_1, X_2, \ldots, X_n$ be a random sample of size n from a normal distribution with parameters μ and σ^2. Let μ be considered as a random variable having a prior distribution which is also normal but with parameters υ, γ^2. The loss function to be considered is

$$(\hat{\mu} - \mu)^2.$$

(*a*) Show that the posterior distribution of μ, given $\overline{X} = \bar{x}$, is normal.
(*b*) Find the Bayes' estimator for μ.

9.* The following data relate road width x and accident frequency Y. Road width (in feet) was treated as a constant; and values of the random variable Y, in accidents per 10^8 vehicle miles, were observed.

Number of observations = 7.

$$\sum_{i=1}^{7} x_i = 354 \qquad \sum_{i=1}^{7} Y_i = 481$$

$$\sum_{i=1}^{7} x_i^2 = 19{,}956 \qquad \sum_{i=1}^{7} Y_i^2 = 35{,}451$$

$$\sum_{i=1}^{7} x_i Y_i = 22{,}200$$

x	y
26	92
30	85
44	78
50	81
62	54
68	51
74	40

Assume that Y is normally distributed with mean $\alpha + \beta x$ and constant variance for all x, and that the sample is random.

(*a*) Fit a least-squares line to the data.
(*b*) Construct a 95 percent prediction interval for Y_+, a future observation of Y, corresponding to $x_+ = 55$ feet.
(*c*) Suppose that two future observations on Y, both corresponding to $x_+ = 55$ feet, are to be made. Construct prediction intervals for both of these observations so that the probability is *at least* 95 percent that *both* future values of Y will fall into them simultaneously.

Hint: If k predictions are to be made [such as given in (d)], each with probability $1 - \alpha$, then the probability is *at least* $1 - k\alpha$ that all k future observations will fall into their respective intervals.

(d) Construct a simultaneous tolerance interval for the future value of Y corresponding to $x_+ = 55$ feet with $P = 0.90$ and $1 - \alpha = 0.95$.

10. The data below are observations on a dependent random variable Y taken at various levels of an independent variable x. [It is to be assumed that $E(Y_i|x_i) = \alpha + \beta x_i$, and Y_i are independent normal random variables with mean 0 and variance σ^2.] Suppose the data are as follows.

x_i	0	2	4	6	8
y_i	0	4	7	13	16

(a) Estimate the linear relationship by the method of least squares.

(b) Find a 95 percent confidence interval for the expected value of Y at $x_* = 10$.

(c) Find a 95 percent prediction interval for a future observation to be taken at $x_+ = 10$.

(d) Obtain an upper 95 percent confidence limit for σ^2.

(e) Prove that

$$a + bx_+ \pm t_{\alpha/2;n-2}s_{y|x}\sqrt{\left[\frac{1}{N} + \frac{1}{n} + \frac{(x_+ - \bar{x})^2}{\Sigma(x_i - \bar{x})^2}\right]}$$

is a $100(1 - \alpha)$ percent prediction interval for the arithmetic mean $\overline{Y}_+$ of N future observations to be taken at x_+.

(f) For $x_+ = 10$, $P = 0.90$, and $(1 - \alpha) = 0.95$, find a simultaneous tolerance interval for the future value of Y_+.

11. If a particle is dropped at time $t = 0$, physical theory indicates that the relationship between r, the distance traveled, and t, the time elapsed, is $r = gt^k$ for some positive constants g and k. A transformation to linearity can be obtained by taking logarithms:

$$\log r = \log g + k \log t.$$

Letting $y = \log r$, $A = \log g$, and $x = \log t$, this relation becomes $y = A + kx$. Due to random error in measurement, however, it can only be stated that $E(Y|x) = A + kx$. Assume Y is normally distributed with mean $A + kx$ and variance σ^2.

A physicist who wishes to estimate k and g performs the following experiment: At time 0 the particle is dropped. At time t the distance r is measured. She performs this experiment 5 times, obtaining the following data:

y = Logarithm r	x = Logarithm t
−3.95	−2.0
−2.12	−1.0
0.08	0.0
2.20	+1.0
3.87	+2.0

(a) Obtain least-squares estimates for k and $\log g$.

(b) Find a 95 percent confidence interval for $E(\log r|x)$ when $x_* = 0$.

(c) Find a 95 percent prediction interval for r when $x_+ = 0$.

The following data has been calculated:

$$\bar{y} = 0.016$$

$$\Sigma(x_i - \bar{x})^2 = 10$$

$$\Sigma(x_i - \bar{x})(y_i - \bar{y}) = 19.96$$

$$\Sigma(y_i - \tilde{y}_i)^2 = 0.0858$$

$$\frac{s_{y|x}}{\sqrt{\Sigma(x_i - \bar{x})^2}} = 0.053$$

$$s_{y|x}\sqrt{\tfrac{1}{5}} = 0.076.$$

Note: All logarithms are to the base 10.

12. Suppose that the relation between Y and x is given by

$$E(Y|x) = \beta x,$$

where Y is assumed to be normally distributed with mean βx and *known* variance σ^2. n independent pairs of observations are taken and are denoted by $x_1, y_1; x_2, y_2; \ldots; x_n, y_n$.

(*a*) Find the least-squares estimate of β.

(*b*) Find the variance of the least-squares estimate of β.

(*c*) For $X = x_*$, find a 95 percent two-sided confidence interval for

$$E(Y|X = x_*).$$

5

Markov Chains

5.1 Introduction

In decision-making problems, we often are faced with making decisions based upon phenomena that have uncertainty associated with them. This uncertainty is caused by inherent variation due to sources of variation that elude control or due to the inconsistency of natural phenomena. Rather than treat this variability qualitatively, we can incorporate it into the mathematical model and thus handle it quantitatively. This treatment generally can be accomplished if the natural phenomena exhibit some degree of regularity, so that their variation can be described by a probability model. We assume that the reader has a basic knowledge of probability theory. The ensuing sections are concerned with special types of probability models.

5.2 Stochastic Processes

A **stochastic process** is defined to be simply *an indexed collection of random variables* $\{X_t\}$, *where the index t runs through a given set T*. Often T is taken to be the set of nonnegative integers, and X_t represents a measurable characteristic of interest at time t. For example, the stochastic process, $X_1, X_2, X_3, \ldots$, can represent the collection of weekly (or monthly) inventory levels of a given product, or it can represent the collection of weekly (or monthly) demands for this product.

There are many stochastic processes that are of interest. A consideration of the behavior of a system operating for some period of time often leads to the analysis of a stochastic process with the following structure. At particular points of time t labeled $0, 1, \ldots$, the system is found in exactly one of a finite number of mutually exclusive and exhaustive categories or *states* labeled $0, 1, \ldots, M$. The points in time may be spaced equally, or their spacing may depend upon the overall behavior of the physical system in which the stochastic process is *imbedded*, e.g., the time between occurrences of some phenomenon of interest. Although the states may constitute a qualitative as well as a quantitative characterization of the system, no loss of generality is entailed by the numerical labels, $0, 1, \ldots, M$, which are used henceforth to denote the possible states of the system. Thus the mathematical representation of the physical system is that of a stochastic process $\{X_t\}$, where the random variables are observed at $t = 0, 1, 2, \ldots$, and where each random variable may take on any one of the $(M + 1)$ integers $0, 1, \ldots, M$. These integers are a characterization of the $(M + 1)$ states of the process. It must be emphasized that each state that the stochastic process reaches is given a label that denotes the physical state of the system. It is only for notational convenience that this set is labeled $0, 1, \ldots, M$.

As an example, consider the following inventory problem. A camera store stocks a particular model camera that can be ordered weekly. Let $D_1, D_2, \ldots$, represent the demand for this camera during the first week, second week, $\ldots$, respectively. It is assumed that the D_i are independent and identically distributed random variables having a known probability distribution. Let X_0 represent the number of cameras on hand at the outset, X_1 the number of cameras on hand at the end of week one, X_2 the number of cameras on hand at the end of week two, and so on. Assume that $X_0 = 3$. On Saturday night the store places an order that is delivered in time for the opening of the store on Monday. The store uses the following (s, S) ordering policy[1]: If the number of cameras on hand at the end of the week is less than $s = 1$ (no cameras in stock), the store orders (up to) $S = 3$. Otherwise, the store does not order (if there are any cameras in stock, no order is placed). It is assumed that sales are lost when demand exceeds the inventory on hand. Thus $\{X_t\}$ for $t = 0, 1, \ldots$, is a stochastic process of the form just described. The possible states of the process are the integers 0, 1, 2, 3 representing the possible number of cameras on hand at the end of the week. In fact, the random variables X_t are clearly dependent and may be evaluated iteratively by the expression

$$X_{t+1} = \begin{cases} \max\{(3 - D_{t+1}), 0\}, & \text{if } X_t < 1 \\ \max\{(X_t - D_{t+1}), 0\}, & \text{if } X_t \geq 1, \end{cases}$$

[1] In general, an (s, S) policy is a periodic review policy that calls for ordering up to S units whenever the inventory level dips below s $(S \geq s)$. If the inventory level is s or greater, then no order is placed. These policies are discussed in detail in Chap. 9.

for $t = 0, 1, 2, \ldots$. This example is used for illustrative purposes throughout many of the following sections. Section 5.3 further defines the type of stochastic process considered in this chapter.

5.3 Markov Chains

Assumptions regarding the joint distribution of $X_0, X_1, \ldots$, are necessary to obtain analytical results. One assumption that leads to analytical tractability is that the stochastic process is a Markov chain (defined later), which has the following key property: A stochastic process $\{X_t\}$ is said to have the *Markovian property* if $P\{X_{t+1} = j | X_0 = k_0, X_1 = k_1, \ldots, X_{t-1} = k_{t-1}, X_t = i\} = P\{X_{t+1} = j | X_t = i\}$, for $t = 0, 1, \ldots$, and every sequence $i, j, k_0, k_1 \ldots, k_{t-1}$.

This Markovian property can be shown to be equivalent to stating that the conditional probability of any future "event," given any past "event" and the present state $X_t = i$, is *independent* of the past event and depends only upon the present state of the process. The conditional probabilities $P\{X_{t+1} = j | X_t = i\}$ are called transition probabilities. If, for each i and j,

$$P\{X_{t+1} = j | X_t = i\} = P\{X_1 = j | X_0 = i\}, \qquad \text{for all } t = 0, 1, \ldots,$$

then the (one-step) transition probabilities are said to be *stationary* and are usually denoted by p_{ij}. Thus having stationary transition probabilities implies that the transition probabilities do not change in time. The existence of stationary (one-step) transition probabilities also implies that, for each i, j, and n ($n = 0, 1, 2, \ldots$),

$$P\{X_{t+n} = j | X_t = i\} = P\{X_n = j | X_0 = i\},$$

for all $t = 0, 1, \ldots$. These conditional probabilities are usually denoted by $p_{ij}^{(n)}$ and are called n-step transition probabilities.[1] Thus $p_{ij}^{(n)}$ is just the conditional probability that the random variable X, starting in state i, will be in state j after exactly n steps (time units).

Because the $p_{ij}^{(n)}$ are conditional probabilities, they must be nonnegative, and since the process must make a transition into some state, they must satisfy the properties

$$p_{ij}^{(n)} \geq 0, \qquad \text{for all } i \text{ and } j, \qquad \text{and} \qquad n = 0, 1, 2, \ldots$$

$$\sum_{j=0}^{M} p_{ij}^{(n)} = 1, \qquad \text{for all } i, \qquad \text{and} \qquad n = 0, 1, 2, \ldots .$$

A convenient notation for representing the transition probabilities is the matrix form

$$\mathbf{P}^{(n)} = \begin{array}{c|cccc} \text{State} & 0 & 1 & \cdots & M \\ \hline 0 & p_{00}^{(n)} & & \cdots & p_{0M}^{(n)} \\ 1 & \vdots & & & \vdots \\ \vdots & & & & \\ M & p_{M0}^{(n)} & & \cdots & p_{MM}^{(n)} \end{array}, \qquad \text{for } n = 0, 1, 2, \ldots$$

[1] For $n = 0$, $p_{ij}^{(0)}$ is just $P\{X_0 = j | X_0 = i\}$ and hence is 1 when $i = j$ and 0 when $i \neq j$. For $n = 1$, $p_{ij}^{(1)}$ is just the (one-step) transition probability and is denoted by p_{ij}.

or, equivalently,

$$\mathbf{P}^{(n)} = \begin{bmatrix} p_{00}^{(n)} & \cdots & p_{0M}^{(n)} \\ \vdots & & \\ p_{M0}^{(n)} & & p_{MM}^{(n)} \end{bmatrix}.$$

It is now possible to define a Markov chain. A stochastic process $\{X_t\}$ $(t = 0, 1, \ldots)$ is said to be a *finite-state Markov chain*[1] if it has the following properties:

1. A finite number of states.
2. The Markovian property.
3. Stationary transition probabilities.
4. A set of initial probabilities $P\{X_0 = i\}$ for all i.

Returning to the inventory example developed in the preceding section, it is easily seen that $\{X_t\}$, where X_t is the number of cameras in stock at the end of the tth week (before an order is received), is a Markov chain. Now consider how to obtain the (one-step) transition probabilities, i.e., the elements of the (one-step) *transition matrix*

$$\mathbf{P} = \begin{bmatrix} p_{00} & p_{01} & p_{02} & p_{03} \\ p_{10} & p_{11} & p_{12} & p_{13} \\ p_{20} & p_{21} & p_{22} & p_{23} \\ p_{30} & p_{31} & p_{32} & p_{33} \end{bmatrix},$$

assuming that each D_t has a Poisson distribution with parameter $\lambda = 1$.

To obtain p_{00} it is necessary to evaluate $P\{X_t = 0|X_{t-1} = 0\}$. If $X_{t-1} = 0$, then $X_t = \max\{(3 - D_t), 0\}$. Therefore, if $X_t = 0$, then the demand during the week has to be 3 or more. Hence $p_{00} = P\{D_t \geq 3\}$. This transition probability is just the probability that a Poisson random variable with parameter $\lambda = 1$ takes on a value of 3 or more, which is obtained from Table A5.4 of Appendix 5, so that $p_{00} = 0.08$. $p_{10} = P\{X_t = 0|X_{t-1} = 1\}$ can be obtained in a similar way. If $X_{t-1} = 1$, then $X_t = \max\{(1 - D_t), 0\}$. To have $X_t = 0$, the demand during the week has to be 1 or more. Hence $p_{10} = P\{D_t \geq 1\} = 0.632$ (from Table A5.4 of Appendix 5). To find $p_{21} = P\{X_t = 1|X_{t-1} = 2\}$, note that $X_t = \max\{(2 - D_t), 0\}$ if $X_{t-1} = 2$. Therefore, if $X_t = 1$, then the demand during the week has to be exactly 1. Hence $p_{21} = P\{D_t = 1\} = 0.368$ (from Table A5.4 of Appendix 5). The remaining entries are obtained in a similar manner, which yield the following (one-step) transition matrix:

$$\mathbf{P} = \begin{bmatrix} 0.080 & 0.184 & 0.368 & 0.368 \\ 0.632 & 0.368 & 0 & 0 \\ 0.264 & 0.368 & 0.368 & 0 \\ 0.080 & 0.184 & 0.368 & 0.368 \end{bmatrix}.$$

Several other examples of Markov chains follow: Consider the following model for the value of a stock. At the end of a given day, the price is recorded. If the stock

[1] The definitions of Markovian property and Markov chain are more restrictive in this context than the usages of these terms in the literature because the discussion is confined to a discrete time parameter and finite-state space.

has gone up, the probability that it will go up tomorrow is 0.7. If the stock goes down, the probability that it will go up tomorrow is only 0.5. This is a Markov chain, where state 0 represents the stock going up and state 1 represents the stock going down. The transition matrix is given by

$$\mathbf{P} = \begin{bmatrix} 0.7 & 0.3 \\ 0.5 & 0.5 \end{bmatrix}.$$

Suppose now that the stock market model is changed so that whether or not the stock goes up tomorrow depends upon whether or not it increased today *and* yesterday. In particular, if the stock has increased for the past two days, it will increase tomorrow with probability 0.9. If the stock increased today but decreased yesterday, then it will increase tomorrow with probability 0.6. If the stock decreased today but increased yesterday, then it will increase tomorrow with probability 0.5. Finally, if the stock decreased for the past two days, then it will increase tomorrow with probability 0.3. If we define the state as representing whether the stock goes up or down, the system is no longer a Markov chain. However, we can transform the system into a Markov chain by defining the states as follows[1]:

State 0: The stock increased both today and yesterday.
State 1: The stock increased today but decreased yesterday.
State 2: The stock decreased today but increased yesterday.
State 3: The stock decreased both today and yesterday.

This leads to a four-state Markov chain with the following transition matrix:

$$\mathbf{P} = \begin{bmatrix} 0.9 & 0 & 0.1 & 0 \\ 0.6 & 0 & 0.4 & 0 \\ 0 & 0.5 & 0 & 0.5 \\ 0 & 0.3 & 0 & 0.7 \end{bmatrix}.$$

One row in the matrix will be verified—say, the second. This corresponds to state 1, which represents the stock increasing today but decreasing yesterday. The first element in the row represents the probability of the stock increasing tomorrow, having increased today and given that the stock increased today but decreased yesterday. This is just the probability of the stock increasing tomorrow given that it increased today but decreased yesterday, i.e., 0.6. Similarly, the third element in the row represents the probability of the stock decreasing tomorrow, having increased today and given that the stock increased today but decreased yesterday. This is just the probability of the stock decreasing tomorrow given that it increased today but decreased yesterday, i.e., 0.4. The other two elements are 0 because they pertain to events that are contradictory; namely, they correspond to instances where the stock decreased today.

Another example is gambling. Suppose that a player has \$1, and with each play of the game wins a dollar with probability p or loses a dollar with probability $1 - p$. The game ends when the player either accumulates \$3 or goes broke. This model is a Markov chain with the states representing the player's fortune, i.e., 0, \$1,

[1] This example demonstrates that Markov chains are able to incorporate arbitrary amounts of history, but at the cost of significantly increasing the number of states; i.e., to incorporate N periods of history requires N^M states to model such arbitrary dependency as a Markov chain.

$2, or $3, and with the transition matrix given by

$$\mathbf{P} = \begin{bmatrix} 1 & 0 & 0 & 0 \\ 1-p & 0 & p & 0 \\ 0 & 1-p & 0 & p \\ 0 & 0 & 0 & 1 \end{bmatrix}.$$

Note that in both the inventory and gambling examples, the numeric labeling of the states that the process reaches coincides with the physical expression of the system—i.e., actual inventory levels and the player's fortune, respectively—whereas the numeric labeling of the states in the stock examples represents notational convenience.

5.4 Chapman-Kolmogorov Equations

Section 5.3 introduced the n-step transition probability $p_{ij}^{(n)}$. This transition probability can be useful when the process is in state i, and the probability that the process will be in state j after n periods is desired. The *Chapman-Kolmogorov equations* provide a method for computing these n-step transition probabilities:

$$p_{ij}^{(n)} = \sum_{k=0}^{M} p_{ik}^{(v)} p_{kj}^{(n-v)}, \qquad \text{for all } i, j, n, \qquad \text{and} \qquad 0 \le v \le n.$$

These equations merely point out that in going from state i to state j in n steps, the process will be in some state k after exactly v (less than n) steps. Thus $p_{ik}^{(v)} p_{kj}^{(n-v)}$ is just the conditional probability that, starting from state i, the process goes to state k after v steps and then to state j in $n - v$ steps. Therefore, summing these conditional probabilities over all possible k must yield $p_{ij}^{(n)}$. The special cases of $v = 1$ and $v = n - 1$ lead to the expressions

$$p_{ij}^{(n)} = \sum_{k=0}^{M} p_{ik} p_{kj}^{(n-1)}$$

and

$$p_{ij}^{(n)} = \sum_{k=0}^{M} p_{ik}^{(n-1)} p_{kj},$$

for all i, j, n. It then becomes evident that the n-step transition probabilities can be obtained from the one-step transition probabilities recursively. This recursive relationship is best explained in matrix notation (see Appendix 3). For $n = 2$, these expressions become

$$p_{ij}^{(2)} = \sum_{k=0}^{M} p_{ik} p_{kj}, \qquad \text{for all } i, j.$$

Note that the $p_{ij}^{(2)}$ are the elements of the matrix $\mathbf{P}^{(2)}$. However, it must also be noted that these elements,

$$\sum_{k=0}^{M} p_{ik} p_{kj},$$

are obtained by multiplying the matrix of one-step transition probabilities by itself; that is,

$$\mathbf{P}^{(2)} = \mathbf{P} \cdot \mathbf{P} = \mathbf{P}^2.$$

More generally, it follows that the matrix of n-step transition probabilities can be obtained from the expression

$$\begin{aligned} \mathbf{P}^{(n)} &= \mathbf{P} \cdot \mathbf{P} \cdots \mathbf{P} = \mathbf{P}^n \\ &= \mathbf{P}\mathbf{P}^{n-1} = \mathbf{P}^{n-1}\mathbf{P}. \end{aligned}$$

Thus the n-step transition probability matrix can be obtained by computing the nth power of the one-step transition matrix. For values of n that are not too large, the n-step transition matrix can be calculated in the manner just described. However, when n is large, such computations are often tedious, and furthermore, round-off errors may cause inaccuracies.

Returning to the inventory example, the two-step transition matrix is given by[1]

$$\mathbf{P}^{(2)} = \mathbf{P}^2 = \begin{bmatrix} 0.080 & 0.184 & 0.368 & 0.368 \\ 0.632 & 0.368 & 0 & 0 \\ 0.264 & 0.368 & 0.368 & 0 \\ 0.080 & 0.184 & 0.368 & 0.368 \end{bmatrix} \begin{bmatrix} 0.080 & 0.184 & 0.368 & 0.368 \\ 0.632 & 0.368 & 0 & 0 \\ 0.264 & 0.368 & 0.368 & 0 \\ 0.080 & 0.184 & 0.368 & 0.368 \end{bmatrix}$$

$$= \begin{bmatrix} 0.249 & 0.286 & 0.300 & 0.165 \\ 0.283 & 0.252 & 0.233 & 0.233 \\ 0.351 & 0.319 & 0.233 & 0.097 \\ 0.249 & 0.286 & 0.300 & 0.165 \end{bmatrix}.$$

Thus, given that there is one camera left in stock at the end of a week, the probability is 0.283 that there will be no cameras in stock 2 weeks later; that is, $p_{10}^{(2)} = 0.283$. Similarly, given that there are two cameras left in stock at the end of a week, the probability is 0.097 that there will be three cameras in stock 2 weeks later; that is, $p_{23}^{(2)} = 0.097$.

The four-step transition matrix can also be obtained as follows:

$$\mathbf{P}^{(4)} = \mathbf{P}^4 = \mathbf{P}^{(2)} \cdot \mathbf{P}^{(2)}$$

$$= \begin{bmatrix} 0.249 & 0.286 & 0.300 & 0.165 \\ 0.283 & 0.252 & 0.233 & 0.233 \\ 0.351 & 0.319 & 0.233 & 0.097 \\ 0.249 & 0.286 & 0.300 & 0.165 \end{bmatrix} \begin{bmatrix} 0.249 & 0.286 & 0.300 & 0.165 \\ 0.283 & 0.252 & 0.233 & 0.233 \\ 0.351 & 0.319 & 0.233 & 0.097 \\ 0.249 & 0.286 & 0.300 & 0.165 \end{bmatrix}$$

$$= \begin{bmatrix} 0.289 & 0.286 & 0.261 & 0.164 \\ 0.282 & 0.285 & 0.268 & 0.166 \\ 0.284 & 0.283 & 0.263 & 0.171 \\ 0.289 & 0.286 & 0.261 & 0.164 \end{bmatrix}.$$

Thus, given that there is one camera left in stock at the end of a week, the probability is 0.282 that there will be no cameras in stock 4 weeks later; that is, $p_{10}^{(4)} = 0.282$. Similarly, given that there are two cameras left in stock at the end of a week, the probability is 0.171 that there will be three cameras in stock 4 weeks later; that is, $p_{23}^{(4)} = 0.171$.

[1] Note that round-off errors already appear in the row corresponding to state 1.

It was pointed out that the one- or n-step transition probabilities are conditional probabilities; for example, $P\{X_n = j | X_0 = i\} = p_{ij}^{(n)}$. If the unconditional probability $P\{X_n = j\}$ is desired, it is necessary to have specified the probability distribution of the initial state. Denote this probability distribution by $Q_{X_0}(i)$, where

$$Q_{X_0}(i) = P\{X_0 = i\}, \qquad \text{for } i = 0, 1, \ldots, M.$$

It then follows that

$$P\{X_n = j\} = Q_{X_0}(0)p_{0j}^{(n)} + Q_{X_0}(1)p_{1j}^{(n)} + \cdots + Q_{X_0}(M)p_{Mj}^{(n)}.$$

In the inventory example it was assumed that initially there were 3 units in stock; that is, $X_0 = 3$. Thus

$$Q_{X_0}(0) = Q_{X_0}(1) = Q_{X_0}(2) = 0,$$

and

$$Q_{X_0}(3) = 1.$$

Hence the (unconditional) probability that there will be three cameras in stock 2 weeks after the inventory system began is 0.165; that is, $P\{X_2 = 3\} = (1)p_{33}^{(2)}$. If, instead, it were given that $Q_{X_0}(i) = \frac{1}{4}$, for $i = 0, 1, 2, 3$, then

$$P\{X_2 = 3\} = \tfrac{1}{4}(0.165) + \tfrac{1}{4}(0.233) + \tfrac{1}{4}(0.097) + \tfrac{1}{4}(0.165) = 0.165.$$

The fact that the same answer is obtained using these two initial probability distributions is purely coincidental.

5.5 Classification of States of a Markov Chain

It is evident that the transition probabilities associated with the states play an important role in the study of Markov chains. In order to further describe the properties of Markov chains it is necessary to present some concepts and definitions concerning these states.

State j is said to be **accessible** from state i if $p_{ij}^{(n)} > 0$ for some $n \geq 0$. Recall that $p_{ij}^{(n)}$ is just the conditional probability of being in state j after n steps, starting in state i. It is easily shown then that state j is accessible from state i if, and only if, it is possible for the system to enter state j starting from state i. In the inventory example, $p_{ij}^{(2)} > 0$ for all i, j, so that every state is accessible from every other state. Obviously, a sufficient condition for *all* states to be accessible is that there exists a value of n, not dependent upon i and j, for which $p_{ij}^{(n)} > 0$ for all i and j. In the inventory example, all states are accessible since $p_{ij}^{(2)} > 0$ for all i and j. In the gambling example, state 2 is not accessible from state 3. This can be deduced from the context of the game (once the player reaches state 3, the player never leaves this state) or by noting that the n-step transition matrix for the game, $\mathbf{P}^{(n)}$, is of the form

$$\mathbf{P}^{(n)} = \begin{bmatrix} 1 & 0 & 0 & 0 \\ * & * & * & * \\ * & * & * & * \\ 0 & 0 & 0 & 1 \end{bmatrix},$$

for all n, where the symbol $*$ represents nonnegative numbers. Note, however, that state 3 is accessible from state 2.

If state j is accessible from state i, and, in addition, state i is accessible from state j, then states i and j are said to **communicate**. In the inventory example, all states communicate. In the gambling example, states 2 and 3 do not. In general, (1) any state communicates with itself [because $p_{ii}^{(0)} = P\{X_0 = i | X_0 = i\} = 1$]; (2) if state i communicates with state j, then state j communicates with state i; and furthermore, (3) if state i communicates with state j, and state j communicates with state k, then state i communicates with state k. Properties (1) and (2) follow from the definition of states communicating, whereas property (3) follows from the Chapman-Kolmogorov equations.

As a result of these three properties of communication, the state space may be partitioned into disjoint classes, with two communicating states said to belong to the same class. Thus the states of a Markov chain may consist of one or more disjoint classes (a class may consist of a single state). If there is only one class, i.e., all the states communicate, the Markov chain is said to be **irreducible**. In the inventory example, the Markov chain is irreducible. In the first stock example, the Markov chain is irreducible. The gambling example contains three classes. State 0 forms a class, state 3 forms a class, and states 1 and 2 form a class.

It is often useful to talk about whether or not a process, starting in state i, will ever return to this state. Let f_{ii} denote the probability that the process will ever return to state i given that it starts in state i. State i is called a **recurrent** state if $f_{ii} = 1$ and **transient** if $f_{ii} < 1$. A special case of a recurrent state is an *absorbing* state. A state i is said to be an **absorbing** state if the (one-step) transition probability p_{ii} equals 1 (which results in a value of 1 appearing as an element of the diagonal of the one-step transition matrix).

Determining whether or not a state is recurrent or transient by evaluating f_{ii} is not generally simple. Hence, it is not always evident whether a state should be classified as recurrent or transient. For example, although all states in the inventory example are recurrent (as shown later in this section), it is not simple to prove that f_{ii} equals 1 for all i.

In the gambling example, state 0 and state 3 are absorbing states (each belonging to a separate class), and hence are recurrent, so f_{00} and f_{33} must equal 1. However, states 1 and 2 are transient, and it is not simple to show that f_{11} and f_{22} are less than 1.

It is possible to determine some properties of f_{ii} that may be useful in determining its value. If a Markov process is in state i, and i is recurrent, the probability is 1 that the process will return to that state. Since the process is a Markov chain, this is equivalent to the process beginning once more from state i, and with probability 1, it will once again return to that state. Repeating this argument leads to the conclusion that state i will be entered infinitely often. Hence, a recurrent state has the property that the expected number of time periods that the process is in state i is infinite.

If a Markov process is in state i, and i is transient, then the probability of reentering state i is f_{ii} (less than 1) and the probability of not reentering state i is $1 - f_{ii}$. It is simple to show that the expected number of time periods that the process is in state i is finite and given by

$$\frac{1}{1 - f_{ii}}.$$

Therefore, it follows that state i is recurrent if, and only if, the expected number of time periods that the process is in state i is infinite given that the process started in state i.

In order to calculate the expected number of time periods that the process is in state i given that $X_0 = i$, define

$$B_n = \begin{cases} 1 & \text{if } X_n = i \\ 0 & \text{if } X_n \neq i. \end{cases}$$

The quantity

$$\sum_{n=1}^{\infty} B_n | X_0 = i$$

represents the number of time periods that the process is in state i given that $X_0 = i$. Therefore, its expectation is given by

$$\begin{aligned} E\left(\sum_{n=1}^{\infty} B_n | X_0 = i\right) &= \sum_{n=1}^{\infty} E(B_n | X_0 = i) \\ &= \sum_{n=1}^{\infty} P\{X_n = i | X_0 = i\} \\ &= \sum_{n=1}^{\infty} p_{ii}^{(n)}. \end{aligned}$$

Thus, it has been shown that a state i is recurrent if, and only if,

$$\sum_{n=1}^{\infty} p_{ii}^{(n)} = \infty.$$

This result can be used to show that recurrence is a class property. That is, all states in a class are either recurrent or transient. Furthermore, in a finite-state Markov chain, not all states can be transient. Otherwise, after a finite amount of time has elapsed, every one of the states will never be reentered (since each state is assumed to be transient), which is impossible since the process must be in some state after this elapsed time. Therefore, all states in an irreducible finite-state Markov chain are recurrent. Indeed, one can identify an irreducible finite-state Markov chain (and therefore conclude that all states are recurrent) by showing that all states of the process communicate. It has already been pointed out that a sufficient condition for *all* states to be accessible (and therefore communicate with each other) is that there exists a value of n, not dependent upon i and j, for which $p_{ij}^{(n)} > 0$ for all i and j. Thus, all states in the inventory example are recurrent, since $p_{ij}^{(2)}$ is positive for all i and j. Similarly, the first stock example contains only recurrent states, since p_{ij} is positive for all i and j. By calculating $p_{ij}^{(2)}$ for all i and j in the second stock example, it follows that all states are recurrent since $p_{ij}^{(2)} > 0$.

As another example, suppose that a Markov process has the following transition matrix:

$$\mathbf{P} = \begin{array}{c} \\ 0 \\ 1 \\ 2 \\ 3 \\ 4 \end{array} \overset{\text{State}}{\begin{array}{c} \begin{array}{ccccc} 0 & 1 & 2 & 3 & 4 \end{array} \\ \begin{bmatrix} \frac{1}{4} & \frac{3}{4} & 0 & 0 & 0 \\ \frac{1}{2} & \frac{1}{2} & 0 & 0 & 0 \\ 0 & 0 & 1 & 0 & 0 \\ 0 & 0 & \frac{1}{3} & \frac{2}{3} & 0 \\ 1 & 0 & 0 & 0 & 0 \end{bmatrix} \end{array}}.$$

It is evident that state 2 is an absorbing state (and hence a recurrent state) because once the process enters state 2 (third row of the matrix), it will never leave. States 3 and 4 are transient states because once the process is in state 3, there is a positive probability that it will never return. The probability is $\frac{1}{3}$ that the process will go from state 3 to state 2 on the first step. Once the process is in state 2, it remains in state 2. Once a process leaves state 4, it can never return. States 0 and 1 are recurrent states. As indicated earlier, to show that states 0 and 1 are recurrent it is sufficient to show that $f_{00} = 1$ and $f_{11} = 1$, which is not a simple task. Hence, an alternative method is desirable. Observe that the n-step transition matrix of the preceding example always has the appearance

$$\mathbf{P}^{(n)} = \begin{bmatrix} * & * & 0 & 0 & 0 \\ * & * & 0 & 0 & 0 \\ 0 & 0 & 1 & 0 & 0 \\ 0 & 0 & * & * & 0 \\ * & * & 0 & 0 & 0 \end{bmatrix},$$

where the symbol $*$ represents positive numbers. Hence it is intuitively evident that once the process is in state 0, it will return to state 0 (possibly passing through state 1) after some number of steps. A similar argument holds for state 1.

Another property of Markov chains that is to be considered is the property of *periodicities*. The **period** of state i is defined to be the integer t ($t > 1$), such that $p_{ii}^{(n)} = 0$ for all values of n other than t, $2t$, $3t$, . . . , and t is the largest integer with this property. In the gambling example, starting in state 1, it is possible for the process to enter state 1 only at times 2, 4, . . . , in which case state 1 is said to have period 2. This is evident by noting that the player can break even (neither be winning or losing) only at times 2, 4, . . . , or by calculating $p_{11}^{(n)}$ for all n and noting that $p_{11}^{(n)} = 0$ for n odd.

If there are two consecutive numbers, s and $(s + 1)$, such that the process can be in state i at times s and $(s + 1)$, the state is said to have period 1 and is called an **aperiodic** state.

Just as recurrence is a class property, it can be shown that periodicity is a class property. That is, if state i in a class has period t, then all states in that class have period t. In the gambling example, state 2 also has period 2.

A final property of Markov chains pertains to further classifying recurrent states. A recurrent state i is said to be **positive recurrent** if, starting in state i, the expected time for the process to reenter state i is finite. Similarly, a recurrent state i is said to be **null recurrent** if, starting in state i, the expected time for the process to reenter state i is infinite. It can be shown that for a finite-state Markov chain all recurrent states are positive recurrent states. Positive recurrent states that are aperiodic are called **ergodic** states.

5.6 First Passage Times

Section 5.4 dealt with finding n-step transition probabilities [i.e., given that the process is in state i, determining the (conditional) probability that the process will be in state j after n periods]. It is often desirable to make probability statements about the number of transitions made by the process in going from state i to state j *for the first time*. This length of time is called the *first passage time* in going from state i to state j. When $j = i$, this first passage time is just the number of transitions until the process returns to the initial state i. In this case, the first passage time is called the *recurrence time* for state i.

To illustrate these definitions, reconsider the inventory example developed in the preceding sections. Recall that the initial inventory (X_0) contains three cameras. Suppose that it turns out that there are two cameras at the end of the first week (X_1 takes on the value 2), one camera at the end of the second week (X_2 takes on the value 1), no cameras in stock at the end of the third week (X_3 takes on the value 0), three cameras at the end of the fourth week (X_4 takes on the value 3), and one camera at the end of the fifth week (X_5 takes on the value 1). In this case, the first passage time in going from state 3 to state 1 is 2 weeks, the first passage time in going from state 3 to state 0 is 3 weeks, and the recurrence time of state 3 is 4 weeks.

In general, the first passage times are random variables and hence have probability distributions associated with them. These probability distributions depend upon the transition probabilities of the process. In particular, let $f_{ij}^{(n)}$ denote the probability that the first passage time from state i to j is equal to n. It can be shown that these probabilities satisfy the following recursive relationships:

$$f_{ij}^{(1)} = p_{ij}^{(1)} = p_{ij},$$
$$f_{ij}^{(2)} = p_{ij}^{(2)} - f_{ij}^{(1)}p_{jj},$$
$$\vdots$$
$$f_{ij}^{(n)} = p_{ij}^{(n)} - f_{ij}^{(1)}p_{jj}^{(n-1)} - f_{ij}^{(2)}p_{jj}^{(n-2)} \cdots - f_{ij}^{(n-1)}p_{jj}.$$

Thus the probability of a first passage time from state i to state j in n steps can be computed recursively from the one-step transition probabilities. In the inventory example, the probability distribution of the first passage time in going from state 3 to state 0 is obtained as follows:

$$f_{30}^{(1)} = 0.080$$
$$f_{30}^{(2)} = (0.249) - (0.080)(0.080) = 0.243.$$
$$\vdots$$

For fixed i and j, the $f_{ij}^{(n)}$ are nonnegative numbers such that

$$\sum_{n=1}^{\infty} f_{ij}^{(n)} \leq 1.$$

Unfortunately, this sum may be strictly less than 1, which implies that a process initially in state i may never reach state j. When the sum does equal 1, $f_{ij}^{(n)}$ (for $n = 1, 2, \ldots$) can be considered as a probability distribution for the random variable, the first passage time.

Whereas calculating $f_{ij}^{(n)}$ for all n may be difficult, it is relatively simple to obtain the expected first passage time from state i to state j. Denote this expectation by μ_{ij}, which is defined by the expressions

$$\mu_{ij} = \begin{cases} \infty, & \text{if } \sum_{n=1}^{\infty} f_{ij}^{(n)} < 1 \\ \sum_{n=1}^{\infty} n f_{ij}^{(n)}, & \text{if } \sum_{n=1}^{\infty} f_{ij}^{(n)} = 1. \end{cases}$$

Whenever
$$\sum_{n=1}^{\infty} f_{ij}^{(n)} = 1,$$

then μ_{ij} satisfies uniquely the equation

$$\mu_{ij} = 1 + \sum_{k \neq j} p_{ik}\mu_{kj}.$$

When $i = j$, μ_{ii} is called the **expected recurrence time**.

For the inventory example, these equations can be used to compute the expected time until the cameras are out of stock, assuming the process is started when three cameras are available; i.e., the expected first passage time, μ_{30}, can be obtained. Since all the states are recurrent, the system of equations leads to the expressions

$$\mu_{30} = 1 + p_{31}\mu_{10} + p_{32}\mu_{20} + p_{33}\mu_{30},$$

$$\mu_{20} = 1 + p_{21}\mu_{10} + p_{22}\mu_{20} + p_{23}\mu_{30},$$

$$\mu_{10} = 1 + p_{11}\mu_{10} + p_{12}\mu_{20} + p_{13}\mu_{30},$$

or

$$\mu_{30} = 1 + 0.184\mu_{10} + 0.368\mu_{20} + 0.368\mu_{30},$$

$$\mu_{20} = 1 + 0.368\mu_{10} + 0.368\mu_{20},$$

$$\mu_{10} = 1 + 0.368\mu_{10}.$$

The simultaneous solution to this system of equations is

$$\mu_{10} = 1.58 \text{ weeks},$$

$$\mu_{20} = 2.51 \text{ weeks},$$

$$\mu_{30} = 3.50 \text{ weeks},$$

so that the expected time until the cameras are out of stock is 3.50 weeks. In making these calculations, we also obtain μ_{20} and μ_{10}.

5.7 Long-Run Properties of Markov Chains

Steady-State Probabilities

In Sec. 5.4 the four-step transition matrix for the inventory example was obtained. It will now be instructive to examine the eight-step transition probabilities given by the matrix

$$\mathbf{P}^{(8)} = \mathbf{P}^8 = \mathbf{P}^4 \cdot \mathbf{P}^4 = \begin{bmatrix} 0.286 & 0.285 & 0.264 & 0.166 \\ 0.286 & 0.285 & 0.264 & 0.166 \\ 0.286 & 0.285 & 0.264 & 0.166 \\ 0.286 & 0.285 & 0.264 & 0.166 \end{bmatrix}.$$

Notice the rather remarkable fact that each of the four rows has identical entries. This implies that the probability of being in state j after 8 weeks appears to be independent of the initial level of inventory. In other words, it appears that there is a limiting probability that the system will be in state j after a large number of transitions, and this probability is independent of the initial state. An important result related to the long-run behavior of finite-state Markov processes follows.

For an irreducible ergodic Markov chain, it can be shown that $\lim_{n\to\infty} p_{ij}^{(n)}$ exists and is independent of i. Furthermore,

$$\lim_{n\to\infty} p_{ij}^{(n)} = \pi_j,$$

where the π_j's uniquely satisfy the following steady-state equations:

$$\pi_j > 0,$$

$$\pi_j = \sum_{i=0}^{M} \pi_i p_{ij}, \qquad \text{for } j = 0, 1, \ldots, M,$$

$$\sum_{j=0}^{M} \pi_j = 1.$$

The π_j's are called the **steady-state probabilities** of the Markov chain and are equal to the reciprocal of the expected recurrence time; that is,

$$\pi_j = \frac{1}{\mu_{jj}}, \qquad \text{for } j = 0, 1, \ldots, M.$$

The term *steady-state* probability means that the probability of finding the process in a certain state, say j, after a large number of transitions tends to the value π_j, independent of the initial probability distribution defined over the states. It is important to note that steady-state probability does *not* imply that the process settles down into one state. On the contrary, the process continues to make transitions from state to state, and at any step n the transition probability from state i to state j is still p_{ij}.

The π_j's can also be interpreted as stationary probabilities (not to be confused with stationary transition probabilities). If the initial absolute probability of being in state j is given by π_j (that is, $P\{X_0 = j\} = \pi_j$) for all j, then the absolute probability of finding the process in state j at time $n = 1, 2, \ldots,$ is also given by π_j (that is, $P\{X_n = j\} = \pi_j$).

It should be noted that the steady-state equations consist of $(M + 2)$ equations in $(M + 1)$ unknowns. Because it has a unique solution, at least one equation must be redundant and can, therefore, be deleted. It cannot be the equation

$$\sum_{j=0}^{M} \pi_j = 1,$$

because $\pi_j = 0$ for all j will satisfy the other $(M + 1)$ equations. Furthermore, the solutions to the other $(M + 1)$ steady-state equations have a unique solution up to a multiplicative constant, and it is the final equation that forces the solution to be a probability distribution.

Returning to the inventory example, the steady-state equations can be expressed as

$$\begin{aligned}
\pi_0 &= \pi_0 p_{00} + \pi_1 p_{10} + \pi_2 p_{20} + \pi_3 p_{30},\\
\pi_1 &= \pi_0 p_{01} + \pi_1 p_{11} + \pi_2 p_{21} + \pi_3 p_{31},\\
\pi_2 &= \pi_0 p_{02} + \pi_1 p_{12} + \pi_2 p_{22} + \pi_3 p_{32},\\
\pi_3 &= \pi_0 p_{03} + \pi_1 p_{13} + \pi_2 p_{23} + \pi_3 p_{33},\\
1 &= \pi_0 + \pi_1 + \pi_2 + \pi_3.
\end{aligned}$$

Substituting values for p_{ij} into these equations leads to the equations

$$\begin{aligned}
\pi_0 &= (0.080)\pi_0 + (0.632)\pi_1 + (0.264)\pi_2 + (0.080)\pi_3, \\
\pi_1 &= (0.184)\pi_0 + (0.368)\pi_1 + (0.368)\pi_2 + (0.184)\pi_3, \\
\pi_2 &= (0.368)\pi_0 \qquad\qquad\qquad\quad + (0.368)\pi_2 + (0.368)\pi_3, \\
\pi_3 &= (0.368)\pi_0 \qquad\qquad\qquad\qquad\qquad\qquad + (0.368)\pi_3, \\
1 &= \pi_0 + \pi_1 + \pi_2 + \pi_3.
\end{aligned}$$

Solving the last four equations provides the simultaneous solutions

$$\begin{aligned}
\pi_0 &= 0.285, \\
\pi_1 &= 0.285, \\
\pi_2 &= 0.264, \\
\pi_3 &= 0.166,
\end{aligned}$$

which are essentially the results that appear in the matrix $\mathbf{P}^{(8)}$. Thus after many weeks the probability of finding zero, one, two, and three cameras in stock tends to 0.285, 0.285, 0.264, and 0.166, respectively. The corresponding expected recurrence times are

$$\begin{aligned}
\mu_{00} &= \frac{1}{\pi_0} = 3.51 \text{ weeks}, \\
\mu_{11} &= \frac{1}{\pi_1} = 3.51 \text{ weeks}, \\
\mu_{22} &= \frac{1}{\pi_2} = 3.79 \text{ weeks}, \\
\mu_{33} &= \frac{1}{\pi_3} = 6.02 \text{ weeks}.
\end{aligned}$$

There are other important results concerning steady-state probabilities. In particular, if i and j are recurrent states belonging to different classes, then

$$P_{ij}^{(n)} = 0, \qquad \text{for all } n.$$

This result follows from the definition of a class.

Similarly, if j is a transient state, then

$$\lim_{n\to\infty} p_{ij}^{(n)} = 0, \qquad \text{for all } i.$$

This result implies that the probability of finding the process in a transient state after a large number of transitions tends to zero.

Expected Average Cost Per Unit Time

The previous subsection dealt with Markov chains whose states were ergodic (positive recurrent and aperiodic). If the requirement that the states be aperiodic is relaxed, then the limit

$$\lim_{n\to\infty} p_{ij}^{(n)}$$

may not exist. To illustrate this point, consider the two-state transition matrix

$$\mathbf{P} = \begin{bmatrix} 0 & 1 \\ 1 & 0 \end{bmatrix}.$$

If the process starts in state 0 at time 0, it will be in state 0 at times 2, 4, 6, . . . , and in state 1 at times 1, 3, 5, ... Thus $p_{00}^{(n)} = 1$ if n is even and $p_{00}^{(n)} = 0$ if n is odd, so that

$$\lim_{n\to\infty} p_{00}^{(n)}$$

does not exist. However, the following limit always exists: For an irreducible Markov chain with positive recurrent states, e.g., a finite-state chain, then

$$\lim_{n\to\infty} \left\{ \frac{1}{n} \sum_{k=1}^{n} p_{ij}^{(k)} \right\} = \pi_j,$$

where the π_j's satisfy the steady-state equations presented on p. 115.

This result is extremely important in computing the long-run average cost per unit time associated with a Markov chain. Suppose that a cost (or other penalty function) $C(X_t)$ is incurred when the process is in state X_t at time t, for $t = 0, 1, 2, \ldots$. Note that $C(X_t)$ is a random variable that takes on any one of the values $C(0), C(1), \ldots, C(M)$, and the function $C(\cdot)$ is independent of t. The expected average cost incurred over the first n periods is given by the expression

$$E\left[\frac{1}{n} \sum_{t=1}^{n} C(X_t)\right].$$

Using the result that

$$\lim_{n\to\infty} \left\{ \frac{1}{n} \sum_{k=1}^{n} p_{ij}^{(k)} \right\} = \pi_j,$$

it is simple to show that the (long-run) *expected average cost per unit time* is given by

$$\lim_{n\to\infty} \left\{ E\left[\frac{1}{n} \sum_{t=1}^{n} C(X_t)\right]\right\} = \sum_{j=0}^{M} \pi_j C(j).$$

As an example, suppose the camera store finds that a storage charge is being allocated for each camera remaining on the shelf at the end of the week. The cost is charged as follows: If $X_t = 0$, then $C(0) = 0$. If $X_t = 1$, then $C(1) = 2$. If $X_t = 2$, then

$C(2) = 8$. Finally, if $X_t = 3$, then $C(3) = 18$. The long-run expected average holding cost per week can then be obtained from the preceding equation; that is,

$$\lim_{n\to\infty} \left\{ E\left[\frac{1}{n} \sum_{t=1}^{n} C(X_t) \right] \right\} = 0(0.285) + 2(0.285) + 8(0.264) + 18(0.166) = 5.67.$$

It should be noted that an alternative measure to the (long-run) expected average cost per unit time is the (long-run) *actual average cost per unit time*. It can be shown that this latter measure is given by

$$\lim_{n\to\infty} \left\{ \frac{1}{n} \sum_{t=1}^{n} C(X_t) \right\} = \sum_{j=0}^{M} \pi_j C(j)$$

for almost all paths of the process. Thus either measure leads to the same result. These results can also be used to interpret the meaning of the π_j's. To interpret them, let

$$C(X_k) = \begin{cases} 1, & \text{if } X_k = j \\ 0, & \text{if } X_k \neq j. \end{cases}$$

The (long-run) expected fraction of times the system is in state j is then given by

$$\lim_{n\to\infty} \left\{ E\left[\frac{1}{n} \sum_{t=1}^{n} C(X_t) \right] \right\} = \lim_{n\to\infty} \{E\,[\text{fraction of times system is in state } j]\} = \pi_j.$$

Similarly, π_j can also be interpreted as the (long-run) actual fraction of times that the system is in state j.

Expected Average Cost Per Unit Time for Complex Cost Functions

In the preceding subsection, the cost function was based solely on the state that the process is in at time t. In many important problems encountered in practice, the cost may depend upon another random variable as well as upon the state that the process is in. For example, in the inventory example developed in this chapter, suppose that the costs to be considered are the ordering cost and the penalty cost for unsatisfied demand (storage costs will be ignored). It is reasonable to assume that the number of cameras ordered depends only upon the state of the process (the number of cameras in stock) when the order is placed. The cost for unsatisfied demand may be assumed to depend upon the demand during the week as well as upon the state of the process at the beginning of the week. The charges for period t will be made at the end of the week and will include the cost of the order delivered on the Monday of that week and the cost of unsatisfied demand during the week. Thus the cost incurred for period t can be described as a function of X_{t-1} and D_t, that is, $C(X_{t-1}, D_t)$. Note that the demands $D_t, D_{t+1}, \ldots$, during successive weeks are assumed to be independent and identically distributed random variables. Furthermore, recall that the (s, S) policy $(1, 3)$ is being used. X_{t-1}, the stock level at the end of period $t - 1$ (before ordering), is defined iteratively by the expression given in Sec. 5.2. Thus it follows that $(X_0, X_1, X_2, \ldots, X_{t-1})$ and D_t are independent random variables because $X_0, X_1, X_2, \ldots, X_{t-1}$ are functions only of $X_0, D_1, \ldots, D_{t-1}$, which are independent of D_t. Under these conditions, it can be shown that the (long-run) *expected average cost per unit time* is given by

$$\lim_{n\to\infty} \left\{ E\left[\frac{1}{n} \sum_{t=1}^{n} C(X_{t-1}, D_t) \right] \right\} = \sum_{j=0}^{M} k(j)\pi_j,$$

where
$$k(j) = E[C(j, D_t)],$$

and this latter (conditional) expectation is taken with respect to the probability distribution of the random variable D_t (given the state). Similarly, the (long-run) actual average cost per unit time is given by

$$\lim_{n\to\infty} \left\{ \frac{1}{n} \sum_{t=1}^{n} C(X_{t-1}, D_t) \right\} = \sum_{j=0}^{M} k(j)\pi_j.$$

Suppose that the following costs are associated with the (s, S) inventory policy given earlier (storage charges are now neglected). If $z > 0$ cameras are ordered, the cost incurred is $10 + 25z$ dollars. If no cameras are ordered, no ordering cost is incurred. For each unit of unsatisfied demand (lost sales), there is a penalty of \$50 per unit. If the $(s = 1, S = 3)$ ordering policy is followed, then the cost in week t is given by $C(X_{t-1}, D_t)$, where

$$C(X_{t-1}, D_t) = \begin{cases} 10 + (25)(3) + 50 \max\{(D_t - 3), 0\}, & \text{if } X_{t-1} < 1 \\ 50 \max\{(D_t - X_{t-1}), 0\}, & \text{if } X_{t-1} \geq 1 \end{cases}$$

for $t = 1, 2, \ldots$. Hence

$$C(0, D_t) = 85 + 50 \max\{(D_t - 3), 0\},$$

so that
$$k(0) = E[C(0, D_t)] = 85 + 50E[\max\{(D_t - 3), 0\}]$$
$$= 85 + 50[1P_D(4) + 2P_D(5) + 3P_D(6) + \cdots],$$

where $P_D(i)$ is the probability that the demand equals i and has been assumed to have a Poisson distribution with parameter $\lambda = 1$. Hence $k(0) = 86.2$. Similar calculations lead to the results

$$\begin{aligned} k(1) = E[C(1, D_t)] &= 50E[\max\{(D_t - 1), 0\}] \\ &= 50[1P_D(2) + 2P_D(3) + 3P_D(4) + \cdots] \\ &= 18.4, \end{aligned}$$

$$\begin{aligned} k(2) = E[C(2, D_t)] &= 50E[\max\{(D_t - 2), 0\}] \\ &= 50[1P_D(3) + 2P_D(4) + 3P_D(5) + \cdots] \\ &= 5.2, \end{aligned}$$

and
$$\begin{aligned} k(3) = E[C(3, D_t)] &= 50E[\max\{(D_t - 3), 0\}] \\ &= 50[1P_D(4) + 2P_D(5) + \cdots] \\ &= 1.2. \end{aligned}$$

Thus the (long-run) expected average inventory cost per week is given by

$$\sum_{j=0}^{3} k(j)\pi_j = (86.2)(0.285) + (18.4)(0.285) + (5.2)(0.264) + (1.2)(0.166) = 31.4.$$

This cost is the cost associated with the (s, S) policy, $(s, S) = (1, 3)$. The cost of other (s, S) policies can be evaluated in a similar way to identify the policy that minimizes the expected average inventory cost per week.

The results of this section were presented only in terms of the inventory example. However, the (nonnumerical) results still hold for other problems as long as the following conditions are satisfied:

1. $\{X_t\}$ is an irreducible Markov chain whose states are positive recurrent.
2. Associated with this Markov chain is a sequence of random variables $\{D_t\}$ each of which is independent and identically distributed.
3. For a fixed $m = 0, \pm 1, \pm 2, \ldots$, a cost $C(X_t, D_{t+m})$ is incurred at time t, for $t = 0, 1, 2, \ldots$.
4. The sequence $(X_0, X_1, X_2, \ldots, X_t)$ must be independent of D_{t+m}.

In particular, if these conditions are satisfied then

$$\lim_{n \to \infty} \left\{ E\left[\frac{1}{n} \sum_{t=1}^{n} C(X_t, D_{t+m}) \right] \right\} = \sum_{j=0}^{M} k(j)\pi_j,$$

where

$$k(j) = E[C(j, D_{t+m})],$$

and this latter conditional expectation is taken with respect to the probability distribution of the random variable D_t (given the state). Furthermore,

$$\lim_{n \to \infty} \left\{ \frac{1}{n} \sum_{t=1}^{n} C(X_t, D_{t+m}) \right\} = \sum_{j=0}^{M} k(j)\pi_j$$

for almost all paths of the process.

5.8 Absorption States

It was pointed out that a state k is called an *absorbing state* if $P_{kk} = 1$, so that once the chain visits k it remains there forever. If k is an absorbing state, the first passage probability from i to k is called the probability of absorption into k, having started at i. When there are two or more absorbing states in a chain, and when it is evident that the process will be absorbed into one of these states, it is desirable to find these probabilities of absorption. These probabilities can be obtained by solving a system of linear equations. Suppose that the Markov chain is such that ultimately one of the absorbing states will be reached. If the state k is an absorbing state, then the set of absorption probabilities f_{ik} satisfies the system of equations

$$f_{ik} = \sum_{j=0}^{M} p_{ij} f_{jk}, \qquad \text{for } i = 0, 1, \ldots, M,$$

subject to the conditions

$$f_{kk} = 1,$$

$$f_{ik} = 0, \qquad \text{if state } i \text{ is recurrent and } i \neq k.$$

Absorption probabilities are important in "random walks." A **random walk** is a Markov chain with the property that if the system is in a state i, then in a single transition the system either remains at i or moves to one of the states immediately adjacent to i. For example, a random walk often is used as a model for situations involving gambling. To illustrate, consider a gambling example similar to that presented in Sec. 5.3. However, suppose now that two players, each having \$2, agree to keep playing a game and betting \$1 at a time until one is broke. The amount of money that player A has forms a Markov chain with the states representing player A's fortune, i.e., 0, 1, 2, 3, 4, and with transition matrix

$$\mathbf{P} = \begin{bmatrix} 1 & 0 & 0 & 0 & 0 \\ 1-p & 0 & p & 0 & 0 \\ 0 & 1-p & 0 & p & 0 \\ 0 & 0 & 1-p & 0 & p \\ 0 & 0 & 0 & 0 & 1 \end{bmatrix}.$$

If p represents the probability of A winning a single encounter, then the probability of absorption into state 0 (A losing all his money) can be obtained from the preceding system of equations. It can be shown that these equations then result in the alternate expressions (for general M rather than $M = 4$ as in this example),

$$\begin{aligned} 1 - f_{i0} &= \frac{\sum_{m=0}^{i-1} \rho^m}{\sum_{m=0}^{M-1} \rho^m}, && \text{for } i = 1, 2, \ldots, M, \\ &= \frac{1-\rho^i}{1-\rho^M}, && \text{for } p \neq \frac{1}{2}, \\ &= \frac{i}{M}, && \text{for } p = \frac{1}{2}, \end{aligned}$$

where $\rho = (1 - p)/p$.

For $M = 4$, $i = 2$, and $p = \frac{2}{3}$, the probability of A going broke is given by

$$f_{20} = 1 - \left[\frac{1-\rho^2}{1-\rho^4}\right] = \frac{1}{5},$$

and the probability of A winning \$4 (B going broke) is given by

$$f_{24} = 1 - f_{20} = \frac{4}{5}.$$

There are many other situations where absorbing states play an important role. Consider a department store that classifies the balance of a customer's bill as fully paid (state 0), 1–30 days in arrears (state 1), 31–60 days in arrears (state 2), or bad debt (state 3). The accounts are checked monthly, and the state of each customer is determined. In general, credit is not extended and customers are expected to pay their bills within 30 days. Occasionally, customers pay only portions of their bill. If this occurs when the balance is within 30 days in arrears (state 1), the store views the customer as remaining in state 1. If this occurs when the balance is between 31 and 60 days in arrears, the store views the customer as moving to state 1 (1–30 days in arrears). Customers that are more than 60 days in arrears are put into the bad debt category (state 3), and then bills are sent to a collection agency. After examining data over the past several years, the store has developed the following transition matrix[1]:

State \ State	0: Fully Paid	1: 1–30 Days in Arrears	2: 31–60 Days in Arrears	3: Bad Debt
0: fully paid	1	0	0	0
1: 1–30 days in arrears	0.7	0.2	0.1	0
2: 31–60 days in arrears	0.5	0.1	0.2	0.2
3: bad debt	0	0	0	1

Although each customer ends up in state 0 or 3, the store is interested in determining the probability that a customer will end up as a bad debt given that the account belongs to the 1–30 days in arrears state, and similarly, given that the account belongs to the 31–60 days in arrears state.

In order to obtain this information, the set of equations presented at the beginning of this section must be solved. In particular, f_{13} and f_{23} are to be obtained. Substituting, the following two equations are obtained:

$$f_{13} = p_{10}f_{03} + p_{11}f_{13} + p_{12}f_{23} + p_{13}f_{33},$$

$$f_{23} = p_{20}f_{03} + p_{21}f_{13} + p_{22}f_{23} + p_{23}f_{33}.$$

Noting that $f_{03} = 0$ and $f_{33} = 1$, there are now two equations in two unknowns, namely,

$$(1 - p_{11})f_{13} = p_{13} + p_{12}f_{23},$$

$$(1 - p_{22})f_{23} = p_{23} + p_{21}f_{13}.$$

Substituting the values from the transition matrix leads to

$$0.8f_{13} = 0.1f_{23},$$

$$0.8f_{23} = 0.2 + 0.1f_{13},$$

and the solution is

$$f_{13} = 0.032,$$

$$f_{23} = 0.254.$$

Thus, approximately 3 percent of the customers whose accounts are 1–30 days in arrears end up as bad debts, where 25 percent of the customers whose accounts are 31–60 days in arrears end up as bad debts.

[1] Customers that are fully paid (in state 0) may be viewed as ''new'' customers if they return for repeat purchases.

5.9 Continuous Time Markov Chains

All the previous sections assumed that the time parameter t was discrete (that is, $t = 0, 1, 2, \ldots$). Such an assumption is suitable for many problems, but there are certain cases (such as for some queueing models) where a continuous time parameter is required.

Following the definitions for discrete time Markov chains, let $\{X(t)\}$, where $t \geq 0$, be a stochastic process taking on values $0, 1, \ldots, M$. This process is called a **continuous time Markov chain** if the transition probabilities can be expressed as

$$P\{X(t + s) = j|X(s) = i \text{ and } X(r) = x(r)\} = P\{X(t + s) = j|X(s) = i\},$$

for all $i, j = 0, 1, \ldots, M$ and $0 \leq r < s$. Furthermore, if these transition probabilities are independent of s, they can be expressed as

$$P_{ij}(t) = P\{X(t + s) = j|X(s) = i\},$$

and are called *stationary transition probabilities*. This function is assumed to be continuous at $t = 0$, with

$$\lim_{t \to 0} p_{ij}(t) = \begin{cases} 1, & \text{if } i = j \\ 0, & \text{if } i \neq j. \end{cases}$$

Note that a continuous time Markov chain has the property (similar to that of discrete time Markov chains) that the probability of any future "event" occurring, given any past "event" and the present state, is *independent* of the past event and depends only upon the present state of the process. Furthermore, if the continuous time Markov chain is stationary (as will be assumed herein), the transition probability $p_{ij}(t)$ is independent of the time at which the present state occurred.

It is worthwhile exploring the implications of the Markovian property. The Markovian property (with stationary transition probabilities) implies that the probability of the system being in state j at time $t + s$, given that the system is in state i at time s, that is,

$$P\{X(t + s) = j|X(s) = i\},$$

depends only on i and j and the incremental value of t and is independent of s; that is, it can be expressed as $p_{ij}(t)$. Similarly, when $i = j$, the probability of the system remaining in state i at time $t + s$, given that it is in state i at time s, depends only on i and the incremental value of t and is independent of s; that is, it can be expressed as $p_{ii}(t)$. Now consider the random variable T_i, which represents the time required for the system to transit out of state i. The system remains in state i at time $t + s$, given that the system is in state i at time s, that is, $X(t + s) = i|X(s) = i$, if, and only if, the time required for the system to transit out of state i exceeds $t + s$ when the time the system has spent in state i exceeds s; that is, $T_i > t + s|T_i > s$. Therefore,

$$P\{X(t + s) = i|X(s) = i\} = P\{T_i > t + s|T_i > s\}.$$

Similarly, the system remains in state i at time t given that the system starts in state i at time 0, that is, $X(t) = i|X(0) = i$, if, and only if, the time required for the system to transit out of state i exceeds t when the system starts in state i at time 0, that is, $T_i > t|T_i > 0$, or, equivalently, $T_i > t$. Therefore,

$$P\{X(t) = i|X(0) = i\} = P\{T_i > t\}.$$

The Markovian property (with stationary transition probabilities) implies that

$$P\{X(t+s)=i|X(s)=i\} = P\{X(t)=i|X(0)=i\},$$

so that

$$P\{T_i > t+s|T_i > s\} = P\{T_i > t\}.$$

This is a rather unusual condition for a probability distribution to possess. The probability distribution of the time required for the system to transit out of a given state always is the same, regardless of how much time the system has already spent in that state. In effect, the random variable is memoryless; the process forgets its history. There is only one (continuous) probability distribution that possesses this property, the exponential distribution with, say, mean $1/q$ (see Sec. 6.4 for a complete discussion of this distribution), i.e.,

$$P\{T_i < t\} = 1 - e^{-tq} \qquad \text{for } t \geq 0.$$

This result leads to an equivalent way of defining a continuous time Markov chain. A stochastic process $\{X(t)\}$, where $t \geq 0$ taking on values $0, 1, \ldots, M$, is a continuous time Markov chain if:

1. Each time the process enters a state i, the amount of time it spends in that state before transiting to a different state has an exponential distribution with mean time $1/q_i$;
2. When leaving state i, the process moves to a state j, with probabilities p_{ij}, where the p_{ij} satisfy the conditions

$$p_{ii} = 0, \qquad \text{for all } i,$$

and

$$\sum_{j=0}^{M} p_{ij} = 1, \qquad \text{for all } i;$$

and

3. The next state visited after transiting from state i is independent of the time spent in state i.

Just as the one-step transition probabilities played a major role in describing the Markov process for a discrete time parameter chain, the analogous role for a continuous time parameter chain is played by the transition intensities. The **transition intensity** is defined by

$$q_j = -\frac{d}{dt}p_{jj}(0) = \lim_{t\to 0}\frac{1-p_{jj}(t)}{t}, \qquad \text{for } j = 0, 1, \ldots, M,$$

and

$$q_{ij} = \frac{d}{dt}p_{ij}(0) = \lim_{t\to 0}\frac{p_{ij}(t)}{t} = q_i p_{ij}, \qquad \text{for all } i \neq j.$$

(For finite-state continuous Markov chains these limits exist and are finite.) Here q_j is called the *intensity of passage*, given that the Markov chain is in state j, and q_{ij} is called the *intensity of transition to state j from the state i*. The quantity p_{ij} has already been defined as the probability, when leaving state i, that the process moves to state j. The intensity of passage, q_i, is just the reciprocal of the expected value of the exponentially distributed random variable that represents the time required for the process to transit out of state i; that is, q_i is the *rate* at which the process leaves state

i when it is in state i. Similarly, the transition intensity, q_{ij}, can be interpreted as the transition rate into state j when the process is in state i. Thus, the process, starting in state i, spends an amount of time in that state before transiting to a different state that has an exponential distribution with rate q_i. The process then moves to state j with probability $p_{ij} = q_{ij}/q_i$ and spends an amount of time in state j that has an exponential distribution with rate q_j, and so on. Finally, note that the intensity of passage and the intensity of transition are related in that

$$q_i = \sum_{j=0, j\neq i}^{M} q_{ij}.$$

These results suggest still another equivalent way of defining a continuous time Markov chain. A stochastic process $\{X(t)\}$, where $t \geq 0$, taking on values $0, 1, \ldots, M$, is a continuous time Markov chain if the process, when in state i, is poised to move to one of the M other alternative states, and the times required to transit from state i to each possible state j ($j = 0, 1, \ldots, M$ and $j \neq i$), denoted by the random variables T_{ij}, have independent exponential distributions with means equal to $1/q_{ij}$. From state i, the particular state to which the process ultimately transits, denoted by j^*, is that state j whose corresponding T_{ij} is smallest. Once the process enters the new state j^*, a new set of independent exponential random variables is utilized in a similar fashion to determine the time the chain resides in the state j^*, and the next state visited from j^*. The q_{ij} are the transition rates alluded to previously. If T_{ij^*} is defined as the minimum of the T_{ij} ($j \neq i$), it is easily shown (see Sec. 6.4) that T_{ij^*} has an exponential distribution with mean time $1/q_i$ (where q_i is the intensity of passage defined earlier) and, as previously noted,[1]

$$q_i = \sum_{j=0, j\neq i}^{M} q_{ij}.$$

Just as the discrete time parameter models satisfy the Chapman-Kolmogorov equations, the continuous time transition probability function also satisfies these equations; i.e., for any states i and j, and positive numbers t and v ($0 \leq v \leq t$),

$$p_{ij}(t) = \sum_{k=0}^{M} p_{ik}(v)p_{kj}(t - v).$$

A pair of states i and j are said to communicate if there are times t_1 and t_2 such that $p_{ij}(t_1) > 0$ and $p_{ji}(t_2) > 0$. All states that communicate are said to form a class. If all states in a chain form a single class, i.e., an irreducible chain (which will be assumed), then

$$p_{ij}(t) > 0, \qquad \text{for all } t > 0 \qquad \text{and} \qquad \text{all states } i \text{ and } j.$$

Furthermore,

$$\lim_{t\to\infty} p_{ij}(t) = \pi_j$$

always exists and is independent of the initial state of the chain, for $j = 0, 1, \ldots, M$.

The π_j satisfy the equations

$$\pi_j = \sum_{i=0}^{M} \pi_i p_{ij}(t), \qquad \text{for } j = 0, 1, \ldots, M, \qquad \text{and} \qquad \text{for every } t \geq 0.$$

[1] This useful result is used again in the final illustrative example in the chapter.

The limiting probabilities also satisfy the equations

$$\pi_j q_j = \sum_{i=0, i \neq j}^{M} \pi_i q_{ij}, \qquad \text{for } j = 0, 1, \ldots, M,$$

and

$$\sum_{j=0}^{M} \pi_j = 1,$$

which can be used to obtain the values of these limiting probabilities.

It has already been noted that q_j is the rate at which the process leaves state j given that it is in state j so that $q_j \pi_j$ is just the rate (in the long run) at which the process leaves state j. Similarly, q_{ij} is just the transition rate into state j given that the process is in state i so that $\Sigma_{i \neq j} \pi_i q_{ij}$ is just the overall rate of transition into state j. Since the long-term rate at which a process leaves a state j must equal the long-term rate at which a process arrives at state j, in order to maintain a stable system,

$$\pi_j q_j = \sum_{i=0, i \neq j}^{M} \pi_i q_{ij}, \qquad \text{for } j = 0, 1, \ldots, M,$$

which is the equation given above for obtaining limiting probabilities and is sometimes called the *balance equations*.

As an example, consider the following repairer problem, which is discussed in detail in Chap. 6 (''Queueing Theory''). There are two machines serviced by a single repairer. A machine that breaks down is serviced immediately unless the repairer is servicing another machine, in which case a waiting line is formed. A system is said to be in state n if n machines are not working. If the process is in state n, where $1 \leq n \leq 2$, this state means that one machine is being serviced and $(n - 1)$ are in the waiting line waiting to be repaired. If the process is in state 0, then all machines are working, and the repairer is idle. Let $X(t)$ be the number of machines not working at time t.

Assume that the time to failure of a working machine has an exponential distribution with mean $1/\lambda$ and that the service time for a failed machine has an independent exponential distribution with mean $1/\mu$. The stochastic process $\{X(t)\}$ is a continuous parameter Markov process with possible states 0, 1, 2. The steady-state probabilities of being in these states are of interest.

Before solving the balance equations, it is important to note some characteristics of the q_{ij} for this model. A process in, say, state 1 (one machine not working) can move only to an adjacent state, that is, zero machines not working or two machines not working. Hence, $q_{ij} = 0$ whenever $|i - j| \geq 2$. The balance equations become

$$\begin{aligned}
\pi_0 q_0 &= \pi_1 q_{10}, \\
\pi_1 q_1 &= \pi_0 q_{01} + \pi_2 q_{21}, \\
\pi_2 q_2 &= \pi_1 q_{12}.
\end{aligned}$$

Solving these equations in terms of π_0 leads to

$$\pi_1 = \pi_0 \frac{q_0}{q_{10}}$$

$$\pi_2 = \pi_0 \frac{q_0}{q_{10}} \frac{q_{12}}{q_2}.$$

Noting that $\pi_0 + \pi_1 + \pi_2 = 1$ results in

$$\pi_0 = \frac{1}{1 + (q_0/q_{10}) + (q_0/q_{10})(q_{12}/q_2)}.$$

Values for q_0, q_2, q_{10}, and q_{12} can be obtained from the model as functions of μ and λ.

In order to find these values, the following general result is utilized (see Sec. 6.4). If $T_1, T_2, \ldots, T_k$ are k independent random variables having exponential distributions with rates $v_1, v_2, \ldots, v_k$, respectively, then Z, the minimum of these k random variables, also has an exponential distribution but with rate $v_1 + v_2 + \cdots + v_k$. Returning to the machine example, q_0 is the rate at which the process leaves state 0 (all machines working) when it is in state 0. Since there are two machines, both working, and each breaks down according to an exponential distribution with rate λ, the process leaves state 0 when the first machine breaks down, i.e., when the minimum of the two failure times occur. Hence,

$$q_0 = \lambda + \lambda = 2\lambda.$$

Similarly, q_2 is the rate at which the process leaves state 2 (both machines are down) when it is in state 2. When both machines are down, one is being repaired by the serviceperson, so that the process leaves state 2 when the machine is repaired. Therefore, the rate at which the process leaves state 2 is given by

$$q_2 = \mu.$$

The rate q_{10} is the transition rate into state 0 (both machines working) when the process is in state 1 (one machine down). The rate at which this occurs is the rate at which the machine is repaired, i.e.,

$$q_{10} = \mu.$$

Similarly, q_{12} is the transition rate into state 2 (both machines down) when the process is in state 1 (one machine down). The rate at which this occurs is the rate at which a single machine fails, i.e.,

$$q_{12} = \lambda.$$

Although it is not necessary to calculate q_1 to get the steady-state probabilities, it is instructive to do so. The rate q_1 is the rate at which the process leaves state 1 (one machine down and one machine operating) when it is in state 1. The process leaves state 1 when either the operating machine goes down or the down machine is repaired, i.e., at the minimum of the time until the operating machine fails and the time until the down machine is repaired. The rate q_1 is then given by

$$q_1 = \lambda + \mu.$$

Another way of viewing this result for q_1 is as follows: Suppose there are two clocks, one set to give the time for the operating machine to break down, and the second set to give the time for the down machine to be repaired. These times have independent exponential distributions with respective rates λ and μ. The clocks are set and an alarm sounds as soon as either the operating machine breaks down or the down machine is repaired. When the alarm sounds, the process leaves state 1. The rate at which the process leaves state 1, q_1, then is as shown above.

Using the derived values for q_0, q_2, q_{10}, and q_{12} results in

$$\pi_0 = \frac{1}{1 + 2\lambda/\mu + 2(\lambda/\mu)^2},$$

$$\pi_1 = \frac{\pi_0 2\lambda}{\mu},$$

$$\pi_2 = \pi_0\, 2\left(\frac{\lambda}{\mu}\right)^2.$$

The results for M machines are given in Sec. 6.6 (basic model with a limited source).

SELECTED REFERENCES

1. Chung, K. L.: *Markov Chains with Stationary Transition Probabilities,* Springer-Verlag, Berlin, 1960.
2. Feller, W.: *An Introduction to Probability Theory and Its Applications,* vol. 1, 2d ed., Wiley, New York, 1957.
3. Heyman, D., and M. Sobel: *Stochastic Models in Operations Research,* vol. 1, McGraw-Hill, New York, 1982.
4. Karlin, S., and H. Taylor: *First Course in Stochastic Processes,* 2d ed., Academic Press, New York, 1975.
5. Ross, S.: *Introduction to Probability Models,* 3d ed., Academic Press, New York, 1985.
6. Ross, S.: *Stochastic Processes,* Wiley, New York, 1983.
7. Taylor, H., and S. Karlin: *An Introduction to Stochastic Modeling,* Academic Press, New York, 1984.

PROBLEMS

1. Assume that the probability of rain tomorrow is 0.5 if it is raining today, and assume that the probability of its being clear tomorrow is 0.9 if it is clear today

(*a*) Determine the one-step transition matrix of the Markov chain.
(*b*) Find the two-step transition matrix.
(*c*) Find the steady-state probabilities.

2. Assume that the probability of rain tomorrow is α if it is raining today, and assume that the probability of its being clear tomorrow is β if it is clear today.

(*a*) Determine the one-step transition matrix of the Markov chain.
(*b*) Find the two-step transition matrix.
(*c*) Find the steady-state probabilities.

3. Consider the stock market model presented in Sec. 5.3. Whether or not the stock goes up tomorrow depends upon whether or not it increased today *and* yesterday. If the stock has increased for the past two days, it will increase tomorrow with probability α_1. If the stock increased today but decreased yesterday, it will increase tomorrow with probability α_2. If the stock decreased today but increased yesterday, it will increase tomorrow with probability α_3. Finally, if the stock decreased for the past two days, it will increase tomorrow with probability α_4.

(*a*) Determine the one-step transition matrix of the Markov chain.
(*b*) Find the steady-state probabilities.

4. Consider the stock market example given in Prob. 3. Suppose, however, that whether or not the stock goes up tomorrow depends upon whether or not it increased today, yesterday, and the day before yesterday. Can this problem be formulated as a Markov chain? If so, what are the possible states?

5.* Determine the classes of the Markov chains and whether or not they are recurrent.

$$\mathbf{A} = \begin{bmatrix} 0 & 0 & \frac{1}{3} & \frac{2}{3} \\ 1 & 0 & 0 & 0 \\ 0 & 1 & 0 & 0 \\ 0 & 1 & 0 & 0 \end{bmatrix}, \quad \mathbf{B} = \begin{bmatrix} 1 & 0 & 0 & 0 \\ 0 & \frac{1}{2} & \frac{1}{2} & 0 \\ 0 & \frac{1}{2} & \frac{1}{2} & 0 \\ \frac{1}{2} & 0 & 0 & \frac{1}{2} \end{bmatrix}.$$

6. Determine the classes of the Markov chains and whether or not they are recurrent.

$$\mathbf{A} = \begin{bmatrix} 0 & \frac{1}{3} & \frac{1}{3} & \frac{1}{3} \\ \frac{1}{3} & 0 & \frac{1}{3} & \frac{1}{3} \\ \frac{1}{3} & \frac{1}{3} & 0 & \frac{1}{3} \\ \frac{1}{3} & \frac{1}{3} & \frac{1}{3} & 0 \end{bmatrix}, \quad \mathbf{B} = \begin{bmatrix} 0 & 0 & 1 \\ \frac{1}{2} & \frac{1}{2} & 0 \\ 0 & 1 & 0 \end{bmatrix}.$$

7. Determine the classes of the Markov chain and whether or not they are recurrent.

$$\begin{bmatrix} \frac{1}{4} & \frac{3}{4} & 0 & 0 & 0 \\ \frac{3}{4} & \frac{1}{4} & 0 & 0 & 0 \\ \frac{1}{3} & \frac{1}{3} & \frac{1}{3} & 0 & 0 \\ 0 & 0 & 0 & \frac{3}{4} & \frac{1}{4} \\ 0 & 0 & 0 & \frac{1}{4} & \frac{3}{4} \end{bmatrix}.$$

8. Suppose that a communications network transmits binary digits, 0 or 1. In passing through the network, there is a probability q that the binary digit will be received incorrectly at the next stage. If X_0 denotes the binary digit entering the system, X_1 the binary digit recorded after the first transmission, X_2 the binary digit recorded after the second transmission, . . . , then $\{X_n\}$ is a Markov chain. Find the one-step and steady-state transition matrices.

9. A transition matrix $\mathbf{P}$ is said to be doubly stochastic if the sum over each column equals 1; that is,

$$\sum_{i=0}^{M} p_{ij} = 1, \qquad \text{for all } j.$$

If such a chain is irreducible, aperiodic, and consists of $M + 1$ states, show that

$$\pi_j = \frac{1}{M+1}, \qquad \text{for} \qquad j = 0, 1, \ldots, M.$$

10.* A particle moves on a circle through points that have been marked 0, 1, 2, 3, 4 (in a clockwise order). The particle starts at point 0. At each step it has probability q of moving one point clockwise (0 follows 4) and $1 - q$ of moving one point counterclockwise. Let X_n $(n \geq 0)$ denote its location on the circle. $\{X_n\}$ is a Markov chain.

(*a*) Find the transition probability matrix.

(*b*) Find the steady-state probabilities.

11. The leading brewery on the West Coast (labeled A) has hired an operations research analyst to analyze its market position. It is particularly concerned about its major competitor (labeled B). The analyst believed that brand switching can be modeled as a Markov chain using three states, with states A and B representing customers drinking beer produced from the

aforementioned breweries and state C representing all other brands. Data are taken monthly, and the analyst has constructed the following transition matrix from past data.

	A	B	C
A	0.7	0.2	0.1
B	0.2	0.75	0.05
C	0.1	0.1	0.8

What are the steady-state market shares for the two major breweries?

12. A manufacturer has a machine that, when it breaks down, requires a day to repair. The machine breaks down with probability p. Denote the state of the system as 0 when the machine is found to be operating at the end of the day; 1 when the machine is found to have broken down during the day; and 2 when the machine, having broken down the previous day, has spent the day being repaired. Show that the process forms a Markov chain and find the steady-state probabilities.

	0	1	2
0	$1-p$	p	0
1	0	0	1
2	1	0	0

13. Suppose that the manufacturer in Prob. 12 keeps a spare machine that is used when the primary machine fails. Denote the state of the system by (x, y), where x and y take on the values 1 or 0 depending upon whether or not the machines are working at the end of the day.

(*a*) Show that the process forms a Markov chain and find the transition matrix.

(*b*) Find the steady-state probabilities.

Hint: Note that (0, 0) is not a possible state.

14. A computer is inspected at the end of every hour. It is found to be either working (up) or failed (down). If the computer is found to be up, the probability of it remaining up for the next hour is 0.90. If it is down, repair action, which may require more than an hour, is taken. Whenever the machine is down (regardless of how long it has been down), the probability of it still being down an hour later is 0.35.

(*a*) Show that this is a Markov chain and find the transition matrix.

(*b*) Find the steady-state probabilities of the machine being up and down.

15. Consider the following blood inventory problem facing a hospital. Suppose there is need for a rare blood type, e.g., type AB, Rh negative blood. Suppose the demand over a three-day period is given by

$$P\{D = 0\} = 0.4, \quad P\{D = 1\} = 0.3, \quad P\{D = 2\} = 0.2, \quad \text{and} \quad P\{D = 3\} = 0.1.$$

Note that the expected demand is then 1 unit. Suppose that there are three days between deliveries. The hospital proposes a policy of receiving one pint at each delivery and uses the oldest blood first, i.e., it uses a FIFO policy (first in, first out). If more blood is required than is on hand, an expensive emergency delivery is made. Blood is discarded if it is still on the shelf after 21 days. Denote the state of the system as the number of pints on hand just after a delivery. Noting that the largest state is 7:

(*a*) Find the transition matrix.

(*b*) Find the steady-state probabilities.

16. A soap company specializes in a luxury type of bath soap. The sales of this soap fluctuate between two levels—"Low" and "High"—depending upon two factors: (1) whether they advertise or not, and (2) the advertising and marketing of new products being done by competitors. The second factor is out of the company's control, but they are trying to determine what their own advertising policy should be. For example, the marketing manager's proposal is to advertise when sales are low but not to advertise when sales are high. Advertising in any quarter of a year has its primary impact on sales in the *following* quarter. Therefore, at the beginning of each quarter, the needed information is available to forecast accurately whether sales will be low or high that quarter and to decide whether to advertise that quarter.

The cost of advertising is \$1,000,000 for each quarter of a year in which it is done. When advertising during a quarter, the probability of having high sales the next quarter is $\frac{1}{2}$ or $\frac{3}{4}$, depending upon whether the current quarter's sales are low or high. These probabilities go down to $\frac{1}{4}$ or $\frac{1}{2}$ when advertising is not done during the current quarter. The company's quarterly profits (excluding advertising costs) are \$4,000,000 when sales are high, but only \$2,000,000 when sales are low. (Hereafter, use units of millions of dollars.)

(*a*) Assuming the company always chooses not to advertise, determine the transition matrix. Assuming the company always chooses to advertise, determine the transition matrix. Assuming the company follows the marketing manager's proposal, determine the transition matrix.

(*b*) Determine the steady-state probabilities for each of the three cases in part (*a*).

17. Consider the camera inventory problem presented in Sec. 5.2 except that demand now has the following probability distribution:

$$P\{D = 0\} = \tfrac{1}{4},$$
$$P\{D = 1\} = \tfrac{1}{2},$$
$$P\{D = 2\} = \tfrac{1}{4},$$
$$P\{D \geq 3\} = 0.$$

The inventory policy is still an (s, S) policy with $s = 1$ and $S = 2$. Assume that there is one camera in stock at the time the policy is instituted. Noting that there are three states:

(*a*) Find the transition matrix.

(*b*) Find the steady-state probabilities.

(*c*) Assuming that the store pays a storage cost for each camera remaining on the shelf at the end of the week according to the function $C(0) = 0$, $C(1) = \$2$, and $C(2) = \$8$, find the long-run expected average holding cost per week.

18. A production process contains a machine that deteriorates rapidly in both quality and output under heavy usage, so that it is inspected at the end of each day. Immediately after inspection, the condition of the machine is noted and classified into one of four possible states:

State	Condition
0	Good as new
1	Operable—minimum deterioration
2	Operable—major deterioration
3	Inoperable and replaced by a good-as-new machine

The process can be modeled as a Markov chain with transition matrix given by

State	0	1	2	3
0	0	$\frac{7}{8}$	$\frac{1}{16}$	$\frac{1}{16}$
1	0	$\frac{3}{4}$	$\frac{1}{8}$	$\frac{1}{8}$
2	0	0	$\frac{1}{2}$	$\frac{1}{2}$
3	1	0	0	0

(*a*) Find the steady-state probabilities.

(*b*) If the costs of being in states 0, 1, 2, 3 are 0, \$1,000, \$3,000, and \$6,000, respectively, what is the long-run expected average cost per day?

19. Using ordering costs and unsatisfied demand costs, evaluate the expected average inventory cost per week for the inventory example introduced in Sec. 5.7, using the (s, S) policy, $(s, S) = (2, 3)$.

20.* Consider the inventory example introduced in Sec. 5.2. Instead of following an (s, S) policy, we use a (q, Q) policy. If the stock level at the end of each period is less than $q = 2$ units, $Q = 2$ additional units will be ordered. Otherwise, no ordering will take place. This policy is a (q, Q) policy with $q = 2$ and $Q = 2$. Let X_t denote the number of units on hand at the end of the tth period. Assume that demand that is not filled results in lost sales. $\{X_n\}$ is a Markov chain (assume $X_0 = 0$). Using the cost values and demand distribution given for the inventory example in the text:

(*a*) Find the steady-state probabilities.

(*b*) Find the long-run expected average cost per unit time.

21. Consider the following (k, Q) inventory policy. Let $D_1, D_2, \ldots,$ be the demand for a product in periods $1, 2, \ldots,$ respectively. If the demand during a period exceeds the number of items available, this unsatisfied demand is backlogged; i.e., it is filled when the next order is received. Let Z_n $(n = 0, 1, \ldots)$ denote the amount of inventory on hand minus the number of units backlogged before ordering at the end of period n $(Z_0 = 0)$. If Z_n is zero or positive, no orders are backlogged. If Z_n is negative, then $-Z_n$ represents the number of backlogged units and no inventory is on hand. If at the end of period n, $Z_n < k = 1$, an order is placed for $2m$ (Qm in general) units, where m is the smallest integer such that $Z_n + 2m \geq 1$. (The amount ordered is the smallest integral multiple of 2, which brings the level to at least 1 unit.) Let D_n be independent and identically distributed random variables taking on the values, 0, 1, 2, 3, 4, each, with probability $\frac{1}{5}$. Let X_n denote the amount of stock on hand *after* ordering at the end of period n $(X_0 = 2)$. It is evident that

$$X_n = \begin{cases} X_{n-1} - D_n + 2m, & \text{if } X_{n-1} - D_n < 1 \\ X_{n-1} - D_n, & \text{if } X_{n-1} - D_n \geq 1 \end{cases} \qquad (n = 1, 2, \ldots),$$

and $\{X_n\}$ $(n = 0, 1, \ldots)$ is a Markov chain with only two states: 1 and 2. [The only time that ordering will take place is when $Z_n = 0, -1, -2,$ or -3, in which case 2, 2, 4, and 4 units are ordered, respectively, leaving $X_n = 2, 1, 2, 1$, respectively. In general, for any (k, Q) policy, the possible states are $k, k+1, k+2, \ldots, k+Q-1$.]

(*a*) Find the one-step transition matrix.

(*b*) Find the stationary probabilities (see Prob. 5).

(*c*) Suppose that the ordering cost is given by $(2 + 2m)$ if an order is placed and zero otherwise. The holding cost per period is Z_n if $Z_n \geq 0$ and zero otherwise. The shortage cost per period is $-4Z_n$ if $Z_n < 0$ and zero otherwise. Find the (long-run) expected average cost per unit time.

22. An important unit consists of two components placed in parallel. The unit performs satisfactorily if one of the two components is operating. A component breaks down in a given period with probability q. Assume that the component breaks down only at the end of a period. When this occurs, the parallel component takes over, if available, beginning at the next period. Only one servicewoman is assigned to service each component in need of repair, and it takes two periods to complete the servicing. Let X_t be a vector consisting of two elements U and V. U represents the number of components operating at the end of the tth period. V takes on the value 1 if the servicewoman requires only one additional period to complete a repair, if she is so engaged, and zero otherwise. Thus the state space consists of the four states (2, 0), (1, 0), (0, 1), and (1, 1). For example, the state (1, 1) implies that one component is operative and the other component needs an additional period for repair before becoming operative. Denote these four states by 0, 1, 2, 3, respectively. $\{X_t\}$ $(t = 0, 1, \ldots)$ is a Markov chain [assume that X_0 is the vector (2, 0); that is, $X_0 = 0$] with transition matrix,

$$
\mathbf{P} = \begin{bmatrix} 1-q & q & 0 & 0 \\ 0 & 0 & q & 1-q \\ 0 & 1 & 0 & 0 \\ 1-q & q & 0 & 0 \end{bmatrix}.
$$

(*a*) What is the probability that there is a waiting line of length 1 (a component needing service but not being worked on) at the end of a current period?

(*b*) What are the steady-state probabilities?

(*c*) If it costs \$30,000 per period when the unit is inoperable (both components down) and zero otherwise, what is the (long-run) expected average cost per period?

23. Consider the following gambler's ruin problem. A gambler bets one unit on each play of a game. She has a probability p of winning and $q = 1 - p$ of losing. She will continue to play until she goes broke or nets a fortune of T units. Let X_n denote the gambler's fortune on the nth play of the game. Then

$$
X_{n+1} = \begin{cases} X_n + 1, & \text{with probability } p \\ X_n - 1, & \text{with probability } q = 1 - p, \end{cases} \qquad \text{for } 0 < X_n < T.
$$

$$
X_{n+1} = X_n, \qquad \text{for } X_n = 0 \text{ or } T.
$$

$\{X_n\}$ is a Markov chain. Assume that successive plays of the game are independent and that the gambler has an initial fortune of X_0.

(*a*) Determine the one-step transition matrix of the Markov chain.

(*b*) Find the classes of the Markov chain.

(*c*) Let $T = 3$ and $p = 0.3$. Find f_{10}, f_{1T}, f_{20}, f_{2T}.

(*d*) Let $T = 3$ and $p = 0.7$. Find f_{10}, f_{1T}, f_{20}, f_{2T}.

What can you conclude from (*c*) and (*d*)?

24. A video recorder manufacturer is so certain of its quality control that it is offering a complete replacement warranty if the set fails within two years. Based upon compiled data, the company has noted that only 1 percent of its recorders fail during the first year and 5 percent fail during the second year. The warranty does not cover replaced recorders.

(*a*) Formulate this problem as a Markov chain and determine the transition matrix.

(*b*) Find the probability that the manufacturer will have to honor the warranty.

25. Consider the repairer problem presented in Sec. 5.9. Derive the expressions for the steady-state probabilities when there are M machines in the system.

26. Consider a single-server queueing system, in which customers arrive according to a Poisson input process with parameter λ (see Sec. 6.6), and the service times for the respective calling units are independent and identically distributed random variables. For $n = 1, 2, \ldots$, let X_n denote the number of calling units in the system at the moment t_n when the nth calling unit to be served (over a certain time interval) has finished being served. The sequence of time $\{t_n\}$ corresponding to the moments when successive calling units depart service are called *regeneration points*. Furthermore, $\{X_n\}$, which represents the number of calling units in the system at the corresponding sequence of time $\{t_n\}$, is a Markov chain and is known as an *imbedded Markov chain*. Imbedded Markov chains are useful for studying the properties of continuous time parameter stochastic processes.

Now consider the particular special case where the service time of successive calling units is a fixed constant, say, 10 minutes, and the mean arrival rate is one every 50 minutes. To obtain a *finite* number of states, assume as an approximation that, if there are four calling units in the system, the system becomes saturated so that additional arrivals are turned away. Therefore, $\{X_n\}$ is an imbedded Markov chain with state 0, 1, 2, or 3. (Because there are never more than four calling units in the system, there can never be more than three in the system at a regeneration point.) Because the system is observed at successive departures, X_n can never decrease by more than 1. Furthermore, the probabilities of transitions that result in increases in X_n are obtained directly from the Poisson distribution.

(*a*) Find the one-step transition matrix. (In obtaining the transition probability from state 3 to state 3, use the probability of one or more arrivals rather than just one arrival, and similarly for other transitions to state 3.)

(*b*) Find the steady-state probabilities for the number of calling units in the system at regeneration points.

(*c*) Compute the expected number of calling units in the queueing system at regeneration points, and compare it to the value of L for the single-server model in Sec. 6.7.

6

Queueing Theory

Queueing theory involves the mathematical study of queues, or waiting lines. The formation of waiting lines is, of course, a common phenomenon that occurs whenever the current demand for a service exceeds the current capacity to provide that service. Decisions regarding the amount of capacity to provide must be made frequently in industry and elsewhere. However, because it is often impossible to predict accurately when units will arrive to seek service and/or how much time will be required to provide that service, these decisions often are difficult ones. Providing too much service involves excessive costs. On the other hand, not providing enough service capacity causes the waiting line to become excessively long at times. Excessive waiting also is costly in some sense, whether it be a social cost, the cost of lost customers, the cost of idle employees, or some other important cost. Therefore, the ultimate goal is to achieve an economic balance between the cost of service and the cost associated with waiting for that service. Queueing theory itself does not solve this problem directly; however, it does contribute vital information required for such decisions by predicting such various characteristics of the waiting line as the average waiting time.

Queueing theory provides a large number of alternative mathematical models for describing a waiting-line situation. Mathematical results that predict some of the characteristics of the waiting line often are available for these models. After some general discussion, this chapter presents most of the more elementary models and their basic results. Chapter 7 discusses how the information provided by queueing theory might be used for making decisions.

6.1 Prototype Example

The emergency room of COUNTY HOSPITAL provides quick medical care for emergency cases brought to the hospital by ambulance or private automobile. At any hour there is always one doctor on duty in the emergency room. However, because of a growing tendency for emergency cases to use these facilities rather than go to a private physician, the hospital has been experiencing a continuing increase in the number of emergency room visits each year. As a result, it has become quite common for patients arriving during peak usage hours (the early evening) to have to wait until it is their turn to be treated by the doctor. Therefore, a proposal has been made that a second doctor should be assigned to the emergency room during these hours, so that two emergency cases can be treated simultaneously. The hospital's Management Engineer has been assigned to study this question.[1]

The Management Engineer began by gathering the relevant historical data and then projecting these data into the next year. Recognizing that the emergency room is a queueing system, he applied several alternative queueing theory models to predict the waiting characteristics of the system with one doctor and with two doctors—as you will see in the latter sections of this chapter (see Tables 6.2, 6.3, and 6.4).

6.2 Basic Structure of Queueing Models

The Basic Queueing Process

The basic process assumed by most queueing models is the following. *Customers* requiring service are generated over time by an *input source*. These customers enter the *queueing system* and join a *queue*. At certain times a member of the queue is selected for service by some rule known as the *queue discipline* (or *service discipline*). The required service is then performed for the customer by the *service mechanism*, after which the customer leaves the queueing system. This process is depicted in Fig. 6.1.

Many alternative assumptions can be made about the various elements of the queueing process; they are discussed next.

Input Source (Calling Population)

One characteristic of the input source is its size. The *size* is the total number of customers that might require service from time to time, i.e., the total number of

[1] For one actual case study of this kind, see Bolling, W. Blaker: "Queueing Model of a Hospital Emergency Room," *Industrial Engineering*, pp. 26–31, September 1972.

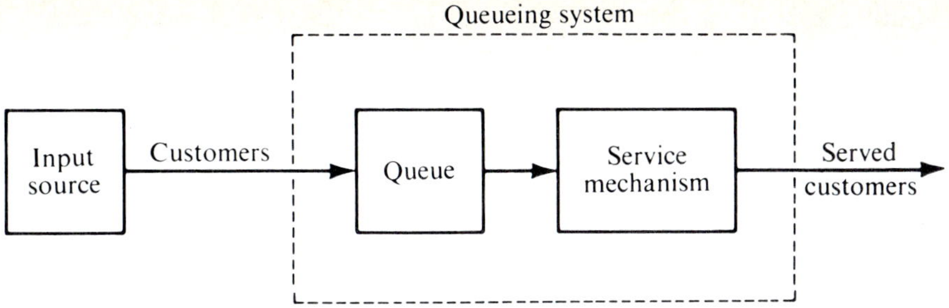

Figure 6.1 The basic queueing process.

distinct potential customers. This population from which arrivals come is referred to as the **calling population**. The size may be assumed to be either *infinite* or *finite* (so that the input source also is said to be either *unlimited* or *limited*). Because the calculations are far easier for the infinite case, this assumption often is made even when the actual size is some relatively large finite number, and it should be taken to be the implicit assumption for any queueing model that does not state otherwise. The finite case is more difficult analytically because the number of customers in the queueing system affects the number of potential customers outside the system at any time. However, the finite assumption must be made if the rate at which the input source generates new customers is significantly affected by the number of customers in the queueing system.

The statistical pattern by which customers are generated over time must also be specified. The common assumption is that they are generated according to a *Poisson process;* i.e., the number of customers generated until any specific time has a Poisson distribution. As we discuss in Sec. 6.4, this case is the one where arrivals to the queueing system occur randomly but at a certain fixed mean rate, regardless of how many customers already are there (so the *size* of the input source is *infinite*). An equivalent assumption is that the probability distribution of the time between consecutive arrivals is an *exponential* distribution. (The properties of this distribution are described in Sec. 6.4.) The time between consecutive arrivals is referred to as the **interarrival time**.

Any unusual assumptions about the behavior of the customers must also be specified. One example is *balking,* where the customer refuses to enter the system and is lost if the queue is too long.

Queue

A queue is characterized by the maximum permissible number of customers that it can contain. Queues are called *infinite* or *finite,* according to whether this number is infinite or finite. The assumption of an *infinite queue* is the standard one for most queueing models, even for situations where there actually is a (relatively large) finite upper bound on the permissible number of customers, because dealing with such an upper bound would be a complicating factor in the analysis. However, for queueing systems where this upper bound is small enough that it actually would be reached with some frequency, it becomes necessary to assume a *finite queue*.

Queue Discipline

The queue discipline refers to the order in which members of the queue are selected for service. For example, it may be first-come-first-served, random, according to some priority procedure, or some other order. First-come-first-served usually is assumed by queueing models unless stated otherwise.

Service Mechanism

The service mechanism consists of one or more *service facilities,* each of which contains one or more *parallel service channels,* called **servers**. If there is more than one service facility, the customer may receive service from a sequence of these (*service channels in series*). At a given facility, the customer enters one of the parallel service channels and is completely serviced by that server. A queueing model must specify the arrangement of the facilities and the number of servers (parallel channels) at each one. Most elementary models assume one service facility with either one or a finite number of servers.

The time elapsed from the commencement of service to its completion for a customer at a service facility is referred to as the **service time** (or *holding time*). A model of a particular queueing system must specify the probability distribution of service times for each server (and possibly for different types of customers), although it is common to assume the *same* distribution for all servers (all models in this chapter make this assumption). The service-time distribution that is most frequently assumed in practice (largely because it is far more tractable than any other) is the *exponential* distribution discussed in Sec. 6.4, and most of our models will be of this type. Other important service-time distributions are the *degenerate* distribution (constant service time) and the *Erlang* (gamma) distribution, as illustrated by models in Sec. 6.7.

An Elementary Queueing Process

As we have already suggested, queueing theory has been applied to many different types of waiting-line situations. However, the most prevalent type of situation is the following: A single waiting line (which may be empty at times) forms in the front of a single service facility, within which are stationed one or more servers. Each customer generated by an input source is serviced by one of the servers, perhaps after some waiting in the queue (waiting line). The queueing system involved is depicted in Fig. 6.2.

Notice that the queueing process in the illustrative example of Sec. 6.1 is of this type. The input source generates customers in the form of emergency cases requiring medical care. The emergency room is the service facility, and the doctors are the servers.

A server need not be a single individual; it may be a group of persons, e.g., a repair crew that combines forces to perform simultaneously the required service for a customer. Furthermore, servers need not even be people. In many cases, a server may be a machine or a piece of equipment, e.g., a forklift that performs a given service on call (although probably with human guidance). By the same token, the customers in the waiting line need not be people. For example, they may be items waiting for a certain operation by a given type of machine, or they may be cars waiting in front of a toll booth.

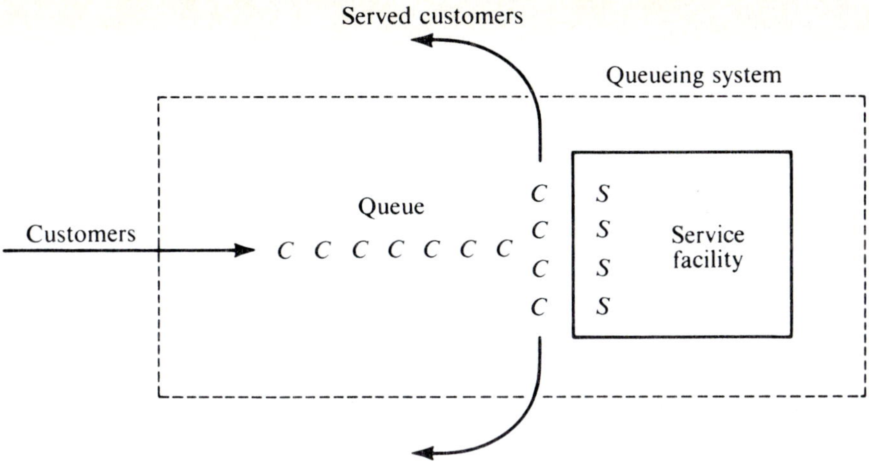

Figure 6.2 An elementary queueing system (each customer is indicated by a C and each server by an S).

It is not necessary that there actually be a physical waiting line forming in front of a physical structure that constitutes the service facility; that is, the members of the queue may be scattered throughout an area waiting for a server to come to them, e.g., machines waiting to be repaired. The server or group of servers assigned to a given area constitute the service facility for that area. Queueing theory still gives the average number waiting, the average waiting time, and so on, because it is irrelevant whether or not the customers wait together in a group. The only essential requirement for queueing theory to be applicable is that changes in the number of customers waiting for a given service occur just as though the physical situation described in Fig. 6.2 (or a legitimate counterpart) prevails.

Except for Sec. 6.9, all the queueing models discussed in this chapter are of the elementary type depicted in Fig. 6.2. Many of these models further assume that all *interarrival times* are independent and identically distributed and that all *service times* are independent and identically distributed. Such models conventionally are labeled as follows:

Distribution of service times

–/–/– Number of servers

Distribution of interarrival times,

where M = exponential distribution (Markovian), as described in Sec. 6.4,
D = degenerate distribution (constant times), as discussed in Sec. 6.7,
E_k = Erlang distribution (shape parameter = k), as described in Sec. 6.7,
G = general distribution (any arbitrary distribution allowed),[1] as discussed in Sec. 6.7.

For example, the M/M/s model discussed in Sec. 6.6 assumes that both interarrival times and service times have an exponential distribution and that the number of servers

[1] When referring to interarrival times, it is conventional to replace the symbol G by GI = general *i*ndependent distribution.

is s (any positive integer). The M/G/1 model discussed again in Sec. 6.7 assumes that interarrival times have an exponential distribution, but it places no restriction on what the distribution of service times must be, whereas the number of servers is restricted to be exactly one. Various other models that fit this labeling scheme also are introduced in Sec. 6.7.

Terminology and Notation

Unless otherwise noted, the following standard terminology and notation will be used:

State of the system = number of customers in queueing system.
Queue length = number of customers waiting for service
= state of system minus number of customers being served.
$N(t)$ = number of customers in queueing system at time t ($t \geq 0$).
$P_n(t)$ = probability that exactly n customers are in queueing system at time t, given number at time 0.
s = number of servers (parallel service channels) in queueing system.
λ_n = mean arrival rate (expected number of arrivals per unit time) of new customers when n customers are in system.
μ_n = mean service rate for overall system (expected number of customers completing service per unit time) when n customers are in system. *Note:* μ_n represents *combined* rate at which all *busy* servers (those serving customers) achieve service completions.
λ, μ, ρ: see following paragraph.

When λ_n is a constant for all n, this constant is denoted by λ. When the mean service rate *per busy server* is a constant for all $n \geq 1$, this constant is denoted by μ. (In this case, $\mu_n = s\mu$ when $n \geq s$, that is, when all s servers are busy.) Under these circumstances, $1/\lambda$ and $1/\mu$ are the *expected interarrival time* and the *expected service time,* respectively. Also, $\rho = \lambda/s\mu$ is the **utilization factor** for the service facility, i.e., the expected fraction of time the individual servers are busy, because $\lambda/s\mu$ represents the fraction of the system's service capacity ($s\mu$) that is being *utilized* on the average by arriving customers (λ).

Certain notation also is required to describe *steady-state* results. When a queueing system has recently begun operation, the state of the system (number of customers in the system) will be greatly affected by the initial state and the time that has since elapsed. The system is now said to be a **transient condition**. However, after sufficient time has elapsed, the state of the system becomes essentially independent of the initial state and the elapsed time (except under unusual circumstances).[1] The system has now essentially reached a **steady-state condition**, where the probability distribution of the state of the system remains the same (the *steady-state* or *stationary* distribution) over time. Queueing theory has tended to focus largely on the steady-state condition, partially because the transient case is more difficult analyti-

[1] When λ and μ are defined, these unusual circumstances are that $\rho \geq 1$, in which case the state of the system tends to grow continually larger as time goes on.

cally. (Some transient results exist, but they are generally beyond the technical scope of this book.) The following notation assumes that the system is in a *steady-state condition*:

P_n = probability that exactly n customers are in queueing system.
L = expected number of customers in queueing system.
L_q = expected queue length (excludes customers being served).
$\mathcal{W}$ = waiting time in system (includes service time) for each individual customer.
$W = E(\mathcal{W})$.
$\mathcal{W}_q$ = waiting time in queue (excludes service time) for each individual customer.
$W_q = E(\mathcal{W}_q)$.

Relationships between L, W, L_q, and W_q

Assume that λ_n is a constant λ for all n. It has been proven that in a steady-state queueing process,

$$L = \lambda W.$$

(Because John D. C. Little[1] provided the first rigorous proof, this equation sometimes is referred to as **Little's formula**.) Furthermore, the same proof also shows that

$$L_q = \lambda W_q.$$

If the λ_n are not equal, then λ can be replaced in these equations by $\bar{\lambda}$, the *average* arrival rate over the long run. (We shall show later how $\bar{\lambda}$ can be determined for some basic cases.)

Now assume that the mean service time is a constant, $1/\mu$, for all $n \geq 1$. It then follows that

$$W = W_q + \frac{1}{\mu}.$$

These relationships are extremely important because they enable all four of the fundamental quantities—L, W, L_q, and W_q—to be immediately determined as soon as one of them is found analytically. This situation is fortunate because some of these quantities often are much easier to find than others when solving a queueing model from basic principles.

6.3 Examples of Real Queueing Systems

It may appear that our description of queueing systems in the preceding section is relatively abstract and applicable to only rather special practical situations. On the contrary, queueing systems are surprisingly prevalent in a wide variety of contexts. To broaden your horizons on the applicability of queueing theory, we shall briefly mention various examples of real queueing systems.

One important class of queueing systems that we all encounter in our daily lives is **commercial service systems**, where outside customers receive service from com-

[1] Little, John D. C.: "A Proof for the Queueing Formula: $L = \lambda W$," *Operations Research,* **9**(3):383–387, 1961; also see Stidham, Shaler, Jr.: "A Last Word on $L = \lambda W$," *Operations Research,* **22**(2):417–421, 1974.

mercial organizations. Many of these involve person-to-person service at a fixed location, such as a *barber shop* (the barbers are the servers), *bank teller* service, *check-out stands* at a grocery store, and a *cafeteria line* (service channels in series). However, many others do not, such as *home appliance repairs* (the server travels to the customers), a *vending machine* (the server is a machine), and a *gas station* (the cars are the customers).

Another important class is **transportation service systems**. For some of these systems the vehicles are the customers, such as *cars* waiting at a *toll booth* or *traffic light* (the server), a *truck* or *ship* waiting to be loaded or unloaded by a *crew* (the server), and *airplanes* waiting to land or take off from a *runway* (the server). (An unusual example of this kind is a *parking lot,* where the cars are the customers and the parking spaces are the servers, but there is no queue because arriving customers go elsewhere to park if the lot is full.) In other cases, the vehicles, such as *taxicabs, fire trucks,* and *elevators,* are the servers.

In recent years, queueing theory probably has been applied most to **business-industrial internal service systems**, where the customers receiving service are *internal* to the organization. Examples include *materials-handling systems,* where materials-handling units (the servers) move loads (the customers); *maintenance systems,* where maintenance crews (the servers) repair machines (the customers); and *inspection stations,* where quality control inspectors (the servers) inspect items (the customers). *Employee facilities* and *typing pools* also fit into this category. In addition, *machines* can be viewed as servers whose customers are the jobs being processed. A related example of great importance is a *computer facility,* where the computer is viewed as the server.

There is now growing recognition that queueing theory also is applicable to **social service systems**. For example, a *judicial system* is a queueing network, where the courts are service facilities, the judges (or panels of judges) are the servers, and the cases waiting to be tried are the customers. A *legislative system* is a similar queueing network, where the customers are the bills waiting to be processed. Various *health-care systems* also are queueing systems. You already have seen one example in Sec. 6.1 (a hospital emergency room), but you can also view *ambulances, x-ray machines,* and *hospital beds* as servers in their own queueing systems. Similarly, families waiting for low- and moderate-income housing, or other social services, can be viewed as customers in a queueing system.

Although these are four broad classes of queueing systems, they still do not exhaust the list. In fact, queueing theory first began early in this century with applications to *telephone engineering* (the founder of queueing theory, A. K. Erlang, was an employee of the Danish Telephone Company in Copenhagen), and telephone engineering still is an important application. Furthermore, we all have our own *personal queues*—homework assignments, books to be read, and so forth. However, these examples are sufficient to suggest that queueing systems do indeed pervade many areas of society.

6.4 The Role of the Exponential Distribution

The operating characteristics of queueing systems are determined largely by two statistical properties, namely, the probability distribution of *interarrival times* (see "Input Source" in Sec. 6.2) and the probability distribution of *service times* (see "Service

Mechanism'' in Sec. 6.2). For real queueing systems, these distributions can take on almost any form. (The only restriction is that negative values cannot occur.) However, to formulate a queueing theory *model* as a representation of the real system, it is necessary to specify the assumed form of each of these distributions. To be useful, the assumed form should be *sufficiently realistic,* so that the model provides *reasonable predictions* while, at the same time, being *sufficiently simple,* so that the model is *mathematically tractable*. Based on these considerations, the most important probability distribution in queueing theory is the *exponential distribution*.

Suppose that a random variable T represents either *interarrival* or *service times*. (We shall refer to the occurrences marking the end of these times—arrivals or service completions—as *events*.) T is said to have an *exponential distribution with parameter* α if its probability density function is

$$f_T(t) = \begin{cases} \alpha e^{-\alpha t}, & \text{for } t \geq 0 \\ 0, & \text{for } t < 0, \end{cases}$$

as shown in Fig. 6.3. In this case, the cumulative probabilities are

$$\begin{aligned} P\{T \leq t\} &= 1 - e^{-\alpha t}, \\ P\{T > t\} &= e^{-\alpha t}, \end{aligned} \qquad (t \geq 0)$$

and the expected value and variance of T are

$$E(T) = \frac{1}{\alpha},$$

$$\text{var}(T) = \frac{1}{\alpha^2}.$$

What are the implications of assuming that T has an exponential distribution for a queueing model? To explore this question, let us examine six key properties of the exponential distribution.

Property 1: $f_T(t)$ is a strictly *decreasing* function of t $(t \geq 0)$.

One consequence of Property 1 is that

$$P\{0 \leq T \leq \Delta t\} > P\{t \leq T \leq t + \Delta t\}$$

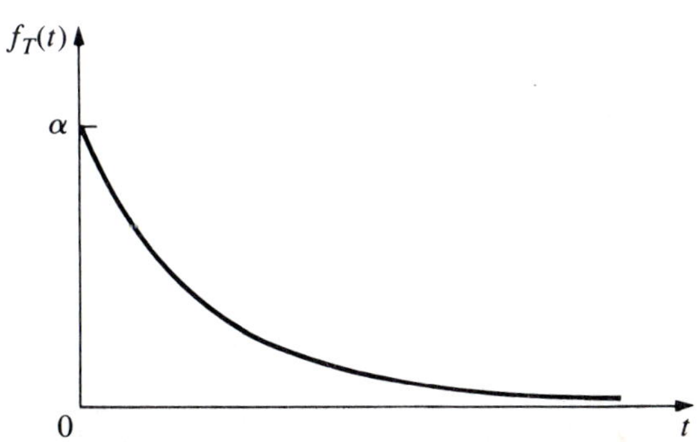

Figure 6.3 Probability density function for the exponential distribution.

for any strictly positive values of Δt and t. [This consequence follows from the fact that these probabilities are the area under the $f_T(t)$ curve over the indicated interval of length Δt, and the average height of the curve is less for the second probability than for the first.] Therefore, it is not only possible but relatively likely that T will take on a small value near zero. In fact,

$$P\left\{0 \le T \le \frac{1}{2}\frac{1}{\alpha}\right\} = 0.393,$$

whereas

$$P\left\{\frac{1}{2}\frac{1}{\alpha} \le T \le \frac{3}{2}\frac{1}{\alpha}\right\} = 0.383,$$

so that the value T takes on is more likely to be ''small'' [i.e., less than half of $E(T)$] than ''near'' its expected value [i.e., no further away than half of $E(T)$], even though the second interval is twice as wide as the first.

Is this really a reasonable property for T in a queueing model? If T represents *service times,* the answer depends upon the general nature of the service involved, as discussed next.

If the service required is essentially identical for each customer, with the server always performing the same sequence of service operations, then the actual service times tend to be near the expected service time. Small deviations from the mean may occur, but usually because of only minor variations in the efficiency of the server. A small service time far below the mean is essentially impossible, because a certain minimum amount of time is needed to perform the required service operations even when the server is working at top speed. The exponential distribution clearly does not provide a close approximation to the service-time distribution for this type of situation.

On the other hand, consider the type of situation where the specific tasks required of the server differ among the customers. The broad nature of the service may be the same, but the specific type and amount of service differ. For example, this would be the case in the *County Hospital* emergency room problem discussed in Sec. 6.1. The doctors encounter a wide variety of medical problems. In most cases, they can provide the required treatment rather quickly, but an occasional patient requires extensive care. Similarly, bank tellers and grocery store checkout clerks are other servers of this general type, where the required service is often brief but must occasionally be extensive. An exponential service-time distribution would seem quite plausible for this type of service situation.

If T represents *interarrival times,* Property 1 rules out situations where potential customers approaching the queueing system tend to postpone their entry if they see another customer entering ahead of them. On the other hand, it is entirely consistent with the common phenomenon of arrivals occurring ''randomly,'' described by subsequent properties.

Property 2: Lack of memory.

This property can be stated mathematically as

$$P\{T > t + \Delta t | T > \Delta t\} = P\{T > t\}$$

for any positive quantities t and Δt. In other words, the probability distribution of the *remaining* time until the event (arrival or service completion) occurs always is the same, regardless of how much time (Δt) already has passed. In effect, the process "forgets" its history. This surprising phenomenon occurs with the exponential distribution because

$$P\{T > t + \Delta t | T > \Delta t\} = \frac{P\{T > \Delta t,\ T > t + \Delta t\}}{P\{T > \Delta t\}}$$

$$= \frac{P\{T > t + \Delta t\}}{P\{T > \Delta t\}}$$

$$= \frac{e^{-\alpha(t+\Delta t)}}{e^{-\alpha\Delta t}}$$

$$= e^{-\alpha t}$$

For *interarrival times,* this property describes the common situation where the time until the next arrival is completely uninfluenced by when the last arrival occurred. For *service times,* the property is more difficult to interpret. We should not expect it to hold in a situation where the server must perform the same fixed sequence of operations for each customer, because then a long elapsed service should imply that probably little remains to be done. On the other hand, in the type of situation where the required service operations differ among the customers, the mathematical statement of the property may be quite realistic. For this case, if considerable service has already elapsed for a customer, the only implication may be that this particular customer requires more extensive service than most.

Property 3: The *minimum* of several independent exponential random variables has an exponential distribution.

To state this property mathematically, let $T_1, T_2, \ldots, T_n$ be *independent* exponential random variables with parameters $\alpha_1, \alpha_2, \ldots, \alpha_n$, respectively. Also let U be the random variable that takes on the value equal to the *minimum* of the values actually taken on by $T_1, T_2, \ldots, T_n$; that is,

$$U = \text{minimum}\ \{T_1, T_2, \ldots, T_n\}.$$

Thus, if T_i represents the time until a particular kind of event will occur, then U represents the time until the *first* of the n different events will occur. Now note that for any $t \geq 0$,

$$P\{U > t\} = P\{T_1 > t,\ T_2 > t,\ \ldots,\ T_n > t\}$$

$$= P\{T_1 > t\}P\{T_2 > t\} \cdots P\{T_n > t\}$$

$$= e^{-\alpha_1 t}e^{-\alpha_2 t} \cdots e^{\alpha_n t}$$

$$= \exp\left\{-\sum_{i=1}^{n} \alpha_i t\right\},$$

so that U indeed has an exponential distribution with parameter

$$\alpha = \sum_{i=1}^{n} \alpha_i.$$

This property has some implications for *interarrival times* in queueing models. In particular, suppose that there are several (n) *different* types of customers, but the interarrival times for *each* type (type i) have an exponential distribution with parameter α_i ($i = 1, 2, \ldots, n$). By Property 2, the *remaining* time from any specified instant until the next arrival of a customer of type i would have this same distribution. Therefore, let T_i be this remaining time measured from the instant a customer of *any* type arrives. Property 3 then tells us that U, the interarrival times for the queueing system as a whole, has an exponential distribution with parameter α defined by the last equation. As a result, you can choose to ignore the distinction between customers and still have exponential interarrival times for the queueing model.

However, the implications are even more important for *service times* in queueing models having more than one server than they are for interarrival times. For example, consider the situation where all the servers have the same exponential service-time distribution with parameter μ. For this case, let n be the number of servers *currently* providing service, and let T_i be the *remaining* service time for server i ($i = 1, 2, \ldots, n$), which also has an exponential distribution with parameter $\alpha_i = \mu$. It then follows that U, the time until the *next* service completion from any of these servers, has an exponential distribution with parameter $\alpha = n\mu$. In effect, the queueing system *currently* would be performing just like a *single*-server system, where service times have an exponential distribution with parameter $n\mu$. We shall make frequent use of this implication for analyzing multiple-server models later in the chapter.

Property 4: Relationship to the Poisson distribution.

Suppose that the *time* between consecutive occurrences of some particular kind of event (e.g., arrivals or service completions by a continuously busy server) has an *exponential distribution* with parameter α. Property 4 then has to do with the resulting implication about the probability distribution of the *number* of times this kind of event occurs over a specified length of time. In particular, let $X(t)$ be the number of occurrences by time t ($t > 0$), where time 0 designates the instant at which the count begins. The implication is that

$$P\{X(t) = n\} = \frac{(\alpha t)^n e^{-\alpha t}}{n!}, \qquad \text{for } n = 0, 1, 2, \ldots;$$

that is, $X(t)$ has a *Poisson distribution* with parameter αt. For example, with $n = 0$,

$$P\{X(t) = 0\} = e^{-\alpha t},$$

which is just the probability from the exponential distribution that the *first* event occurs after time t. The mean of this Poisson distribution is

$$E\{X(t)\} = \alpha t,$$

so that the expected number of events *per unit time* is α. Thus α is said to be the *mean rate* at which the events occur. When the events are counted on a continuing basis, the counting process $\{X(t); t > 0\}$ is said to be a **Poisson process** with parameter α (the mean rate).

This property provides useful information about *service completions* when service times have an exponential distribution with parameter μ. We obtain this information by defining $X(t)$ as the number of service completions achieved by a *continuously busy* server in elapsed time t, where $\alpha = \mu$. For *multiple*-server queueing models, $X(t)$ can also be defined as the number of service completions achieved by n continuously busy servers in elapsed time t, where $\alpha = n\mu$.

The property is particularly useful for describing the probabilistic behavior of *arrivals* when interarrival times have an exponential distribution with parameter λ. In this case, $X(t)$ would be the *number* of arrivals in elapsed time t, where $\alpha = \lambda$ is the *mean arrival rate*. Therefore, arrivals occur according to a **Poisson input process** with parameter λ. Such queueing models also are described as assuming a *Poisson input*.

Arrivals sometimes are said to occur *randomly,* meaning that they occur in accord with a Poisson input process. One intuitive interpretation of this phenomenon is that every time period of fixed length has the *same* chance of having an arrival regardless of when the preceding arrival occurred, as suggested by the following property.

Property 5: For all positive values of t, $P\{T \leq t + \Delta t | T > t\} \approx \alpha \Delta t$, for small Δt.

Continuing to interpret T as the time from the last event of a certain type (arrival or service completion) until the *next* such event, suppose that a time t already has elapsed without the event occurring. We know from Property 2 that the probability that the event will occur within the next time interval of fixed length Δt is a *constant* (identified in the next paragraph), regardless of how large or small t is. Property 5 goes further to say that, when the value of Δt is small, this constant probability can be approximated very closely by $\alpha\,\Delta t$. Furthermore, when considering different small values of Δt, this probability is essentially *proportional* to Δt, with proportionality factor α. In fact, α is the *mean rate* at which the events occur (see Property 4), so that the *expected number* of events in the interval of length Δt is *exactly* $\alpha\,\Delta t$. The only reason that the probability of an event occurring differs slightly from this value is the possibility that *more than one* event will occur, which has negligible probability when Δt is small.

To see why Property 5 holds mathematically, note that the constant value of our probability (for a fixed value of $\Delta t > 0$) is just

$$\begin{aligned} P\{T \leq t + \Delta t | T > t\} &= P\{T \leq \Delta t\} \\ &= 1 - e^{-\alpha \Delta t}, \end{aligned}$$

for any $t \geq 0$. Therefore, because the series expansion of e^x for any exponent x is

$$e^x = 1 + x + \sum_{n=2}^{\infty} \frac{x^n}{n!},$$

it follows that

$$\begin{aligned} P\{T \leq t + \Delta t | T > t\} &= 1 - 1 + \alpha\,\Delta t - \sum_{n=2}^{\infty} \frac{(-\alpha\,\Delta t)^n}{n!} \\ &\approx \alpha\,\Delta t, \text{ for small } \Delta t,^1 \end{aligned}$$

because the summation terms become relatively negligible for sufficiently small values of $\alpha\,\Delta t$.

[1] More precisely,

$$\lim_{\Delta t \to 0} \frac{P\{T \leq t + \Delta t | T > t\}}{\Delta t} = \alpha.$$

Because T can represent either *interarrival* or *service times* in queueing models, this property provides a convenient approximation of the probability that the event of interest occurs in the next small interval (Δt) of time. An analysis based on this approximation also can be made exact by taking appropriate limits as $\Delta t \to 0$.

Property 6: Unaffected by aggregation or disaggregation.

This property is relevant primarily for verifying that the *input process* is *Poisson*. Therefore, we shall describe it in these terms, although it also applies directly to the exponential distribution (exponential interarrival times) because of Property 4.

Suppose that there are several (n) *different* types of customers, where the customers of *each* type (type i) arrive according to a *Poisson input process* with parameter λ_i ($i = 1, 2, \ldots, n$). Assuming that these are *independent* Poisson processes, the property says that the *aggregate* input process (arrival of *all* customers without regard to type) also must be *Poisson,* with parameter (arrival rate) $\lambda = \lambda_1 + \lambda_2 + \cdots + \lambda_n$. In other words, having a Poisson process is *unaffected by aggregation*.

This part of the property follows directly from Properties 3 and 4. The latter property implies that the *interarrival times* for customers of type i have an exponential distribution with parameter λ_i. For this identical situation, we already discussed for Property 3 that it implies that the interarrival times for *all* customers also must have an exponential distribution, with parameter $\lambda = \lambda_1 + \lambda_2 + \cdots + \lambda_n$. Using Property 4 again then implies that the aggregate input process is Poisson.

The second part of Property 6 ("unaffected by disaggregation") refers to the reverse case, where the *aggregate* input process is known to be Poisson with parameter λ, but the question now is the nature of the *disaggregated* input processes for the individual customer types. Assuming that each arriving customer has a *fixed* probability p_i of being of type i ($i = 1, 2, \ldots, n$), with

$$\lambda_i = p_i \lambda \qquad \text{and} \qquad \sum_{i=1}^{n} p_i = 1,$$

the property says that the input process for customers of type i also must be *Poisson,* with parameter λ_i. In other words, having a Poisson process is *unaffected by disaggregation*.

As one example of the usefulness of this second part of the property, consider the following situation. Indistinguishable customers arrive according to a Poisson process with parameter λ. Each arriving customer has a fixed probability p of *balking* (leaving without entering the queueing system), so the probability of entering the system is $(1 - p)$. Thus there are two types of customers—those that balk and those that enter the system. The property says that *each* type arrives according to a Poisson process, with parameters $p\lambda$ and $(1 - p)\lambda$, respectively. Therefore, by using the latter Poisson process, queueing models that assume a Poisson input process can still be used to analyze the performance of the queueing system for those customers that enter the system.

6.5 The Birth-and-Death Process

Most elementary queueing models assume that the inputs (arriving customers) and outputs (leaving customers) of the queueing system occur according to the *birth-and-death process*. This important process in probability theory has applications in various

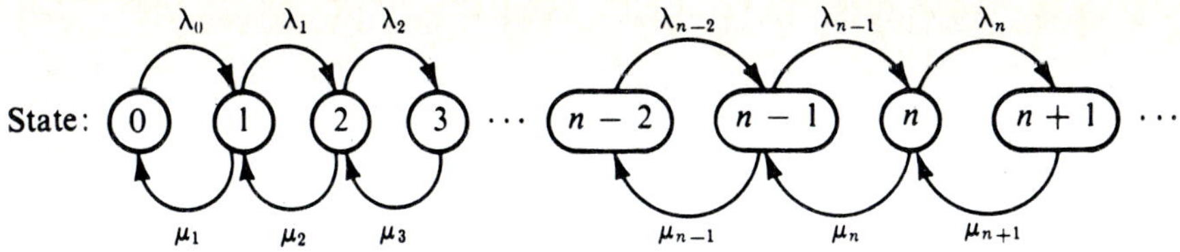

Figure 6.4 Rate diagram for the birth-and-death process.

areas. However, in the context of queueing theory, the term **birth** refers to the *arrival* of a new customer into the queueing system, and **death** refers to the *departure* of a served customer. The *state* of the system at time t ($t \geq 0$), $N(t)$, is the number of customers in the queueing system at time t. The birth-and-death process describes *probabilistically* how $N(t)$ changes as t increases. Broadly speaking, it says that *individual* births and deaths occur *randomly,* where their mean occurrence rates depend only upon the current state of the system. More precisely, the assumptions of the birth-and-death process are the following:

ASSUMPTION 1: Given $N(t) = n$, the current probability distribution of the *remaining* time until the next *birth* (arrival) is *exponential* with parameter λ_n ($n = 0, 1, 2, \ldots$).

ASSUMPTION 2: Given $N(t) = n$, the current probability distribution of the *remaining* time until the next *death* (service completion) is *exponential* with parameter μ_n ($n = 1, 2, \ldots$).

ASSUMPTION 3: Only *one* birth or death can occur at a time.

Because of Assumptions 1 and 2, the birth-and-death process is a special type of *continuous time Markov chain.* (See Sec. 5.9 for a description of continuous time Markov chains and their properties.) Queueing models that can be represented by a continuous time Markov chain are far more tractable analytically than any other, and Assumption 3 simplifies the analysis considerably further.

Because Property 4 for the exponential distribution (see Sec. 6.4) implies that the λ_n and μ_n are *mean rates,* we can summarize these assumptions by the *rate diagram* shown in Fig. 6.4. The arrows in this diagram show the only possible *transitions* in the state of the system (as specified by Assumption 3), and the entry for each arrow gives the mean rate for that transition (as specified by Assumptions 1 and 2) when the system is in the state at the base of the arrow.

Except for a few special cases, analysis of the birth-and-death process is very difficult when the system is in a *transient* condition. Some results about the probability distribution of $N(t)$[1] have been obtained, but they are too complicated to be of much practical use. On the other hand, it is relatively straightforward to derive this distribution *after* the system has reached a *steady-state* condition (assuming that this condition can be reached). This derivation can be done directly from the rate diagram, as outlined next.

Consider any particular state of the system n ($n = 0, 1, 2, \ldots$). Starting at time 0, suppose that a count is made of the number of times that the process *enters*

[1] Karlin, S., and J. McGregor: "Many Server Queueing Processes with Poisson Input and Exponential Service Times," *Pacific Journal of Mathematics,* **8**:87–118, 1958.

this state and the number of times it *leaves* this state, as denoted below:

$$E_n(t) = \text{number of times that the process } \textit{enters} \text{ state } n \text{ by time } t.$$
$$L_n(t) = \text{number of times that the process } \textit{leaves} \text{ state } n \text{ by time } t.$$

Because the two types of events (entering and leaving) must alternate, these two numbers must always be either *equal* or differ by just 1; that is,

$$|E_n(t) - L_n(t)| \leq 1.$$

Dividing through both sides by t and then letting $t \to \infty$ gives

$$\left| \frac{E_n(t)}{t} - \frac{L_n(t)}{t} \right| \leq \frac{1}{t}, \quad \text{so} \quad \lim_{t\to\infty} \left| \frac{E_n(t)}{t} - \frac{L_n(t)}{t} \right| = 0.$$

Dividing $E_n(t)$ and $L_n(t)$ by t gives the actual *rate* (number of events per unit time) at which these two kinds of events have occurred, and letting $t \to \infty$ then gives the *mean rate* (expected number of events per unit time).

$$\lim_{t\to\infty} \frac{E_n(t)}{t} = \text{mean rate at which the process enters state } n.$$

$$\lim_{t\to\infty} \frac{L_n(t)}{t} = \text{mean rate at which the process leaves state } n.$$

These results yield the following key principle:

RATE IN = RATE OUT PRINCIPLE: For any state of the system n (n = 0, 1, 2, . . .), mean entering rate = mean leaving rate.

The *equation* expressing this principle is called the **balance equation** for state n. After constructing the balance equations for *all* the states in terms of the *unknown* P_n probabilities, we can solve this system of equations to find these probabilities.

To illustrate a balance equation, consider state 0. The process enters this state *only* from state 1. Thus the steady-state probability of being in state 1 (P_1) represents the proportion of time that it would be *possible* for the process to enter state 0. *Given* that the process is in state 1, the mean rate of entering state 0 is μ_1. (In other words, for each cumulative unit of time that the process spends in state 1, the expected number of times that it would leave state 1 to enter state 0 is μ_1.) From any *other* state, this mean rate is 0. Therefore, the overall mean rate at which the process leaves its current state to *enter* state 0 (the *mean entering rate*) is

$$\mu_1 P_1 + 0(1 - P_1) = \mu_1 P_1.$$

By the same reasoning, the *mean leaving rate* must be $\lambda_0 P_0$, so the balance equation for state 0 is

$$\mu_1 P_1 = \lambda_0 P_0.$$

For every other state there are *two* possible transitions both into and out of the state. Therefore, each side of the balance equations for these states represents the *sum* of the mean rates for the two transitions involved. Otherwise, the reasoning is just the same as for state 0. These balance equations are summarized in Table 6.1.

Notice that the first balance equation contains two variables for which to solve (P_0 and P_1), the first two equations contain three variables (P_0, P_1, and P_2), and so

Table 6.1 **Balance Equations for Birth-and-Death Process**

State	Rate In = Rate Out
0	$\mu_1 P_1 = \lambda_0 P_0$
1	$\lambda_0 P_0 + \mu_2 P_2 = (\lambda_1 + \mu_1) P_1$
2	$\lambda_1 P_1 + \mu_3 P_3 = (\lambda_2 + \mu_2) P_2$
⋮	⋮
$n-1$	$\lambda_{n-2} P_{n-2} + \mu_n P_n = (\lambda_{n-1} + \mu_{n-1}) P_{n-1}$
n	$\lambda_{n-1} P_{n-1} + \mu_{n+1} P_{n+1} = (\lambda_n + \mu_n) P_n$
⋮	⋮

on, so that there always is one "extra" variable. Therefore, the procedure in solving these equations is to solve in terms of one of the variables, the most convenient one being P_0. Thus the first equation is used to solve for P_1 in terms of P_0; this result and the second equation are then used to solve for P_2 in terms of P_0, and so forth. At the end, the requirement that the sum of all the probabilities must equal 1 can be used to evaluate P_0.

Applying this procedure yields the following results:

State

$$0: \quad P_1 = \frac{\lambda_0}{\mu_1} P_0$$

$$1: \quad P_2 = \frac{\lambda_1}{\mu_2} P_1 + \frac{1}{\mu_2}(\mu_1 P_1 - \lambda_0 P_0) = \frac{\lambda_1}{\mu_2} P_1 = \frac{\lambda_1 \lambda_0}{\mu_2 \mu_1} P_0$$

$$2: \quad P_3 = \frac{\lambda_2}{\mu_3} P_2 + \frac{1}{\mu_3}(\mu_2 P_2 - \lambda_1 P_1) = \frac{\lambda_2}{\mu_3} P_2 = \frac{\lambda_2 \lambda_1 \lambda_0}{\mu_3 \mu_2 \mu_1} P_0$$

$$\vdots \qquad \vdots$$

$$n-1: \quad P_n = \frac{\lambda_{n-1}}{\mu_n} P_{n-1} + \frac{1}{\mu_n}(\mu_{n-1} P_{n-1} - \lambda_{n-2} P_{n-2}) = \frac{\lambda_{n-1}}{\mu_n} P_{n-1} = \frac{\lambda_{n-1} \lambda_{n-2} \cdots \lambda_0}{\mu_n \mu_{n-1} \cdots \mu_1} P_0$$

$$n: \quad P_{n+1} = \frac{\lambda_n}{\mu_{n+1}} P_n + \frac{1}{\mu_{n+1}}(\mu_n P_n - \lambda_{n-1} P_{n-1}) = \frac{\lambda_n}{\mu_{n+1}} P_n = \frac{\lambda_n \lambda_{n-1} \cdots \lambda_0}{\mu_{n+1} \mu_n \cdots \mu_1} P_0$$

$$\vdots \qquad \vdots$$

To simplify notation, let

$$C_n = \frac{\lambda_{n-1} \lambda_{n-2} \cdots \lambda_0}{\mu_n \mu_{n-1} \cdots \mu_1}, \quad \text{for } n = 1, 2, \ldots .$$

Thus the steady-state probabilities are

$$P_n = C_n P_0, \qquad \text{for } n = 1, 2, \ldots .$$

The requirement that

$$\sum_{n=0}^{\infty} P_n = 1$$

implies that

$$\left[1 + \sum_{n=1}^{\infty} C_n\right] P_0 = 1,$$

so that

$$P_0 = \frac{1}{1 + \sum_{n=1}^{\infty} C_n}.$$

Given this information,

$$L = \sum_{n=0}^{\infty} n P_n.$$

Also, because the number of servers s represents the number of customers that can be served (and thus are not in the queue) simultaneously,

$$L_q = \sum_{n=s}^{\infty} (n - s) P_n.$$

Furthermore, the relationships given in Sec. 6.2 yield

$$W = \frac{L}{\bar{\lambda}}, \qquad W_q = \frac{L_q}{\bar{\lambda}},$$

where $\bar{\lambda}$ is the *average* arrival rate over the long run. Because λ_n is the mean arrival rate while the system is in state n ($n = 0, 1, 2, \ldots$), and P_n is the proportion of time that the system is in this state,

$$\bar{\lambda} = \sum_{n=0}^{\infty} \lambda_n P_n.$$

Several of the expressions just given involve summations with an infinite number of terms. Fortunately, these summations have analytic solutions for a number of interesting special cases,[1] as seen in the next section. Otherwise, they can be approximated by summing a finite number of terms on a computer.

These steady-state results have been derived under the assumption that the λ_n

[1] These solutions are based on the following known results for the sum of any geometric series:

$$\sum_{n=0}^{N} x^n = \frac{1 - x^{N+1}}{1 - x}, \qquad \text{for any } x,$$

$$\sum_{n=0}^{\infty} x^n = \frac{1}{1 - x}, \qquad \text{if } |x| < 1.$$

and μ_n parameters have values such that the process actually can *reach* a steady-state condition. This assumption *always* holds if $\lambda_n = 0$ for some value of n greater than the initial state, so that only a finite number of states (those less than this n) are possible. It also *always* holds when λ and μ are defined (see "Terminology and Notation" in Sec. 6.2) and $\rho = \lambda/s\mu < 1$. It does *not* hold if $\Sigma_{n=1}^{\infty} C_n = \infty$.

The following section describes several queueing models that are just *special cases* of the birth-and-death process. Therefore, the *general* steady-state results just given in boxes will be used over and over again to obtain the *specific* steady-state results for these models.

6.6 Queueing Models Based on the Birth-and-Death Process

Because each of the mean rates $\lambda_0, \lambda_1, \ldots$ and $\mu_1, \mu_2, \ldots$ for the birth-and-death process can be assigned any nonnegative value, we have great flexibility in modeling a queueing system. Probably the most widely used models in queueing theory are based directly upon this process. Because of Assumptions 1 and 2 (and Property 4 for the exponential distribution), these models are said to have a **Poisson input** and **exponential service times**. The models differ only in their assumptions about how the λ_n and the μ_n change with n. We present four of these models in this section for four important types of queueing systems.

The M/M/s Model

As described in Sec. 6.2, the M/M/s model assumes that all *interarrival times* are independently and identically distributed according to an exponential distribution (i.e., the input process is Poisson), that all *service times* are independent and identically distributed according to another exponential distribution, and that the number of servers is s (any positive integer). Consequently, this model is just the special case of the birth-and-death process where the queueing system's *mean arrival rate* and *mean service rate per busy server* are constant (λ and μ, respectively) regardless of the state of the system. When the system has just a *single server* ($s = 1$), the implication is that the parameters for the birth-and-death process are $\lambda_n = \lambda$ ($n = 0, 1, 2, \ldots$) and $\mu_n = \mu$ ($n = 1, 2, \ldots$). The resulting *rate diagram* is shown in Fig. 6.5*a*.

(*a*) *Single-server case* ($s = 1$) $\lambda_n = \lambda$, for $n = 0, 1, 2, \ldots$
$\mu_n = \mu$, for $n = 1, 2, \ldots$

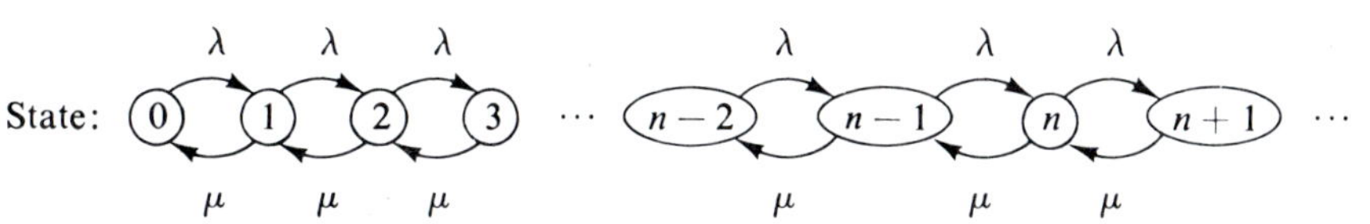

(*b*) *Multiple-server case* ($s > 1$) $\lambda_n = \lambda$, for $n = 0, 1, 2, \ldots$

$$\mu_n = \begin{cases} n\mu, & \text{for } n = 1, 2, \ldots, s \\ s\mu, & \text{for } n = s, s+1, \ldots \end{cases}$$

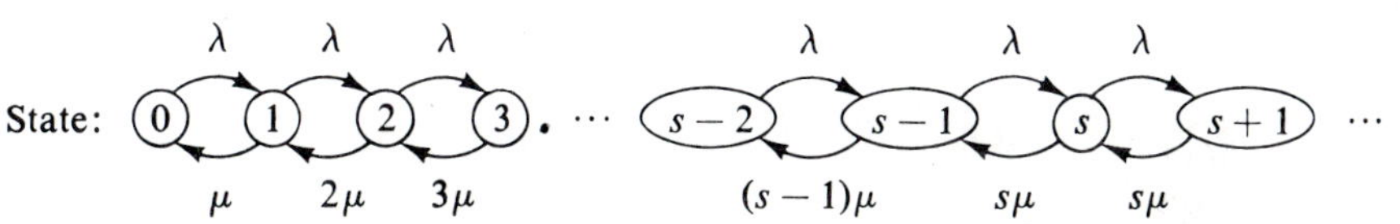

Figure 6.5 Rate diagrams for the M/M/s model.

However, when the system has *multiple servers* ($s > 1$), the μ_n cannot be expressed this simply. Keep in mind that μ_n represents the mean service rate for the *overall* queueing system (i.e., the mean rate at which service completions occur, so that customers leave the system) when there are n customers currently in the system. As mentioned for Property 4 of the exponential distribution (see Sec. 6.4), when the mean service rate per busy server is μ, the overall mean service rate for n busy servers must be $n\mu$. Therefore, $\mu_n = n\mu$ when $n \leq s$, whereas $\mu_n = s\mu$ when $n \geq s\mu$ so that all s servers are busy. The rate diagram for this case is shown in Fig. 6.5*b*.

When the maximum mean service rate ($s\mu$) exceeds the mean arrival rate (λ), that is, when

$$\rho = \frac{\lambda}{s\mu} < 1,$$

a queueing system fitting this model will eventually reach a steady-state condition. In this situation, the steady-state results derived in Sec. 6.5 for the general birth-and-death process are directly applicable. However, these results simplify considerably for this model and yield closed-form expressions for the P_n, L, L_q, and so forth, as shown next.

RESULTS FOR THE SINGLE-SERVER CASE (M/M/1): For $s = 1$, the C_n factors for the birth-and-death process reduce to

$$C_n = \left(\frac{\lambda}{\mu}\right)^n = \rho^n, \qquad \text{for } n = 1, 2, \ldots .$$

Therefore,

$$P_n = \rho^n P_0, \qquad \text{for } n = 1, 2, \ldots ,$$

where

$$P_0 = \frac{1}{1 + \sum_{n=1}^{\infty} \rho^n}$$

$$= \left(\sum_{n=0}^{\infty} \rho^n\right)^{-1}$$

$$= \left(\frac{1}{1-\rho}\right)^{-1}$$

$$= 1 - \rho.$$

Thus

$$P_n = (1-\rho)\rho^n, \qquad \text{for } n = 0, 1, 2, \ldots .$$

Consequently,

$$L = \sum_{n=0}^{\infty} n(1-\rho)\rho^n$$

$$= (1-\rho)\rho \sum_{n=0}^{\infty} \frac{d}{d\rho}(\rho^n)$$

$$= (1-\rho)\rho \frac{d}{d\rho}\left(\sum_{n=0}^{\infty} \rho^n\right)$$

$$= (1-\rho)\rho \frac{d}{d\rho}\left(\frac{1}{1-\rho}\right)$$

$$= \frac{\rho}{1-\rho} = \frac{\lambda}{\mu - \lambda}.$$

Similarly,
$$L_q = \sum_{n=1}^{\infty} (n-1)P_n$$
$$= L - 1(1 - P_0)$$
$$= \frac{\lambda^2}{\mu(\mu - \lambda)}.$$

When $\lambda \geq \mu$, so that the mean arrival rate exceeds the mean service rate, the preceding solution "blows up" (because the summation for computing P_0 diverges). For this case, the queue would "explode" and grow without bound.

Assuming again that $\lambda < \mu$, we now can derive the probability distribution of the waiting time in the system (*including* service) $\mathcal{W}$ for a random arrival when the queue discipline is first-come-first-served. If this arrival finds n customers already in the system, he will have to wait through $(n + 1)$ exponential service times, including his own. (For the customer currently being served, recall the lack of memory property for the exponential distribution discussed in Sec. 6.4.) Therefore, let $T_1, T_2, \ldots$ be independent service-time random variables having an exponential distribution with parameter μ, and let

$$S_{n+1} = T_1 + T_2 + \cdots + T_{n+1}, \qquad \text{for } n = 0, 1, 2, \ldots,$$

so that S_{n+1} represents the *conditional* waiting time given n customers already in the system. As discussed in Sec. 6.7, S_{n+1} is known to have an *Erlang distribution.*[1] Because the probability that the random arrival will find n customers in the system is P_n, it follows that

$$P\{\mathcal{W} > t\} = \sum_{n=0}^{\infty} P_n P\{S_{n+1} > t\},$$

which reduces after considerable algebraic manipulation to

$$P\{\mathcal{W} > t\} = e^{-\mu(1-\rho)t}, \qquad \text{for } t \geq 0.$$

The surprising conclusion is that $\mathcal{W}$ has an *exponential* distribution with parameter $\mu(1 - \rho)$. Therefore,

$$W = E(\mathcal{W}) = \frac{1}{\mu(1-\rho)}$$
$$= \frac{1}{\mu - \lambda}.$$

These results *include* service time in the waiting time. In some contexts (e.g., the *County Hospital* emergency room problem), the more relevant waiting time is just until service begins. Thus consider the waiting time in the queue (so *excluding* service time) $\mathcal{W}_q$ for a random arrival when the queue discipline is first-come-first-served. If this arrival finds *no* customers already in the system, he is served immediately, so

[1] Outside of queueing theory, this distribution is known as the *gamma distribution*.

that

$$P\{\mathcal{W}_q = 0\} = P_0 = 1 - \rho.$$

If he finds $n > 0$ customers already there instead, then he has to wait through n exponential service times until his own service begins, so that

$$\begin{aligned}
P\{\mathcal{W}_q > t\} &= \sum_{n=1}^{\infty} P_n P\{S_n > t\} \\
&= \sum_{n=1}^{\infty} (1 - \rho)\rho^n P\{S_n > t\} \\
&= \rho \sum_{n=0}^{\infty} P_n P\{S_{n+1} > t\} \\
&= \rho P\{\mathcal{W} > t\} \\
&= \rho e^{-\mu(1-\rho)t}, \qquad \text{for } t \geq 0.
\end{aligned}$$

By deriving the mean of this distribution [or applying either $L_q = \lambda W_q$ or $W_q = W - (1/\mu)$],

$$W_q = E(\mathcal{W}_q) = \frac{\lambda}{\mu(\mu - \lambda)}.$$

Results for the Multiple-Server Case ($s > 1$): When $s > 1$, the C_n factors become

$$C_n = \frac{(\lambda/\mu)^n}{n!}, \qquad \text{for } n = 1, 2, \ldots, s,$$

and

$$C_n = \frac{(\lambda/\mu)^s}{s!}\left(\frac{\lambda}{s\mu}\right)^{n-s} = \frac{(\lambda/\mu)^n}{s!\,s^{n-s}}, \qquad \text{for } n = s, s+1, \ldots.$$

Consequently, if $\lambda < s\mu$, then

$$\begin{aligned}
P_0 &= 1 \Bigg/ \left[\sum_{n=0}^{s-1} \frac{(\lambda/\mu)^n}{n!} + \frac{(\lambda/\mu)^s}{s!} \sum_{n=s}^{\infty} \left(\frac{\lambda}{s\mu}\right)^{n-s}\right] \\
&= 1 \Bigg/ \left[\sum_{n=0}^{s-1} \frac{(\lambda/\mu)^n}{n!} + \frac{(\lambda/\mu)^s}{s!} \frac{1}{1 - (\lambda/s\mu)}\right]
\end{aligned}$$

and

$$P_n = \begin{cases} \dfrac{(\lambda/\mu)^n}{n!} P_0, & \text{if } 0 \leq n \leq s \\[2ex] \dfrac{(\lambda/\mu)^n}{s!\,s^{n-s}} P_0, & \text{if } n \geq s. \end{cases}$$

Using the notation $\rho = \lambda/\mu s$,

$$
\begin{aligned}
L_q &= \sum_{n=s}^{\infty} (n - s)P_n \\
&= \sum_{j=0}^{\infty} jP_{s+j} \\
&= \sum_{j=0}^{\infty} j \frac{(\lambda/\mu)^s}{s!} \rho^j P_0 \\
&= P_0 \frac{(\lambda/\mu)^s}{s!} \rho \sum_{j=0}^{\infty} \frac{d}{d\rho}(\rho^j) \\
&= P_0 \frac{(\lambda/\mu)^s}{s!} \rho \frac{d}{d\rho}\left(\sum_{j=0}^{\infty} \rho^j\right) \\
&= P_0 \frac{(\lambda/\mu)^s}{s!} \rho \frac{d}{d\rho}\left(\frac{1}{1-\rho}\right) \\
&= \frac{P_0(\lambda/\mu)^s \rho}{s!(1-\rho)^2}.
\end{aligned}
$$

$$
W_q = \frac{L_q}{\lambda};
$$

$$
W = W_q + \frac{1}{\mu};
$$

$$
L = \lambda\left(W_q + \frac{1}{\mu}\right) = L_q + \frac{\lambda}{\mu}.
$$

Figures 6.6 and 6.7 show how P_0 and L change with ρ for various values of s.

The single-server method for finding the probability distribution of waiting times also can be extended to the multiple-server case. This yields[1] (for $t \geq 0$)

$$
P\{\mathcal{W} > t\} = e^{-\mu t}\left[1 + \frac{P_0(\lambda/\mu)^s}{s!(1-\rho)}\left(\frac{1 - e^{-\mu t(s-1-\lambda/\mu)}}{s - 1 - \lambda/\mu}\right)\right]
$$

and
$$
P\{\mathcal{W}_q > t\} = [1 - P\{\mathcal{W}_q = 0\}]e^{-s\mu(1-\rho)t},
$$

where
$$
P\{\mathcal{W}_q = 0\} = \sum_{n=0}^{s-1} P_n.
$$

If $\lambda \geq s\mu$, so that the mean arrival rate exceeds the maximum mean service rate, the queue grows without bound, so the preceding steady-state solutions are not applicable.

Example: For the *County Hospital* emergency room problem (see Sec. 6.1), the Management Engineer has concluded that the emergency cases arrive pretty much at

[1] When $s - 1 - \lambda/\mu = 0$, $(1 - e^{-\mu t(s-1-\lambda/\mu)})/(s - 1 - \lambda/\mu)$ should be replaced by μt.

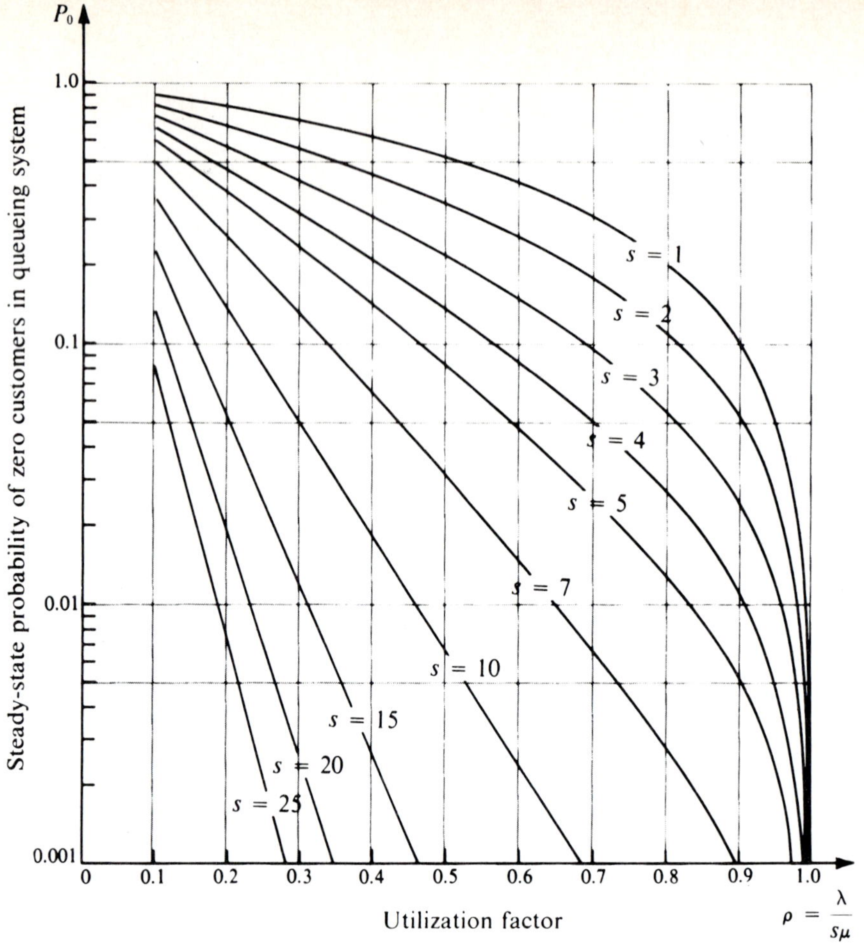

Figure 6.6 Values of P_0 for the M/M/s model (Sec. 6.6).

random (a *Poisson input process*), so that interarrival times have an exponential distribution. He also has concluded that the time spent by a doctor treating the cases approximately follows an *exponential distribution*. Therefore, he has chosen the M/M/s model for a preliminary study of this queueing system.

By projecting the available data for the early evening shift into next year, he estimates that patients will arrive at an *average* rate of one every half hour. A doctor requires an average of 20 minutes to treat each patient. Thus, using an hour as the unit of time,

$$\frac{1}{\lambda} = \frac{1}{2} \text{ hour per customer}$$

$$\frac{1}{\mu} = \frac{1}{3} \text{ hour per customer,}$$

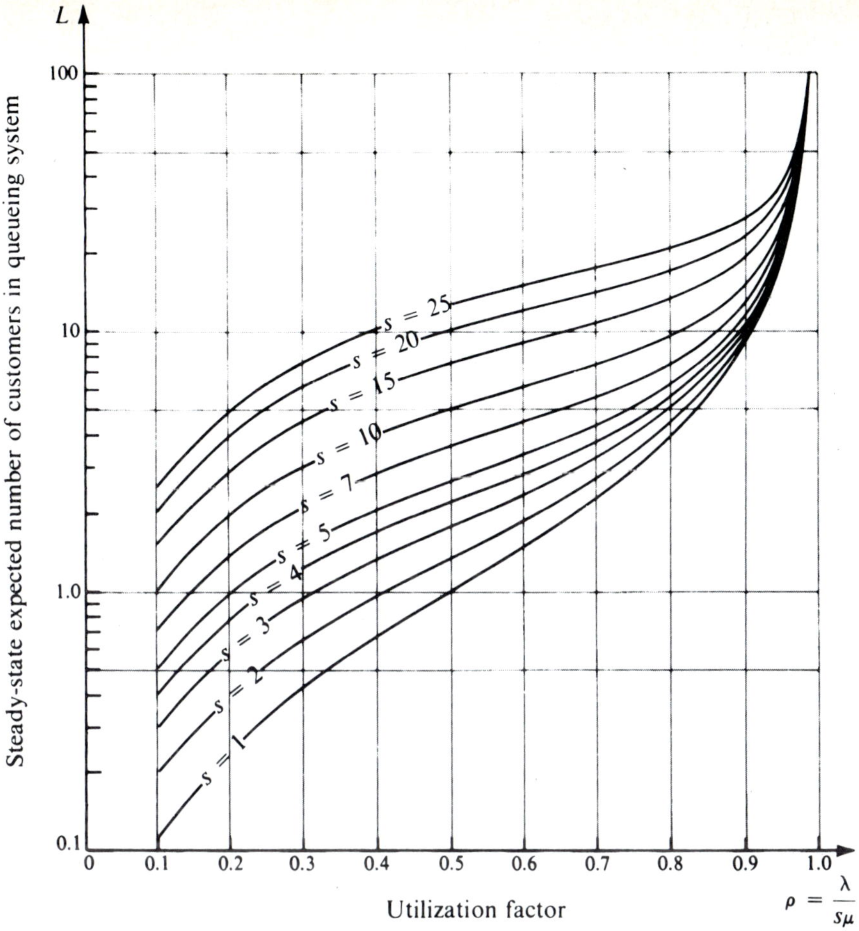

Figure 6.7 Values of L for the M/M/s model (Sec. 6.6).

so that

$$\lambda = 2 \text{ customers per hour}$$

$$\mu = 3 \text{ customers per hour.}$$

The two alternatives being considered are to continue having just one doctor during this shift ($s = 1$) or to add a second doctor ($s = 2$). In both cases,

$$\rho = \frac{\lambda}{s\mu} < 1,$$

so that the system should approach a steady-state condition. (Actually, because λ is somewhat different during other shifts, the system will never truly reach a steady-state condition, but the Management Engineer feels that steady-state results will provide a good approximation.) Therefore, the preceding equations are used to obtain the results shown in Table 6.2.

On the basis of these results, he tentatively concluded that a single doctor would be inadequate next year for providing the relatively prompt treatment needed in a

Table 6.2 **Steady-State Results from M/M/s Model for County Hospital Problem**

	$s = 1$	$s = 2$
ρ	$\frac{2}{3}$	$\frac{1}{3}$
P_0	$\frac{1}{3}$	$\frac{1}{2}$
P_1	$\frac{2}{9}$	$\frac{1}{3}$
P_n for $n \geq 2$	$\frac{1}{3}(\frac{2}{3})^n$	$(\frac{1}{3})^n$
L_q	$\frac{4}{3}$	$\frac{1}{12}$
L	2	$\frac{3}{4}$
W_q	$\frac{2}{3}$	$\frac{1}{24}$ (in hours)
W	1	$\frac{3}{8}$ (in hours)
$P\{\mathcal{W}_q > 0\}$	0.667	0.167
$P\{\mathcal{W}_q > \frac{1}{2}\}$	0.404	0.022
$P\{\mathcal{W}_q > 1\}$	0.245	0.003
$P\{\mathcal{W}_q > t\}$	$\frac{2}{3}e^{-t}$	$\frac{1}{6}e^{-4t}$
$P\{\mathcal{W} > t\}$	e^{-t}	$\frac{1}{2}e^{-3t}(3 - e^{-t})$

hospital emergency room. You will see later how the Management Engineer checked this conclusion by applying two other queueing models that provide better representations of the real queueing system in some ways.

The Finite Queue Variation of the M/M/s Model

We mentioned in the discussion of queues in Sec. 6.2 that queueing systems sometimes have a *finite queue*; i.e., the number of customers in the system is not permitted to exceed some specified number (denoted by K). Any customer that arrives while the queue is "full" is refused entry into the system and so leaves forever. From the viewpoint of the birth-and-death process, the *mean input rate* into the system becomes zero at these times. Therefore, the one modification needed in the M/M/s model to introduce a finite queue is to change the λ_n parameters to

$$\lambda_n = \begin{cases} \lambda, & \text{for } n = 0, 1, 2, \ldots, K - 1 \\ 0, & \text{for } n \geq K. \end{cases}$$

Because $\lambda_n = 0$ for some values of n, a queueing system that fits this model will eventually reach a steady-state condition.

This model commonly is labeled as M/M/s/K, where the presence of the fourth symbol distinguishes it from the M/M/s model. The single difference in the formulation of these two models is that K is finite for the M/M/s/K model and $K = \infty$ for the M/M/s model.

The usual physical interpretation for the M/M/s/K model is that there is only *limited waiting room* that will accommodate a maximum of K customers in the system. For example, for the *County Hospital* emergency room problem, this system actually would have a finite queue if there were only K cots for the patients and if the policy were to send arriving patients to another hospital whenever there were no empty cots.

Another possible interpretation is that arriving customers will leave and "take their business elsewhere" whenever they find too many customers (K) ahead of them in the system because they are not willing to incur a long wait. This *balking* phenomenon is quite common in *commercial service systems*. However, there are other models available (e.g., see Prob. 5) that fit this interpretation even better.

The *rate diagram* for this model is identical to that shown in Fig. 6.5 for the M/M/s model, *except* that it stops with state K.

Results for the Single-Server Case (M/M/1/K): For this case,

$$C_n = \begin{cases} \left(\dfrac{\lambda}{\mu}\right)^n = \rho^n, & \text{for } n = 1, 2, \ldots, K \\ 0, & \text{for } n > K. \end{cases}$$

Therefore, for $\rho \neq 1$,[1]

$$P_0 = \frac{1}{\sum_{n=0}^{K} (\lambda/\mu)^n}$$

$$= 1 \bigg/ \left[\frac{1 - (\lambda/\mu)^{K+1}}{1 - (\lambda/\mu)}\right]$$

$$= \frac{1 - \rho}{1 - \rho^{K+1}},$$

so that $$P_n = \left(\frac{1 - \rho}{1 - \rho^{K+1}}\right) \rho^n, \qquad \text{for } n = 0, 1, 2, \ldots, K.$$

Hence $$L = \sum_{n=0}^{K} nP_n$$

$$= \frac{1 - \rho}{1 - \rho^{K+1}} \rho \sum_{n=0}^{K} \frac{d}{d\rho} (\rho^n)$$

$$= \frac{1 - \rho}{1 - \rho^{K+1}} \rho \frac{d}{d\rho} \left(\sum_{n=0}^{K} \rho^n\right)$$

$$= \left(\frac{1 - \rho}{1 - \rho^{K+1}}\right) \rho \frac{d}{d\rho} \left(\frac{1 - \rho^{K+1}}{1 - \rho}\right)$$

$$= \rho \frac{-(K + 1)\rho^K + K\rho^{K+1} + 1}{(1 - \rho^{K+1})(1 - \rho)}$$

$$= \frac{\rho}{1 - \rho} - \frac{(K + 1)\rho^{K+1}}{1 - \rho^{K+1}}.$$

As usual (when $s = 1$),

$$L_q = L - (1 - P_0).$$

Notice that the preceding results do not require that $\lambda < \mu$ (i.e., that $\rho < 1$).

When $\rho < 1$, it can be verified that the second term in the final expression for L converges to zero as $K \to \infty$, so that *all* of the preceding results do indeed converge to the corresponding results given earlier for the M/M/1 model.

[1] If $\rho = 1$, then $P_n = 1/(K + 1)$ for $n = 0, 1, 2, \ldots, K$, so that $L = K/2$.

The waiting-time distributions can be derived by using the same reasoning as for the M/M/1 model (see Prob. 16). However, no simple expressions are obtained in this case, so computer calculations are required. Fortunately, even though $L \neq \lambda W$ and $L_q \neq \lambda W_q$ for the current model because the λ_n are not equal for all n (see the end of Sec. 6.2), the *expected* waiting times for customers entering the system still can be obtained directly from the expressions given at the end of Sec. 6.5,

$$W = \frac{L}{\bar{\lambda}}, \qquad W_q = \frac{L_q}{\bar{\lambda}},$$

where

$$\bar{\lambda} = \sum_{n=0}^{\infty} \lambda_n P_n = \sum_{n=0}^{K-1} \lambda P_n = \lambda(1 - P_K).$$

Results for the Multiple-Server Case ($s > 1$): Because this model does not allow more than K customers in the system, K is the maximum number of servers that could ever be used. Therefore, assume that $s \leq K$. In this case, C_n becomes

$$C_n = \begin{cases} \dfrac{(\lambda/\mu)^n}{n!}, & \text{for } n = 1, 2, \ldots, s \\ \dfrac{(\lambda/\mu)^s}{s!}\left(\dfrac{\lambda}{s\mu}\right)^{n-s} = \dfrac{(\lambda/\mu)^n}{s!s^{n-s}}, & \text{for } n = s, s+1, \ldots, K, \\ 0, & \text{for } n > K. \end{cases}$$

Hence

$$P_n = \begin{cases} \dfrac{(\lambda/\mu)^n}{n!} P_0, & \text{for } n = 1, 2, \ldots, s \\ \dfrac{(\lambda/\mu)^n}{s!s^{n-s}} P_0, & \text{for } n = s, s+1, \ldots, K \\ 0, & \text{for } n > K, \end{cases}$$

where

$$P_0 = 1 \Bigg/ \left[\sum_{n=1}^{s} \frac{(\lambda/\mu)^n}{n!} + \frac{(\lambda/\mu)^s}{s!} \sum_{n=s+1}^{K} \left(\frac{\lambda}{s\mu}\right)^{n-s}\right].$$

Adapting the derivation of L_q for the M/M/s model to this case (see Prob. 22) yields

$$L_q = \frac{P_0(\lambda/\mu)^s \rho}{s!(1-\rho)^2}[1 - \rho^{K-s} - (K-s)\rho^{K-s}(1-\rho)],$$

where $\rho = \lambda/s\mu$.[1] It can then be shown (see Prob. 33) that

$$L = \sum_{n=0}^{s-1} nP_n + L_q + s\left(1 - \sum_{n=0}^{s-1} P_n\right).$$

W and W_q are obtained from these quantities just as shown for the single-server case.

[1] If $\rho = 1$, it is necessary to apply L'Hôpital's rule twice to this expression for L_q. Otherwise, all of these multiple-server results hold for all $\rho > 0$. The reason that this queueing system can reach a steady-state condition even when $\rho \geq 1$ is that $\lambda_n = 0$ for $n \geq K$, so that the number of customers in the system cannot continue growing indefinitely.

Now assume that the only deviation from the M/M/s model is that (as defined in Sec. 6.2) the *input source* is *limited*; i.e., the size of the *calling population* is *finite*. For this case, let N denote the size of the calling population. Thus, when the number of customers in the queueing system is n ($n = 0, 1, 2, \ldots, N$), there are only ($N - n$) *potential* customers remaining in the input source.

The most important application of this model has been to the *machine repair problem,* where one or more repairmen are assigned the responsibility of maintaining in operational order a certain group of N machines by repairing each one that breaks down. The repairmen are considered to be individual servers in the queueing system if they work individually on different machines, whereas the entire crew is considered to be a single server if they work together on each machine. The machines constitute the calling population. Each one is considered to be a customer in the queueing system when it is down waiting to be repaired, whereas it is outside the queueing system while it is operational.

Note that each member of the calling population alternates between being *inside* and *outside* the queueing system. Therefore, the analog of the M/M/s model that fits this situation assumes that *each* member's *outside time* (i.e., the elapsed time from leaving the system until returning for the next time) has an *exponential distribution* with parameter λ. When n of the members are *inside,* and so ($N - n$) members are *outside*, the current probability distribution of the *remaining* time until the next arrival to the queueing system is the distribution of the *minimum* of the *remaining outside times* for the latter ($N - n$) members. Properties 2 and 3 for the exponential distribution imply that this distribution must be exponential with parameter $\lambda_n = (N - n)\lambda$. Hence this model is just the special case of the birth-and-death process that has the rate diagram shown in Fig. 6.8.

Because $\lambda_n = 0$ for $n = N$, any queueing system that fits this model will eventually reach a steady-state condition. The available steady-state results are sum-

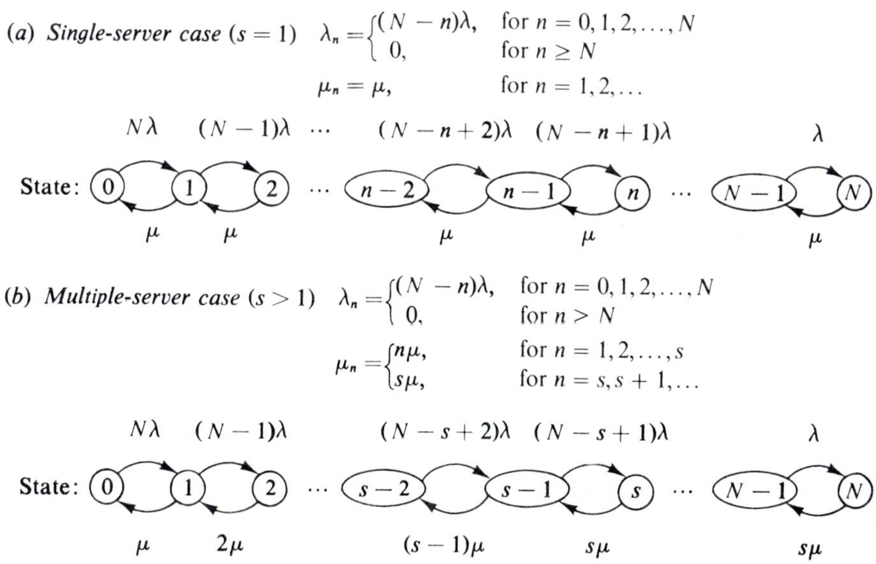

Figure 6.8 Rate diagrams for the finite calling population variation of the M/M/s model.

marized as follows:

Results for the Single-Server Case ($s = 1$): When $s = 1$, the C_n factors in Sec. 6.5 reduce to

$$C_n = N(N-1)\cdots(N-n+1)\left(\frac{\lambda}{\mu}\right)^n = \frac{N!}{(N-n)!}\left(\frac{\lambda}{\mu}\right)^n, \qquad \text{for } n \le N$$

$$= 0, \qquad \text{for } n > N,$$

for this model. Therefore,

$$P_0 = 1 \bigg/ \sum_{n=0}^{N} \left[\frac{N!}{(N-n)!}\left(\frac{\lambda}{\mu}\right)^n\right];$$

$$P_n = \frac{N!}{(N-n)!}\left(\frac{\lambda}{\mu}\right)^n P_0, \qquad \text{if } n = 1, 2, \ldots, N;$$

$$L_q = \sum_{n=1}^{N} (n-1)P_n,$$

which can be reduced to

$$L_q = N - \frac{\lambda + \mu}{\lambda}(1 - P_0);$$

$$L = \sum_{n=0}^{N} nP_n = L_q + (1 - P_0)$$

$$= N - \frac{\mu}{\lambda}(1 - P_0).$$

Finally,

$$W = \frac{L}{\bar{\lambda}}, \qquad W_q = \frac{L_q}{\bar{\lambda}},$$

where

$$\bar{\lambda} = \sum_{n=0}^{\infty} \lambda_n P_n = \sum_{n=0}^{N} (N-n)\lambda P_n = \lambda(N - L).$$

Results for the Multiple-Server Case ($s > 1$): For $N \ge s > 1$,

$$C_n = \begin{cases} \dfrac{N!}{(N-n)!n!}\left(\dfrac{\lambda}{\mu}\right)^n, & \text{for } n = 1, 2, \ldots, s \\ \dfrac{N!}{(N-n)!s!s^{n-s}}\left(\dfrac{\lambda}{\mu}\right)^n, & \text{for } n = s, s+1, \ldots, N \\ 0, & \text{for } n > N. \end{cases}$$

Hence

$$P_n = \begin{cases} \dfrac{N!}{(N-n)!n!}\left(\dfrac{\lambda}{\mu}\right)^n P_0, & \text{if } 0 \le n \le s \\ \dfrac{N!}{(N-n)!s!s^{n-s}}\left(\dfrac{\lambda}{\mu}\right)^n P_0, & \text{if } s \le n \le N \\ 0, & \text{if } n > N, \end{cases}$$

where $P_0 = 1 \Bigg/ \left[\sum_{n=0}^{s-1} \frac{N!}{(N-n)!n!} \left(\frac{\lambda}{\mu}\right)^n + \sum_{n=s}^{N} \frac{N!}{(N-n)!s!s^{n-s}} \left(\frac{\lambda}{\mu}\right)^n \right].$

Finally,

$$L_q = \sum_{n=s}^{N} (n-s)P_n,$$

$$L = \sum_{n=0}^{s-1} nP_n + L_q + s\left(1 - \sum_{n=0}^{s-1} P_n\right),$$

which then yield W and W_q by the same equations as in the single-server case.

Extensive tables of computational results are available[1] for this model for both the single-server and multiple-server cases.

For both cases, it has been shown[2] that the preceding formulas for P_n and P_0 (and so for L_q, L, W, and W_q) *also* hold for a generalization of this model. In particular, we can *drop* the assumption that the times spent *outside* the queueing system by the members of the calling population have an *exponential distribution*, even though this takes the model outside the realm of the birth-and-death process. As long as these times are identically distributed with mean $1/\lambda$ (and the assumption of exponential service times still holds), these outside times can have *any* probability distribution!

A Model with State-Dependent Service Rate and/or Arrival Rate

All the models thus far have assumed that the mean service rate is always a constant, regardless of how many customers are in the system. Unfortunately, this rate often is not a constant in real queueing systems, particularly when the servers are people. When there is a large backlog of work (i.e., a long queue), it is quite likely that such servers will tend to work faster than they do when the backlog is small or nonexistent. This increase in the service rate may result merely because the servers increase their effort when they are under the pressure of a long queue. However, it may also result partly because the quality of the service is compromised or because assistance is obtained on certain service phases.

Given that the mean service rate does increase as the queue size increases, it would be desirable to develop a theoretical model that seems to describe the pattern by which it increases. This model not only should be a reasonable approximation of the actual pattern but also should be simple enough to be practical for implementation. One such model is formulated next. (You have the flexibility to formulate many similar models within the framework of the birth-and-death process.) We then show how the same results apply when the *arrival rate* is affected by the queue size in an analogous way.

FORMULATION FOR THE SINGLE-SERVER CASE ($s = 1$): Let

$$\mu_n = n^c \mu_1, \qquad \text{for } n = 1, 2, \ldots,$$

where n = number of customers in system,

μ_n = mean service rate when there are n customers in system,

[1] Peck, L. G., and R. N. Hazelwood: *Finite Queueing Tables*, Wiley, New York, 1958.

[2] Bunday, B. D., and R. E. Scraton: "The G/M/r Machine Interference Model," *European Journal of Operational Research*, **4**:399–402, 1980.

$1/\mu_1$ = expected "normal" service time—expected time to service customer when that customer is the only one in system,

c = "pressure coefficient"—positive constant that indicates degree to which service rate of system is affected by system state.

Thus, selecting $c = 1$, for example, hypothesizes that the mean service rate is directly proportional to n; $c = \frac{1}{2}$ implies that the mean service rate is proportional to the square root of n; and so on. The preceding queueing models in this section have implicitly assumed that $c = 0$.

Now assume additionally that the queueing system has a Poisson input with $\lambda_n = \lambda$ (for $n = 0, 1, 2, \ldots$) and exponential service times with μ_n as just given. This case is now a special case of the birth-and-death process, where

$$C_n = \frac{(\lambda/\mu_1)^n}{(n!)^c}, \qquad \text{for } n = 1, 2, \ldots .$$

Thus all the steady-state results given in Sec. 6.5 are applicable to this model. (A steady-state condition always can be reached when $c > 0$.) Unfortunately, analytical expressions are not available for the summations involved. However, nearly exact values of P_0 and L have been tabulated[1] for various values of c and λ/μ_1 by summing a finite number of terms on a computer. A small portion of these results also is shown in Figs. 6.9 and 6.10.

A queueing system may react to a long queue by decreasing the arrival rate instead of increasing the service rate. (The arrival rate may be decreased, for example, by diverting some of the customers requiring service to another service facility.) The corresponding model for describing mean arrival rates for this case is to let

$$\lambda_n = (n + 1)^{-b}\lambda_0, \qquad \text{for } n = 0, 1, 2, \ldots ,$$

where b is a constant whose interpretation is analogous to that for c. The C_n values for the birth-and-death process with these λ_n (and with $\mu_n = \mu$ for $n = 1, 2, \ldots$) are *identical* to those just shown (replacing λ by λ_0) for the state-dependent service rate model when $c = b$ and $\lambda/\mu_1 = \lambda_0/\mu$, so the steady-state results also are the same.

A more general model that combines these two patterns can also be used when both the mean arrival rates and the mean service rates are state-dependent. Thus let

$$\mu_n = n^a\mu_1, \qquad \text{for } n = 1, 2, \ldots ,$$

$$\lambda_n = (n + 1)^{-b}\lambda_0, \qquad \text{for } n = 0, 1, 2, \ldots .$$

Once again, the C_n values for the birth-and-death process with these parameters are identical to those shown for the state-dependent service rate model when $c = a + b$ and $\lambda/\mu_1 = \lambda_0/\mu_1$, so the tabulated steady-state results actually are applicable to this general model.

Formulation for the Multiple-Server Case ($s > 1$): To generalize this combined model further to the multiple-server case, it would seem natural to have the μ_n and λ_n vary with the number of customers *per server* (n/s) in essentially the same

[1] Conway, Richard W., and William L. Maxwell: "A Queueing Model with State Dependent Service Rate," *Journal of Industrial Engineering*, **12**:132–136, 1961.

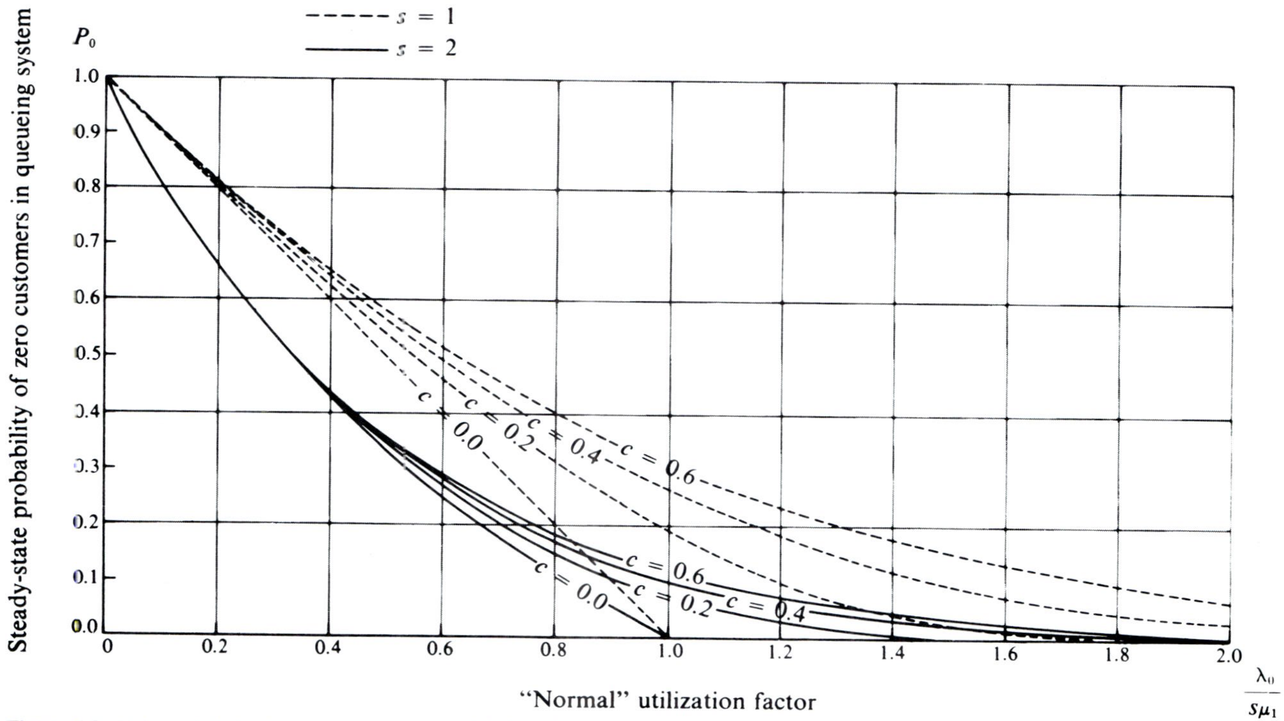

Figure 6.9 Values of P_0 for the state-dependent model (Sec. 6.6).

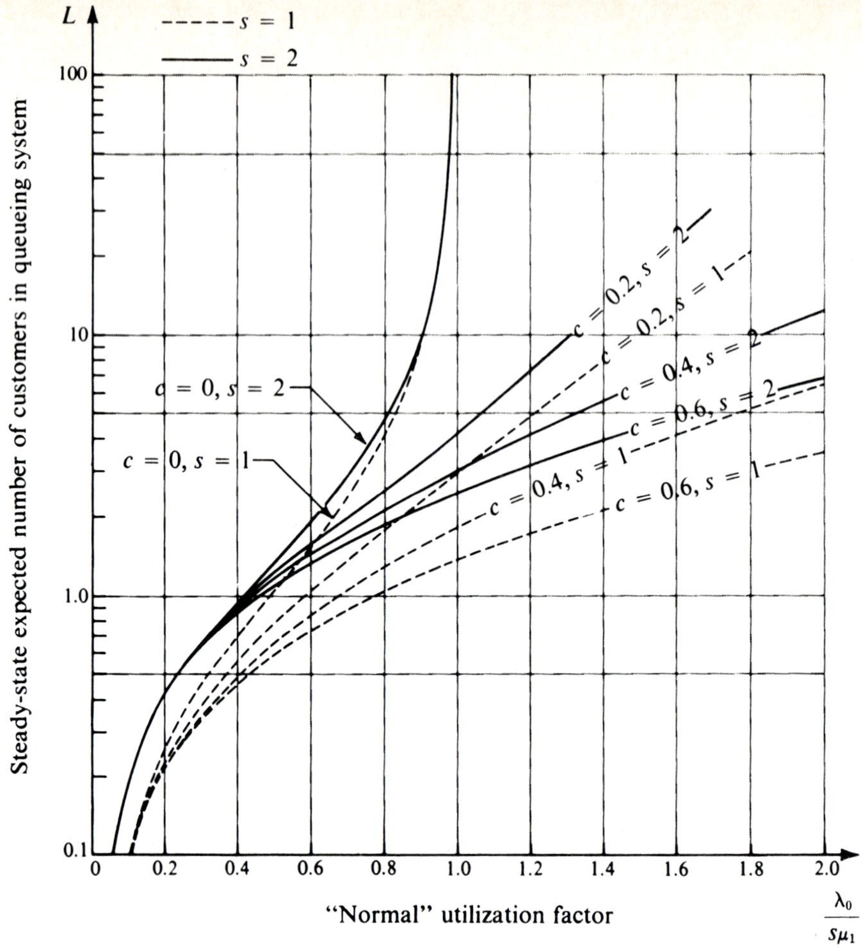

Figure 6.10 Values of L for the state-dependent model (Sec. 6.6).

way they vary with n for the single-server case. Thus let

$$\mu_n = \begin{cases} n\mu_1, & \text{if } n \le s \\ \left(\dfrac{n}{s}\right)^a s\mu_1, & \text{if } n \ge s, \end{cases}$$

$$\lambda_n = \begin{cases} \lambda_0, & \text{if } n \le s - 1 \\ \left(\dfrac{s}{n+1}\right)^b \lambda_0, & \text{if } n \ge s - 1. \end{cases}$$

Therefore, the birth-and-death process with these parameters has

$$C_n = \begin{cases} \dfrac{(\lambda_0/\mu_1)^n}{n!}, & \text{for } n = 1, 2, \ldots, s \\ \dfrac{(\lambda_0/\mu_1)^n}{s!(n!/s!)^c s^{(1-c)(n-s)}}, & \text{for } n = s, s+1, \ldots, \end{cases}$$

where $c = a + b$.

Computational results for P_0, L_q, and L have been tabulated[1] for various values of c, (λ_0/μ_1), and s. Some of these results also are given in Figs. 6.9 and 6.10.

Example: After gathering additional data for the *County Hospital* emergency room, the Management Engineer found that the time a doctor spends with a patient tends to decrease as the number of patients waiting increases. Part of the explanation is simply that the doctor works faster, but the main reason is that more of the treatment is turned over to a nurse for completion. The pattern of the μ_n (the mean rate at which a doctor treats patients while there are a total of n patients to be treated in the emergency room) seems to fit reasonably the *state-dependent service rate model* presented here. Therefore, the Management Engineer has decided to apply this model.

The new data indicate that the average time a doctor spends treating a patient is 24 minutes if no other patients are waiting, whereas this average becomes 12 minutes when each doctor has six patients (so five are waiting their turn). Thus, with a single doctor on duty,

$$\mu_1 = 2\tfrac{1}{2} \text{ customers per hour,}$$
$$\mu_6 = 5 \text{ customers per hour.}$$

Therefore, the pressure coefficient c (or a in the general model) must satisfy the relationship

$$\mu_6 = 6^c \mu_1, \qquad \text{so } 6^c = 2.$$

Using logarithms to solve for c yields $c = 0.4$. Because $\lambda = 2$ from before, this solution for c completes the specification of parameter values for this model.

To compare the two alternatives of having *one* doctor ($s = 1$) or *two* doctors ($s = 2$) on duty, the Management Engineer developed the various measures of performance shown in Table 6.3. The values of P_0, L, and (for $s = 2$) L_q were obtained directly from the tabulated results for this model. (Except for this L_q, you can approximate the same values from Figs. 6.9 and 6.10.) These values were then used to calculate

$$P_1 = C_1 P_0,$$

$$L_q = L - (1 - P_0), \qquad \text{if } s = 1,$$

$$[L_q = (L - P_1) - 2(1 - P_0 - P_1), \qquad \text{if } s = 2]$$

$$W_q = \frac{L_q}{\lambda}, \qquad W = \frac{L}{\lambda},$$

$$P\{\mathscr{W}_q > 0\} = 1 - \sum_{n=0}^{s-1} P_n.$$

The fact that some of the results in Table 6.3 do not deviate substantially from those in Table 6.2 reinforce the tentative conclusion that a single doctor will be inadequate next year.

[1] Hillier, F. S., R. W. Conway, and W. L. Maxwell: "A Multiple Server Queueing Model with State Dependent Service Rate," *Journal of Industrial Engineering*, **15**:153–157, 1964.

Table 6.3 **Steady-State Results from State-Dependent Service Rate Model for County Hospital Problem**

	$s = 1$	$s = 2$	
$\frac{\lambda}{s\mu_1}$	0.8	0.4	
$\frac{\lambda}{s\mu_{6s}}$	0.4	0.2	
P_0	0.367	0.440	
P_1	0.294	0.352	
L_q	0.618	0.095	
L	1.251	0.864	
W_q	0.309	0.048	(in hours)
W	0.626	0.432	(in hours)
$P\{W_q > 0\}$	0.633	0.208	

6.7 Queueing Models Involving Nonexponential Distributions

Because all the queueing theory models in the preceding section (except for one generalization) are based on the birth-and-death process, both their interarrival and service times are required to have *exponential* distributions. As discussed in Sec. 6.4, this type of probability distribution has many convenient properties for queueing theory, but it provides a reasonable fit for only certain kinds of queueing systems. In particular, the assumption of exponential *interarrival times* implies that arrivals occur *randomly* (a Poisson input process), which is a reasonable approximation in many situations but *not* when the arrivals are carefully scheduled or regulated. Furthermore, the *actual service-time distribution* frequently deviates greatly from the exponential form, particularly when the service requirements of the customers are quite similar. Therefore, it is important to have available other queueing models that use alternative distributions.

Unfortunately, the mathematical analysis of queueing models with nonexponential distributions is much more difficult. However, it has been possible to obtain some useful results for a few such models. This analysis is beyond the level of this book, but in this section we shall summarize the models and describe their results.

The M/G/1 Model

As introduced in Sec. 6.2, the M/G/1 model assumes that the queueing system has a *single server* and a *Poisson input process* (exponential interarrival times) with a *fixed* mean arrival rate λ. As usual, it is assumed that the customers have *independent* service times with the *same* probability distribution. However, *no* restrictions are imposed on what this service-time distribution can be. In fact, it is only necessary to know (or estimate) the mean $1/\mu$ and variance σ^2 of this distribution.

Any such queueing system can eventually reach a *steady-state condition* if $\rho = \lambda/\mu < 1$. The readily available steady-state results[1] for this general model are the following:

[1] A recursion formula also is available for calculating the probability distribution of the number of customers in the system; see Hordijk, A., and H. C. Tijms: *Statistica Neerlandica*, **30**:97–100, 1976.

$$P_0 = 1 - \rho,$$

$$L_q = \frac{\lambda^2\sigma^2 + \rho^2}{2(1 - \rho)},$$

$$L = \rho + L_q,$$

$$W_q = \frac{L_q}{\lambda},$$

$$W = W_q + \frac{1}{\mu}.$$

Considering the complexity involved in analyzing a model that permits *any* service-time distribution, it is remarkable that such a simple formula can be obtained for L_q. This formula is one of the most important results in queueing theory because of its ease of use and the prevalence of M/G/1 queueing systems in practice. This equation for L_q (or its counterpart for W_q) commonly is referred to as the **Pollaczek-Khintchine formula**, named after the Frenchman and the Russian who derived it more than 50 years ago.

For any fixed expected service time $1/\mu$, notice that L_q, L, W_q, and W all increase as σ^2 is increased. This result is important because it indicates that the consistency of the server has a major bearing on the performance of the service facility—not just his average speed. This key point is illustrated in the next subsection.

When the service-time distribution is exponential, $\sigma^2 = 1/\mu^2$, and the preceding results will reduce to the corresponding results for the M/M/1 model given at the beginning of Sec. 6.6.

The complete flexibility in the service-time distribution provided by this model is extremely useful, so it is unfortunate that efforts to derive similar results for the *multiple-server* case have been unsuccessful. However, some multiple-server results have been obtained for the important special cases described by the following two models.

The M/D/s Model

When the service consists of essentially the same routine task to be performed for all customers, there tends to be little variation in the service time required. The M/D/s model often provides a reasonable representation for this kind of situation, because it assumes that all service times actually equal some fixed *constant* (the *degenerate* service-time distribution) and that we have a *Poisson* input process with a fixed mean arrival rate λ.

When there is just a *single* server, the M/D/1 model is just the special case of the M/G/1 model where $\sigma^2 = 0$, so that the *Pollaczek-Khintchine formula* reduces to

$$L_q = \frac{\rho^2}{2(1 - \rho)},$$

where L, W_q, and W are obtained from L_q as just shown. Notice that this L_q and W_q are exactly *half* as large as for the exponential service-time case of Sec. 6.6 (the M/M/1 model), where $\sigma^2 = 1/\mu^2$, so decreasing σ^2 can *greatly* improve the measures of performance of a queueing system.

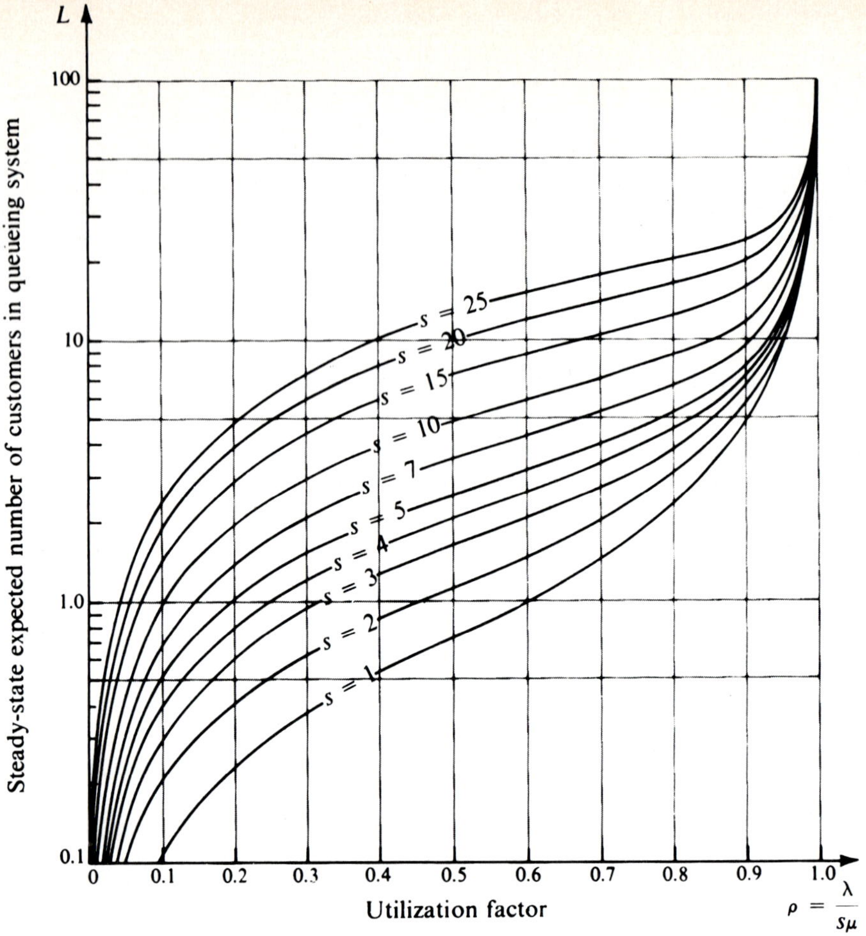

Figure 6.11 Values of L for the M/D/s model (Sec. 6.7).

For the multiple-server version of this model (M/D/s), a complicated method is available[1] for deriving the steady-state probability distribution of the number of customers in the system and its mean (assuming $\rho = \lambda/s\mu < 1$). However, these results have been tabulated for numerous cases,[2] and the means (L) also are given graphically in Fig. 6.11.

The M/E_k/s Model

The M/D/s model assumes *zero* variation in the service times ($\sigma = 0$), whereas the *exponential* service-time distribution assumes a very large variation ($\sigma = 1/\mu$). Between these two rather extreme cases lies a long middle ground ($0 < \sigma < 1/\mu$), where most *actual* service-time distributions fall. Another kind of theoretical service-time distribution that fills this middle ground is the **Erlang distribution** (named after the founder of queueing theory).

[1] See Prabhu, N.U.: *Queues and Inventories*, pp. 32–34, Wiley, New York, 1965; also see pp. 344–346 in Selected Reference 3.

[2] Hillier, F. S., and O. S. Yu, with D. Avis, L. Fossett, F. Lo, and M. Reiman: *Queueing Tables and Graphs*, Elsevier North-Holland, New York, 1981.

The probability density function for the Erlang distribution is

$$f(t) = \frac{(\mu k)^k}{(k-1)!} t^{k-1} e^{-k\mu t}, \qquad \text{for } t \geq 0,$$

where μ and k are strictly positive parameters of the distribution, and k is further restricted to be *integer*. (Except for this integer restriction, and the definition of the parameters, this distribution is *identical* to the *gamma* distribution.) Its mean and standard deviation are

$$\text{Mean} = \frac{1}{\mu},$$

$$\text{Standard deviation} = \frac{1}{\sqrt{k}} \frac{1}{\mu}.$$

Thus k is the parameter that specifies the degree of variability of the service times relative to the mean. It usually is referred to as the *shape parameter*.

The Erlang distribution is a very important distribution in queueing theory for two reasons. To describe the first one, suppose that $T_1, T_2, \ldots, T_k$ are k independent random variables with an identical *exponential* distribution whose mean is $1/k\mu$. Then their sum,

$$T = T_1 + T_2 + \cdots + T_k,$$

has an *Erlang* distribution with parameters μ and k. The discussion of the exponential distribution in Sec. 6.4 suggested that the time required to perform certain kinds of tasks might well have an exponential distribution. However, the total service required by a customer may involve the server performing not just one specific task but a sequence of k tasks. If the respective tasks have an identical exponential distribution for their duration, the total service time would have an Erlang distribution, which would be the case, for example, if the server must perform the *same* exponential task k times for each customer.

The Erlang distribution also is very useful because it is a large (two-parameter) family of distributions permitting only nonnegative values. Hence empirical service-time distributions can usually be reasonably approximated by an Erlang distribution. In fact, both the *exponential* and the *degenerate* (constant) distributions are special cases of the Erlang distribution, with $k = 1$ and $k = \infty$, respectively. Intermediate values of k provide intermediate distributions with mean $= 1/\mu$, mode $= (k-1)/k\mu$, and variance $= 1/k\mu^2$, as suggested by Fig. 6.12.

Now consider the $M/E_k/1$ model, which is just the special case of the M/G/1 model where service times have an *Erlang* distribution with shape parameter $= k$. Applying the *Pollaczek-Khintchine formula* with $\sigma^2 = 1/k\mu^2$ (and the accompanying results given for M/G/1) yields

$$L_q = \frac{\lambda^2/k\mu^2 + \rho^2}{2(1-\rho)} = \frac{1+k}{2k} \frac{\lambda^2}{\mu(\mu-\lambda)},$$

$$W_q = \frac{1+k}{2k} \frac{\lambda}{\mu(\mu-\lambda)},$$

$$W = W_q + \frac{1}{\mu},$$

$$L = \lambda W.$$

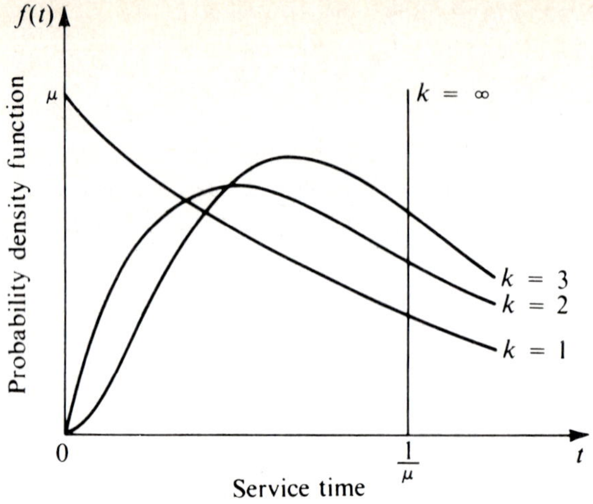

Figure 6.12 A family of Erlang distributions with constant mean $1/\mu$.

With *multiple* servers ($M/E_k/s$), the relationship of the Erlang distribution to the exponential distribution just described can be exploited to formulate a *modified* birth-and-death process (continuous time Markov chain) in terms of individual exponential service phases (k per customer) rather than complete customers. However, it has not been possible to derive a general steady-state solution (when $\rho = \lambda/s\mu < 1$) for the probability distribution of the number of customers in the system as we did in Sec. 6.5. Instead, advanced theory is required to solve individual cases numerically. Once again, these results have been obtained and tabulated for numerous cases.[1] The means (L) also are given graphically in Fig. 6.13 for some cases where $s = 2$.

Models without a Poisson Input

All the queueing models presented thus far have assumed a Poisson input process (exponential interarrival times). However, this assumption is violated if the arrivals are scheduled or regulated in some way that prevents them from occurring randomly, in which case another model is needed.

As long as the service times have an *exponential* distribution with a fixed parameter, three such models are readily available. These models are obtained by merely *reversing* the assumed distributions of the *interarrival* and *service times* in the preceding three models. Thus the first new model (GI/M/s) imposes no restriction on what the *interarrival time* distribution can be. In this case, there are some steady-state results available[2] (particularly in regard to waiting-time distributions) for *both* the single-server and multiple-server versions of the model, but these results are not nearly as convenient as the simple expressions given for the M/G/1 model. The second new model (D/M/s) assumes that all interarrival times equal some fixed *constant*, which would represent a queueing system where arrivals are *scheduled* at regular intervals. The third new model ($E_k/M/s$) assumes an *Erlang* interarrival time distribution, which provides a middle ground between *regularly scheduled* (constant) and *completely random* (exponential) arrivals. Extensive computational results have

[1] Ibid.

[2] For example, see pp. 304–320 of Selected Reference 3.

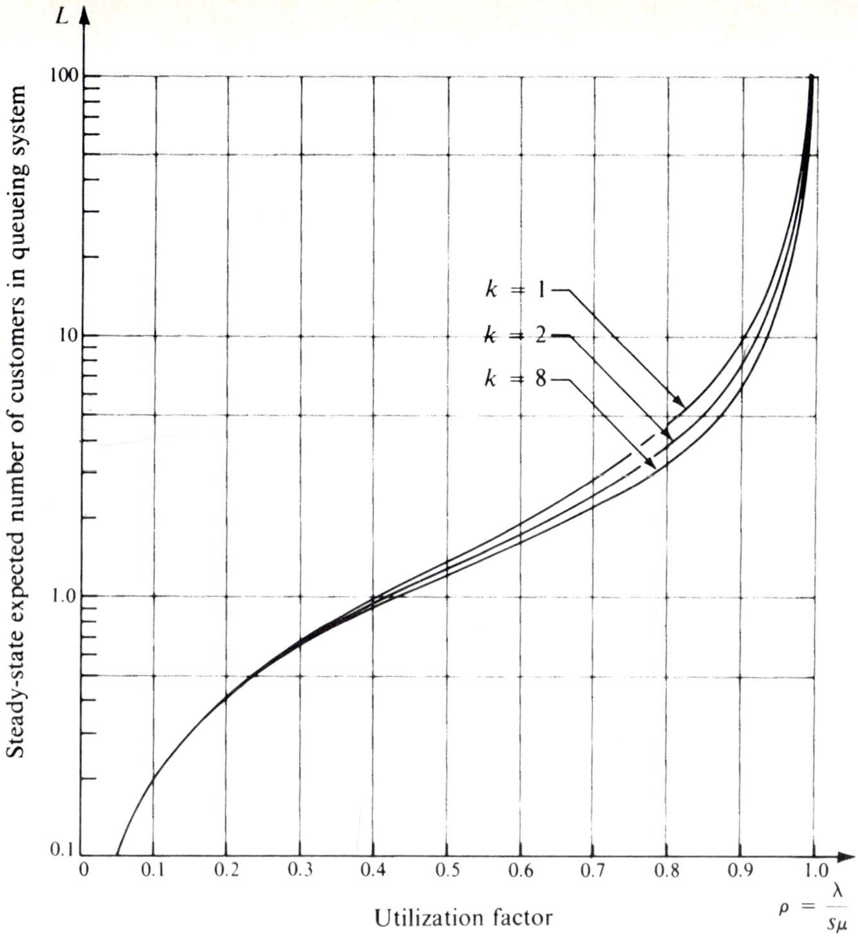

Figure 6.13 Values of L for the $M/E_k/2$ model (Sec. 6.7).

been tabulated[1] for these latter two models, including the values of L given graphically in Figs. 6.14 and 6.15.

If *neither* the interarrival times nor the service times for a queueing system have an *exponential* distribution, then there are three additional queueing models for which computational results also are available.[2] One of these models ($E_m/E_k/s$) assumes an *Erlang* distribution for *both* these times. The other two models ($E_k/D/s$ and $D/E_k/s$) assume that *one* of these times has an *Erlang* distribution and the *other* time equals some fixed *constant*.

Other Models

Although you have seen in this section a large number of queueing models that involve nonexponential distributions, we have far from exhausted the list. For example, another distribution that occasionally is used for either interarrival times or service times

[1] F. S. Hillier and O. S. Yu, op. cit.

[2] Ibid.

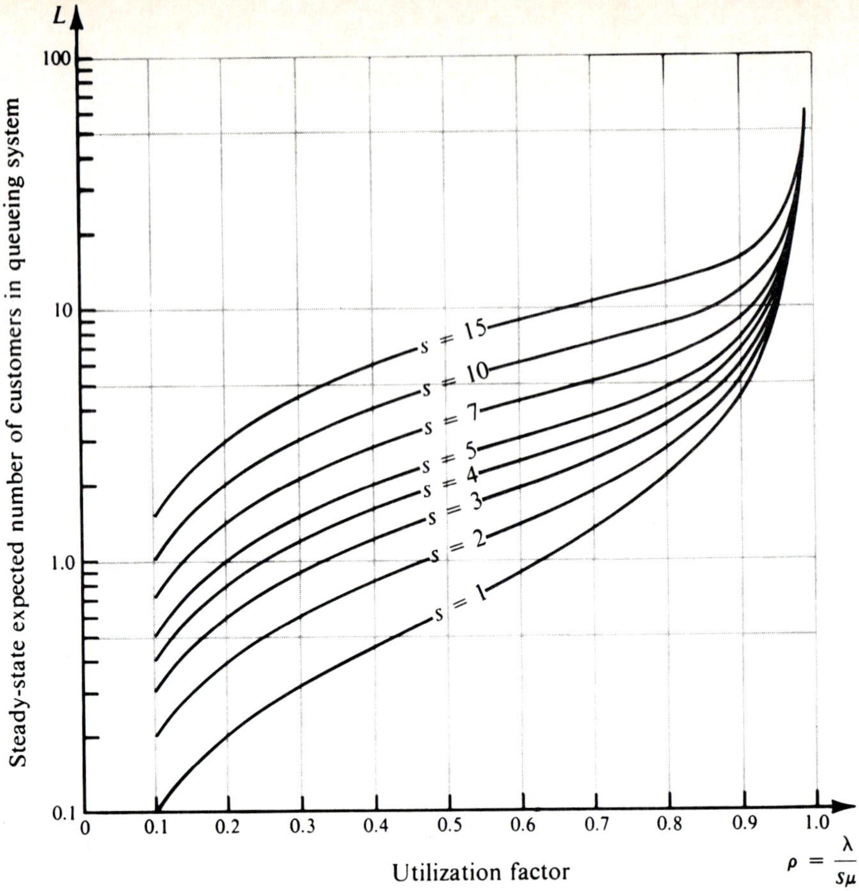

Figure 6.14 Values of L for the D/M/s model (Sec. 6.7).

is the **hyperexponential distribution**. The key characteristic of this distribution is that, even though only nonnegative values are allowed, its standard deviation, σ, actually is larger than its mean, $1/\mu$. This characteristic is in contrast to the *Erlang* distribution, where $\sigma < 1/\mu$ in *every* case except $k = 1$ (exponential distribution), which has $\sigma = 1/\mu$. To illustrate a typical situation where $\sigma > 1/\mu$ can occur, we suppose that the service involved in the queueing system is the repair of some kind of machine or vehicle. If many of the repairs turn out to be routine (small service times) but occasional repairs require an extensive overhaul (very large service times), then the standard deviation of service times will tend to be quite large relative to the mean, in which case the hyperexponential distribution may be used to represent the service-time distribution.

Another family of distributions coming into general use consists of **phase-type distributions** (some of which also are called *generalized Erlangian distributions*). These distributions are obtained by breaking down the total time into a number of *phases*, each having an *exponential* distribution, where the parameters of these exponential distributions may be different and the phases may be either *in series* or *in parallel* (or both). A group of phases being *in parallel* means that the process randomly selects *one* of the phases to go through each time according to specified probabilities.

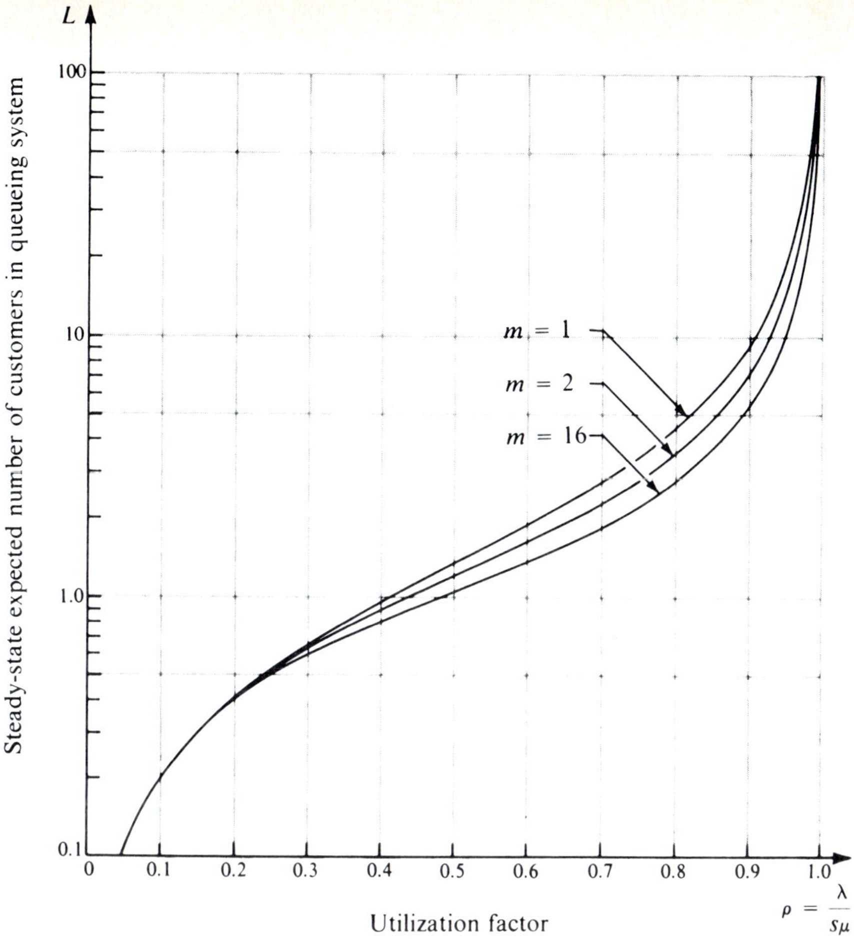

Figure 6.15 Values of L for the $E_k/M/2$ model (Sec. 6.7).

This approach is, in fact, how the *hyperexponential* distribution is derived, so this distribution is a special case of the phase-type distributions. Another special case is the *Erlang* distribution, which has the restrictions that all of its k phases are *in series* and that these phases have the *same* parameter for their exponential distributions. Removing these restrictions means that phase-type distributions in general can provide considerably more flexibility than the Erlang distribution in fitting the *actual* distribution of interarrival times or service times observed in a real queueing system. This flexibility is especially valuable when using the actual distribution directly in the model is not analytically tractable, whereas the ratio of the *mean* to the *standard deviation* for the actual distribution does not closely match the available ratios ($\sqrt{k}$ for $k = 1, 2, \ldots$) for the Erlang distribution.

Because they are built up from combinations of exponential distributions, queueing models using phase-type distributions still can be represented by a *continuous time Markov chain*. This Markov chain generally will have an infinite number of states, so solving for the steady-state distribution of the state of the system requires solving an infinite system of linear equations that has a relatively complicated structure. Solving such a system is far from a routine thing, but recent theoretical advances have enabled us to solve these queueing models numerically in some cases. An extensive tabulation of these results for models with various phase-type distributions (including the hypergeometric distribution) now is available.[1]

6.8 A Priority-Discipline Queueing Model

Priority-discipline queueing models are those where the queue discipline is based on a *priority system*. Thus the order in which members of the queue are selected for service is on the basis of their assigned priorities.

Many real queueing systems fit these priority-discipline models much more closely than other available models. Rush jobs are taken ahead of other jobs, and important customers may be given precedence over others. Therefore, the use of priority-discipline models often provides a very welcome refinement over the more usual queueing models.

Unfortunately, the inclusion of priorities makes the mathematical analysis fairly complicated, so that only limited results are available. Almost all these results are for the single-server case. However, usable results are available for one *multiple-server* model. This model assumes that there are *N priority classes* (class 1 has the highest priority and class N the lowest) and that, whenever a server becomes free to begin serving a new customer from the queue, the one selected is that member of the *highest* priority class represented in the queue who has waited longest. In other words, customers are selected to begin service in the order of their priority classes, but on a first-come-first-served basis within each priority class. A *Poisson* input process and *exponential* service times are assumed for each priority class. The model also makes the somewhat restrictive assumption that the mean service time is the *same* for all priority classes. However, it does permit the mean arrival rate to differ among the priority classes.

Notice that if this distinction between customers in different priority classes is ignored, this model actually fits the M/M/s model studied in Sec. 6.6. Therefore, when we count just the *total* number of customers in the system, the steady-state distribution given there still applies here. Consequently, the formulas for L and L_q also carry over, as do the expected waiting-time results (by Little's formula), W and W_q, for a randomly selected customer. What changes is the *distribution* of waiting times, which was derived in Sec. 6.6 under the assumption of a first-come-first-served queue discipline. With a priority discipline, this distribution has a much larger *variance*, because the waiting times of customers in the highest priority classes tend to be much smaller than under a first-come-first-served discipline, whereas the waiting times in the lowest priority classes tend to be much larger. By the same token, the breakdown of the total number of customers in the system tends to be disproportionately weighted toward the lower priority classes. But this condition is just the reason for imposing priorities on the queueing system in the first place. We want to *improve* the *measures of performance* for each of the higher priority classes at the expense of performance for the lower priority classes. To determine how much improvement is being made, we need to obtain such measures as *expected waiting time in the system*

[1] Seelen, L. P., H. C. Tijms, and M. H. Van Hoorn: *Tables for Multi-Server Queues*, North-Holland, Amsterdam, 1985.

and *expected number of customers in the system* for the individual priority classes. We obtain these measures next for this priority-discipline model.

First consider the case of **nonpreemptive priorities**, where a customer being served cannot be ejected back into the queue (preempted) if a higher priority customer enters the queueing system. Therefore, once a server has begun serving a customer, the service must be completed without interruption. Under this assumption for the priority-discipline model, W_k, the steady-state *expected waiting time in the system* (including service time) for a member of priority class k, is

$$W_k = \frac{1}{A \cdot B_{k-1} \cdot B_k} + \frac{1}{\mu}, \qquad \text{for } k = 1, 2, \ldots, N,$$

where $$A = s! \left(\frac{s\mu - \lambda}{r^s} \right) \sum_{j=0}^{s-1} \frac{r^j}{j!} + s\mu,$$

$$B_0 = 1,$$

$$B_k = 1 - \frac{\sum_{i=1}^{k} \lambda_i}{s\mu}, \qquad \text{for } k = 1, 2, \ldots, N,$$

and s = number of servers,

μ = mean service rate per busy server,

λ_i = mean arrival rate for priority class i, for $i = 1, 2, \ldots, N$,

$$\lambda = \sum_{i=1}^{N} \lambda_i,$$

$$r = \frac{\lambda}{\mu}.$$

(These results assume that

$$\sum_{i=1}^{k} \lambda_i < s\mu,$$

so that priority class k can reach a steady-state condition.) *Little's formula* still applies to individual priority classes, so L_k, the steady-state *expected number of members of priority class k in the queueing system* (including those being served), is

$$L_k = \lambda_k W_k, \qquad \text{for } k = 1, 2, \ldots, N.$$

To determine the *expected waiting time in the queue* (excluding service time) for priority class k, merely subtract $1/\mu$ from W_k; the corresponding *expected queue length* is again obtained by multiplying by λ_k. For the special case where $s = 1$, the expression for A reduces to $A = \mu^2/\lambda$.

Now consider the case of **preemptive priorities**, whereby the lowest priority customer being served is *preempted* (ejected back into the queue) whenever a higher priority customer enters the queueing system. A server is thereby freed to begin servicing the new arrival immediately. (When a server does succeed in *finishing* a service, the next customer to begin receiving service is selected in just the same way as described earlier, so a preempted customer normally will get back into service again and, after enough tries, eventually finish.) If the other assumptions of the preceding model are retained, this preemption feature changes the *total* expected waiting time in the system (including total service time) to

$$W_k = \frac{1/\mu}{B_{k-1} \cdot B_k}, \qquad \text{for } k = 1, 2, \ldots, N,$$

for the *single-server* case ($s = 1$). When $s > 1$, the W_k can be calculated by an iterative procedure that will be illustrated soon by the County Hospital example. The L_k just defined continue to satisfy the relationship

$$L_k = \lambda_k W_k, \qquad \text{for } k = 1, 2, \ldots, N.$$

The corresponding results for the *queue* (excluding customers in service) also can be obtained from W_k and L_k as just described for the case of nonpreemptive priorities. Because of the lack of memory property of the exponential distribution (see Sec. 6.4), preemptions do not affect the service process (occurrence of service completions) in any way. The expected *total* service time for any customer still is $1/\mu$. This lack of memory property also implies that we don't need to worry about defining the point at which service begins when a preempted customer returns to service; the distribution of the *remaining* service time *always* is the same. (For any *other* service-time distribution, it becomes important to distinguish between *preemptive-resume* systems, where service for a preempted customer resumes at the point of interruption, and *preemptive-repeat* systems, where service must start at the beginning again.)

Some results are also available[1] for a few other single-server, priority-discipline models involving other service-time distributions and/or unequal expected service times.

EXAMPLE: For the *County Hospital* emergency room problem, the Management Engineer has noticed that the patients are not treated on a first-come-first-served basis. Rather, the admitting nurse seems to divide the patients into roughly three categories: (1) *critical* cases, where prompt treatment is vital for survival; (2) *serious* cases, where early treatment is important to prevent further deterioration; and (3) *stable* cases, where treatment can be delayed without adverse medical consequences. Patients are then treated in this order of priority, where those in the same category are normally taken on a first-come-first-served basis. A doctor will interrupt treatment of a patient if a new case in a higher priority category arrives. Approximately 10 percent of the patients fall into the first category, 30 percent into the second, and 60 percent into the third. Because the more serious cases will be sent into the hospital for further care after receiving emergency treatment, the average treatment time by a doctor in the emergency room actually does not differ greatly among these categories.

The Management Engineer has decided to use the priority-discipline queueing model just described as a reasonable representation of this queueing system, where the three categories of patients constitute the three priority classes in the model. Because treatment is interrupted by the arrival of a higher priority case, the *preemptive priorities* version of this model is the appropriate one. Given the previously available data ($\mu = 3$ and $\lambda = 2$), the preceding percentages yield $\lambda_1 = 0.2$, $\lambda_2 = 0.6$, $\lambda_3 = 1.2$. Table 6.4 gives the resulting expected waiting times in the queue (so *excluding*

[1] See Selected Reference 4.

Table 6.4 **Steady-State Results from Priority-Discipline Model for County Hospital Problem**

	Preemptive Priorities		*Nonpreemptive Priorities*	
	$s = 1$	$s = 2$	$s = 1$	$s = 2$
A	—	—	4.5	36
B_1	0.933	—	0.933	0.967
B_2	0.733	—	0.733	0.867
B_3	0.333	—	0.333	0.667
$W_1 - \frac{1}{\mu}$	0.024	0.00037	0.238	0.029
$W_2 - \frac{1}{\mu}$	0.154	0.00793	0.325	0.033
$W_3 - \frac{1}{\mu}$	1.033	0.06542	0.889	0.048

treatment time) in *hours* for the respective priority classes[1] when there is one doctor ($s = 1$) or two doctors ($s = 2$) on duty. (The corresponding results for the *non-preemptive priorities* version of the model also are given in Table 6.4 to show the effect of preempting.)

These preemptive priority results for $s = 2$ were obtained as follows. Because the waiting times for priority class 1 customers are completely unaffected by the presence of customers in lower priority classes, W_1 would be the same for any other values of λ_2 and λ_3, including $\lambda_2 = 0$, $\lambda_3 = 0$. Therefore, W_1 must equal W for the corresponding *one-class* model (the M/M/s model in Sec. 6.6) with $s = 2$, $\mu = 3$, and $\lambda = \lambda_1 = 0.2$, which yields

$$\begin{aligned} W_1 &= W \qquad \text{for } \lambda = 0.2 \\ &= 0.33370, \end{aligned}$$

so

$$W_1 - \frac{1}{\mu} = 0.33370 - 0.33333 = 0.00037.$$

Now consider the first *two* priority classes. Again note that customers in these classes are completely unaffected by lower priority classes (just priority class 3 in this case), which can therefore be ignored in the analysis. Let $\overline{W}_{1-2}$ be the expected waiting time in the system (so including service time) of a *random arrival* in *either* of these two classes, so the probability is $\lambda_1/(\lambda_1 + \lambda_2) = \frac{1}{4}$ that this arrival is in class 1 and $\lambda_2/(\lambda_1 + \lambda_2) = \frac{3}{4}$ that it is in class 2. Therefore,

$$\overline{W}_{1-2} = \tfrac{1}{4}W_1 + \tfrac{3}{4}W_2.$$

Furthermore, because *expected* waiting time is the same for *any* queue discipline, $\overline{W}_{1-2}$ must also equal W for the M/M/s model in Sec. 6.6, with $s = 2$, $\mu = 3$, and $\lambda = \lambda_1 + \lambda_2 = 0.8$, which yields

$$\begin{aligned} \overline{W}_{1-2} &= W \qquad \text{for } \lambda = 0.8 \\ &= 0.33937. \end{aligned}$$

[1] Note that these expected times can no longer be interpreted as the expected time before treatment begins when $k > 1$, because treatment may be interrupted at least once, causing additional waiting time, before being completed.

Combining these facts gives

$$W_2 = \frac{4}{3}\left[0.33937 - \frac{1}{4}(0.33370)\right] = 0.34126.$$

$$\left(W_2 - \frac{1}{\mu} = 0.00793.\right)$$

Finally, let $\overline{W}_{1-3}$ be the expected waiting time in the system (so including service time) for a *random arrival* in *any* of the three priority classes, so the probabilities are 0.1, 0.3, and 0.6 that it is in classes 1, 2, and 3, respectively. Therefore,

$$\overline{W}_{1-3} = 0.1W_1 + 0.3W_2 + 0.6W_3.$$

Furthermore, $\overline{W}_{1-3}$ must also equal W for the M/M/s model in Sec. 6.6, with $s = 2$, $\mu = 3$, and $\lambda = \lambda_1 + \lambda_2 + \lambda_3 = 2$, so that (from Table 6.2)

$$\overline{W}_{1-3} = W \qquad \text{for } \lambda = 2$$

$$= 0.375.$$

Consequently,

$$W_3 = \frac{1}{0.6}[0.375 - 0.1(0.33370) - 0.3(0.34126)]$$

$$= 0.39875.$$

$$\left(W_3 - \frac{1}{\mu} = 0.06542.\right)$$

The corresponding W_q results for the M/M/s model in Sec. 6.6 also could have been used in exactly the same way to derive the $[W_k - (1/\mu)]$ quantities directly.

When $s = 1$, the $[W_k - (1/\mu)]$ values in Table 6.4 for the preemptive priorities case indicate that providing just a single doctor would cause *critical* cases to wait about $1\frac{1}{2}$ minutes (0.024 hour) on the average, *serious* cases to wait more than 9 minutes, and *stable* cases to wait more than an hour. (Contrast these results with the average wait of $W_q = \frac{2}{3}$ hour for all patients that was obtained in Table 6.2 under the first-come-first-served queue discipline.) However, these values represent *statistical expectations*, so some patients have to wait considerably longer than the average for their priority class. This wait would not be tolerable for the critical and serious cases, where a few minutes can be vital. By contrast, the $s = 2$ results in Table 6.4 (preemptive priorities case) indicate that adding a second doctor would virtually eliminate waiting for all but the stable cases. Therefore, the Management Engineer recommended that there be two doctors on duty in the emergency room during the early evening hours next year. The Board of Directors for County Hospital adopted this recommendation and simultaneously raised the charge for using the emergency room!

6.9 Queueing Networks

Thus far we have considered only queueing systems that have a *single* service facility with one or more servers. However, queueing systems encountered in operations research studies are sometimes actually *queueing networks*, i.e., networks of service facilities where customers must receive service at some or all of these facilities. For example, orders being processed through a job shop must be routed through a sequence of machine groups (service facilities). It is therefore necessary to study the entire network to obtain such information as expected total waiting time, expected number of customers in the entire system, and so forth.

Because of the importance of queueing networks, research into this area has been very active. However, this is a difficult area, and most of the work has been confined to cases with a Poisson input process and exponential service times. Many of the results that have been obtained[1] have necessarily been quite involved and unsuitable for general routine use.

However, there is available one simple result that is of such fundamental importance for queueing networks that this finding and its implications warrant special attention here. This fundamental result is the following *equivalence property* for the *input process* of arriving customers and the *output process* of departing customers for certain queueing systems.

> **Equivalence property:** Assume that a service facility with s servers and an infinite queue has a Poisson input with parameter λ and the same exponential service-time distribution with parameter μ for each server (the M/M/s model), where $s\mu > \lambda$. Then the steady-state *output* of this service facility is also a Poisson process with parameter λ.[2]

Notice that this property makes no assumption about the type of queue discipline used. Whether it be first-come-first-served, random, or even a priority discipline as in Sec. 6.8, the served customers will leave the service facility according to a Poisson process. The crucial implication of this fact for queueing networks is that if these customers must then go to another service facility for further service, this second facility *also* will have a *Poisson* input. With an exponential service-time distribution, the *equivalence property* will hold for this facility as well, which can then provide a *Poisson input* for a third facility, etc. We discuss the consequences for two basic kinds of networks next.

Infinite Queues in Series

Suppose that customers must all receive service at a series of m service facilities in a fixed sequence. Assume that each facility has an *infinite queue* (no limitation on the number of customers allowed in the queue), so that the series of facilities form a system of *infinite queues in series*. Assume further that the customers arrive at the first facility according to a Poisson process with parameter λ and that each facility i ($i = 1, 2, \ldots, m$) has the same exponential service-time distribution with parameter μ_i for its servers, where $s\mu_i > \lambda$. It then follows from the *equivalence property* that (under steady-state conditions) *each* service facility has a Poisson input with parameter λ. Therefore, the elementary M/M/s model of Sec. 6.6 (or its priority-discipline counterpart in Sec. 6.8) can be used to analyze each service facility independently of the others!

[1] For a survey, see Lemoine, Austin J.: "Networks of Queues—A Survey of Equilibrium Analysis," *Management Science*, **24**(4):464–481, 1977.

[2] For a proof, see Burke, P. J.: "The Output of a Queueing System," *Operations Research*, **4**(6):699–704, 1956.

Being able to use the M/M/s model to obtain all measures of performance for each facility independently, rather than analyzing interactions between facilities, is a tremendous simplification. For example, the probability of having n customers at a given facility is given by the formula for P_n given in Sec. 6.6 for the M/M/s model. The *joint probability* of n_1 customers at facility 1, n_2 customers at facility 2, etc., then, is the *product* of the individual probabilities obtained in this simple way. In particular, this joint probability can be expressed as

$$P\{(N_1, N_2, \ldots, N_m) = (n_1, n_2, \ldots, n_m)\} = P_{n_1} P_{n_2} \cdots P_{n_m}.$$

(This simple form for the solution is called the **product form solution**.) Similarly, the expected total waiting time and the expected number of customers in the entire system can be obtained by merely summing the corresponding quantities obtained at the respective facilities.

Unfortunately, the equivalence property and its implications do not hold for the case of *finite* queues discussed in Sec. 6.6. This case is actually quite important in practice, because there is often a definite limitation on the queue length in front of service facilities in networks. For example, only a small amount of buffer storage space is typically provided in front of each facility (station) in a production-line system. For such systems of *finite queues in series*, no simple *product form solution* is available. The facilities must be analyzed jointly instead, and only limited results have been obtained.

Jackson Networks

Systems of infinite queues in series are not the only queueing networks where the M/M/s model can be used to analyze each service facility independently of the others. Another prominent kind of network with this property (a product form solution) is the *Jackson network*, named after the individual who first characterized the network and showed that this property holds.[1]

The characteristics of a Jackson network are the same as assumed above for the system of infinite queues in series, except now the customers visit the facilities in different orders (and may not visit them all). For each facility, its arriving customers come from *both* outside the system (according to a Poisson process) and from the other facilities. These characteristics are summarized below.

> A **Jackson network** is a system of m service facilities where facility i ($i = 1, 2, \ldots, m$) has:
>
> 1. An infinite queue;
> 2. Customers arriving from outside the system according to a Poisson input process with parameter a_i; and
> 3. s_i servers with the same exponential service-time distribution with parameter μ_i.
>
> A customer leaving facility i is routed next to facility j ($j = 1, 2, \ldots, m$, but $j \neq i$) with probability p_{ij} or departs the system with probability
>
> $$q_i = 1 - \sum_{\substack{j=1 \\ j \neq i}}^{m} p_{ij}.$$

[1] See Jackson, James R.: "Jobshop-Like Queueing Systems," *Management Science*, **10**(1):131–142, 1963.

Any such network has the following key property.

> Under steady-state conditions, each facility j ($j = 1, 2, \ldots, m$) in a Jackson network behaves as if it were an *independent* M/M/s queueing system with arrival rate
>
> $$\lambda_j = a_j + \sum_{\substack{i=1 \\ i \neq j}}^{m} \lambda_i p_{ij},$$
>
> where $s_j \mu_j > \lambda_j$.

This key property cannot be *proven* directly from the *equivalence property* this time (the reasoning would become circular), but its *intuitive underpinning* is still provided by the latter property. The intuitive viewpoint (not quite technically correct) is that, for each facility i, its input processes from the various sources (outside and other facilities) are *independent Poisson processes*, so the *aggregate* input process is Poisson with parameter λ_i (Property 6 in Sec. 6.4). The *equivalence property* then says that the *aggregate output* process for facility i must be Poisson with parameter λ_i. Disaggregating this output process (Property 6 again), the process for customers going from facility i to facility j must be Poisson with parameter $\lambda_i p_{ij}$. This process becomes one of the Poisson *input* processes for facility j, thereby helping to maintain the series of Poisson processes in the overall system.

The above equation for obtaining λ_j is based on the fact that λ_i is the *departure rate* as well as the arrival rate for all customers using facility i. Because p_{ij} is the proportion of customers departing from facility i that go next to facility j, the *rate* at which customers from facility i arrive at facility j is $\lambda_i p_{ij}$. Summing this product over all $i \neq j$, and then adding this sum to a_j, gives the *total arrival rate* to facility j from all sources.

To calculate λ_j from this equation requires knowing the λ_i for $i \neq j$, but these λ_i also are unknowns given by the corresponding equations. Therefore, the procedure is to solve *simultaneously* for $\lambda_1, \lambda_2, \ldots, \lambda_m$ by obtaining the simultaneous solution of the entire system of linear equations for λ_j for $j = 1, 2, \ldots, m$.

To illustrate these calculations, consider a Jackson network with three service facilities that have the parameters shown in Table 6.5. Plugging into the formula for λ_j for $j = 1, 2, 3$, the resulting system of equations is

$$\begin{aligned}
\lambda_1 &= 1 \qquad\qquad\qquad + 0.1\lambda_2 + 0.4\lambda_3 \\
\lambda_2 &= 4 + 0.6\lambda_1 \qquad\qquad\quad + 0.4\lambda_3 \\
\lambda_3 &= 3 + 0.3\lambda_1 + 0.3\lambda_2.
\end{aligned}$$

(Reason through each equation to see why it gives the total arrival rate.) The simultaneous solution for this system is

$$\lambda_1 = 5, \qquad \lambda_2 = 10, \qquad \lambda_3 = 7\tfrac{1}{2}.$$

Table 6.5 **Data for the Example of a Jackson Network**

				p_{ij}		
Facility j	s_j	μ_j	a_j	$i = 1$	$i = 2$	$i = 3$
$j = 1$	1	10	1	—	0.1	0.4
$j = 2$	2	10	4	0.6	—	0.4
$j = 3$	1	10	3	0.3	0.3	—

Given this simultaneous solution, each of the three service facilities now can be analyzed *independently* by using the formulas for the M/M/s model given in Sec. 6.6. For example, to obtain the distribution of the number of customers, $N_i = n_i$, at facility i, note that

$$\rho_i = \frac{\lambda_i}{s_i \mu_i} = \begin{cases} \frac{1}{2}, & \text{for } i = 1 \\ \frac{1}{2}, & \text{for } i = 2 \\ \frac{3}{4}, & \text{for } i = 3. \end{cases}$$

Plugging these values (and the parameters in Table 6.5) into the formula for P_n gives

$$P_{n_1} = \tfrac{1}{2}\left(\tfrac{1}{2}\right)^{n_1}, \qquad \text{for facility 1}$$

$$P_{n_2} = \begin{cases} \frac{1}{3}, & \text{for } n_2 = 0 \\ \frac{1}{3}, & \text{for } n_2 = 1, \\ \frac{1}{3}\left(\frac{1}{2}\right)^{n_2 - 1}, & \text{for } n_2 \geq 2 \end{cases} \qquad \text{for facility 2}$$

$$P_{n_3} = \tfrac{1}{4}\left(\tfrac{3}{4}\right)^{n_3}, \qquad \text{for facility 3.}$$

The *joint probability* of (n_1, n_2, n_3) then is given simply by the *product form solution*,

$$P\{(N_1, N_2, N_3) = (n_1, n_2, n_3)\} = P_{n_1} P_{n_2} P_{n_3}.$$

In a similar manner, the *expected* number of customers L_i at facility i can be calculated from Sec. 6.6 as

$$L_1 = 1, \qquad L_2 = \tfrac{4}{3}, \qquad L_3 = 3.$$

The expected *total* number of customers in the entire system then is

$$L = L_1 + L_2 + L_3 = 5\tfrac{1}{3}.$$

Obtaining W, the expected *total* waiting time in the system (including service times) for a customer, is a little trickier. You cannot simply add the expected waiting times at the respective facilities, because a customer does not necessarily visit each facility exactly once. However, Little's formula can still be used, where the system arrival rate λ is the sum of the arrival rates *from outside* to the facilities, $\lambda = a_1 + a_2 + a_3 = 8$. Thus

$$W = \frac{L}{a_1 + a_2 + a_3} = \frac{2}{3}.$$

In conclusion, we should point out that there do exist other (more complicated) kinds of queueing networks as well, where the individual service facilities can be analyzed independently from the others. In fact, finding queueing networks with a *product form solution* has been the Holy Grail for research on queueing networks. One source of additional information is Selected Reference 5.

6.10 Conclusions

Queueing systems are prevalent throughout society. The adequacy of these systems can have an important effect on the quality of life and productivity.

Queueing theory studies queueing systems by formulating mathematical models of their operation and then using these models to derive measures of performance.

This analysis provides vital information for effectively designing queueing systems that achieve an appropriate balance between the cost of providing a service and the cost associated with waiting for that service.

This chapter presented the most basic models of queueing theory for which particularly useful results are available. However, many other interesting models could be considered if space permitted. In fact, *several thousand* research papers formulating and/or analyzing queueing models have already appeared in the technical literature, and many more are being published each year!

The *exponential distribution* plays a fundamental role in queueing theory for representing the distribution of interarrival and service times, because this assumption enables us to represent the queueing system as a *continuous time Markov chain*. For the same reason, *phase-type distributions* such as the *Erlang distribution*, where the total time is broken down into individual phases having an exponential distribution, are very useful. Useful analytical results have been obtained for only a relatively few queueing models making other assumptions.

Priority-discipline queueing models are useful for the common situation where some categories of customers are given priority over others for receiving service.

Another common situation is where customers must receive service at several different service facilities. Models for queueing networks are gaining widespread use for such situations. This is an area of especially active ongoing research.

When no tractable model that provides a reasonable representation of the queueing system under study is available, a common approach is to obtain relevant performance data by developing a computer program for *simulating* the operation of the system. This technique is discussed in Chap. 15.

Chapter 7 describes how queueing theory can be used to help design effective queueing systems.

SELECTED REFERENCES

1. Asmussen, S.: *Applied Probability and Queues,* Wiley, New York, 1987.
2. Cooper, Robert B.: *Introduction to Queueing Theory,* 2d ed., Elsevier North-Holland, New York, 1981.
3. Gross, Donald, and Carl M. Harris: *Fundamentals of Queueing Theory,* 2d ed., Wiley, New York, 1985.
4. Jaiswell, N. K.: *Priority Queues,* Academic Press, New York, 1968.
5. Kelly, F. P.: *Reversibility and Stochastic Networks,* Wiley, New York, 1979.
6. Kleinrock, Leonard: *Queueing Systems, Vol. I: Theory,* Wiley, New York, 1975.
7. Neuts, M. F.: *Matrix-Geometric Solutions in Stochastic Models,* Johns Hopkins University Press, Baltimore, 1981.

PROBLEMS[1]

1.* Consider a typical barber shop. Demonstrate that it is a queueing system by describing its components.

2. Identify the customers and the servers in the queueing system in each of the following situations:

(*a*) The checkout stand in a grocery store.
(*b*) A fire station.

[1] See also the end of Chap. 7 for problems involving the application of queueing theory.

(*c*) The toll booth for a bridge.
(*d*) A bicycle repair shop.
(*e*) A shipping dock.
(*f*) A group of semiautomatic machines assigned to one operator.
(*g*) The materials-handling equipment in a factory area.
(*h*) A plumbing shop.
(*i*) A job shop producing custom orders.
(*j*) A secretarial typing pool.

3. Suppose that a queueing system has *two* servers, an *exponential* interarrival time distribution with a mean of 2 hours, and an *exponential* service-time distribution with a mean of 2 hours. Furthermore, a customer has just arrived at 12:00 noon.

(*a*) What is the probability that the *next* arrival will come before 1:00 P.M.? Between 1:00 and 2:00 P.M.? After 2:00 P.M.?
(*b*) Suppose that *no* additional customers arrive before 1:00 P.M. Now what is the probability that the next arrival will come between 1:00 and 2:00 P.M.?
(*c*) What is the probability that the *number* of arrivals between 1:00 and 2:00 P.M. will be *zero*? *One*? *Two or more*?
(*d*) Suppose that both servers are serving customers at 1:00 P.M. What is the probability that *neither* customer will have service completed before 2:00 P.M.? Before 1:10 P.M.? Before 1:01 P.M.?

4.* The jobs to be performed on a particular machine arrive according to a Poisson input process with a mean rate of two per hour. Suppose that the machine breaks down and will require 1 hour to be repaired. What is the probability that the number of new jobs that will arrive during this time is (*a*) zero, (*b*) two, (*c*) five or more?

5. A service station has one gasoline pump. Cars wanting gasoline arrive according to a *Poisson* process at a mean rate of 15 per hour. However, if the pump already is being used, these potential customers may *balk* (drive on to another service station). In particular, if there are n cars already at the service station, the probability that an arriving potential customer will balk is $n/3$ for $n = 1, 2, 3$. The time required to service a car has an *exponential* distribution with a mean of 4 minutes.

(*a*) Construct the rate diagram for this queueing system.
(*b*) Develop the balance equations.
(*c*) Solve these equations to find the steady-state probability distribution of the number of cars at the station. Verify that this solution is the same as that given by the general solution for the *birth-and-death process*.
(*d*) Find the expected waiting time (including service) for those cars that stay.

6. Consider a variation of the M/M/1 model where customers *renege* (leave the queueing system without being served) if their waiting time in the queue grows too large. In particular, assume that the time each customer is willing to wait in the queue before reneging has an exponential distribution with a mean of $1/\theta$.

(*a*) Construct the rate diagram for this queueing system.
(*b*) Develop the balance equations.

7.* A certain small grocery store has a single checkout stand with a full-time cashier. Customers arrive at the stand "randomly" (i.e., a Poisson input process) at a mean rate of 30 per hour. When there is only one customer at the stand, he is processed by the cashier alone, with an expected service time of 1.5 minutes. However, the stock boy has been given standard instructions that whenever there is more than one customer at the stand, he is to help the cashier by bagging the groceries. This help reduces the expected time required to process a customer to 1 minute. In both cases, the service-time distribution is exponential.

(*a*) Construct the rate diagram for this queueing system.

(*b*) What is the steady-state probability distribution of the number of customers at the checkout stand?
(*c*) Derive L for this system. Use this information to determine L_q, W, and W_q.

8. Consider a self-service model in which the customer is also the server. Note that this corresponds to having an infinite number of servers available. Customers arrive according to a Poisson process with parameter λ, and service times have an exponential distribution with parameter μ.
(*a*) Find L_q and W_q.
(*b*) Construct the rate diagram for this queueing system.
(*c*) Use the balance equations to find the expression for P_n in terms of P_0.
(*d*) Find P_0.
(*e*) Find L and W.

9. For each of the following models, write the *balance equations* and show that they are satisfied by the solution given in Sec. 6.6.
(*a*) The M/M/1 model.
(*b*) The *finite queue variation* of the M/M/1 model, with $K = 2$.
(*c*) The *finite calling population variation* of the M/M/1 model, with $N = 2$.

10. Consider the M/M/s model.
(*a*) Suppose there is *one* server and the expected service time is exactly 1 minute. Compare L for the cases where the mean arrival rate is 0.5, 0.9, and 0.99 customers per minute, respectively. Do the same for L_q, W, W_q, and $P\{\mathcal{W} > 5\}$.
(*b*) Now suppose there are *two* servers and the expected service time is exactly 2 minutes. Make the same comparisons as for part (*a*).

11. A bank employs four tellers to serve its customers. Customers arrive according to a *Poisson* process at a mean rate of three per minute. If a customer finds all tellers busy, he joins a queue that is serviced by all tellers; that is, there are no lines in front of each teller, but rather one line waiting for the first available teller. The transaction time between the teller and customer has an exponential distribution with a mean of 1 minute.
(*a*) Construct the rate diagram for this queueing system.
(*b*) Find the steady-state probability distribution of the number of customers in the bank.
(*c*) Find L_q, W_q, W, and L.

12.* Jobs arrive at a particular work center according to a Poisson input process at a mean rate of two per day, and the operation time has an exponential distribution with a mean of $\frac{1}{4}$ day. Enough in-process storage space is provided at the work center to accommodate three jobs in addition to the one being processed, whereas excess jobs are stored temporarily in a less convenient location. For what proportion of the time will this storage space at the work center be adequate to accommodate all waiting jobs?

13. It is necessary to determine how much in-process storage space to allocate to a particular work center in a new factory. Jobs arrive at this work center according to a *Poisson* process with a mean rate of three per hour, and the time required to perform the necessary work has an *exponential* distribution with a mean of 0.25 hour. Whenever the waiting jobs require more in-process storage space than has been allocated, the excess jobs are stored temporarily in a less convenient location. If each job requires 1 square foot of floor space while it is in in-process storage at the work center, how much space must be provided to accommodate all waiting jobs (*a*) 50 percent of the time? (*b*) 90 percent? (*c*) 99 percent? *Hint:* The sum of a geometric series is

$$\sum_{n=0}^{N} x^n = \frac{1 - x^{N+1}}{1 - x}.$$

14. Section 6.6 gives the following equations for the M/M/1 model:

(1) $$P\{\mathcal{W} > t\} = \sum_{n=0}^{\infty} P_n P\{S_{n+1} > t\}.$$

(2) $$P\{\mathcal{W} > t\} = e^{-\mu(1-\rho)t}.$$

Show that Eq. (1) reduces algebraically to Eq. (2).

15. Derive W_q directly for the following cases by developing and reducing an expression analogous to Eq. (1), Prob. 14. (*Hint:* Use the *conditional* expected waiting time in the queue given that a random arrival finds n customers already in the system.)

(*a*) The M/M/1 model.
(*b*) The M/M/s model.

16. For the *finite queue variation* of the M/M/1 model, develop an expression analogous to Eq. (1) in Prob. 14 for the following probabilities:

(*a*) $P\{\mathcal{W} > t\}$.
(*b*) $P\{\mathcal{W}_q > t\}$.

[*Hint:* Arrivals can occur only when the system is not full, so the probability that a random arrival finds n customers already there is $P_n/(1 - P_K)$.]

17. An airline ticket office has two ticket agents answering incoming phone calls for flight reservations. In addition, one caller can be put on hold until one of the agents is available to take the call. If all three phone lines (both agent lines and the hold line) are busy, a potential customer gets a busy signal, and it is assumed that the call goes to another ticket office and that the business is lost. The calls and attempted calls occur *randomly* (i.e., according to a *Poisson* process) at a mean rate of 15 per hour. The length of a telephone conversation has an *exponential* distribution with a mean of 4 minutes.

(*a*) Construct the rate diagram for this queueing system.
(*b*) Find the steady-state probability that
 (i) A caller will get to talk to an agent immediately,
 (ii) The caller will be put on hold,
 (iii) The caller will get a busy signal.

18. The Copy Shop is open 5 days per week for copying materials that are brought to the shop. It has three identical copying machines, but only two operators are kept on duty to run the machines, so the third machine is a spare that is used only when one of the other machines breaks down. When a machine is being used, the time until it breaks down has an exponential distribution with a mean of 2 weeks. If a machine breaks down while the other two are operational, a repairman is called in to repair it, in which case the total time from the breakdown until the repair is completed has an exponential distribution with a mean of 0.2 week. However, if a second machine breaks down before the first one has been repaired, the third machine is shut off while the two operators work together to repair this second machine quickly, in which case its repair time has an exponential distribution with a mean of only $\frac{1}{15}$ week. If the repairman finishes repairing the first machine before the two operators complete the repair of the second, the operators go back to running the two operational machines while the repairman finishes the second repair, in which case the remaining repair time has an exponential distribution with a mean of 0.2 week.

(*a*) Letting the state of the system be the number of machines not working, construct the rate diagram for this queueing system.
(*b*) Find the steady-state distribution of the number of machines not working.
(*c*) What is the expected number of operators available for copying?

19. Consider a single-server queueing system with a finite queue that can hold a maximum of two customers *excluding* any being served. The server can provide *batch service* to two customers simultaneously, where the service time has an exponential distribution with a mean of one unit of time regardless of the number being served. Whenever the queue is not full, customers arrive individually according to a Poisson process at a mean rate of one per unit of time.

(*a*) Assume that the server *must* serve two customers simultaneously. Thus, if the server is idle when only one customer is in the system, the server must wait for another arrival before beginning service. Formulate the queueing model as a *continuous time Markov chain* by defining the states and then constructing the rate diagram. Give the balance equations, but do not solve further.

(*b*) Now assume that the batch size for a service is two only if two customers are in the queue when the server finishes the preceding service. Thus, if the server is idle when only one customer is in the system, the server must serve this single customer, and any subsequent arrivals must wait in the queue until service is completed for this customer. Formulate the resulting queueing model as a *continuous time Markov chain* by defining the states and then constructing the rate diagram. Give the balance equations, but do not solve further.

20. Consider a queueing system that has two classes of customers, two clerks providing service, and *no queue*. Potential customers from each class arrive according to a Poisson process, with a mean arrival rate of 10 customers per hour for class 1 and 5 customers per hour for class 2, but these arrivals are lost to the system if they cannot immediately enter service.

Each customer of class 1 that enters the system will receive service from either one of the clerks that is free, where the service times have an exponential distribution with a mean of 5 minutes.

Each customer of class 2 that enters the system requires the *simultaneous use of both clerks* (the two clerks would work together as a single server), where the service times have an exponential distribution with a mean of 5 minutes. Thus, an arriving customer of this kind would be lost to the system unless both clerks are free to begin service immediately.

(*a*) Formulate the queueing model as a *continuous time Markov chain* by defining the states and constructing the rate diagram.

(*b*) Construct the balance equations and use them to solve for the steady-state joint distribution of the number of customers of each class in the system.

(*c*) For each of the two classes of customers, what is the expected fraction of arrivals that are unable to enter the system?

21.* Plans are being made to open a small car-wash operation, and the owner must decide how much space to provide for waiting cars. It is estimated that customers would arrive randomly (i.e., a *Poisson* input process) with a mean rate of one every 4 minutes, unless the waiting area is full, in which case the arriving customers would take their cars elsewhere. The time that can be attributed to washing one car has an exponential distribution with a mean of 3 minutes. Compare the expected fraction of potential customers that would be *lost* because of inadequate waiting space if (*a*) zero spaces (not including the car being washed) were to be provided. (*b*) Two spaces. (*c*) Four spaces.

22. Consider the *finite queue variation* of the M/M/s model. Derive the expression for L_q given in Sec. 6.6 for this model.

23.* Suppose that one repairman has been assigned the responsibility of maintaining three machines. For each machine, the probability distribution of the running time before a breakdown is *exponential,* with a mean of 9 hours. The repair time also has an *exponential* distribution, with a mean of 2 hours.

(*a*) Calculate the steady-state probability distribution and the expected number of machines that are not running.

(*b*) As a crude approximation, assume that the calling population is infinite, so that the input process is *Poisson* with a mean arrival rate of three every 9 hours. Compare the result from part (*a*) with that obtained by making this approximation using (1) the corresponding infinite queue model and (2) the corresponding finite queue model.

(*c*) Now suppose that a second repairman is available whenever more than one of these three machines requires repair. Calculate the information specified in part (*a*).

24. Plans are currently being developed for a new factory. One department has been allocated a large number of automatic machines of a certain type, and we need to determine how many machines should be assigned to each operator for servicing (loading, unloading, adjusting, setup, and so on). For the purpose of this analysis, the following information has been provided.

The running time (time between completing service and the machine requiring service again) of each machine has an *exponential* distribution, with a mean of 150 minutes. The service time has an *exponential* distribution, with a mean of 15 minutes. Each operator attends to her own machine; she does not give help to or receive help from other operators. For the department to achieve the required production rate, the machines must be running at least 89 percent of the time on the average.

(*a*) What is the maximum number of machines that can be assigned to an operator while still achieving the required production rate?

(*b*) Given that the maximum number found in part (*a*) is assigned to each operator, what is the expected fraction of time that the operators will be busy servicing machines?

25. Consider a single-server queueing system. It has been observed that this server seems to speed up as the number of customers in the system increases, and that the pattern of acceleration seems to fit the *state-dependent model* presented at the end of Sec. 16.6. Furthermore, it is estimated that the expected service time is 8 minutes when there is only *one* customer in the system. Determine the pressure coefficient c for this model for the following cases:

(*a*) The expected service time is estimated to be 4 minutes when there are *four* customers in the system.

(*b*) The expected service time is estimated to be 5 minutes when there are *four* customers in the system.

26. For the *state-dependent model* presented at the end of Sec. 6.6, show the effect of the *pressure coefficient* c by using Fig. 6.10 to construct a table giving the *ratio* (expressed as a decimal number) of L for this model to L for the corresponding M/M/s model (that is, with $c = 0$). Tabulate these ratios for $\lambda_0/s\mu_1 = 0.5, 0.9, 0.99$ when $c = 0.2, 0.4, 0.6$, and $s = 1, 2$.

27.* Consider the M/G/1 model.

(*a*) Compare the expected waiting time in the queue if the service-time distribution is (i) exponential, (ii) constant, (iii) Erlang with the amount of variation (i.e., the standard deviation) halfway between the constant and exponential cases.

(*b*) What is the effect on the expected waiting time in the queue and on the expected queue length if both λ and μ are doubled and the scale of the service-time distribution is changed accordingly?

28. Consider a queueing system with a *Poisson* input, where the server must perform two distinguishable tasks in sequence for each customer, so the total service time is the sum of the two task times (which are statistically independent).

(*a*) Suppose that the first task time has an *exponential* distribution with a mean of 3 minutes and the second task time has an *Erlang* distribution with a mean of 9 minutes and with the shape parameter $k = 3$. Which queueing theory model should be used to represent this system?

(*b*) Suppose that part (*a*) is modified so that the first task time also has an *Erlang* distribution with the shape parameter $k = 3$ (but with the mean still equal to 3 minutes). Which queueing theory model should be used to represent this system?

29.* An airline maintenance base has facilities for overhauling only one airplane engine at a time. Therefore, to return the airplanes to use as soon as possible, the policy has been to stagger the overhauling of the four engines of each airplane. In other words, only one engine is overhauled each time an airplane comes into the shop. Under this policy, airplanes have arrived according to a *Poisson* process at a mean rate of one per day. The time required for an engine overhaul (once work has begun) has an *exponential* distribution with a mean of $\frac{1}{2}$ day.

A proposal has been made to change the policy so that all four engines are overhauled consecutively each time an airplane comes into the shop. Although this would quadruple the expected service time, each plane would need to come into the shop only one-fourth as often.

Use queueing theory to compare the two alternatives on a meaningful basis.

30. Consider a shoe repair shop with a single repairman. Pairs of shoes are brought in to be repaired (on a first-come-first-served basis) according to a Poisson process at a mean rate of one pair per hour. The time required to repair each individual shoe has an exponential distribution with a mean of 15 minutes.

(*a*) Consider the formulation of this queueing system where the *individual* shoes (not pairs of shoes) are considered to be the customers. For this formulation, construct the rate diagram and develop the balance equations, but do not solve further.

(*b*) Now consider the formulation of this queueing system where the *pairs* of shoes are considered to be the customers. Identify the specific queueing model that fits this formulation, and then use the available results for this model to calculate the expected waiting time W. (Interpret W to be the expected waiting time until *both* shoes in a pair have been repaired.)

31. Airplanes arrive at a certain maintenance base for overhaul according to a Poisson process with parameter $\lambda = 3$ (per week). Service times have an Erlang distribution with parameters $\mu = 4$ (per week) and $k = 2$.

(*a*) How should the states of the system be defined in order to formulate the queueing model as a *continuous time Markov chain*?

(*b*) Construct the corresponding rate diagram.

(*c*) What are L_q, L, W_q, and W for this queueing system? [*Hint:* Do not use parts (*a*) and (*b*).]

32. Consider a single-server queueing system with *any* service-time distribution and *any* distribution of interarrival times (the G1/G/1 model). Use only basic definitions and the relationships given in Sec. 6.2 to verify the following general relationships:

(*a*) $L = L_q + (1 - P_0)$.

(*b*) $L = L_q + \rho$.

(*c*) $P_0 = 1 - \rho$.

33. Show that

$$L = \sum_{n=0}^{s-1} nP_n + L_q + s\left(1 - \sum_{n=0}^{s-1} P_n\right)$$

by using the statistical definitions of L and L_q in terms of the P_n.

34. A company currently has *two* tool cribs, each with a *single* clerk, in its manufacturing area. One tool crib handles only the tools for the heavy machinery; the second one handles all other tools. However, for each crib the mechanics arrive to obtain tools at a mean rate of 24 per hour, and the expected service time is 2 minutes.

Because of complaints that the mechanics coming to the tool cribs have to wait too long, it has been proposed that the two tool cribs be combined so that either clerk can handle either kind of tool as the demand arises. It is believed that the mean arrival rate to the combined two-clerk tool crib would double to 48 per hour, and the expected service time would continue to be 2 minutes. However, information is not available on the *form* of the probability distributions for *interarrival* and *service times,* so it is not clear which queueing model would be most appropriate.

Compare the status quo and the proposal with respect to the total expected number of mechanics at the tool crib(s) and the expected waiting time (including service) for each mechanic. Do this by tabulating these data for the four queueing models considered in Figs. 6.7, 6.11, 6.13, and 6.14 (use $k = 2$ when an *Erlang* distribution is appropriate).

35. Consider the model with *nonpreemptive priorities* presented in Sec. 6.8. Suppose there are just two priority classes, with $\lambda_1 = 4$ and $\lambda_2 = 4$. In designing this queueing system, you are offered the choice between the following two alternatives: (1) one fast server ($\mu = 10$) and (2) two slow servers ($\mu = 5$).

Compare these alternatives with the usual four mean measures of performance (W, L, W_q, L_q) for the individual priority classes (W_1, W_2, L_1, L_2, and so forth). Which alternative is preferred if your primary concern is expected waiting time in the *system* for priority class 1 (W_1)? Which is preferred if your primary concern is expected waiting time in the *queue* for priority class 1?

36.* A particular work center in a job shop can be represented as a single-server queueing system, where jobs arrive according to a *Poisson* process, with a mean rate of eight per day. Although the arriving jobs are of three distinct types, the time required to perform any of these jobs has the same *exponential* distribution, with a mean of 0.1 working day. The practice has been to work on arriving jobs on a first-come-first-served basis. However, it is important that jobs of type 1 do not have to wait very long, whereas the wait is only moderately important for jobs of type 2 and relatively unimportant for jobs of type 3. These three types arrive with a mean rate of two, four, and two per day, respectively. Because all three types have experienced rather long delays on the average, it has been proposed that the jobs be selected according to an appropriate priority discipline instead.

Compare the expected waiting time (including service) for each of the three types of jobs if the queue discipline is (*a*) first-come-first-served, (*b*) nonpreemptive priority, or (*c*) preemptive priority.

37. Reconsider the *County Hospital* emergency room problem as analyzed in Sec. 6.8. Suppose that the definitions of the three categories of patients are tightened somewhat in order to move marginal cases into a lower category. Consequently, only 5 percent of the patients will qualify as *critical* cases, 20 percent as *serious* cases, and 75 percent as *stable* cases. Develop a table showing the data presented in Table 6.4 for this revised problem.

38. One inspector has been assigned the full-time task of inspecting the output from a group of 10 identical machines. Jobs to be done by any one of the machines arrive according to a *Poisson* process at a mean rate of 70 per hour. The time required by a machine to perform each job has an *exponential* distribution with a mean of 6 minutes. Thus, whenever all 10 machines are busy, the jobs are being completed ready for inspection at a mean rate of 100 per hour. Unfortunately, the inspector is able to inspect them at a mean rate of only 80 per hour. (In particular, his inspection time has an *Erlang* distribution with a mean of 0.75 minute and a shape parameter $k = 25$.) This inspection rate has resulted in a substantial average amount of in-process inventory at the inspection station (i.e., the expected number of jobs waiting to be inspected is fairly large), in addition to that already found at the group of machines. Management feels that there is too much capital tied up in in-process inventory, so it has instructed the production manager to cut down on such inventory. Therefore, the production manager has made two alternative proposals to reduce the average level of in-process inventory. Proposal 1 is to use slightly less power for the machines (which would increase their expected time to perform a job to 7 minutes), so that the inspector can keep up with their output better. Proposal 2 is to substitute a certain younger inspector for this task. He is somewhat faster (albeit more variable in his inspection times because of less experience), so he should keep up better. (His inspection time would have an *Erlang* distribution with a mean of 0.72 minute and a shape parameter $k = 2$.)

The production manager has asked you to "use the latest OR techniques to see how much each proposal would cut down on in-process inventory."

(*a*) What would be the effect of proposal 1? Why? How would you explain this outcome to the production manager?

(*b*) Determine the effect of proposal 2. How would you explain this outcome to the production manager?

(*c*) What suggestions would you make for reducing the average level of in-process inventory at the inspection station? At the group of machines?

39. Consider the queueing-theory model with a *preemptive priority* queue discipline presented in Sec. 6.8. Suppose that $s = 1$, $N = 2$, and $(\lambda_1 + \lambda_2) < \mu$, and let P_{ij} be the steady-state probability that there are i members of the higher priority class and j members of the lower priority class in the queueing system ($i = 0, 1, 2, \ldots; j = 0, 1, 2, \ldots$). Use a method analogous to that presented in Sec. 6.5 to derive a system of linear equations whose simultaneous solution is the P_{ij}. Do not actually obtain this solution.

40. Consider a queueing system with two servers, where the customers arrive from two different sources. From Source 1, the customers always arrive two at a time, where the time between consecutive arrivals of pairs of customers has an exponential distribution with a mean of 20 minutes. Source 2 is itself a two-server queueing system, which has a Poisson input process with a mean rate of 7 customers per hour, and the service time from each of these two servers has an exponential distribution with a mean of 15 minutes. When a customer completes service at Source 2, it immediately enters the queueing system under consideration for another type of service. In the latter queueing system, the queue discipline is preemptive priority where customers from Source 1 always have preemptive priority over customers from Source 2. However, service times are independent and identically distributed for both types of customers according to an exponential distribution with a mean of 6 minutes.

(*a*) First focus on the problem of deriving the steady-state distribution of *only* the number of Source 1 customers in the queueing system under consideration. Using a *continuous time Markov chain* formulation, define the states and construct the rate diagram for most efficiently deriving this distribution (but don't actually derive it).

(*b*) Now focus on the problem of deriving the steady-state distribution of the *total* number of customers of both types in the queueing system under consideration. Using a *continuous time Markov chain* formulation, define the states and construct the rate diagram for most efficiently deriving this distribution (but don't actually derive it).

(*c*) Now focus on the problem of deriving the steady-state *joint* distribution of the number of customers of each type in the queueing system under consideration. Using a *continuous time Markov chain* formulation, define the states and construct the rate diagram for deriving this distribution (but don't actually derive it).

41.* Consider a single-server queueing system with a *Poisson* input, *Erlang* service times, and a *finite* queue. In particular, suppose that $k = 2$, the mean arrival rate is two customers per hour, the expected service time is 0.25 hour, and the maximum permissible number of customers in the system is two. Derive the steady-state probability distribution of the number of customers in the system, and then calculate the expected number. Compare this result with the corresponding results when the service-time distribution is *exponential*.

42. Consider the $E_2/M/1$ model with $\lambda = 4$ and $\mu = 5$.

(*a*) How should the states of the system be defined in order to formulate this model as a *continuous time Markov chain*?

(*b*) Construct the corresponding rate diagram.

43. A shop contains three identical machines that are subject to a failure of a certain kind. Therefore, a maintenance system is provided to perform the maintenance operation (recharging) required by a failed machine. The time required by each operation has an exponential distribution with a mean of 30 minutes. However, with probability $\frac{1}{3}$, the operation must be performed a second time (with the same distribution of time) in order to bring the failed machine back to a satisfactory operational state. The maintenance system works on only one failed machine at a time, performing all the operations (one or two) required by that machine, on a first-come-first-served basis. After a machine is repaired, the time until its next failure has an exponential distribution with a mean of 3 hours.

(*a*) How should the states of the system be defined in order to formulate this queueing system as a *continuous time Markov chain*? (*Hint:* Given that a first operation is being performed on a failed machine, completing this operation *successfully* and completing it *unsuccessfully* are two separate events of interest. Then use Property 6 regarding disaggregation for the exponential distribution.)

(*b*) Construct the corresponding rate diagram.

(*c*) Develop the balance equations.

44. A company has one repairman to keep a large group of machines in running order. Treating this group as an infinite calling population, individual breakdowns occur according to a *Poisson* process at a mean rate of one per hour. For each breakdown, the probability is 0.9 that only a *minor repair* is needed, in which case the repair time has an *exponential* distribution with a mean of $\frac{1}{2}$ hour. Otherwise, a *major repair* is needed, in which case the repair time has an *exponential* distribution with a mean of 5 hours. Because both of these *conditional* distributions are exponential, the *unconditional* (combined) distribution of repair times is *hyperexponential*.

(*a*) Compute the mean and standard deviation of this hyperexponential distribution. [*Hint:* Use the general relationships from probability theory that, for any random variable X and any pair of mutually exclusive events, E_1 and E_2, $E(X) = E(X|E_1)P(E_1) + E(X|E_2)P(E_2)$ and $\text{VAR}(X) = E(X^2) - E(X)^2$.] Compare this standard deviation with that for an *exponential* distribution having this mean.

(*b*) What are P_0, L_q, L, W_q, and W for this queueing system?

(*c*) What is the *conditional* value of W, *given* that the machine involved requires a *major repair*? A *minor repair*? What is the division of L between machines requiring the two types of repairs? (*Hint: Little's formula* still applies for the individual categories of machines.)

(*d*) How should the states of the system be defined in order to formulate this queueing system as a *continuous time Markov chain*? (*Hint:* Consider what additional information must be given, besides the number of machines down, for the conditional distribution of the time remaining until the next event of each kind to be exponential.)

(*e*) Construct the corresponding rate diagram.

45. Consider a system of two *infinite queues in series,* where each of the two service facilities has a single server. All service times are independent and have an exponential distribution, with a mean of 3 minutes at facility 1 and 4 minutes at facility 2. Facility 1 has a Poisson input process with a mean rate of 10 per hour.

(*a*) Find the steady-state distribution of the number of customers at facility 1, and then at facility 2. Then show the *product form solution* for the *joint* distribution of the number at the respective facilities.

(*b*) What is the probability that *both* servers are idle?

(*c*) Find the expected *total* number of customers in the system and the expected *total* waiting time (including service times) for a customer.

46. Under the assumptions specified in Sec. 6.9 for a system of *infinite queues in series,* this kind of queueing network actually is a *special case* of a *Jackson network.* Demonstrate that this is true by describing this system as a Jackson network, including specifying the values of the a_j and the p_{ij} given λ for this system.

47. Consider a *Jackson network* with three service facilities having the parameter values shown below.

Facility j	s_j	μ_j	a_j	p_{ij} $i = 1$	$i = 2$	$i = 3$
$j = 1$	1	40	10	—	0.3	0.4
$j = 2$	1	50	15	0.5	—	0.5
$j = 3$	1	30	3	0.3	0.2	—

(*a*) Find the *total* arrival rate at each of the facilities.

(*b*) Find the steady-state distribution of the number of customers at facility 1. At facility 2. At facility 3. Then show the *product form solution* for the *joint* distribution of the number at the respective facilities.

(*c*) What is the probability that *all* the facilities have *empty queues* (no customers waiting to begin service)?

(*d*) Find the expected *total* number of customers in the system.

(*e*) Find the expected *total* waiting time (including service times) for a customer.

7

The Application of Queueing Theory

Queueing theory has enjoyed a prominent place among the modern analytical techniques of operations research. However, the emphasis thus far has been on developing a *descriptive* mathematical theory. Thus queueing theory is not directly concerned with achieving the goal of operations research: *optimal decision making*. Rather, it develops *information* on the *behavior* of queueing systems. This theory provides *part* of the information needed to conduct an operations research study attempting to find the *best design* for a queueing system.

This chapter discusses the *application* of queueing theory in the broader context of an overall operations research study. It begins by introducing three examples that will be used for illustration throughout the chapter. Section 7.2 discusses the basic considerations for *decision making* in this context. The following two sections then develop *decision models* for the *optimal* design of queueing systems. The last model requires the incorporation of *travel-time models,* which are presented in Sec. 7.5. Section 7.6 then focuses on the issue of obtaining point estimates and confidence intervals for both the parameters and measures of performance of the queueing system being studied.

7.1 Examples

Example 1—How Many Repairers?

SIMULATION, INC., a small company that makes gidgets for analog computers, has 10 gidget-making machines. However, because these machines break down and require repair frequently, the company has only enough operators to operate *eight* machines at a time, so two machines are available on a standby basis for use while other machines are down. Thus eight machines are always operating whenever no more than two machines are waiting to be repaired, but the number of operating machines is reduced by one for each additional machine waiting to be repaired.

The time until any given operating machine breaks down has an exponential distribution, with a mean of 20 days. The time required to repair a machine also has an exponential distribution, with a mean of 2 days. Until now the company has had just *one* repairer to repair these machines, which has frequently resulted in reduced productivity because *fewer than eight* machines are operating. Therefore, the company is considering hiring a *second* repairer, so that *two* machines can be repaired simultaneously.

Thus the queueing system to be studied has the repairers as its servers and the machines requiring repair as its customers, where the problem is to choose between having *one* or *two* servers. (Notice the analogy between this problem and the *County Hospital* emergency room problem described in Sec. 6.1.) With one slight exception, this system fits the *finite calling population variation* of the M/M/s model presented in Sec. 6.6, where $N = 10$ machines, $\lambda = 1/20$ customer per day (for each operating machine), and $\mu = 1/2$ customer per day. The exception is that the λ_0 and λ_1 parameters of the birth-and-death process are changed from $\lambda_0 = 10\lambda$ and $\lambda_1 = 9\lambda$ to $\lambda_0 = 8\lambda$ and $\lambda_1 = 8\lambda$. (All the other parameters are the same as those given in Sec. 6.6.) Therefore, the C_n factors for calculating the P_n probabilities change accordingly (see Sec. 6.5).

Each repairer costs the company approximately \$70/day. However, the estimated *lost profit* from having fewer than eight machines operating to produce gidgets is \$100/day for each machine down. (The company can sell the full output from eight operating machines, but not much more.)

The analysis of this problem will be pursued in Secs. 7.3 and 7.4.

Example 2—Which Computer?

EMERALD UNIVERSITY currently has one large computer that is shared by everyone on campus. Because students have been experiencing long turnaround times, the university now is planning to lease an additional small batch-processing computer for the exclusive use of its students, while reserving the large computer for the other users. Two models are being considered: one from the MBI Corporation and the other from the EG Company. The MBI computer costs more but is somewhat faster than the EG computer. In particular, if a sequence of typical student programs were run continuously for 1 hour, the number completed would have a Poisson distribution with a mean of 30 and 25 for the MBI and the EG computers, respectively. A statistical study has shown that the student population actually submits programs to be run every 3 minutes on the average during all operating hours, and that the time from one submission to the next has an exponential distribution with this mean. The leasing cost per operating hour would be \$100 for the MBI computer and \$75 for the EG computer.

Thus the queueing system of concern has the new small computer as its (single) server and the students' programs as its customers. Furthermore, this system fits the M/M/1 model presented at the beginning of Sec. 6.6. With 1 hour as the unit of time, λ = (60 minutes per hour)/(3 minutes per customer) = 20 customers per hour, and μ = 30 and 25 customers per hour with the MBI and the EG computers, respectively. You will see in Secs. 7.3 and 7.4 how the decision was made between the two computers.

Example 3—How Many Tool Cribs?

The MECHANICAL COMPANY is designing a new plant. This plant will need to include one or more tool cribs in the factory area to store tools required by the shop mechanics. The tools will be handed out by clerks as the mechanics arrive and request them and will be returned to the clerks when they are no longer needed. In existing plants, there have been frequent complaints from foremen that their mechanics have had to waste too much time traveling to tool cribs and waiting to be served, so it appears that there should be *more* tool cribs and *more* clerks in the new plant. On the other hand, management is exerting pressure to reduce overhead in the new plant, and this reduction would lead to *fewer* tool cribs and *fewer* cribs. To resolve these conflicting pressures, an operations research study is to be conducted to determine just how many tool cribs and clerks the new plant should have.

Each tool crib constitutes a queueing system, with the clerks as its servers and the mechanics as its customers. Based on previous experience, it is estimated that the time required by a tool crib clerk to service a mechanic has an exponential distribution, with a mean of $\frac{1}{2}$ minute. Judging from the anticipated number of mechanics in the entire factory area, it is also predicted that they would require this service *randomly* but at a mean rate of two mechanics per minute. Therefore, it was decided to use the M/M/s model of Sec. 6.6 to represent each queueing system. With 1 hour as the unit of time, μ = 120. If only one tool crib were to be provided, λ also would be 120. With more than one tool crib, this mean arrival rate would be divided among the different queueing systems.

The total cost to the company of each tool crib clerk is about \$10/hour. The capital recovery costs, upkeep costs, and so forth, associated with each tool crib provided are estimated to be \$8/working hour. While a mechanic is busy, the value to the company of his output averages about \$24/hour.

Sections 7.3 and 7.5 include discussions of how this (and additional) information was used to make the required decisions.

7.2 Decision Making

Queueing-type situations that require decision making arise in a wide variety of contexts. For this reason, it is not possible to present a meaningful decision-making procedure that is applicable to all these situations. Instead, this section attempts to give a broad conceptual picture of the general approach to a predominant group of waiting-line problems.

A large proportion of waiting-line problems that arise in practice involve making one or a combination of the following decisions:

1. Number of servers at a service facility.
2. Efficiency of the servers.
3. Number of service facilities.

When such problems are formulated in terms of a queueing model, the corresponding decision variables usually would be s (number of servers at each facility), μ (mean service rate per busy server), and λ (mean arrival rate at each facility). The *number of service facilities* is directly related to λ because, assuming a uniform workload among the facilities, λ equals the total mean arrival rate to all facilities divided by the number of facilities.

Refer back to Sec. 7.1 and note how the three examples there respectively illustrate situations involving these three decisions. In particular, the decision facing *Simulation, Inc.*, is *how many repairers* (servers) to provide. The problem for *Emerald University* is *how fast a computer* (server) is needed. The problem facing *Mechanical Company* is *how many tool cribs* (service facilities) to install, as well as *how many clerks* (servers) to provide at each facility.

The first kind of decision is particularly common in practice. However, the other two also arise frequently, particularly for the *business-industrial internal service systems* described in Sec. 6.3. One example illustrating a decision on the *efficiency of the servers* is the selection of the type of materials-handling equipment (the servers) to purchase to transport certain kinds of loads (the customers). Another such example is the determination of the size of a maintenance crew (where the entire crew is one server). Other decisions concern the *number of service facilities,* such as restrooms, first-aid centers, drinking fountains, storage areas, and so on, to distribute throughout an area.

All the specific decisions discussed here involve the general question of the *appropriate level of service* to provide in a queueing system. As mentioned at the beginning of Chap. 6, decisions regarding the amount of service capacity to provide usually are based primarily on two considerations: (1) the cost incurred by providing the service, as shown in Fig. 7.1, and (2) the amount of waiting for that service, as suggested in Fig. 7.2. Figure 7.2 can be obtained by using the appropriate waiting-time equation from queueing theory.

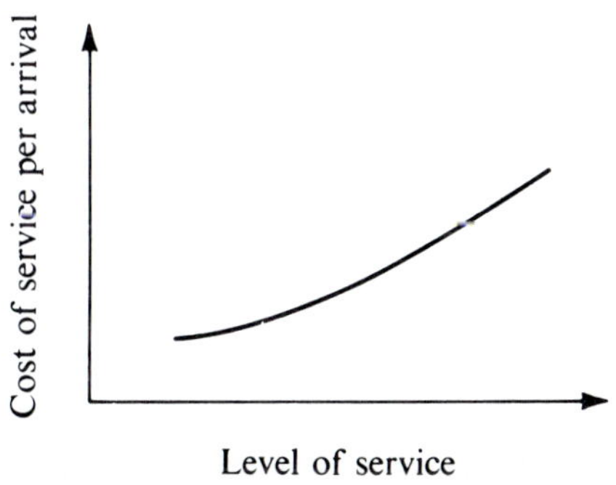

Figure 7.1 Service cost as a function of service level.

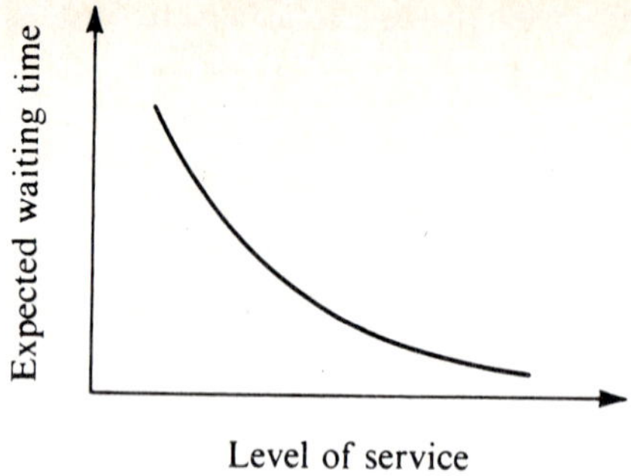

Figure 7.2 Expected waiting time as a function of service level.

It is readily apparent that these two considerations create conflicting pressures on the decision maker. The objective of reducing service costs recommends a minimal level of service. On the other hand, long waiting times are undesirable, which recommends a high level of service. Therefore, it is necessary to strive for some type of compromise. To assist us in finding this compromise, Figs. 7.1 and 7.2 may be combined, as shown in Fig. 7.3. The problem is thereby reduced to selecting the point on the curve of Fig. 7.3 that gives the best balance between the average delay in being serviced and the cost of providing that service. Reference to Figs. 7.1 and 7.2 indicates the corresponding level of service.

Unfortunately, it is all too easy to terminate further analysis and make a quick subjective judgment on the basis of Fig. 7.3. Actually, the most crucial portion of the analysis still lies ahead. An intelligent decision on the proper balance between delays and service costs can be made only after the relative seriousness of delays and service costs has been established. Obtaining the proper balance requires answers to such questions as "How much expenditure on service is equivalent (in its detrimental impact) to a customer being delayed one unit of time?" Thus, to compare service costs and waiting times, it is necessary to adopt (explicitly or implicitly) a common measure of their impact. The natural choice for this common measure is cost, so that it becomes necessary to estimate the cost of waiting. This cost probably cannot be identified entirely with expenditures on the accounting books. Nevertheless, a given amount of waiting can be considered to be equivalent in its long-run impact (from the viewpoint of the decision maker) to an expenditure of a certain amount. If it is reasonable to assume that this cost of waiting is proportional to the total amount of waiting, it is sufficient to estimate the cost of waiting per unit time per arrival.

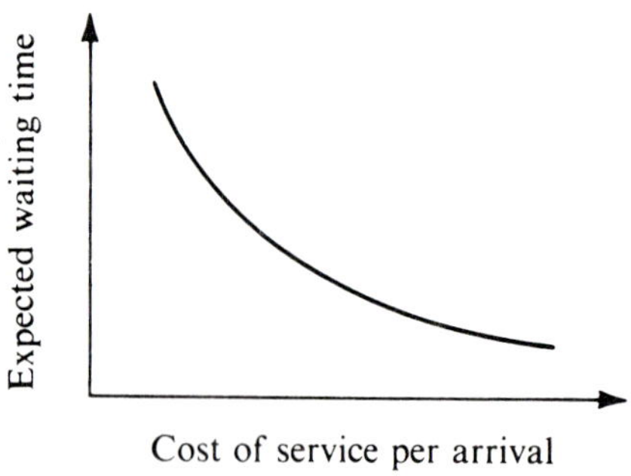

Figure 7.3 Relationship between average delay and service cost.

A common viewpoint in practice is that the cost of waiting is often too intangible to be amenable to estimation; hence the decision must instead be based on more tangible, even if less fundamental, criteria, such as the desired expected waiting time in Fig. 7.3. The fallacy in this viewpoint is that it is impossible to avoid the equivalent of estimating waiting costs when analyzing the problem rationally. Any comparison of waiting times and service costs must inevitably reduce to estimating the cost that is equivalent to the waiting. The only question is whether the estimation should be done explicitly or implicitly. The answer to this question depends, in part, upon the time and cost required to develop a reasonable explicit estimate. Will the time and cost incurred be more or less than the potential savings? It seems evident that performing the penetrating analysis required to obtain this explicit estimate should provide a sounder basis for the required decision than superficial criteria that are ultimately based on these same cost considerations in a very imprecise and intuitive way. In addition to using a better estimate, this explicit procedure permits using rigorous mathematical analysis to identify accurately the decision that minimizes the total estimated expected cost.

Granted that an explicit estimate of the cost of waiting is desirable, the next question is how to develop this estimate. Because of the diversity of waiting-line situations, no single estimating process is generally applicable. However, we shall discuss the basic considerations involved for several types of situations.

One broad category is where the customers are *external* to the organization providing the service; i.e., they are *outsiders* bringing their business to the organization. To discuss this category meaningfully, we need to divide it further in terms of whether the service is being provided *for profit* or *not for profit* (on a nonprofit basis). Consider first the case of *profit-making* organizations (typified by the *commercial service systems* described in Sec. 6.3). From the viewpoint of the decision maker, the cost of waiting probably consists primarily of the *lost profit* from *lost business*. This loss of business may occur immediately (because the customer grows impatient and leaves) and/or in the future (because the customer is sufficiently irritated that he or she does not come again). This kind of cost is quite difficult to estimate, and it may be necessary to revert to other criteria, such as a tolerable probability distribution of waiting times. When the customer is not a human being, but a job being performed on order, there may be more readily identifiable costs incurred, such as those caused by idle in-process inventories or increased expediting and administrative effort.

Now consider the type of situation where service is provided on a *nonprofit* basis to customers *external* to the organization (typical of *social service systems* and some *transportation service systems* described in Sec. 6.3). In this case, the cost of waiting usually is a *social cost* of some kind. Thus it is necessary to evaluate the consequences of the waiting for the individuals involved and/or for society as a whole and to try to impute a monetary value to avoiding these consequences. (We shall illustrate this process in the next section with Example 2—the Emerald University computer problem—the example that most closely fits this type of situation.) Once again, this kind of cost is quite difficult to estimate, and it may be necessary to revert to other criteria.

A situation may be more amenable to estimating waiting costs if the customers are *internal* to the organization providing the service (as for *business-industrial internal service systems*). For example, the customers may be machines (as in Example 1) or employees (as in Example 3) of a firm. Therefore, it may be possible to identify

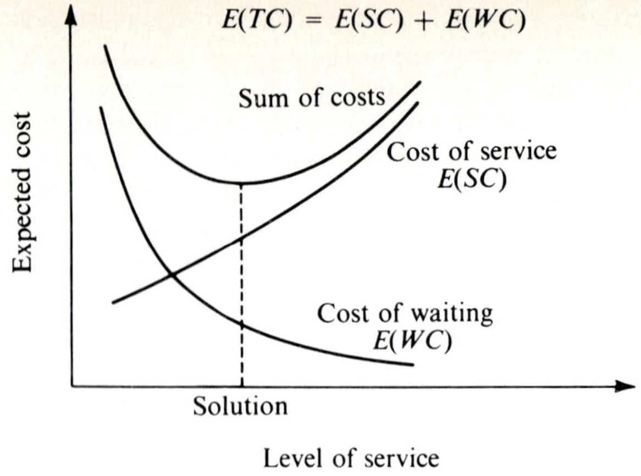

Figure 7.4 Conceptual solution procedure for many waiting-line problems.

directly some or all of the costs associated with the idleness of these customers. (A detailed discussion of the estimating process for this situation is available elsewhere.[1]) To illustrate the underlying rationale (and to warn against a common pitfall), we consider the case where the customers are machine operators. At first glance, it is easy to jump to the conclusion that the relevant cost to the firm if such an employee waits in a queue is her wage during the waiting time. However, this conclusion implies that the net reduction in the earnings of the firm because an operator has to wait is equal to her wage. There is no reason why this particular relationship should hold in general. Perhaps one reason this approach is sometimes used in practice is the misconception that what is being wasted by the waiting is the operator's wage. The operator actually will receive the same wage regardless of the waiting, and what is really lost is the contribution to the firm's earnings that the operator would have made otherwise. Furthermore, the worker does not work in a vacuum; rather, she is the catalyst that causes the efforts of all the economic resources involved (machinery and equipment, materials, managerial skill, capital, and so on) to result in one of the changes in the product necessary to make it salable. Thus, although the output of the machine operator and her colleagues is essential if a salable product is to be produced, the sale of that product must not only pay their wages but also pay for the other necessary economic resources. Therefore, when the operator is idle because she is waiting in a queue, what is being wasted is productive output that would have helped pay the fixed expenses of the firm in addition to the operator's wage. In short, rather than focusing solely on the value of the one economic resource that physically waits in the queue, the emphasis instead should be on finding the value of all the economic resources that would be idled as a consequence of this waiting. This approach frequently boils down to evaluating the *lost profit* from *all lost productivity*.

Given that the cost of waiting has been evaluated explicitly, the remainder of the analysis is conceptually straightforward. The objective is to determine the level of service that minimizes the total of the expected cost of service and the expected cost of waiting for that service. This concept is depicted in Fig. 7.4, where *WC* denotes *waiting cost, SC* denotes *service cost,* and *TC* denotes *total cost.* Thus the

[1] Hillier, Frederick S.: "Cost Models for the Application of Priority Waiting Line Theory to Industrial Problems," *Journal of Industrial Engineering,* **16**(3):178–185, 1965.

mathematical statement of the objective is to

$$\text{Minimize} \qquad E(TC) = E(SC) + E(WC).$$

Sections 7.3 to 7.5 are concerned with the *application* of this concept to various types of problems. Thus Sec. 7.3 describes how $E(WC)$ can be expressed mathematically. Section 7.4 then focuses on $E(SC)$ to formulate the overall objective function $E(TC)$ for several basic design problems (including some with multiple decision variables, so that the level of the service axis in Fig. 7.4 actually requires more than one dimension). This section also introduces the fact that when a decision on the number of service facilities is required, time spent in traveling to and from a facility should be included in the analysis (as part of the total time waiting for service). Section 7.5 discusses how to determine the expected value of this travel time.

7.3 Formulation of Waiting-Cost Functions

To express $E(WC)$ mathematically, we must first formulate a *waiting-cost function* that describes how the actual waiting cost being incurred varies with the current behavior of the queueing system. The form of this function depends on the context of the individual problem. However, most situations can be represented by one of the two basic forms described next.

The $g(N)$ Form

Consider first the situation discussed in the preceding section where the queueing system *customers* are *internal* to the organization providing the service, and so the primary cost of waiting may be the *lost profit from lost productivity*. The *rate* at which productive output is lost sometimes is essentially *proportional* to the number of customers in the queueing system. However, in many cases there is not enough productive work available to keep all the members of the calling population continuously busy. Therefore, little productive output may be lost by having just a few members idle waiting for service in the queueing system, whereas the loss may increase greatly if a few more members are made idle because they require service. Consequently, the primary property of the queueing system that determines the *current rate* at which waiting costs are being incurred is N, the *number of customers* in the system. Thus the form of the waiting-cost function for this kind of situation is that illustrated in Fig. 7.5, namely, a function of N. We shall denote this form by $g(N)$.

The $g(N)$ function would be constructed for a particular situation by estimating $g(n)$, the waiting-cost rate incurred when $N = n$, for $n = 1, 2, \ldots,$ where $g(0) = 0$. After computing the P_n probabilities for a given design of the queueing system, we can calculate

$$E(WC) = E\{g(N)\}.$$

Because N is a random variable, this calculation is made by using the equation for the expected value of a *function* of a *discrete* random variable,

$$E(WC) = \sum_{n=0}^{\infty} g(n)P_n.$$

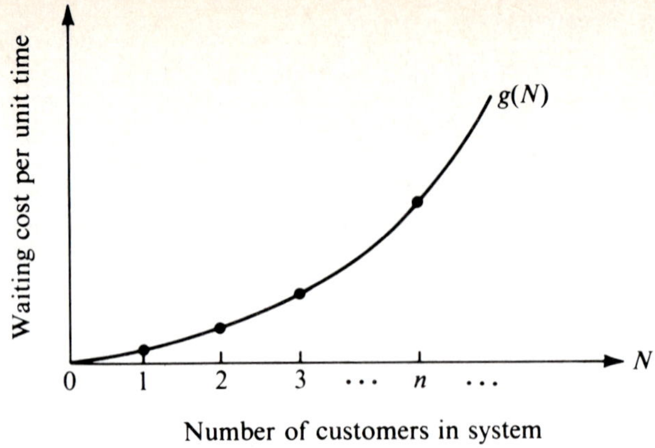

Figure 7.5 The waiting-cost function as a function of N.

When $g(N)$ is a *linear* function (i.e., when the waiting-cost rate is proportional to N), then

$$g(N) = C_w N,$$

where C_w is the cost of waiting per unit time for each customer. In this case, $E(WC)$ reduces to

$$E(WC) = C_w \sum_{n=0}^{\infty} nP_n = C_w L.$$

Example 1—How Many Repairers? For Example 1 of Sec. 7.1, *Simulation, Inc.*, has two *standby* gidget-making machines, so there is *no* lost productivity as long as the number of customers (machines requiring repair) in the system does not exceed *two*. However, for each *additional* customer (up to the maximum of 10 total), the estimated lost profit is \$100/day. Therefore,

$$g(n) = \begin{cases} 0, & \text{for } n = 0, 1, 2 \\ 100(n-2), & \text{for } n = 3, 4, \ldots, 10, \end{cases}$$

as shown in Table 7.1. Consequently, $E(WC)$ is calculated by summing the last column of Table 7.1 for each of the two cases of interest, namely, having *one* repairer ($s = 1$) or *two* repairers ($s = 2$).

The $h(\mathcal{W})$ Form

Now consider the cases discussed in Sec. 7.2 where the queueing system *customers* are *external* to the organization providing the service. Three major types of queueing systems described in Sec. 6.3—commercial service systems, transportation service systems, and social service systems—typically fall into this category. In the case of *commercial service systems*, the primary cost of waiting may be the *lost profit from lost future business*. For *transportation service systems* and *social systems*, the primary cost of waiting may be in the form of a *social cost*. However, for either type of cost, its magnitude tends to be affected greatly by the size of the waiting times experienced by the customers. Thus the primary property of the queueing system that determines the waiting cost currently being incurred is $\mathcal{W}$, the waiting time in the system for the

Table 7.1 **Calculation of $E(WC)$ for Example 1**

		s = 1		s = 2	
$N = n$	$g(n)$	P_n	$g(n)P_n$	P_n	$g(n)P_n$
0	0	0.271	0	0.433	0
1	0	0.217	0	0.346	0
2	0	0.173	0	0.139	0
3	100	0.139	14	0.055	6
4	200	0.097	19	0.019	4
5	300	0.058	17	0.006	2
6	400	0.029	12	0.001	0
7	500	0.012	6	3×10^{-4}	0
8	600	0.003	2	4×10^{-5}	0
9	700	7×10^{-4}	0	4×10^{-6}	0
10	800	7×10^{-5}	0	2×10^{-7}	0
	$E(WC)$		\$70/day		\$12/day

individual customers. Consequently, the form of the waiting-cost function for this kind of situation is that illustrated in Fig. 7.6, namely, a function of $\mathcal{W}$. We shall denote this form by $h(\mathcal{W})$.

One way of constructing the $h(\mathcal{W})$ function is to estimate $h(w)$ (the waiting cost incurred when a customer's waiting time $\mathcal{W} = w$) for several different values of w and then to fit a polynomial to these points. The expectation of this *function* of a *continuous* random variable is then defined as

$$E\{h(\mathcal{W})\} = \int_0^\infty h(w) f_{\mathcal{W}}(w)\, dw,$$

where $f_{\mathcal{W}}(w)$ is the probability density function of $\mathcal{W}$. However, because $E\{h(\mathcal{W})\}$ is the expected waiting cost *per customer* and $E(WC)$ is the expected waiting cost *per unit time*, these two quantities are not equal in this case. To relate them it is necessary to multiply $E\{h(\mathcal{W})\}$ by the expected *number of customers per unit time* entering the queueing system. In particular, if the mean arrival rate is a constant λ, then

$$E(WC) = \lambda E\{h(\mathcal{W})\} = \lambda \int_0^\infty h(w) f_{\mathcal{W}}(w)\, dw.$$

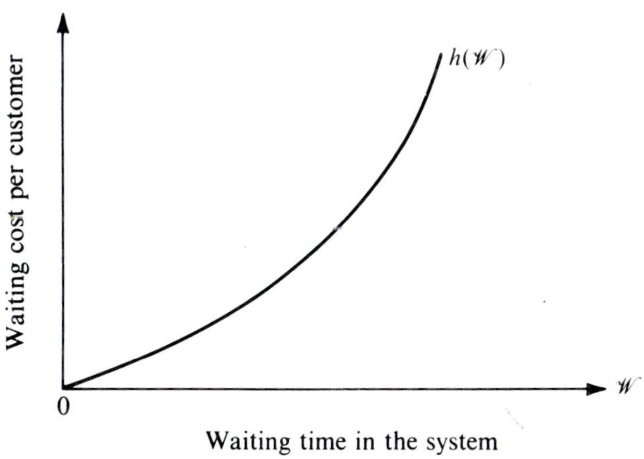

Figure 7.6 The waiting-cost function as a function of $\mathcal{W}$.

Example 2—Which Computer? Because the students of *Emerald University* would experience different turnaround times with the two computers under consideration (see Sec. 7.1), the choice between the computers required an evaluation of the *consequences* of making students wait for their programs to be run. This evaluation led to the following conclusions.

From the university's viewpoint, there are two primary consequences of making students wait. First, it decreases the time available for the student to pursue other academic endeavors because little effective studying can be done during the wait. Second, it detracts from the efficiency and academic value of the computer assignment because it breaks the continuity in dealing with the problem, and it may prevent the student from completing the assignment satisfactorily.

To evaluate the first consequence, an estimate was made that an average student studies at only one-third of full efficiency during the wait. Furthermore, by calculating the total expenditures from all sources for a student's college education, a monetary figure of \$15/hour was placed on the value of being able to study at *full* efficiency. Therefore, this component of waiting cost was estimated to be \$10/hour, that is, $10\mathcal{W}$, where $\mathcal{W}$ is expressed in units of hours.

To evaluate the second consequence, two groups of students were interviewed just after experiencing turnaround times of $\frac{1}{2}$ hour and 1 hour, respectively. They were asked to estimate how much additional time working on the assignment had been caused by the break in the continuity, as well as the effect of the wait on their ability to complete the assignment satisfactorily. On this basis, averages of \$2 and \$8 were imputed to the value of a student being able to avoid this consequence entirely rather than having a wait of $\frac{1}{2}$ hour and 1 hour, respectively. Therefore, this component of the waiting cost was estimated to be $8\mathcal{W}^2$.

This analysis yields

$$h(\mathcal{W}) = 10\mathcal{W} + 8\mathcal{W}^2.$$

Because

$$f_{\mathcal{W}}(w) = \mu(1 - \rho)e^{-\mu(1-\rho)w}$$

for the M/M/1 model (see Sec. 6.6) fitting this single-server queueing system,

$$E\{h(\mathcal{W})\} = \int_0^\infty (10w + 8w^2)\mu(1 - \rho)e^{-\mu(1-\rho)w}\,dw.$$

Using the fact that $\mu(1 - \rho) = \mu - \lambda$ for a single-server system, the values of μ and λ presented in Sec. 7.1 give

$$\mu(1 - \rho) = \begin{cases} 10, & \text{for the MBI computer} \\ 5, & \text{for the EG computer.} \end{cases}$$

Evaluating the integral for these two cases yields

$$E\{h(\mathcal{W})\} = \begin{cases} 1.16, & \text{for the MBI computer} \\ 2.64, & \text{for the EG computer.} \end{cases}$$

This result represents the expected waiting cost (in dollars) for each student arriving with a computer program to be run. Because $\lambda = 20$, the total expected waiting cost per hour becomes

$$E(WC) = \begin{cases} \$23.20/\text{hr}, & \text{for the MBI computer} \\ \$52.80/\text{hr}, & \text{for the EG computer.} \end{cases}$$

The Linear Case: When $h(\mathcal{W})$ is a *linear* function,

$$h(\mathcal{W}) = C_w \mathcal{W},$$

$E(WC)$ reduces to

$$E(WC) = \lambda E(C_w \mathcal{W}) = C_w(\lambda W) = C_w L.$$

Note that this result is identical to the result when $g(N)$ is a linear function. Consequently, when the total waiting cost incurred by the queueing system is simply *proportional* to the total waiting time, it does not matter whether the $g(N)$ or the $h(\mathcal{W})$ form is used for the waiting-cost function.

Example 3—How Many Tool Cribs? As indicated in Sec. 7.1, the value to the *Mechanical Company* of a busy mechanic's output averages about \$24/hour. Thus $C_w = 24$. Consequently, *for each tool crib* the expected waiting cost per hour is

$$E(WC) = 24L,$$

where L represents the expected number of mechanics waiting (or being served) at the tool crib.

7.4 Decision Models

We mentioned in Sec. 7.2 that three common decision variables in designing queueing systems are s (number of servers), μ (mean service rate for each server), and λ (mean arrival rate at each service facility). We shall now formulate models for making some of these decisions.

Model 1—Unknown *s*

Model 1 is designed for the case where both μ and λ are fixed at a particular service facility, but where a decision must be made on the number of servers to have on duty at the facility.

Formulation of Model 1

Definition: C_s = marginal cost of a server per unit time.
Given: μ, λ, C_s.
To find: s.
Objective: Minimize $E(TC) = sC_s + E(WC)$.

Because only a few alternative values of s normally need to be considered, the usual way of solving this model is to calculate $E(TC)$ for these values of s and select the minimizing one.

Example 1—How Many Repairers? For Example 1 of Sec. 7.1, each repairer (server) costs *Simulation, Inc.*, approximately \$70/day. Thus, with a day as the unit of time, $C_s = 70$. Using the values of $E(WC)$ calculated in Table 7.1 then yields the results shown in Table 7.2, which indicate that the company should continue having just *one* repairer.

Table 17.2 **Calculation of *E(TC)* for Example 1**

s	sC_s	$E(WC)$	$E(TC)$
1	70	70	\$140/day ← minimum
2	140	12	\$152/day
≥3	≥210	≥ 0	≥\$210/day

Model 2—Unknown μ and s

Model 2 is designed for the case where both the *efficiency* of service, measured by μ, and the *number* of servers s at a service facility need to be selected. Alternative values of μ may be available because there is a choice on the *quality* of the servers. (In one example, both the *type* and *quantity* of materials-handling equipment to transport certain kinds of loads must be selected for purchase.) Another possibility is that the *speed* of the servers can be adjusted mechanically. (For example, the speed of machines frequently can be adjusted by changing the amount of power consumed, which also changes the cost of operation.) Still another type of example is the selection of the number of crews (the servers) and the size of each crew (which determines μ) for jointly performing a certain task, e.g., maintenance work, loading and unloading operations, inspection work, setup of machines, and so forth. In many cases, only a few alternative values of μ are available, e.g., the efficiency of the alternative types of materials-handling equipment or the efficiency of the alternative crew sizes.

Formulation of Model 2

Definitions: $f(\mu)$ = marginal cost of server per unit time when mean service rate is μ.
A = set of feasible values of μ.

Given: λ, $f(\mu)$, A.

To find: μ, s.

Objective: Minimize $E(TC) = sf(\mu) + E(WC)$, subject to $\mu \in A$.

Example 2—Which Computer? As indicated in Sec. 7.1, $\mu = 30$ for the *MBI computer* and $\mu = 25$ for the *EG computer,* where 1 hour is the unit of time. These computers are the only two being considered by Emerald University, so

$$A = \{25, 30\}.$$

Because the leasing cost per operating hour would be \$75 for the EG computer ($\mu = 25$) and \$100 for the MBI computer ($\mu = 30$),

$$f(\mu) = \begin{cases} 75, & \text{for } \mu = 25 \\ 100, & \text{for } \mu = 30. \end{cases}$$

The computer chosen will be the only one available for student use, so the number of servers (computers) for this queueing system is restricted to be $s = 1$. Hence

$$E(TC) = f(\mu) + E(WC),$$

where $E(WC)$ is given in Sec. 7.3 for the two alternatives. Thus

$$\begin{aligned} E(TC) &= 75 + 52.80 \\ &= \$127.80/\text{hr}, \qquad \text{for EG computer,} \end{aligned}$$

$$E(TC) = 100 + 23.20$$
$$= \$123.20/\text{hr}, \qquad \text{for MBI computer.}$$

Consequently, the decision was made to lease the MBI computer.

This example illustrates a case where the number of feasible values of μ is *finite* but the value of s is fixed. If s were not fixed, a two-stage approach could be used to solve such a problem. First, for each individual value of μ, set $C_s = f(\mu)$, and solve for the value of s that minimizes $E(TC)$ for model 1. Second, compare these minimum $E(TC)$ for the alternative values of μ, and select the one giving the overall minimum.

When the number of feasible values of μ is *infinite* (as when mechanically setting the speed of a machine or piece of equipment within some feasible interval), another two-stage approach sometimes can be used to solve the problem. First, for each individual value of s, *analytically* solve for the value of μ that minimizes $E(TC)$. [This approach requires setting to zero the derivative of $E(TC)$ with respect to μ and then solving this equation for μ, which can be done only when analytical expressions are available for both $f(\mu)$ and $E(WC)$.] Second, compare these minimum $E(TC)$ for the alternative values of s, and select the one giving the overall minimum.

This analytical approach frequently is relatively straightforward for the case of $s = 1$ (see Prob. 14). However, because far fewer and less convenient analytical results are available for *multiple-server* versions of queueing models, this approach is either difficult (requiring computer calculations with numerical methods to solve the equation for μ) or completely impossible when $s > 1$. Therefore, a more practical approach is to consider only a relatively small number of representative values of μ and to use available tabulated results for the appropriate queueing model to obtain (or approximate) $E(TC)$ for these μ.

Fortunately, under certain fairly common circumstances described next, $s = 1$ (and its minimizing value of μ) *must* yield the overall minimum $E(TC)$ for model 2, so $s > 1$ cases need not be considered at all.

Optimality of a single server: Under certain conditions, $s = 1$ necessarily is *optimal* for model 2.

The primary conditions[1] are

1. The value of μ minimizing $E(TC)$ for $s = 1$ is feasible,
2. $f(\mu)$ is either a *linear* function or a *concave* function (as defined in Appendix 1).

In effect, this optimality result indicates that it is better to concentrate service capacity into one *fast* server rather than dispersing it among several *slow* servers. Condition 2 says that this concentrating of a given amount of service capacity can be done without increasing the cost of service. Condition 1 says that it must be possible to make μ sufficiently large so that a single server can be used to full advantage.

[1] There also are minor restrictions on the queueing model and the waiting-cost function. However, any of the constant service-rate queueing models presented in Chap. 6 for $s \geq 1$ are allowed. If the $g(N)$ form is used for the waiting-cost function, it can be any *increasing* function. If the $h(\mathcal{W})$ form is used, it can be any *linear* function or any *convex* function, which fits most cases of interest.

Table 7.3 **Comparison of Service Efficiency for Model 2 Solutions**

	Mean Rate of Service Completions
$N = n$	(s^*, μ^*) vs. $(1, s^*\mu^*)$ for (s, μ), where $s^* > 1$
$n = 0$	$0 = 0$
$n = 1, 2, \ldots, s^* - 1$	$n\mu^* < s^*\mu^*$
$n \geq s^*$	$s^*\mu^* = s^*\mu^*$

To understand why this result holds, consider *any* other solution to model 2, $(s, \mu) = (s^*, \mu^*)$, where $s^* > 1$. The service capacity of this system (as measured by the mean rate of service completions when all servers are working) would be $s^*\mu^*$. We shall now compare this solution with the corresponding *single-server* solution $(s, \mu) = (1, s^*\mu^*)$ having the *same* service capacity. In particular, Table 7.3 compares the *mean rate* at which service completions occur for each given number of customers in the system $N = n$. Table 7.3 shows that the service efficiency of the (s^*, μ^*) solution sometimes is worse but never is better than for the $(1, s^*\mu^*)$ solution because it can use the full service capacity only when there are at least s^* customers in the system, whereas the single-server solution uses the full capacity whenever there are *any* customers in the system. Because this lower service efficiency can only increase waiting in the system, $E(WC)$ must be *larger* for (s^*, μ^*) than $(1, s^*\mu^*)$. Furthermore, the expected service cost must be *at least as large* because condition 2 [and $f(0) = 0$] implies that

$$s^*f(\mu^*) \geq f(s^*\mu^*).$$

Therefore, $E(TC)$ is *larger* for (s^*, μ^*) than $(1, s^*\mu^*)$. Finally, note that condition 1 implies that there is a *feasible* solution with $s = 1$ that is *at least as good* as $(1, s^*\mu^*)$. The conclusion is that *any* $s > 1$ solution *cannot* be optimal for model 2, so $s = 1$ must be optimal.[1]

This result is still of some use even when one or both conditions fail to hold. If μ cannot be made sufficiently large to permit a single server, it still suggests that a *few* fast servers should be preferred to many slow ones. If condition 2 does not hold, we still know that $E(WC)$ is minimized by concentrating any given amount of service capacity into a single server, so the best $s = 1$ solution must be at least nearly optimal unless it causes a *substantial* increase in service cost.

Model 3—Unknown λ and *s*

Model 3 is designed especially for the case where it is necessary to select both the *number of service facilities* and the *number of servers* (s) at each facility. The typical situation would be where a population (such as the employees in an industrial building) must be provided with a certain service, so a decision must be made as to what proportion of the population (and therefore what value of λ) should be assigned to each service facility. Examples of such facilities include *employee facilities* (drinking fountains, vending machines, and restrooms), *storage facilities,* and *reproduction equipment facilities*. It may sometimes be clear that only a single server should be provided at each facility (e.g., one drinking fountain or one copy machine), but s often is also a decision variable.

[1] For a rigorous proof of this result, see Stidham, Shaler, Jr.: "On the Optimality of Single-Server Queueing Systems," *Operations Research,* **18**:708–732, 1970.

To simplify our presentation, we shall require in model 3 that λ and s must be the same for all service facilities. However, it should be recognized that a slight improvement in the indicated solution might be achieved by permitting minor deviations in these parameters at individual facilities. This should be investigated as part of the detailed analysis that generally follows the application of the mathematical model.

Formulation of Model 3

Definitions: C_s = marginal cost of server per unit time.
C_f = fixed cost of service per service facility per unit time.
λ_p = mean arrival rate for entire population.
n = number of service facilities = λ_p/λ.
Given: μ, C_s, C_f, λ_p.
To find: λ, s.
Objective: Minimize $E(TC)$, subject to $\lambda = \lambda_p/n$, where $n = 1, 2, \ldots$.

It might appear at first glance that the appropriate expression for the expected total cost per unit time of all the facilities should be

$$E(TC) \overset{?}{=} n[(C_f + sC_s) + E(WC)],$$

where $E(WC)$ here represents the expected waiting cost per unit time for *each* facility. However, if this expression actually were valid, it would imply that $n = 1$ *necessarily* is optimal for model 3. [The reasoning is completely analogous to that for the optimality of a single-server result for model 2, namely, any solution $(n, s) = (n^*, s^*)$ with $n^* > 1$ has higher service costs than the $(n, s) = (1, n^*s^*)$ solution, and it *also* has a higher expected waiting cost because it sometimes makes less effective use of the available service capacity. In particular, it sometimes has idle servers at one facility while customers are waiting at another facility, so the mean rate of service completions would be less than it would be if the customers had access to *all* the servers at one common facility.] Because there are many situations where it obviously would *not* be optimal to have just *one* service facility (e.g., the number of restrooms in the Pentagon), something must be wrong with this expression. Its deficiency is that it considers only the cost of service and the cost of waiting *at the service facilities* and totally ignores the cost of the time wasted in *traveling to and from* the facilities. Because travel time would be prohibitive with only one service facility for a large population, enough separate facilities must be distributed throughout the population to hold travel time down to a reasonable level.

Thus, letting the *random variable* T be the round-trip *travel time* for a customer coming to and going back from one of the service facilities, the *total time lost* by the customer actually is $(\mathcal{W} + T)$. (Recall from Chap. 6 that $\mathcal{W}$ is the waiting time in the queueing system *after* the customer arrives.) Therefore, a customer's *total* cost for time lost should be based on $(\mathcal{W} + T)$ rather than just $\mathcal{W}$. To simplify the analysis, we shall separate this total cost into the *sum* of the *waiting-time cost* based on $\mathcal{W}$ (or N) and the *travel-time cost* based on T. We shall also assume that the travel-time cost is *proportional* to T, where C_t is the cost of each unit of travel time for each customer. For ease of presentation, suppose that the probability distribution of T is the same for each service facility, so that $C_tE(T)$ is the *expected travel cost* for each arrival at any of the service facilities. The resulting expression for $E(TC)$ would be

$$E(TC) = n[(C_f + sC_s) + E(WC) + \lambda C_t\, E(T)]$$

because λ is the expected number of arrivals *per unit time* at each facility. Consequently, if $E(T)$ could be evaluated for each case of interest, model 3 can be solved by calculating $E(TC)$ for various values of s for each n and then selecting the solution giving the overall minimum. The next section discusses how to evaluate $E(T)$ and also solves an example (Example 3 of Sec. 7.1) fitting model 3.

7.5 The Evaluation of Travel Time

$E(T)$ can be interpreted as the *average travel time* spent by customers in coming both to and from a given service facility. Therefore, the value of $E(T)$ depends very much upon the characteristics of the individual situation. However, we shall illustrate a rather general approach to evaluating $E(T)$ by developing a basic travel-time model and then calculating $E(T)$ for a particular example involving a more complicated situation. In both cases it is assumed that the portion of the population assigned to the service facility under consideration is *distributed uniformly* throughout the assigned area, that each arrival returns to its *original location* after receiving service, and that the average speed of travel does *not* depend upon the distance traveled. Another basic assumption is that all travel is *rectilinear,* i.e., it progresses along a system of *orthogonal* paths (aisles, streets, highways, and so on) that are *parallel* to the main sides of the area under consideration.

A Basic Travel-Time Model

Description: Rectangular area and rectilinear travel, as shown in Fig. 7.7.

Definitions:
- T = travel time (round trip) for an arrival.
- v = average velocity (speed) of customers in traveling to and from facility.
- a, b, c, d = respective distances from facility to boundary of area assigned to facility, as shown in Fig. 7.7.

Given: v, a, b, c, d.

To find: Expected value of T, $E(T)$.

Using an orthogonal (x, y) coordinate system, Fig. 7.7 shows the coordinates (x, y) of the location of a *particular* customer. The x and y coordinates of the location from which a *random* arrival comes actually are *random variables* X and Y, where X

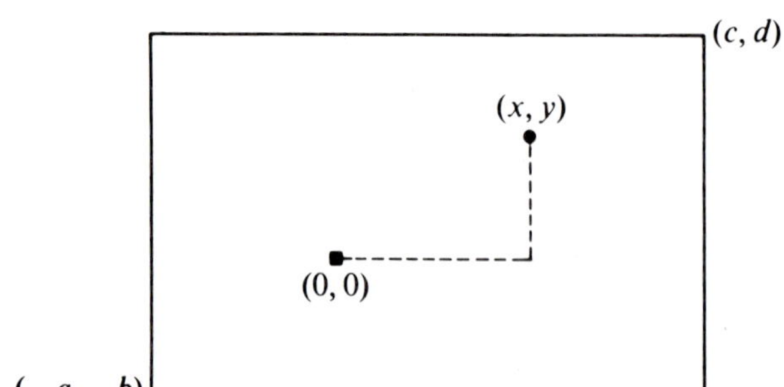

Figure 7.7 Graphical representation of a basic travel-time model, where the service facility is at (0, 0) and a random arrival comes from (and returns to) some location (x, y).

ranges from $-a$ to c and Y ranges from $-b$ to d. Because the total round-trip distance traveled by the random arrival is

$$D = 2(|X| + |Y|)$$

and

$$T = \frac{D}{v},$$

it follows that

$$E(T) = \frac{2}{v}(E\{|X|\} + E\{|Y|\}).$$

Thus the problem is reduced to identifying the probability distributions of $|X|$ and $|Y|$ and then calculating their means.

First consider $|X|$. Its probability distribution can be obtained directly from the distribution of X. Because the customers are assumed to be distributed uniformly throughout the assigned area, and because the *height* of the rectangular area is the *same* for all possible values of $X = x$, X must have a *uniform distribution* between $-a$ and c, as shown in Fig. 7.8*a*. Because $|x| = |-x|$, adding the probability density function values at x and $-x$ then yields the probability distribution of $|X|$ shown in Fig. 7.8*b*.

Therefore, noting that $|x| = x$ for $x \geq 0$,

$$\begin{aligned} E\{|X|\} &= \int_0^{\max\{a,\,c\}} x f_{|X|}(x)\,dx \\ &= \int_0^{\min\{a,\,c\}} \frac{2x}{a+c}\,dx + \int_{\min\{a,\,c\}}^{\max\{a,\,c\}} \frac{x}{a+c}\,dx \\ &= \frac{1}{2}\frac{1}{a+c}[(\min\{a,\,c\})^2 + (\max\{a,\,c\})^2] \\ &= \frac{a^2+c^2}{2(a+c)}. \end{aligned}$$

The analysis for $|Y|$ is completely analogous, where the *width* of the rectangular area for possible values of $Y = y$ now determines the probability distribution of Y.

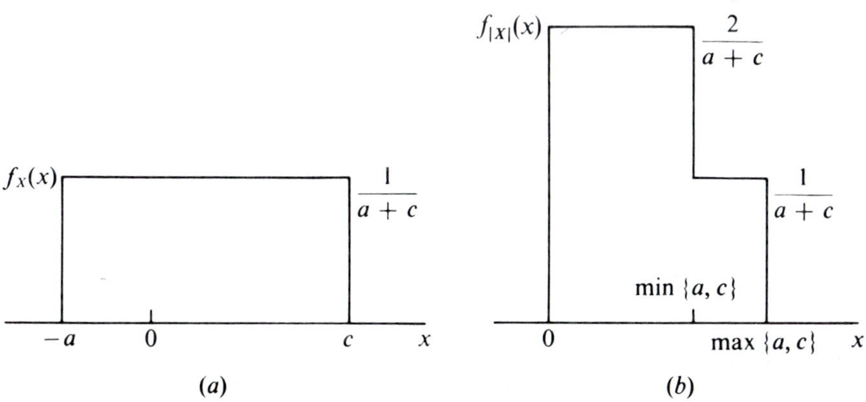

Figure 7.8 Probability density functions of (*a*) X; (*b*) $|X|$.

The result is that

$$E\{|Y|\} = \frac{b^2 + d^2}{2(b + d)}.$$

Consequently,

$$E(T) = \frac{1}{v}\left(\frac{a^2 + c^2}{a + c} + \frac{b^2 + d^2}{b + d}\right).$$

Example 3—How Many Tool Cribs? For the new plant being designed for the *Mechanical Company* (see Sec. 7.1), the layout of the portion of the factory area where the mechanics will work is shown in Fig. 7.9. The three *possible* locations for tool cribs are identified as Locations 1, 2, and 3, where access to these locations will be provided by a system of orthogonal aisles parallel to the sides of the indicated area. The coordinates are given in units of *feet*. The mechanics will be distributed quite uniformly throughout the area shown, and each mechanic will be assigned to the *nearest* tool crib. It is estimated that the mechanics will walk to and from a tool crib at an average speed of slightly less than 3 miles/hour, so v is set at $v = 15{,}000$ feet/hour.

The three basic alternatives being considered are

Alternative 1: Have *three* tool cribs—use Locations 1, 2, and 3;
Alternative 2: Have *one* tool crib—use Location 2;
Alternative 3: Have *two* tool cribs—use Locations 1 and 3.

The calculation of $E(T)$ for each alternative is given next, followed by the use of model 3 to make the choice among them.

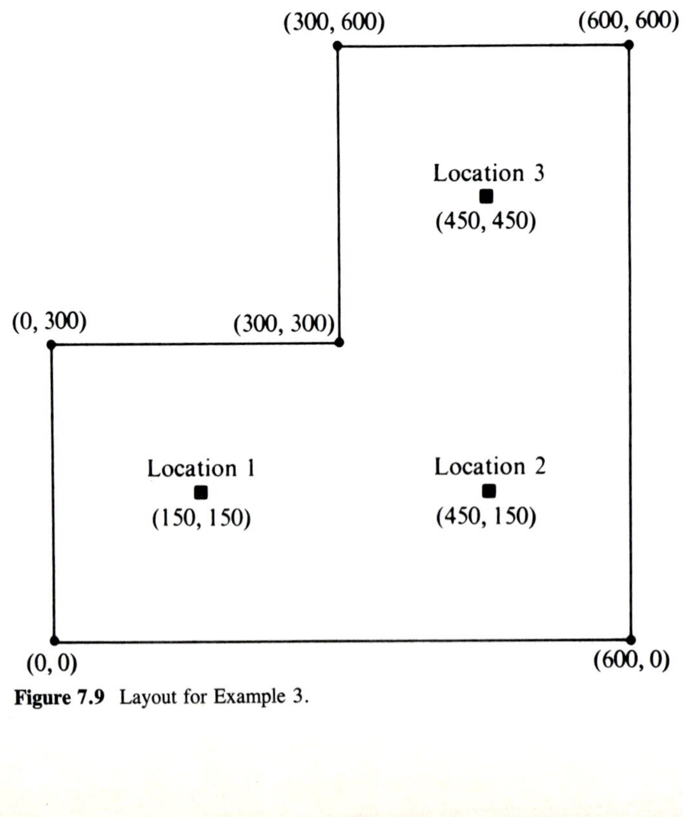

Figure 7.9 Layout for Example 3.

Alternative 1 ($n = 3$): If all three locations were used, *each* tool crib would service a 300×300 foot *square* area. Therefore, this case is just a special case of the basic travel-time model just presented, where $a = c = 150$ and $b = d = 150$. Consequently,

$$E(T) = \frac{1}{15{,}000 \text{ ft/hr}} \left(\frac{150^2 + 150^2}{150 + 150} + \frac{150^2 + 150^2}{150 + 150} \right) \text{ft}$$

$$= \frac{1}{15{,}000 \text{ ft/hr}} (300 \text{ ft})$$

$$= 0.02 \text{ hr}.$$

Alternative 2 ($n = 1$): With just *one* tool crib (in Location 2) to service the entire area shown in Fig. 7.9, the derivation of $E(T)$ is a little more complicated than it is for the basic travel-time model. The first step is to relabel Location 2 as the origin (0, 0) for an (x, y) coordinate system, so that 450 would be subtracted from the first coordinates shown and 150 would be subtracted from the second coordinates. The probability density function for X is then obtained by dividing the *height* for each possible value of $X = x$ by the total area (so that the area under the probability density function curve equals 1), as given in Fig. 7.10*a*. Combining the values for x and $-x$ then yields the probability distribution of $|X|$ shown in Fig. 7.10*b*.

Hence

$$E\{|X|\} = \int_0^{450} x f_{|X|}(x)\, dx$$

$$= \int_0^{150} x \left(\frac{1}{225}\right) dx + \int_{150}^{450} x \left(\frac{1}{900}\right) dx$$

$$= \frac{150^2}{450} + \frac{450^2 - 150^2}{1{,}800} = 150.$$

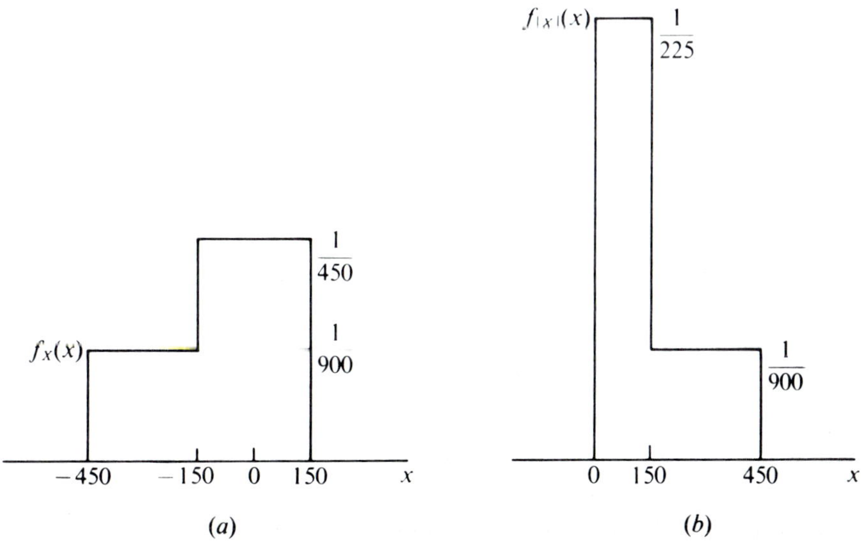

Figure 7.10 Probability density functions of (*a*) X and (*b*) $|X|$ for a tool crib at Location 2 of Fig. 7.9 under Alternative 2 (no other tool cribs).

We suggest that you now try the same approach (using the *width* of the area rather than the height) to derive $E\{|Y|\}$. You will find that the probability distribution of $|Y|$ is *identical* to that for $|X|$, so $E\{|Y|\} = 150$. As a result,

$$E(T) = \frac{2}{15{,}000}(150 + 150)$$
$$= 0.04 \text{ hr.}$$

Alternative 3 ($n = 2$): With tool cribs in just Locations 1 and 3, the areas assigned to them would be divided by a line segment between (300, 300) and (600, 0) in Fig. 7.9. Notice that the two areas and their tool cribs are located symmetrically with respect to this line segment. Therefore, $E(T)$ is the same for both, so we shall derive it just for the tool crib in Location 1. (You might try it for the other tool crib for practice—see Prob. 18.)

Proceeding just as for Alternative 2, relabel Location 1 as the origin (0, 0) for an (x, y) coordinate system, so that 150 would be subtracted from all coordinates shown in Fig. 7.9. This relabeling leads directly to the probability density function of X, and then of $|X|$, shown in Fig. 7.11. As a result,

$$E\{|X|\} = \frac{1}{225}\int_0^{150} x\,dx + \frac{1}{300}\int_{150}^{450}\left(1 - \frac{x}{450}\right)x\,dx$$
$$= \frac{1}{225}\left[\frac{x^2}{2}\right]_0^{150} + \frac{1}{300}\left[\frac{x^2}{2} - \frac{x^3}{1{,}350}\right]_{150}^{450}$$
$$= \frac{1}{225}\,\frac{150^2}{2} + \frac{1}{300}\left(\frac{450^2}{2} - \frac{450^3}{1{,}350}\right) - \frac{1}{300}\left(\frac{150^2}{2} - \frac{150^3}{1{,}350}\right)$$
$$= 133\tfrac{1}{3}.$$

Next, the probability density function of Y is obtained by using the *width* of the area assigned to the tool crib at Location 1 for each possible value of $Y = y$ and then dividing by the size of the area, as given in Fig. 7.12*a*. This result then yields the

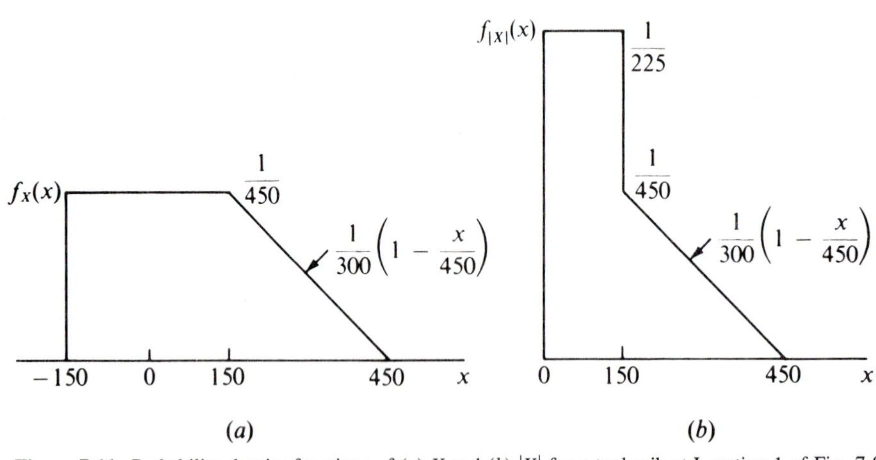

Figure 7.11 Probability density functions of (*a*) X and (*b*) $|X|$ for a tool crib at Location 1 of Fig. 7.9 under Alternative 3 (the only other tool crib is at Location 3).

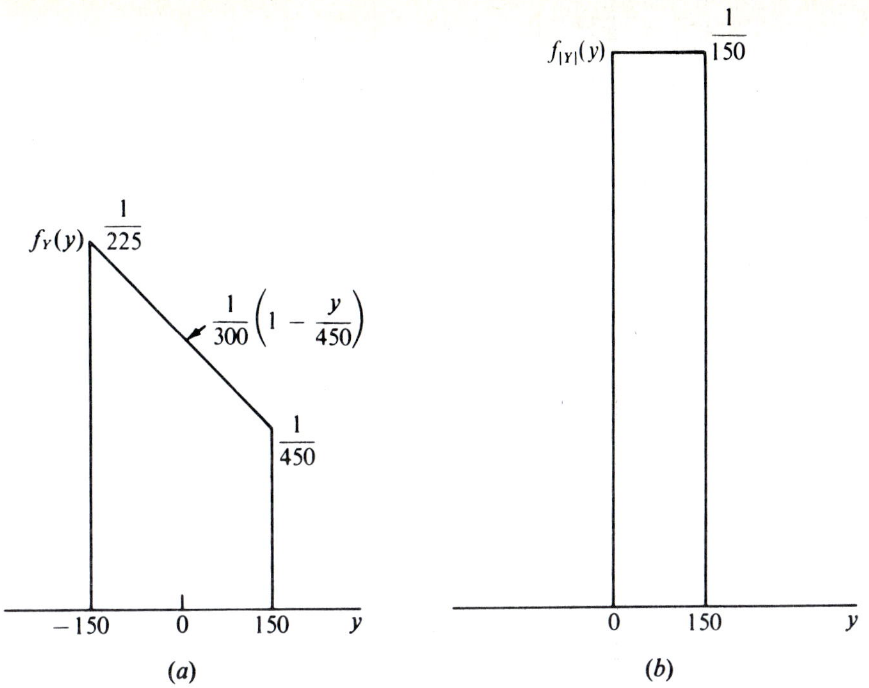

Figure 7.12 Probability density functions of (*a*) Y and (*b*) $|Y|$ for a tool crib at Location 1 of Fig. 7.9 under Alternative 3 (the only other tool crib is at Location 3).

uniform distribution of $|Y|$ shown in Fig. 7.12*b*. Thus

$$E\{|Y|\} = \frac{1}{150}\int_0^{150} y\,dy$$

$$= 75.$$

Consequently,

$$E(T) = \frac{2}{15{,}000}\left(133\tfrac{1}{3} + 75\right)$$

$$= 0.0278 \text{ hr.}$$

Applying Model 3: Because $E(T)$ now has been evaluated for the three alternatives under consideration, the stage is set for using model 3 from Sec. 7.4 to choose among these alternatives. Most of the data required for this model are given in Sec. 7.1, namely,

$$\mu = 120/\text{hr}, \qquad C_f = \$8/\text{hr},$$
$$C_s = \$10/\text{hr},$$
$$\lambda_p = 120/\text{hr}, \qquad C_t = \$24/\text{hr},$$

where the M/M/s model given in Sec. 6.6 would be used to calculate L, and so on. In addition, the end of Sec. 7.3 gives $E(WC) = 24L$ in dollars per hour. Therefore,

$$E(TC) = n\left[(8 + 10s) + 24L + \frac{120}{n}24E(T)\right].$$

Table 7.4 **Calculation of $E(TC)$ in \$/hr for Example 3**

n	λ	s	L	$E(T)$	$C_f + sC_s$	$E(WC)$	$\lambda C_t E(T)$	$E(TC)$
1	120	1	∞	0.04	18.00	∞	115.20	∞
1	120	2	1.333	0.04	28.00	32.00	115.20	175.20
1	120	3	1.044	0.04	38.00	25.06	115.20	178.26
2	60	1	1.000	0.0278	18.00	24.00	40.00	164.00
2	60	2	0.534	0.0278	28.00	12.82	40.00	161.64
3	40	1	0.500	0.02	18.00	12.00	19.20	147.60
3	40	2	0.344	0.02	28.00	8.26	19.20	166.38

The resulting calculation of $E(TC)$ for various s for each n is given in Table 7.4, which indicates that the *overall minimum* $E(TC)$ is obtained by having *three* tool cribs (so $\lambda = 40$ for each), with *one* clerk at *each* tool crib.

7.6 Estimating Model Parameters

All the results derived in Chap. 6 were expressed in terms of the parameters of the interarrival time and service-time distributions. In most practical situations these parameters are unknown, and so estimates must be obtained. This section suggests a few convenient estimating procedures, based on adapting the standard statistical procedures reviewed in Chap. 4 to queueing systems. The assumption is made throughout that statistical data regarding arrivals and service times can be obtained for developing the estimates.

Consider a queueing system where the mean arrival rate and the mean service rate per busy server are both fixed constants (λ and μ, respectively). To estimate λ, one approach is to count the number of customers N that enter the queueing system over a time interval of specified length t; an *unbiased estimator* of λ would then be

$$\hat{\lambda} = \frac{N}{t}.$$

An alternative approach is to instead fix the number of customers observed (call it n) rather than the length of the time interval. Thus, let T be the total elapsed time that the system is observed until the last of the n arrivals has entered the system. T is then the sum of the n interarrival times. An estimator of $1/\lambda$ is given by T/n, so that

$$\hat{\hat{\lambda}} = \frac{n}{T}$$

is an estimator of λ. If the system has a Poisson input, then either $\hat{\lambda}$ or $\hat{\hat{\lambda}}$ is a *maximum likelihood estimator* of λ, depending upon how the data are recorded.

To estimate μ, it might seem reasonable to tabulate both the total number of customers served, M, and the sum of the busy times (time spent servicing customers) of the servers, B, over a time interval of specified length; an estimator of μ would then be

$$\hat{\mu} = \frac{M}{B}.$$

However, it is unclear how to count customers that are in the process of being served at the beginning or the end of the time interval, and it would create a bias to consider only the number and service times of the customers that are completely served during the time interval. Therefore, it is usually preferable to use the alternative approach of fixing the number of customers observed rather than the length of the time interval. Thus, let B be the sum of the service times of the first m customers that begin service after initiating the observation of the system, where m is a prespecified number. (In a multiple-server system it may be more convenient to record the sum of m service times for only *one* of the servers, which is also permissible.) An estimator of $1/\mu$ is then given by B/m, so that

$$\hat{\mu} = \frac{m}{B}$$

is an estimator of μ. If the service-time distribution is exponential, then $\hat{\mu}$ is a *maximum likelihood estimator* of μ.

The *utilization factor* $\rho = \lambda/s\mu$ plays a particularly important role in queueing theory. An estimator $\hat{\rho}$ of this parameter can be obtained from

$$\hat{\rho} = \frac{\hat{\lambda}}{s\hat{\mu}} = \frac{n/T}{sm/B} = \frac{B/m}{sT/n}.$$

For the M/M/s model described in Sec. 6.6, $\hat{\rho}$ is a *maximum likelihood estimator* of ρ.

For this M/M/s model, utilization of distributional properties of the random variables B and T can lead to the calculation of a *confidence interval* for ρ. Since T is the sum of n independent exponentially distributed (each with mean $1/\lambda$) interarrival times, $2\lambda T$ has a chi-square distribution with $2n$ degrees of freedom. Similarly, since B is the sum of m independent exponentially distributed (each with mean $1/\mu$) service times, $2\mu B$ has a chi-square distribution with $2m$ degrees of freedom. Furthermore, B and T are assumed to be independent. Hence it follows that

$$\frac{\mu B/m}{\lambda T/n} = \frac{\hat{\rho}}{\rho}$$

has an F distribution with $2m$ and $2n$ degrees of freedom. Consequently,

$$P\left\{F_{1-\alpha/2;2m,2n} \le \frac{\hat{\rho}}{\rho} \le F_{\alpha/2;2m,2n}\right\} = 1 - \alpha,$$

where $F_{\gamma;v_1,v_2}$ is the 100α percentage point of the F distribution, with v_1 and v_2 degrees of freedom, as tabulated in Table A5.5 of Appendix 5. This probability statement is equivalent[1] to

$$P\left\{\frac{\hat{\rho}}{F_{\alpha/2;2m,2n}} \le \rho \le \hat{\rho}F_{\alpha/2;2n,2m}\right\} = 1 - \alpha,$$

so that

$$\underline{\rho} = \frac{\hat{\rho}}{F_{\alpha/2;2m,2n}}, \qquad \bar{\rho} = \hat{\rho}F_{\alpha/2;2n,2m}$$

are the endpoints of a $100(1 - \alpha)$ percent confidence interval for ρ.

[1] The equivalence is based partially on the fact that

$$F_{1-\gamma;v_1,v_2} = \frac{1}{F_{\gamma,v_2,v_1}}.$$

These results are also useful for obtaining a confidence interval estimator for any *monotone increasing* function of ρ, say, $H(P)$.[1] The appropriate $100(1 - \alpha)$ percent confidence interval for $H(\rho)$ is given by

$$H(\underline{\rho}) \leq H(\rho) \leq H(\bar{\rho}).$$

Some important cases of monotone increasing functions that $H(\rho)$ can represent are L, L_q, and P_0.

For example, when $s = 1$, $L = \rho/(1 - \rho)$. Since $\rho/(1 - \rho)$ is monotone increasing, the endpoints of a $100(1 - \alpha)$ percent confidence interval for L are

$$\frac{\underline{\rho}}{1 - \underline{\rho}}, \qquad \frac{\bar{\rho}}{1 - \bar{\rho}}.$$

The corresponding confidence intervals for L_q and P_0 are shown in Table 7.5. The same approach also can be used for $s > 1$, as illustrated by the two-server results given in Table 7.5. A similar derivation yields the confidence intervals given in the last part of Table 7.5 for the $M/E_k/1$ model described in Sec. 6.7.

Finally, consider the queueing model with state-dependent service rate presented at the end of Sec. 6.6. Suppose that $\lambda_n = \lambda_0$ for $n = 0, 1, \ldots$, so the model indicates that $\mu_n = n\mu_1$ if $n \leq s$ and $\mu_n = (n/s)^c s\mu_1$ if $n \geq s$. Hence, $\log \mu_n = c \log (n/s) + \log (s\mu_1)$, if $n \geq s$. Given that there are n customers in the system, service times are assumed to have an *exponential* distribution with parameter μ_B. Therefore, a *maximum likelihood estimator* of μ_n is the observed number of customers that completed service while n customers were in the system divided by the total observed time n customers were in the system. After obtaining these estimates, $\hat{\mu}_n$ for $n - s = 0, 1, 2, \ldots$, plot $\log \hat{\mu}_n$ versus $\log (n/s)$. The slope c and the intercept $\log (s\mu_1)$ of the $\log \mu_n$ line can then be estimated by the *method of least squares*. Confidence intervals for $\log \mu_n$ can also be obtained from *linear regression theory*.

EXAMPLE: Consider an M/M/1 queueing system. Twenty observations on both the interarrival times and service times are obtained, and a 95 percent confidence interval for L is desired.

Assume the data indicate that

$$\hat{\rho} = \frac{\hat{\hat{\lambda}}}{\hat{\hat{\mu}}} = \frac{B/m}{T/n} = \frac{300/20}{600/20} = \frac{1}{2}.$$

Since $L = \rho/(1 - \rho)$, the maximum likelihood estimate for L is 1. To find the 95 percent confidence interval for L, Table A5.5 of Appendix 5 is used to obtain

$$F_{0.025,40,40} = 1.88.$$

Therefore,

$$\underline{\rho} = \frac{1/2}{1.88} = 0.266$$

and

$$\bar{\rho} = (\tfrac{1}{2})\ 1.88 = 0.94.$$

Hence, the endpoints of a 95 percent confidence interval for L are

$$\frac{0.266}{1 - 0.266} = 0.362, \qquad \frac{0.94}{1 - 0.94} = 16.$$

[1] The results are also applicable for monotone decreasing functions, in which case the limits are reversed.

Table 7.5 **Confidence Intervals for Parameters of Queueing Systems***

Model	Parameter	$100(1-\alpha)$ Percent Confidence Interval
M/M/1	$L = \dfrac{\rho}{1-\rho}$	$\dfrac{\underline{\rho}}{1-\underline{\rho}}, \quad \dfrac{\bar{\rho}}{1-\bar{\rho}}$
	$L_q = \dfrac{\rho^2}{1-\rho}$	$\dfrac{\underline{\rho}^2}{1-\underline{\rho}}, \quad \dfrac{\bar{\rho}^2}{1-\bar{\rho}}$
	$P_0 = 1-\rho$	$1-\bar{\rho}, \quad 1-\underline{\rho}$
M/M/2	$L = \dfrac{2\rho}{1-\rho}$	$\dfrac{2\underline{\rho}}{1-\underline{\rho}}, \quad \dfrac{2\bar{\rho}}{1-\bar{\rho}}$
	$L_q = \dfrac{4\rho^3}{1-2\rho^2}$	$\dfrac{4\underline{\rho}^3}{1-2\underline{\rho}^2}, \quad \dfrac{4\bar{\rho}^3}{1-2\bar{\rho}^2}$
	$P_0 = \dfrac{1-\rho}{1+\rho}$	$\dfrac{1-\bar{\rho}}{1+\bar{\rho}}, \quad \dfrac{1-\underline{\rho}}{1+\underline{\rho}}$
$M/E_k/1$	$L = \dfrac{k-1}{2k}\dfrac{\rho^2}{1-\rho} + \rho$	$\dfrac{k+1}{2k}\dfrac{\underline{\rho}^2}{1-\underline{\rho}} + \underline{\rho}, \quad \dfrac{k+1}{2k}\dfrac{\bar{\rho}^2}{1-\bar{\rho}} + \bar{\rho}$
	$L_q = \dfrac{k+1}{2k}\dfrac{\rho^2}{1-\rho}$	$\dfrac{k+1}{2k}\dfrac{\underline{\rho}^2}{1-\underline{\rho}}, \quad \dfrac{k+1}{2k}\dfrac{\bar{\rho}^2}{1-\bar{\rho}}$

* For all models, $\rho = \lambda/s\mu$. For M/M/s models, $\underline{\rho} = \hat{\rho}/F_{\alpha/2;2m,2n}$ and $\bar{\rho} = \hat{\rho}F_{\alpha/2;2n,2m}$. For the $M/E_k/1$ model, $\underline{\rho} = \hat{\rho}/F_{\alpha/2;2mk,2n}$ and $\bar{\rho} = \hat{\rho}F_{\alpha/2;2n,2mk}$.

This interval is very wide, so little can be concluded yet about how well the queueing system would perform on the average with the current single server and input source. Much more data would be needed to obtain a relatively precise estimator for L.

7.7 Conclusions

This chapter has discussed the application of queueing theory for *designing* queueing systems. Every individual problem has its own special characteristics, so no standard procedure can be prescribed to fit every situation. Therefore, the emphasis has been on introducing fundamental considerations and approaches that can be adapted to most cases. We have focused on three particularly common decision variables (s, μ, and λ) as a vehicle for introducing and illustrating these concepts. However, there are many other possible decision variables (e.g., the size of a waiting room for a queueing system) and many more complicated situations (e.g., designing a *priority* queueing system) that can also be analyzed in a similar way.

The time required to *travel* to and from a service facility sometimes is an important consideration. A rather general approach to evaluating expected travel time has been introduced by applying it to some relatively simple cases. However, once again, many more complicated situations can also be analyzed quite similarly. We have discussed the incorporation of travel-time information into the overall analysis only in the context of determining the *number* of service facilities to provide when *customers* must travel to the nearest facility. But travel-time models also can be very useful when the *servers* must travel to the customer from the service facility (e.g., fire trucks and ambulances), as well as in other contexts.

In order to apply queueing theory, it is necessary to *estimate* the parameters of the model being used. If some version of the queueing system already is in operation, then statistical data can be gathered to develop these estimates. Point estimates and confidence intervals for the model parameters can then be used to derive point estimates and confidence intervals for some measures of performance for contemplated designs of the system.

Another useful area for the application of queueing theory is the development of policies for *controlling* queueing systems, e.g., for *dynamically* adjusting the number of servers or the service rate to compensate for changes in the number of customers in the system. Considerable research is being conducted in this area.

Queueing theory has proven to be a very useful tool, and we anticipate that its use will continue to grow as recognition of the many guises of queueing systems grows.

SELECTED REFERENCES

1. Allen, Arnold O.: *Probability, Statistics, and Queueing Theory with Computer Science Applications,* chaps. 5–6, Academic Press, New York, 1978.
2. Kleinrock, Leonard: *Queueing Systems, Vol. II: Computer Applications,* Wiley, New York, 1976.
3. Kobayashi, Hisashi: *Modeling and Analysis: An Introduction to System Performance Evaluation Methodology,* chap. 3, Addison-Wesley, Reading, Mass., 1978.
4. Lee, Alec M.: *Applied Queueing Theory,* St. Martin's Press, New York, 1966.
5. Newell, Gordon F.: *Applications of Queueing Theory,* 2d ed., Chapman and Hall, London, 1982.

PROBLEMS

1. For each kind of queueing system listed in Chap. 6, Prob. 2, briefly describe the nature of the *cost of service* and the *cost of waiting* that would need to be considered in designing the system.

2.* Suppose that a queueing system fits the M/M/1 model described in Sec. 6.6, with $\lambda = 2$ and $\mu = 4$. Evaluate the expected waiting cost per unit time $E(WC)$ for this system when its waiting-cost function has the form

(*a*) $g(N) = 10N + 2N^2$.

(*b*) $g(N) = \begin{cases} 10N, & \text{for } N = 0, 1, 2 \\ 6N^2, & \text{for } N = 3, 4, 5 \\ N^3, & \text{for } N > 5. \end{cases}$

(*c*) $h(\mathcal{W}) = 25\mathcal{W} + \mathcal{W}^3$.

(*d*) $h(\mathcal{W}) = \begin{cases} \mathcal{W}, & \text{for } 0 \le \mathcal{W} \le 1 \\ \mathcal{W}^2, & \text{for } \mathcal{W} \ge 1. \end{cases}$

3. A certain queueing system has a *Poisson* input, with a mean arrival rate of four customers per hour. The service-time distribution is *exponential,* with a mean of 0.2 hour. The marginal cost of providing each server is \$20/hour, where it is estimated that the cost that is incurred by having each customer *idle* (i.e., in the queueing system) is \$120/hour for the first customer and \$180/hour for each additional customer. Determine the *number of servers* that should be assigned to the system to minimize the expected total cost per hour. [*Hint:* Express $E(WC)$ in terms of L, P_0, and ρ, and then use Figs. 6.6 and 6.7.]

4. A certain small grocery store has a single checkout stand with a full-time cashier. Customers arrive at the stand according to a *Poisson* process at a mean rate of 30 per hour. The service-time distribution is *exponential,* with a mean of 1.5 minutes. This situation has resulted in occasional long lines and complaints from customers. Therefore, because there is no room for a second checkout stand, the proposal has been made that another person be hired to help the cashier by bagging the groceries. This help would reduce the expected time required to process a customer to 1 minute, but the distribution still would be exponential.

The total compensation for the new employee would be \$5 per hour, which is just half that for the cashier. It is estimated that the grocery store incurs lost profit due to lost future business of 5¢ for each minute that each customer has to wait (including service time). The manager wants to determine on an expected total cost basis whether it would be worthwhile to hire the new person.

(*a*) Which decision model presented in Sec. 7.4 applies to this problem? Why?

(*b*) Use this model to determine whether to continue the status quo or to adopt the proposal.

5. The problem is to choose between two types of materials-handling equipment, *A* and *B,* for transporting certain types of goods between certain producing centers in a job shop. Calls for the materials-handling unit to move a load would come essentially at random (i.e., according to a *Poisson* input process) at a mean rate of four per hour. The total time required to move a load has an *exponential* distribution, where the expected time is 12 minutes for *A* and 9 minutes for *B*. The total equivalent uniform hourly cost (capital recovery cost plus operating cost) would be \$50 for *A* and \$150 for *B*. The estimated cost of idle goods (waiting to be moved or in transit) because of increased in-process inventory is \$20/load/hour. Furthermore, the scheduling of the work at the producing centers allows for just 1 hour from the completion of a load at one center to the arrival of that load at the next center. Therefore, an additional \$100/load/hour of delay (including transit time) *after the first hour* is to be charged for lost production because of idle personnel and equipment, extra costs of expediting and supervision, and so forth.

Assuming that only one materials-handling unit is to be purchased, which type of unit should be selected?

6. A railroad company paints its own railroad cars as needed. Alternative 1 is to provide two paint shops, where painting is done by hand (one car at a time in each shop), for a total annual cost of \$300,000. The painting time for a car is 6 hours. Alternative 2 is to provide one spray shop involving an annual cost of \$400,000. In this case, the painting time for a car (again done one at a time) is 3 hours. For both alternatives, the cars arrive according to a *Poisson* input, with a mean arrival rate of one every 5 hours. The cost of idle time per car is \$50/hour. Which alternative should the railroad choose? Assume that the paint shops are always open; i.e., they work (24)(365) = 8,760 hours per year.

7. An airline maintenance base wants to make a change in its overhaul operation. The present situation is that only one airplane can be repaired at a time, and the expected repair time is 36 hours, whereas the expected time between arrivals is 45 hours. This situation has led to frequent and prolonged delays in repairing incoming planes, even though the base operates continuously. The average cost of an idle plane to the airline is \$3,000/hour. It is estimated that each plane goes into the maintenance shop five times per year. It is believed that the input process for the base is essentially *Poisson* and that the probability distribution of repair times is *Erlang,* with shape parameter $k = 2$. Alternative *A* is to provide a duplicate maintenance shop, so that two planes can be repaired simultaneously. The cost, amortized over a period of 5 years, is \$400,000/airplane/year.

Alternative B is to replace the present maintenance equipment by the most efficient (and expensive) equipment available, thereby reducing the expected repair time to 18 hours. The cost, amortized over a period of 5 years, is \$550,000/airplane/year.

Which alternative should the airline choose?

8.* A particular in-process inspection station is used to inspect subassemblies of a certain kind. At present there are two inspectors at the station, and they work together to inspect each subassembly. The inspection time has an *exponential* distribution, with a mean of 15 minutes. The cost of providing this inspection system is \$20/hour.

A proposal has been made to streamline the inspection procedure so that it can be handled by only one inspector. This inspector would begin by visually inspecting the exterior of the subassembly, and he would then use new efficient equipment to complete the inspection. The times required for these two phases of the inspection have independent *Erlang* distributions, with shape parameter $k = 2$ and means of 6 and 12 minutes, respectively. The capitalized cost of providing this inspection system would be \$15/hour.

The subassemblies arrive at the inspection station according to a Poisson process at a mean rate of three per hour. The cost of having the subassemblies wait at the inspection station (thereby increasing in-process inventory and disrupting subsequent production) is estimated to be \$10/hour for each subassembly.

Determine whether to continue the status quo or adopt the proposal in order to minimize expected total cost per hour.

9. A car rental agency has been subcontracting for the maintenance and repair of its cars. However, due to long delays in getting its cars back, the agency has decided to open its own maintenance shop to do this work more quickly. This shop will operate 42 hours per week.

Alternative 1 is to hire two mechanics (at a cost of \$1,000/week each), so that two cars can be worked on at a time. The time required by a mechanic to service a car would have an *exponential* distribution with a mean of 5 hours.

Alternative 2 is to hire just one mechanic but to provide some additional special equipment (at a capitalized cost of \$500/week) to speed up his work. In this case, the maintenance work on each car would be done in stages, where the time required for *each* stage has an *Erlang* distribution with the shape parameter $k = 4$, where the mean is 2 hours for the first stage and 1 hour for the second stage.

For both alternatives, the cars would arrive according to a *Poisson* process, with a mean arrival rate of 0.3 car per hour (during work hours). The agency estimates that its net lost revenue due to having its cars unavailable for rental is \$100/week/car.

Which alternative should the agency choose to minimize expected total cost per week?

10. A certain small car-wash business is currently being analyzed to see if costs can be reduced. Customers arrive according to a *Poisson* process at a mean rate of 15 per hour, and only one car can be washed at a time. At present the time required to wash a car has an *exponential* distribution, with a mean of 4 minutes. It also has been noticed that if there are already four cars waiting (including the one being washed), then any additional arriving customers leave and take their business elsewhere. The lost incremental profit from each such lost customer is \$3.

Two proposals have been made. Proposal 1 is to add certain equipment, at a capitalized cost of \$3/hour, which would reduce the expected washing time to 3 minutes. In addition, each arriving customer would be given a guarantee that if she has to wait longer than $\frac{1}{2}$ hour (according to a time slip she receives upon arrival) before her car is ready, then she receives a free car wash (at a marginal cost of \$2 for the company). This guarantee would be well posted and advertised, so it is believed that no arriving customers would be lost.

Proposal 2 is to obtain the most advanced equipment available, at an increased cost of \$10/hour, where each car would be sent through two cycles of the process in succession. The time required for a cycle has an *exponential* distribution, with a mean of 1 minute, so total expected washing time would be 2 minutes. Because of the increased speed and effectiveness, it is believed that essentially no arriving customers would be lost.

The owner also feels that because of the loss of customer goodwill (and consequent lost future business) when customers have to wait, a cost of 10 cents for each minute that a customer has to wait before her car wash begins should be included in the analysis of all alternatives.

Evaluate the expected total cost per hour $E(TC)$ of the status quo, proposal 1, and proposal 2 to determine which one should be chosen.

11.* A single crew is provided for unloading and/or loading each truck that arrives at the loading dock of a warehouse. These trucks arrive according to a *Poisson* input process at a mean rate of one per hour. The time required by a crew to unload and/or load a truck has an *exponential* distribution (regardless of the crew size). The expected time required by a one-person crew would be 1 hour.

The cost of providing each additional member of the crew is \$10/hour. The cost that is attributable to having a truck not in use (i.e., a truck standing at the loading dock) is estimated to be \$15/hour.

(*a*) Assume that the mean service rate of the crew is proportional to its size. What should the size be to minimize the expected total cost per hour?

(*b*) Assume that the mean service rate of the crew is proportional to the square root of its size. What should the size be to minimize expected total cost per hour?

12. Reconsider Prob. 31, Chap. 6. Suppose that the waiting cost for airplanes at the maintenance base depends linearly on the number in the system with a cost of \$90,000/week/airplane. Suppose that the service rate is continuously adjustable with the cost per week of providing service rate μ given by the expression

$$40{,}000 + 10{,}000\,\mu.$$

Find the optimal value of μ.

13.* A machine shop contains a grinder for sharpening the machine cutting tools. A decision must now be made on the speed at which to set the grinder.

The grinding time required by a machine operator to sharpen his cutting tool has an *exponential* distribution, where the mean $1/\mu$ can be set at anything from $\frac{1}{2}$ minute to 2 minutes, depending upon the speed of the grinder. The running and maintenance costs go up rapidly with the speed of the grinder, so the estimated cost per minute for providing a mean of $1/\mu$ is $\$(0.10\mu^2)$.

The machine operators arrive to sharpen their tools according to a *Poisson* process at a mean rate of one every 2 minutes. The estimated cost of an operator being away from his machine to the grinder is \$0.20/minute.

Plot the expected total cost per minute $E(TC)$ versus μ over the feasible range for μ to solve graphically for the minimizing value of μ.

14. Consider the special case of model 2 where (1) any $\mu > \lambda/s$ is feasible and (2) both $f(\mu)$ and the waiting-cost function are *linear* functions, so that

$$E(TC) = sC_r\mu + C_wL,$$

where C_r is the marginal cost per unit time for each unit of a server's mean service rate and C_w is the cost of waiting per unit time for each customer. The optimal solution is $s = 1$ (by the optimality of a single-server result), and

$$\mu = \lambda + \sqrt{\frac{\lambda C_w}{C_r}}$$

for any queueing system fitting the M/M/1 model presented in Sec. 6.6.

Show that this μ is indeed optimal for the M/M/1 model.

15. Consider a harbor with a single dock for unloading ships. The ships arrive according to a *Poisson* process at a mean rate of λ ships per week, and the service-time distribution is *exponential* with a mean rate of μ unloadings per week. Assume that harbor facilities are owned by the shipping company, so that the objective is to balance the cost associated with idle ships with the cost of running the dock. The shipping company has no control over the arrival rate λ (i.e., λ is fixed); however, by changing the size of the unloading crew, and so on, the shipping company can adjust the value of μ as desired.

Suppose that the expected cost per unit time of running the unloading dock is $D \cdot \mu$. The waiting cost for each idle ship is some constant (C) times the *square* of the total waiting time (including loading time). The shipping company wishes to adjust μ so that the expected *total* cost (including the waiting cost for idle ships) per unit time is minimized. Derive this optimal value of μ in terms of D and C.

16. Reconsider Prob. 24 in Chap. 6.

(*a*) Formulate part (*a*) to fit as closely as possible a special case of one of the decision models presented in Sec. 7.4. (Do not solve.)

(*b*) Because the answer for part (*b*) reveals a very low utilization of the machine operators (who are relatively expensive employees), the department manager has asked you to analyze each of the following alternative ways of organizing the work of the operators: (i) pool the operators so that *any* idle operator can take the next machine needing servicing, (ii) combine the operators into small *crews* to work together on any machine needing servicing within their assigned group of machines, and (iii) keep each operator assigned to her own group of machines but allow idle operators to assist busy operators on their machines. Describe each of these alternatives in queueing theory terms, including their relationship (if any) to the decision models presented in Sec. 7.4. Briefly indicate why each of these alternatives might decrease the total number of operators (thereby increasing their utilization) needed to achieve the required production rate. Also point out any dangers that might prevent this decrease.

17. Consider a factory whose floor area is a square with 600 feet on each side. Suppose that one service facility of a certain kind is provided in the center of the factory. The employees are distributed uniformly throughout the factory, and they walk to and from the facility at an average speed of 3 miles/hour along a system of orthogonal aisles.

Compute the expected travel time $E(T)$ per arrival.

18. Consider *Alternative 3* (tool cribs in Locations 1 and 3) for the example illustrated in Fig. 7.9. Derive $E(T)$ for the tool crib in Location 3 by using the probability density functions of X and Y directly for this tool crib.

19.* Suppose that the calling population for a particular service facility is uniformly distributed over *each* area shown, where the service facility is located at (0, 0). Making the same assumptions as in Sec. 7.5, derive the expected round-trip travel time per arrival $E(T)$ in terms of the average velocity v and the distance r.

(*a*)

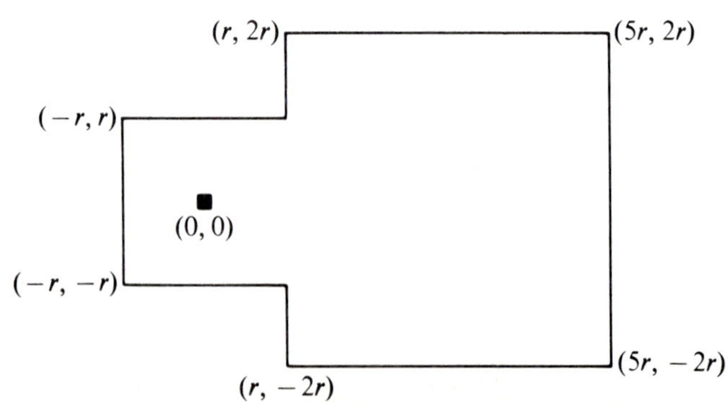

(b)

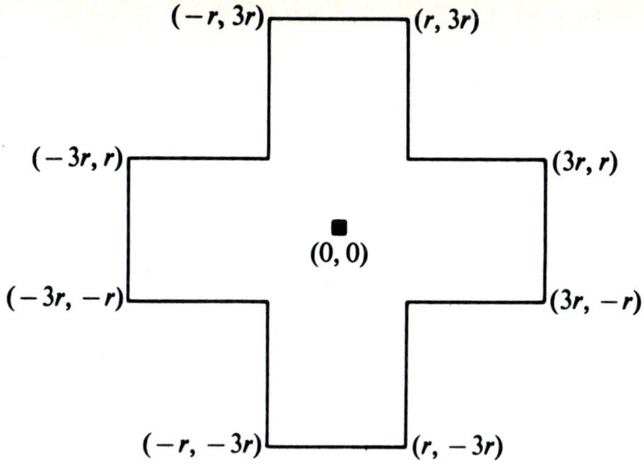
(−r, 3r)
(r, 3r)
(−3r, r)
(3r, r)
(0, 0)
(−3r, −r)
(3r, −r)
(−r, −3r)
(r, −3r)

(c)

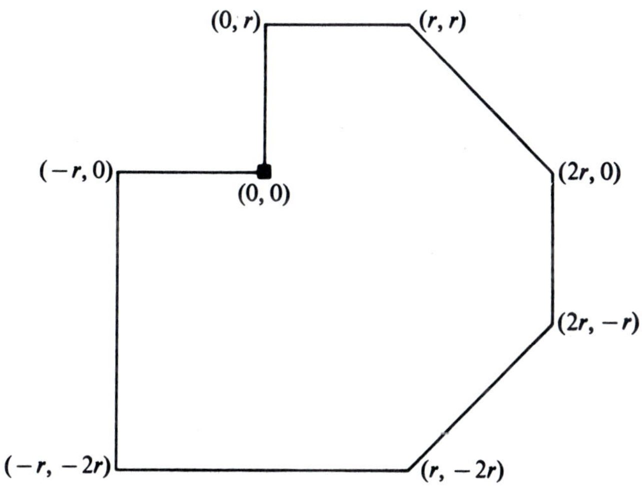
(0, r)
(r, r)
(−r, 0)
(2r, 0)
(0, 0)
(2r, −r)
(−r, −2r)
(r, −2r)

(d)

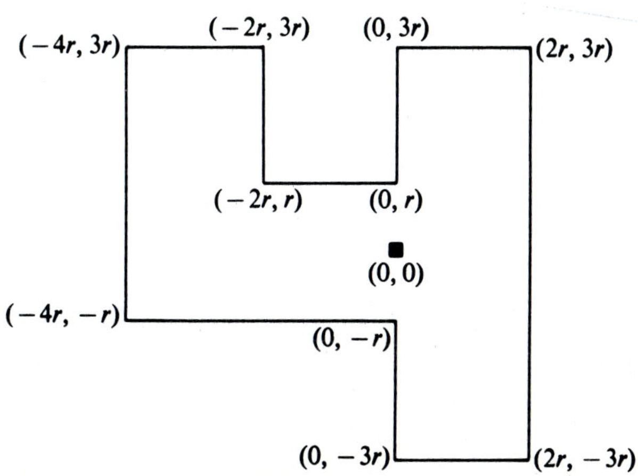
(−2r, 3r)
(0, 3r)
(−4r, 3r)
(2r, 3r)
(−2r, r)
(0, r)
(0, 0)
(−4r, −r)
(0, −r)
(0, −3r)
(2r, −3r)

20. A certain large shop doing light fabrication work uses a single central storage facility (dispatch station) for material in in-process storage. The typical procedure is that each employee personally delivers her finished work (by hand, tote box, or hand cart) and receives new work and materials at the facility. Although this procedure had worked well in earlier years when the shop was smaller, it appears that it may now be advisable to divide the shop into two semi-independent parts, with a separate storage facility for each one. You have been assigned the job of comparing the use of two facilities and of one facility from a cost standpoint.

The factory has the shape of a rectangle 150 by 100 yards. Thus, letting 1 yard be the unit of distance, the (x, y) coordinates of the corners would be (0, 0), (150, 0), (150, 100), and (0, 100). With this coordinate system, the existing facility is located at (50, 50), and the location available for the second facility is (100, 50).

Each facility would be operated by a single clerk. The time required by a clerk to service a caller has an *exponential* distribution, with a mean of 2 minutes. Employees arrive at the present facility according to a *Poisson* input process at a mean rate of 24 per hour. The employees are rather uniformly distributed throughout the shop, and if the second facility were installed, each employee would normally use the nearer of the two facilities. Employees walk at an average speed of about 5,000 yards per hour. All aisles are parallel to the outer walls of the shop. The net cost of providing each facility is estimated to be about \$20/hour, plus \$15/hour for the clerk. The estimated total cost of an employee being idled by traveling or waiting at the facility is \$25/hour.

Given the preceding cost factors, which alternative minimizes expected total cost?

21. A job shop is being laid out in a square area with 600 feet on a side, and one of the decisions to be made is the *number* of facilities for the storage and shipping of final inventory. The capitalized cost associated with providing each facility would be \$10/hour. There are just four potential locations available for these facilities, one in the middle of each of the four sides of the square area as shown in the figure.

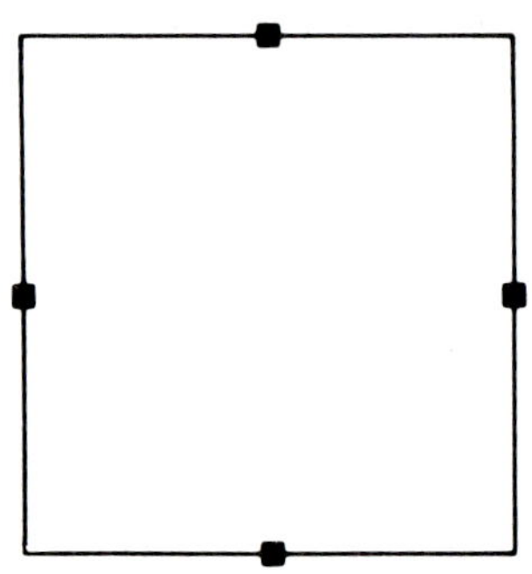

The loads to be moved to a storage and shipping facility would be distributed uniformly throughout the shop area, and they become available according to a *Poisson* process at a mean rate of 90 per hour. Each time a load becomes available, an appropriate materials-handling vehicle would be sent from the *nearest* facility to pick it up (with an expected loading time of 3 minutes) and bring it there, where the cost would be \$40/hour for time spent in traveling, loading, and waiting to be unloaded. The vehicles would travel at a speed of 20,000 feet per hour along a system of orthogonal aisles parallel to the sides of the shop area.

Another decision to be made is the number of men (m) to provide at each storage and shipping facility for unloading an arriving vehicle. These m men would work together on each vehicle, and the time required to unload it would have an *exponential* distribution, with a mean of $2/m$ minutes. The cost of providing each man is \$15/hour.

Determine the number of facilities and the value of m at each that will minimize expected total cost per hour.

22.* Consider the formulation of the *County Hospital* emergency room problem as a preemptive priority queueing system, as presented in Sec. 6.8. Suppose that the following imputed costs are assigned to making patients wait (*excluding* treatment time): \$10/hour for *stable* cases, \$1,000/hour for *serious* cases, and \$100,000/hour for *critical* cases. The cost associated with having an additional doctor on duty would be \$40/hour. Determine on an expected total cost basis whether there should be *one* or *two* doctors on duty.

23. A certain job shop has been experiencing long delays in jobs going through the turret lathe department because of inadequate capacity. The foreman contends that five machines are required, as opposed to the three machines that he now has. However, because of pressure from management to hold down capital expenditures, only one additional machine will be authorized unless there is solid evidence that a second one is necessary.

This shop does three kinds of jobs, namely, government jobs, commercial jobs, and standard products. Whenever a turret lathe finishes a job, it starts a government job if one is waiting; if not, it starts a commercial job if any are waiting; if not, it starts on a standard product if any are waiting. Jobs of the same type are taken on a first-come-first-served basis.

Although much overtime work is required currently, management wants the turret lathe department to operate on an 8-hour, 5-day-a-week basis. The probability distribution of the time required by a turret lathe for a job appears to be approximately *exponential,* with a mean of 10 hours. Jobs come into the shop according to a *Poisson* input process, but at a mean rate of 6 per week for government jobs, 4 per week for commercial jobs, and 2 per week for standard products. (These figures are expected to remain the same for the indefinite future.)

It is worth about \$750, \$450, and \$150 to avoid a delay of one additional (working) day in a government, commercial, and standard job, respectively. The incremental capitalized cost of providing each turret lathe (including the operator and so on) is estimated to be \$250/working day.

Determine the number of *additional* turret lathes that should be obtained to minimize expected total cost.

24. A certain single-server queueing system has been under observation to determine which theoretical model should be used to represent this system. The following service times (in minutes) have been recorded:

0.47, 0.22, 1.42, 0.72, 0.38, 2.85, 0.09, 0.85, 0.72,
0.27, 1.39, 0.32, 0.55, 0.13, 0.25, 2.20, 0.81, 1.03,
0.16, 0.61

Assuming that the service-time distribution is exponential, estimate the mean service rate μ.

25. An M/M/1 queueing system has been under observation. The total elapsed time that the system is observed until the last of 30 arrivals has entered the system is 535 minutes. Similarly, the sum of the service times of these first 30 customers that began service after initiating the observation of the system is 428 minutes.

(*a*) Find the maximum likelihood estimate of the utilization factor.
(*b*) Find a 95 percent confidence interval for the utilization factor.
(*c*) Find a 95 percent confidence interval for L, L_q and P_0.

26. Solve Prob. 25 by using the data given, except that the results are based upon observing 60 customers.

27. Solve Prob. 25 when there are two servers.

28. Solve Prob. 25 when there are two servers and the results are based upon 60 customers.

29. Solve Prob. 25 when the service-time distribution has an Erlang distribution with $k = 2$.

8

Dynamic Programming

Dynamic programming is a useful mathematical technique for making a sequence of interrelated decisions. It provides a systematic procedure for determining the optimal combination of decisions.

In contrast to linear programming, there does not exist a standard mathematical formulation of "the" dynamic programming problem. Rather, dynamic programming is a general type of approach to problem solving, and the particular equations used must be developed to fit each individual situation. Therefore, a certain degree of ingenuity and insight into the general structure of dynamic programming problems is required to recognize when a problem can be solved by dynamic programming procedures and how it can be done. These abilities can best be developed by an exposure to a wide variety of dynamic programming applications and a study of the characteristics that are common to all these situations. A large number of illustrative examples are presented for this purpose.

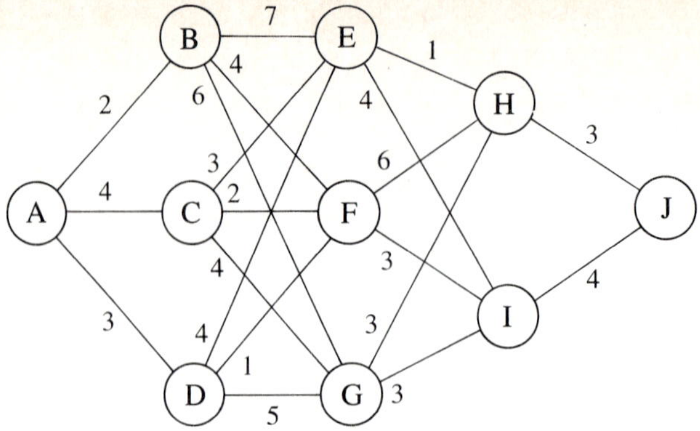

Figure 8.1 The road system and costs for the stagecoach problem.

8.1 Prototype Example

The STAGECOACH PROBLEM is a problem especially constructed[1] to illustrate the features and to introduce the terminology of dynamic programming. It concerns a mythical fortune seeker in Missouri who decided to go west to join the 49'er gold rush in California during the mid-nineteenth century. The journey would require traveling by stagecoach through unsettled country where there was serious danger of attack by marauders. Although his starting point and destination were fixed, he had considerable choice as to which **states** (or territories that subsequently became states) to travel through en route. The possible routes are shown in Fig. 8.1, where each state is represented by a lettered circle. Thus four **stages** (*stage*coach runs) were required to travel from his point of embarkation in state A (Missouri) to his destination in state J (California).

This fortune seeker was a prudent man who was quite concerned about his safety. After some thought, he came up with a rather clever way of determining the safest route. Life insurance policies were offered to stagecoach passengers. Because the cost of the **policy** for taking any given stagecoach run was based on a careful evaluation of the safety of that run, the safest route should be the one with the cheapest total life insurance policy.

The cost for the standard policy on the stagecoach run from state i to state j, which will be denoted by c_{ij}, is

	B	C	D
A	2	4	3

	E	F	G
B	7	4	6
C	3	2	4
D	4	1	5

	H	I
E	1	4
F	6	3
G	3	3

	J
H	3
I	4

These costs are also shown in Fig. 8.1.

[1] This problem was developed by Professor Harvey M. Wagner while he was at Stanford University.

We shall now focus on the question of which route minimizes the total cost of the policy.

Solving the Problem

First note that the shortsighted approach of selecting the cheapest run offered by each successive stage need not yield an overall optimal decision. Following this strategy would give the route A → B → F → I → J, at a total cost of 13. However, sacrificing a little on one stage may permit greater savings thereafter. For example, A → D → F is cheaper overall than A → B → F.

One possible approach to solving this problem is to use trial and error.[1] However, the number of possible routes is large (18) and having to calculate the total cost for each route is not an appealing task.

Fortunately, dynamic programming provides a solution with much less effort than exhaustive enumeration. (The computational savings are enormous for larger versions of this problem.) Dynamic programming starts with a small portion of the original problem and finds the optimal solution for this smaller problem. It then gradually enlarges the problem, finding the current optimal solution from the preceding one, until the original problem is solved in its entirety.

For the stagecoach problem, we start with the smaller problem where the fortune seeker has nearly completed his journey and has only one more stage (stagecoach run) to go. The obvious optimal solution for this smaller problem is to go from his current state (whatever it is) to his ultimate destination (state J). At each subsequent iteration, the problem is enlarged by increasing by one the number of stages left to go to complete the journey. For this enlarged problem, the optimal solution for where to go next from each possible state can be found relatively easily from the results obtained at the preceding iteration. The details involved in implementing this approach follow.

FORMULATION: Let the decision variables x_n ($n = 1, 2, 3, 4$) be the immediate destination on stage n (the nth stagecoach run to be taken). Thus the route selected is $A \to x_1 \to x_2 \to x_3 \to x_4$, where $x_4 = J$.

Let $f_n(s, x_n)$ be the total cost of the best overall *policy* for the *remaining* stages, given that the fortune seeker is in state s, ready to start stage n, and selects x_n as the immediate destination. Given s and n, let x_n^* denote the value of x_n that minimizes $f_n(s, x_n)$, and let $f_n^*(s)$ be the corresponding minimum value of $f_n(s, x_n)$. Thus

$$f_n^*(s) = \min_{x_n} f_n(s, x_n) = f_n(s, x_n^*),$$

where

$$f_n(s, x_n) = \text{immediate cost (stage } n) + \text{minimum future cost (stages } n+1 \text{ onward)}$$

$$= c_{sx_n} + f_{n+1}^*(x_n).$$

[1] This problem also can be formulated as a *shortest path problem* where *costs* here play the role of *distances* in the shortest path problem.

The value of c_{sx_n} is given by the preceding tables for c_{ij} by setting $i = s$ (the current state) and $j = x_n$ (the immediate destination). Because the ultimate destination (state J) is reached at the end of stage 4, $f_5^*(\text{J}) = 0$.

The objective is to find $f_1^*(\text{A})$ and the corresponding route. Dynamic programming finds it by successively finding $f_4^*(s)$, $f_3^*(s)$, $f_2^*(s)$ for each of the possible states s and then using $f_2^*(s)$ to solve for $f_1^*(\text{A})$.[1]

Solution Procedure: When the fortune seeker has only one more stage to go ($n = 4$), his route thereafter is determined entirely by his current state s (either H or I) and his final destination, $x_4 = \text{J}$, so the route for this final stagecoach run is $s \rightarrow \text{J}$. Therefore, since $f_4^*(s) = f_4(s, \text{J}) = c_{s,J}$, the immediate solution to the $n = 4$ problem is

$n = 4$:

s	$f_4^*(s)$	x_4^*
H	3	J
I	4	J

When the fortune seeker has two more stages to go ($n = 3$), the solution procedure requires a few calculations. For example, suppose that the fortune seeker is in state F. Then, as depicted below, he must next go to either state H or I at an immediate cost of $c_{F,H} = 6$ or $c_{F,I} = 3$, respectively. If he chooses state H, the minimum additional cost after he reaches there is given in the preceding table as $f_4^*(\text{H}) = 3$, as shown next to the H box in the above diagram. Therefore, the total cost for this decision would be $6 + 3 = 9$. If he chooses state I instead, the total cost is $3 + 4 = 7$, which is smaller. Therefore, the optimal choice would be this latter one, $x_3^* = \text{I}$, because it gives the minimum cost, $f_3^*(\text{F}) = 7$.

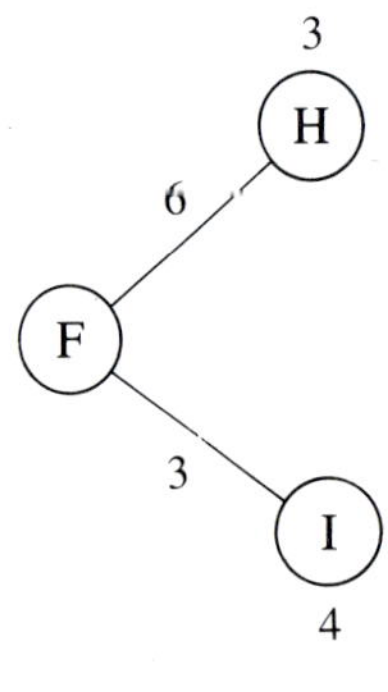

Similar calculations need to be made when starting from the other two possible states, $s = \text{E}$ and $s = \text{G}$, with two stages to go. Try it, proceeding both graphically

[1] Because this procedure involves moving *backward* stage by stage, some writers also count n backward to denote the number of *remaining stages* to the destination. We use the more natural *forward counting* for greater simplicity.

(Fig. 8.1) and algebraically (combining c_{ij} and $f_4^*(s)$ values), to verify the following complete results for the $n = 3$ problem.

$n = 3$:

s \ x_3	$f_3(s,x_3) = c_{sx_3} + f_4^*(x_3)$ H	I	$f_3^*(s)$	x_3^*
E	4	8	4	H
F	9	7	7	I
G	6	7	6	H

The solution for the three-stage problem ($n = 2$) is obtained in a similar fashion. In this case, $f_2(s,x_2) = c_{sx_2} + f_3^*(x_2)$. For example, suppose that the fortune seeker is in state C, as depicted below.

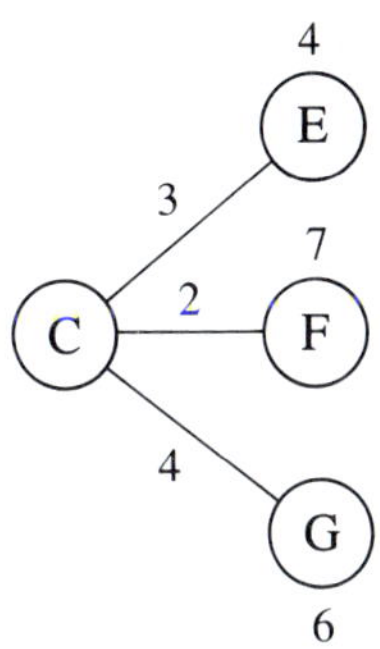

He must next go to state E, F, or G at an immediate cost of $c_{C,E} = 3$, $c_{C,F} = 2$, or $c_{C,G} = 4$, respectively. After getting there, the minimum additional cost for stage 3 to the end is given by the $n = 3$ table as $f_3^*(\text{E}) = 4$, $f_3^*(\text{F}) = 7$, or $f_3^*(\text{G}) = 6$, respectively, as shown next to the E, F, and G states in the above diagram. The resulting calculations for the three alternatives are summarized below.

$x_2 = \text{E}$: $\quad f_2(\text{C, E}) = c_{C,E} + f_3^*(\text{E}) = 3 + 4 = 7.$

$x_2 = \text{F}$: $\quad f_2(\text{C, F}) = c_{C,F} + f_3^*(\text{F}) = 2 + 7 = 9.$

$x_2 = \text{G}$: $\quad f_2(\text{C, G}) = c_{C,G} + f_3^*(\text{G}) = 4 + 6 = 10.$

The minimum of these three numbers is 7, so the minimum total cost from state C to the end is $f_2^*(\text{C}) = 7$ and the immediate destination should be $x_2^* = \text{E}$.

Making similar calculations when starting from state B or D (try it) yields the following results for the $n = 2$ problem:

$n = 2$:

s \ x_2	$f_2(s, x_2) = c_{sx_2} + f_3^*(x_2)$ E	F	G	$f_2^*(s)$	x_2^*
B	11	11	12	11	E or F
C	7	9	10	7	E
D	8	8	11	8	E or F

Moving to the four-stage problem ($n = 1$), the calculations are similar to those just shown for the three-stage problem ($n = 2$), except now there is just *one* possible starting state, $s = \text{A}$, as depicted below.

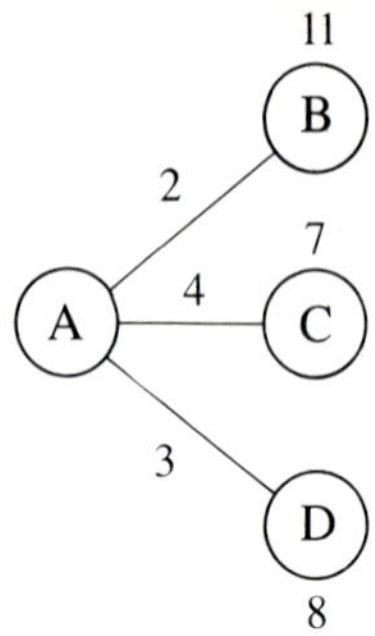

These calculations are summarized next for the three alternatives for the immediate destination:

$$x_1 = \text{B}: \qquad f_1(\text{A}, \text{B}) = c_{A,B} + f_2^*(\text{B}) = 2 + 11 = 13.$$

$$x_1 = \text{C}: \qquad f_1(\text{A}, \text{C}) = c_{A,C} + f_2^*(\text{C}) = 4 + 7 = 11.$$

$$x_1 = \text{D}: \qquad f_1(\text{A}, \text{D}) = c_{A,D} + f_2^*(\text{D}) = 3 + 8 = 11.$$

Since 11 is the minimum, $f_1^*(\text{A}) = 11$ and $x_1^* = \text{C}$ or D, as shown in the following table.

	x_1 / s	$f_1(s, x_1) = c_{sx_1} + f_2^*(x_1)$			$f_1^*(s)$	x_1^*
$n = 1$:		B	C	D		
	A	13	11	11	11	C or D

The optimal solution for the entire problem can now be identified. Results for the $n = 1$ problem indicate that the fortune seeker should go initially to either state C or state D. Suppose that he chooses $x_1^* = \text{C}$. For $n = 2$, the result for $s = \text{C}$ is $x_2^* = \text{E}$. This result leads to the $n = 3$ problem, which gives $x_3^* = \text{H}$ for $s = \text{E}$, and the $n = 4$ problem yields $x_4^* = \text{J}$ for $s = \text{H}$. Hence one optimal route is $\text{A} \rightarrow \text{C} \rightarrow \text{E} \rightarrow \text{H} \rightarrow \text{J}$. Choosing $x_1^* = \text{D}$ leads to the other two optimal routes, $\text{A} \rightarrow \text{D} \rightarrow \text{E} \rightarrow \text{H} \rightarrow \text{J}$ and $\text{A} \rightarrow \text{D} \rightarrow \text{F} \rightarrow \text{I} \rightarrow \text{J}$. They all yield a total cost of $f_1^*(\text{A}) = 11$.

You will see in the next section that the *special terms* describing the particular context of this problem—*stage, state, policy*—actually are part of the *general terminology* of dynamic programming with an analogous interpretation in other contexts.

8.2 Characteristics of Dynamic Programming Problems

The stagecoach problem is a literal prototype of dynamic programming problems. In fact, this example was purposely designed to provide a literal physical interpretation of the rather abstract structure of such problems. Therefore, one way to recognize a situation that can be formulated as a dynamic programming problem is to notice that its basic structure is analogous to that of the stagecoach problem.

These basic features that characterize dynamic programming problems are presented and discussed here.

1. The problem can be divided into **stages**, with a **policy decision** required at each stage.

The stagecoach problem was literally divided into its four *stages* (stagecoaches) that correspond to the four legs of the journey. The policy decision at each stage was which life insurance *policy* to choose (i.e., which destination to select for the next stagecoach ride). Similarly, other dynamic programming problems require making a *sequence of interrelated decisions,* where each decision corresponds to one stage of the problem.

2. Each stage has a number of **states** associated with it.

The states associated with each stage in the stagecoach problem were the states (or territories) in which the fortune seeker could be located when embarking on that particular leg of the journey. In general, the states are the various *possible conditions* in which the system might be at that stage of the problem. The number of states may be either *finite* (as in the stagecoach problem) or *infinite* (as in some subsequent examples).

3. The effect of the policy decision at each stage is to *transform the current state into a state associated with the next stage* (possibly according to a probability distribution).

The fortune seeker's decision as to his next destination led him from his current state to the next state on his journey. This procedure suggests that dynamic programming problems can be interpreted in terms of network theory. Each *node* would correspond to a *state*. The network would consist of columns of nodes, with each *column* corresponding to a *stage,* so that the flow from a node can go only to a node in the next column to the right. The links from a node to nodes in the next column correspond to the possible policy decisions on which state to go to next. The value assigned to each link usually can be interpreted as the *immediate contribution* to the objective function from making that policy decision. In most cases, the objective corresponds to finding either the *shortest* or the *longest route* through the network.

4. The solution procedure is designed to find an **optimal policy** for the overall problem, i.e., a prescription of the optimal policy decision at each stage for *each* of the possible states.

For the stagecoach problem, the solution procedure constructed a table for each stage (n) that prescribes the optimal decision (x_n^*) for *each* possible state (s). Thus, in addition to identifying three *optimal solutions* (optimal routes) for the overall problem, the results also show the fortune seeker how he should proceed if he gets detoured to a state that is not on an optimal route. For any problem, dynamic programming provides this kind of *policy* prescription of what to do under every possible circumstance (which is why the actual decision made upon reaching a particular state at a given stage is referred to as a *policy* decision). Providing this additional information beyond simply specifying an optimal solution (optimal sequence of decisions) can be helpful in a variety of ways, including sensitivity analysis.

5. Given the current state, an *optimal policy for the remaining stages* is *independent* of the policy adopted in *previous stages*. (This is the **principle of optimality** for dynamic programming.)

Given the state in which the fortune seeker is currently located, the optimal life insurance policy (and its associated route) from this point onward is independent of how he got there. For dynamic programming problems in general, knowledge of the current state of the system conveys all the information about its previous behavior necessary for determining the optimal policy henceforth. (This property is the *Markovian property* discussed in Sec. 5.3.) Any problem lacking this property cannot be formulated as a dynamic programming problem.

6. The solution procedure begins by finding the *optimal policy for the last stage*.

The optimal policy for the last stage prescribes the optimal policy decision for *each* of the possible states at that stage. The solution of this one-stage problem is usually trivial, as it was for the stagecoach problem.

7. A **recursive relationship** that identifies the optimal policy for stage n, given the optimal policy for stage $(n + 1)$, is available.

For the stagecoach problem, this recursive relationship was

$$f_n^*(s) = \min_{x_n} \{c_{sx_n} + f_{n+1}^*(x_n)\}.$$

Therefore, finding the *optimal policy decision* when starting in state s at stage n requires finding the minimizing value of x_n. The corresponding minimum cost is achieved by using this value of x_n and then following the optimal policy when starting in state x_n at stage $(n + 1)$.

The precise form of the recursive relationship differs somewhat among dynamic programming problems. However, notation analogous to that introduced in the preceding section will continue to be used here, as summarized below.

N = number of stages.
n = label for current stage ($n = 1, 2, \ldots, N$).
s_n = current *state* for stage n.
x_n = decision variable for stage n.
x_n^* = optimal value of x_n (given s_n).
$f_n(s_n, x_n)$ = contribution of stages $n, n + 1, \ldots, N$ to the objective function if the system starts in state s_n at stage n, the immediate decision is x_n, and optimal decisions are made thereafter.
$f_n^*(s_n) = f_n(s_n, x_n^*)$.

The recursive relationship will always be of the form

$$f_n^*(s_n) = \max_{x_n} \{f_n(s_n, x_n)\} \qquad \text{or} \qquad f_n^*(s_n) = \min_{x_n} \{f_n(s_n, x_n)\},$$

where $f_n(s_n, x_n)$ would be written in terms of s_n, x_n, $f_{n+1}^*(s_{n+1})$ and probably some measure of the first-stage effectiveness (or ineffectiveness) of x_n.

The recursive relationship is given its name because it keeps *recurring* as we move backward stage by stage. When the current stage number n is decreased by one, the new $f_n^*(s_n)$ function is derived by using the $f_{n+1}^*(s_{n+1})$ function that was just derived during the preceding iteration, and then this process keeps repeating. This property is emphasized in our next (and final) characteristic of dynamic programming.

8. When we use this recursive relationship, the solution procedure moves *backward* stage by stage—each time finding the optimal policy for that stage—until it finds the optimal policy starting at the *initial* stage.

This backward movement was demonstrated by the stagecoach problem, where the optimal policy was found successively beginning in each state at stages 4, 3, 2, and 1, respectively.[1] For all dynamic programming problems, a table such as the following one would be obtained for each stage ($n = N, N - 1, \ldots, 1$).

s_n \ x_n	$f_n(s_n, x_n)$	$f_n^*(s_n)$	x_n^*

When this table is finally obtained for the initial stage ($n = 1$), the problem of interest is solved. Because the initial state is known, the initial decision is specified by x_1^* in this table. The optimal value of the other decision variables is then specified by the other tables in turn according to the state of the system that results from the preceding decisions.

8.3 Deterministic Dynamic Programming

This section further elaborates upon the dynamic programming approach to *deterministic* problems, where the *state* at the *next stage* is *completely determined* by the *state* and *policy decision* at the *current stage*. The *probabilistic* case, where there is a probability distribution for what the next state will be, is discussed in the next section.

Deterministic dynamic programming can be described diagrammatically as shown in Fig. 8.2. Thus at stage n the process will be in some state s_n. Making policy decision x_n then moves the process to some state s_{n+1} at stage $(n + 1)$. The contribution *thereafter* to the objective function under an optimal policy has been previously calculated to be $f_{n+1}^*(s_{n+1})$. The policy decision x_n also makes some contribution to the objective function. Combining these two quantities in an appropriate way provides $f_n(s_n, x_n)$, the contribution of stages n onward to the objective function. Optimizing with respect to x_n then gives $f_n^*(s_n) = f_n(s_n, x_n^*)$. After finding x_n^* and $f_n^*(s_n)$ for each possible value of s_n, the solution procedure is ready to move back one stage.

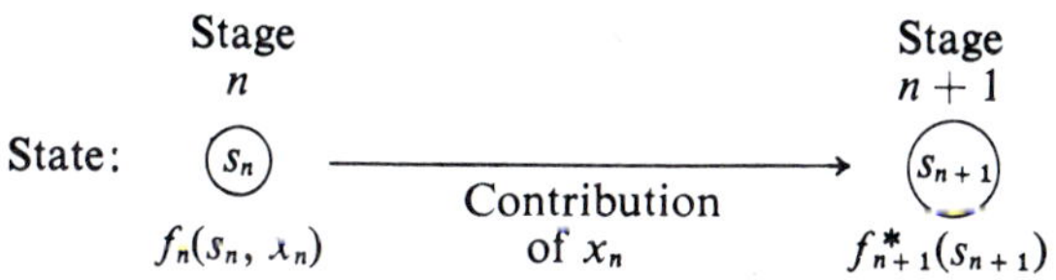

Figure 8.2 The basic structure for deterministic dynamic programming.

[1] Actually, for this problem the solution procedure can move *either* backward or forward. However, for many problems (especially when the stages correspond to *time periods*), the solution procedure *must* move backward.

One way of categorizing deterministic dynamic programming problems is by the *form of the objective function*. For example, the objective might be to *minimize* the *sum* of the contributions from the individual stages (as for the stagecoach problem), or to *maximize* such a sum, or to minimize a *product* of such terms, and so on. Another categorization is in terms of the nature of the *set of states* for the respective stages. In particular, the states s_n might be representable by a *discrete* state variable (as for the stagecoach problem), or by a *continuous* state variable, or perhaps a state *vector* (more than one variable) is required.

Several examples are presented to illustrate these various possibilities. More important, they illustrate that these apparently major differences are actually quite inconsequential (except in terms of computational difficulty) because the underlying basic structure shown in Fig. 8.2 always remains the same.

The first new example arises in a much different context from the stagecoach problem, but it has the same *mathematical formulation* except that the objective is to *maximize* rather than minimize a sum.

Example 2—Distributing Medical Teams to Countries

The WORLD HEALTH COUNCIL is devoted to improving health care in the underdeveloped countries of the world. It now has five *medical teams* available to allocate among three such countries to improve their medical care, health education, and training programs. Therefore, the council needs to determine how many teams (if any) to allocate to each of these countries to maximize the total effectiveness of the five teams. The teams must be kept intact, so the number allocated to each country must be integer.

The measure of performance being used is *additional person-years of life*. (For a particular country, this measure equals the country's *increased life expectancy* in years times its population.) Table 8.1 gives the estimated additional person-years of life (in multiples of 1,000) for each country for each possible allocation of medical teams.

Which allocation maximizes the measure of performance?

FORMULATION: This problem requires making three *interrelated decisions,* namely, how many medical teams to allocate to each of the three countries. Therefore, even though there is no fixed sequence, these three countries can be considered as the three

Table 8.1 **Data for the World Health Council Problem**

	Thousands of Additional Person-Years of Life		
	Country		
No. of Medical Teams	1	2	3
0	0	0	0
1	45	20	50
2	70	45	70
3	90	75	80
4	105	110	100
5	120	150	130

stages in a dynamic programming formulation. The decision variables x_n ($n = 1, 2, 3$) would be the number of teams to allocate to stage (country) n.

The identification of the *states* may not be readily apparent. To determine the states, we ask questions such as the following. What is it that changes from one stage to the next? Given that the decisions have been made at the previous stages, how can the status of the situation at the current stage be described? What information about the current state of affairs is necessary to determine the optimal policy hereafter? On these bases, an appropriate choice for the "state of the system" is

s_n = number of medical teams still available for allocation to the remaining countries ($n, \ldots, 3$).

Thus, at stage 1 (country 1), where all three countries remain under consideration for allocations, $s_1 = 5$. However, at stage 2 or 3 (country 2 or 3), s_n is just 5 minus the number of teams allocated at preceding stages. With the dynamic programming procedure of solving backward stage by stage, when we are solving at stage 2 or 3, we shall not yet have solved for the allocations at the preceding stages. Therefore, we shall consider every possible state we could be in at stage 2 or 3, namely, $s_n = 0$, 1, 2, 3, 4, or 5.

Let $p_i(x_i)$ be the measure of performance from allocating x_i medical teams to country i, as given in Table 8.1. Thus the objective is to choose x_1, x_2, x_3 so as to

$$\text{Maximize} \quad \sum_{i=1}^{3} p_i(x_i),$$

subject to

$$\sum_{i=1}^{3} x_i = 5,$$

and

the x_i are nonnegative integers.

Using the notation presented in Sec. 8.2, $f_n(s_n, x_n)$ is then

$$f_n(s_n, x_n) = p_n(x_n) + \text{maximum} \sum_{i=n+1}^{3} p_i(x_i),$$

where the maximum is taken over $x_{n+1}, \ldots, x_3$ such that

$$\sum_{i=n}^{3} x_i = s_n$$

and the x_i are nonnegative integers, for $n = 1, 2, 3$. In addition,

$$f_n^*(s_n) = \max_{x_n=0,1,\ldots,s_n} f_n(s_n, x_n)$$

Therefore,

$$f_n(s_n, x_n) = p_n(x_n) + f_{n+1}^*(s_n - x_n)$$

(with f_4^* defined to be *zero*). These basic relationships are summarized in Fig. 8.3.

Consequently, the *recursive relationship* relating the f_1^*, f_2^*, and f_3^* functions for this problem is

$$f_n^*(s_n) = \max_{x_n=0,1,\ldots,s_n} \{p_n(x_n) + f_{n+1}^*(s_n - x_n)\}, \qquad \text{for } n = 1, 2.$$

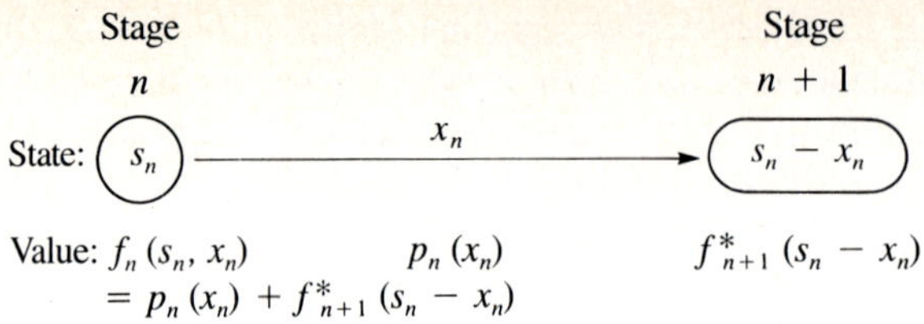

Figure 8.3 The basic structure for the World Health Council problem.

For the last stage ($n = 3$),

$$f_3^*(s_3) = \max_{x_3=0,1,\ldots,s_3} p_3(x_3).$$

The resulting dynamic programming calculations are given next.

Solution Procedure: Beginning with the last stage ($n = 3$), we note that the values of $p_3(x_3)$ are given in the last column of Table 8.1, and that these values keep increasing as we move down the column. Therefore, with s_3 medical teams still available for allocation to country 3, the maximum of $p_3(x_3)$ is automatically achieved by allocating all s_3 teams, so $x_3^* = s_3$ and $f_3^*(s_3) = p_3(s_3)$ as shown in the following table.

$n = 3$:

s_3	$f_3^*(s_3)$	x_3^*
0	0	0
1	50	1
2	70	2
3	80	3
4	100	4
5	130	5

We now move backward to start from the next-to-last stage ($n = 2$). Here, finding x_2^* requires calculating and comparing $f_2(s_2, x_2)$ for the alternative values of x_2, namely, $x_2 = 0, 1, \ldots, s_2$. To illustrate, we depict this situation when $s_2 = 2$ graphically:

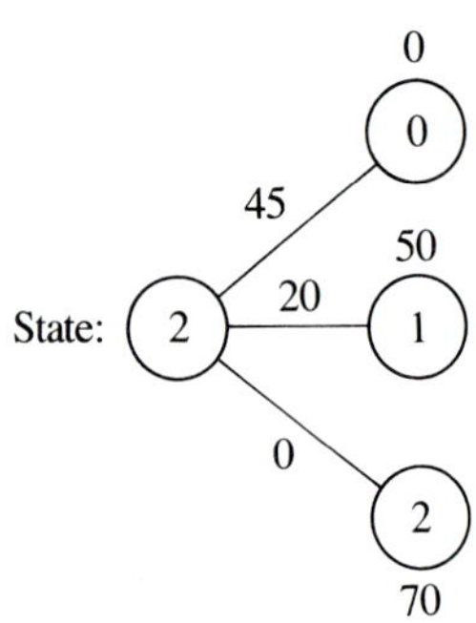

This diagram corresponds to Fig. 8.3 except that all three possible states at stage 3 are shown. Thus, if $x_2 = 0$, the resulting state at stage 3 will be $s_2 - x_2 = 2 - 0 = 2$, whereas $x_2 = 1$ leads to state 1 and $x_2 = 2$ leads to state 0. The corresponding values of $p_2(x_2)$ from the country 2 column of Table 8.1 are shown along the links, and the values of $f_3^*(s_2 - x_2)$ from the $n = 3$ table are given next to the stage 3 nodes. The required calculations for this case of $s_2 = 2$ are summarized below.

$$x_2 = 0: \qquad f_2(2, 0) = p_2(0) + f_3^*(2) = 0 + 70 = 70.$$

$$x_2 = 1: \qquad f_2(2, 1) = p_2(1) + f_3^*(1) = 20 + 50 = 70.$$

$$x_2 = 2: \qquad f_2(2, 2) = p_2(2) + f_3^*(0) = 45 + 0 = 45.$$

Because the objective is *maximization*, $x_2^* = 0$ or 1 with $f_2^*(2) = 70$.

Proceeding in a similar way with the other possible values of s_2 (try it) yields the following table.

$n = 2$:

s_2 \ x_2	$f_2(s_2, x_2) = p_2(x_2) + f_3^*(s_2 - x_2)$						$f_2^*(s_2)$	x_2^*
	0	1	2	3	4	5		
0	0						0	0
1	50	20					50	0
2	70	70	45				70	0 or 1
3	80	90	95	75			95	2
4	100	100	115	125	110		125	3
5	130	120	125	145	160	150	160	4

We now are ready to move backward to solve the original problem where we are starting from stage 1 ($n = 1$). In this case, the only state to be considered is the starting state of $s_1 = 5$, as depicted below.

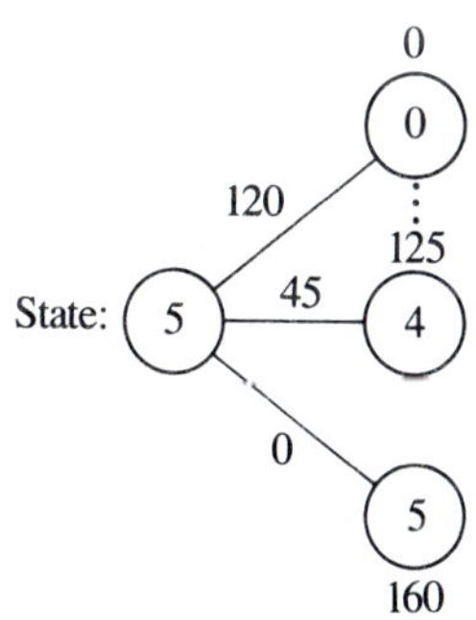

Since allocating x_1 medical teams to country 1 leads to a state of $(5 - x_1)$ at stage 2, a choice of $x_1 = 0$ leads to the bottom node on the right, $x_1 = 1$ leads to the next node up, and so forth up to the top node with $x_1 = 5$. The corresponding $p_1(x_1)$ values from Table 8.1 are shown next to the links. The numbers next to the nodes are obtained from the $f_2^*(s_2)$ column of the $n = 2$ table. As with $n = 2$, the calculation needed for each alternative value of the decision variable involves adding the corresponding link value and node value, as summarized below.

$$x_1 = 0: \qquad f_1(5, 0) = p_1(0) + f_2^*(5) = 0 + 160 = 160.$$

$$x_1 = 1: \qquad f_1(5, 1) = p_1(1) + f_2^*(4) = 45 + 125 = 170.$$

$$\vdots$$

$$x_1 = 5: \qquad f_1(5, 5) = p_1(5) + f_2^*(0) = 120 + 0 = 120.$$

The similar calculations for $x_1 = 2, 3, 4$ (try it) verify that $x_1^* = 1$ with $f_1^*(5) = 170$, as shown in the following table.

	x_1	$f_1(s_1, x_1) = p_1(x_1) + f_2^*(s_1 - x_1)$							
$n = 1$:	s_1	0	1	2	3	4	5	$f_1^*(s_1)$	x_1^*
	5	160	170	165	160	155	120	170	1

Thus the optimal solution has $x_1^* = 1$, which makes $s_2 = 5 - 1 = 4$, so $x_2^* = 3$, which makes $s_3 = 4 - 3 = 1$, so $x_3^* = 1$. Since $f_1^*(5) = 170$, this (1, 3, 1) allocation of medical teams to the three countries will yield an estimated total of 170,000 *additional person-years of life*, which is at least 5,000 more than for any other allocation.

A Prevalent Problem Type—The Distribution of Effort Problem

The preceding example illustrates a particularly common type of dynamic programming problem called the *distribution of effort problem*. For this type of problem, there is just one kind of *resource* that is to be allocated to a number of *activities*. The objective is to determine how to *distribute the effort* (the resource) among the activities most effectively. For the World Health Council example, the resource involved is the *medical teams*, and the three activities are the health care work in the three *countries*.

ASSUMPTIONS: This interpretation of allocating resources to activities also is a common interpretation for *linear programming* problems. However, there also are some key differences between the distribution of effort problem and linear programming that help illuminate the general distinctions between dynamic programming and other areas of mathematical programming.

One key difference is that the distribution of effort problem involves only *one resource* (one functional constraint), whereas linear programming can deal with hundreds or even thousands of resources. (In principle, dynamic programming can handle slightly more than one resource, as we shall illustrate in Example 5 by solving a *three-resource* linear programming problem, but it quickly becomes very inefficient when the number of resources is increased.)

On the other hand, the distribution of effort problem is far more general than linear programming in other ways. The contribution to the overall measure of performance from each activity no longer needs to be *proportional* to the level of the activity. In fact, the relationship between the contribution and level can be anything. Furthermore, the decision variables no longer are required to be *continuous*. In many cases (as in Example 2), the decision variables must be *integer*, which actually simplifies the calculations.

FORMULATION: Because they always involve allocating one kind of resource to a number of activities, distribution of effort problems always have the following dynamic programming formulation (where the ordering of the activities is arbitrary):

Stage n = activity n ($n = 1, 2, \ldots, N$).
x_n = amount of the resource allocated to activity n.
State s_n = amount of the resource still available for allocation to the remaining activities ($n, \ldots, N$).

Therefore, when starting at stage n in state s_n, the choice of x_n always results in the next state at stage $(n + 1)$ being $s_{n+1} = (s_n - x_n)$, as depicted below:

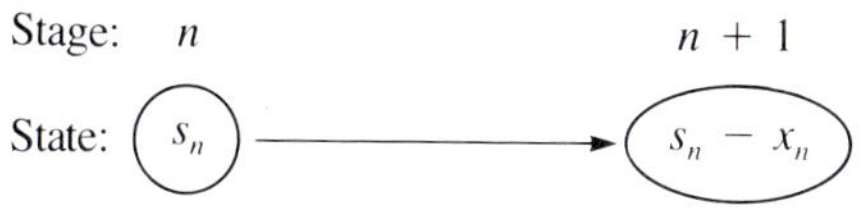

Note how the structure of this diagram corresponds to the one shown in Fig. 8.3 for the World Health Council example of a distribution of effort problem. What will differ from one such example to the next is the *rest* of what is shown in Fig. 8.3, namely, the relationship between $f_n(s_n, x_n)$ and $f^*_{n+1}(s_n - x_n)$, and then the resulting *recursive relationship* between the f^*_n and f^*_{n+1} functions. These relationships depend on the particular objective function for the overall problem.

The *structure* of the next example is similar to the one for the World Health Council because it too is a *distribution of effort problem*. However, its *recursive relationship* differs in that its objective is to *minimize* a *product* of terms for the respective stages.

At first glance this example may appear *not* to be a *deterministic* dynamic programming problem because probabilities are involved. However, it does indeed fit our definition because the state at the next stage is completely determined by the state and policy decision at the current stage.

Example 3—Distributing Scientists to Research Teams

A government space project is conducting research on a certain engineering problem that must be solved before people can fly safely to Mars. Three research teams are currently trying three different approaches for solving this problem. The estimate has been made that, under present circumstances, the probability that the respective teams—call them 1, 2, and 3—will not succeed is 0.40, 0.60, and 0.80, respectively. Thus the current probability that all three teams will fail is $(0.40)(0.60)(0.80) = 0.192$. Because the objective is to minimize the probability of failure, two more top scientists have been assigned to the project.

Table 8.2 gives the estimated probability that the respective teams will fail when 0, 1, or 2 additional scientists are added to that team. Only integer numbers of scientists are considered because each new scientist will need to devote full attention to one team. The problem is to determine how to allocate the two additional scientists to minimize the probability that all three teams will fail.

Formulation: Because both Examples 2 and 3 are *distribution of effort problems*, their underlying structure is actually very similar. In this case, scientists replace medical teams as the kind of resource involved, and research teams replace countries as the activities. Therefore, instead of medical teams being allocated to countries, sci-

Table 8.2 **Data for the Government Space Project Problem**

	Probability of Failure		
	Team		
Number of New Scientists	1	2	3
0	0.40	0.60	0.80
1	0.20	0.40	0.50
2	0.15	0.20	0.30

entists are being allocated to research teams. The only basic difference between the two problems is in their objective functions.

With so few scientists and teams involved, this problem could be solved very easily by a process of exhaustive enumeration. However, the dynamic programming solution is presented for illustrative purposes.

In this case, *stage n* ($n = 1, 2, 3$) corresponds to research team n, and the *state* s_n is the number of new scientists *still available* for allocation to the remaining teams. The decision variables x_n ($n = 1, 2, 3$) are the number of additional scientists allocated to team n.

Let $p_i(x_i)$ denote the probability of failure for team i if it is assigned x_i additional scientists, as given by Table 8.2. Letting Π denote multiplication, the government's objective is to choose x_1, x_2, x_3 so as to

$$\text{Minimize} \quad \prod_{i=1}^{3} p_i(x_i) = p_1(x_1)p_2(x_2)p_3(x_3),$$

subject to

$$\sum_{i=1}^{3} x_i = 2$$

and

the x_i are nonnegative integers.

Consequently, $f_n(s_n, x_n)$ for this problem is

$$f_n(s_n, x_n) = p_n(x_n) \cdot \text{minimum} \prod_{i=n+1}^{3} p_i(x_i),$$

where the minimum is taken over $x_{n+1}, \ldots, x_3$ such that

$$\sum_{i=n}^{3} x_i = s_n$$

and

the x_i are nonnegative integers,

for $n = 1, 2, 3$. Thus

$$f_n^*(s_n) = \min_{x_n \le s_n} f_n(s_n, x_n).$$

Hence

$$f_n(s_n, x_n) = p_n(x_n) \cdot f_{n+1}^*(s_n - x_n)$$

(with f_4^* defined to be *one*). Figure 8.4 summarizes these basic relationships.

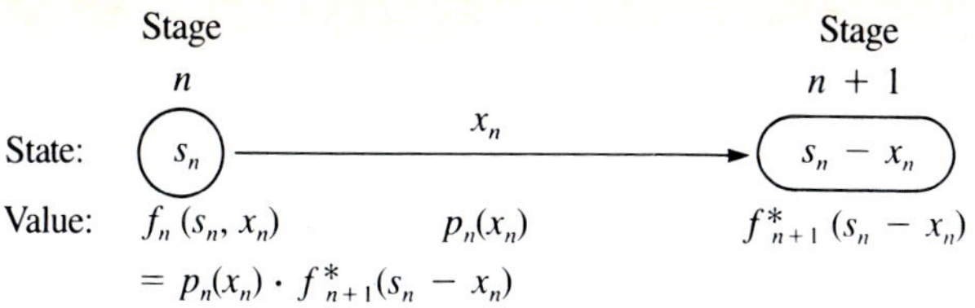

Figure 8.4 The basic structure for the government space project problem.

Thus the *recursive relationship* relating the f_1^*, f_2^*, and f_3^* functions in this case is

$$f_n^*(s_n) = \min_{x_n \le s_n} \{p_n(x_n) \cdot f^*_{n+1}(s_n - x_n)\}, \qquad \text{for } n = 1, 2,$$

and, when $n = 3$,

$$f_3^*(s_3) = \min_{x_3 \le s_3} p_3(x_3).$$

Solution Procedure: The resulting dynamic programming calculations are

$n = 3$:

s_3	$f_3^*(s_3)$	x_3^*
0	0.80	0
1	0.50	1
2	0.30	2

$n = 2$:

s_2 \ x_2	$f_2(s_2, x_2) = p_2(x_2) \cdot f_3^*(s_2 - x_2)$				
	0	1	2	$f_2^*(s_2)$	x_2^*
0	0.48			0.48	0
1	0.30	0.32		0.30	0
2	0.18	0.20	0.16	0.16	2

$n = 1$:

s_1 \ x_1	$f_1(s_1, x_1) = p_1(x_1) \cdot f_2^*(s_1 - x_1)$				
	0	1	2	$f_1^*(s_1)$	x_1^*
2	0.064	0.060	0.072	0.060	1

Therefore, the optimal solution must have $x_1^* = 1$, which makes $s_2 = 2 - 1 = 1$, so that $x_2^* = 0$, which makes $s_3 = 1 - 0 = 1$, so that $x_3^* = 1$. Thus teams 1 and 3 should each receive one additional scientist. The new probability that all three teams will fail would then be 0.060.

All the examples thus far have had a *discrete* state variable s_n at each stage. Furthermore, they all have been *reversible* in the sense that the solution procedure actually could have moved *either* backward or forward stage by stage. (The latter alternative amounts to renumbering the stages in reverse order and then applying the procedure in the standard way.) This reversibility is a general characteristic of *distribution of effort problems* such as Examples 2 and 3, since the activities (stages) can be ordered in any desired manner.

The next example is different in both respects. Rather than being restricted to *integer* values, its state variable s_n at stage n is a *continuous* variable that can take on *any* value over certain intervals. Since s_n now has an infinite number of values, it is no longer possible to consider each of its feasible values individually. Rather, the solution for $f_n^*(s_n)$ and x_n^* must be expressed as *functions* of s_n. Furthermore, this example is *not* reversible because its stages correspond to *time periods,* so the solution procedure *must* proceed backward.

Example 4—Scheduling Employment Levels

The workload for the LOCAL JOB SHOP is subject to considerable seasonal fluctuation. However, machine operators are difficult to hire and costly to train, so the manager is reluctant to lay off workers during the slack seasons. He is likewise reluctant to maintain his peak season payroll when it is not required. Furthermore, he is definitely opposed to overtime work on a regular basis. Since all work is done to custom orders, it is not possible to build up inventories during slack seasons. Therefore, the manager is in a dilemma as to what his policy should be regarding employment levels.

The following estimates are given for the minimum employment requirements during the four seasons of the year for the foreseeable future:

Season	Spring	Summer	Autumn	Winter	Spring
Requirements	255	220	240	200	255

Employment will not be permitted to fall below these levels. Any employment above these levels is wasted at an approximate cost of \$2,000/person/season. It is estimated that the hiring and firing costs are such that the total cost of changing the level of employment from one season to the next is \$200 times the square of the difference in employment levels. Fractional levels of employment are possible because of a few part-time employees, and the cost data also apply on a fractional basis. The manager needs to determine what the employment level should be in each season to minimize total cost.

Formulation: On the basis of the data available, it is not worthwhile to have the employment level go above the peak season requirements of 255. Therefore, spring employment should be at 255, and the problem is reduced to finding the employment level for the other three seasons.

For a dynamic programming formulation, the seasons should be the *stages*. There are actually an indefinite number of stages because the problem extends into the indefinite future. However, each year begins an identical cycle, and because spring employment is known, it is possible to consider only one cycle of four seasons ending with the spring season, as summarized below.

$$\begin{aligned}
\text{Stage 1} &= \text{summer},\\
\text{Stage 2} &= \text{autumn},\\
\text{Stage 3} &= \text{winter},\\
\text{Stage 4} &= \text{spring}.\\
x_n &= \text{employment level for stage } n \ (n = 1, 2, 3, 4).\\
(x_4 &= 255.)
\end{aligned}$$

It is necessary that the spring season be the last stage because the optimal value of the decision variable for each state at the last stage must be either known or obtainable without considering other stages. For every other season, the solution for the optimal employment level must consider the effect on costs in the following season.

Let

$$r_n = \text{minimum employment requirement for stage } n,$$

where these requirements were given earlier as $r_1 = 220$, $r_2 = 240$, $r_3 = 200$, and $r_4 = 255$. Thus the only feasible values for x_n are

$$r_n \le x_n \le 255.$$

Referring to the cost data given in the problem statement,

$$\text{Cost for stage } n = 200(x_n - x_{n-1})^2 + 2{,}000(x_n - r_n).$$

Note that the cost at the current stage depends only upon the current decision x_n and the employment in the preceding season x_{n-1}. Thus the preceding employment level is all the information about the current state of affairs that we need to determine the optimal policy henceforth. Therefore, the *state* s_n for stage n is

$$\text{State } s_n = x_{n-1}.$$

When $n = 1$, $s_1 = x_0 = x_4 = 255$.

For your ease of reference while working through the problem, a summary of the above data is given in Table 8.3.

The objective for the problem is to choose x_1, x_2, x_3 (with $x_4 = 255$) so as to

$$\text{Minimize} \quad \sum_{i=1}^{4} [200(x_i - x_{i-1})^2 + 2{,}000(x_i - r_i)],$$

subject to

$$r_i \le x_i \le 255, \qquad \text{for } i = 1, 2, 3, 4.$$

Table 8.3 **Data for the Local Job Shop Problem**

n	r_n	Feasible x_n	Possible $s_n = x_{n-1}$	Cost
1	220	$220 \le x_1 \le 255$	$s_1 = 255$	$200(x_1 - 255)^2 + 2{,}000(x_1 - 220)$
2	240	$240 \le x_2 \le 255$	$220 \le s_2 \le 255$	$200(x_2 - x_1)^2 + 2{,}000(x_2 - 240)$
3	200	$200 \le x_3 \le 255$	$240 \le s_3 \le 255$	$200(x_3 - x_2)^2 + 2{,}000(x_3 - 200)$
4	255	$x_4 = 255$	$200 \le s_4 \le 255$	$200(255 - x_3)^2$

Thus for stage n onward ($n = 1, 2, 3, 4$),

$$f_n(s_n, x_n) = 200(x_n - s_n)^2 + 2{,}000(x_n - r_n)$$

$$+ \underset{r_i \le x_i \le 255}{\text{minimum}} \sum_{i=n+1}^{4} [200(x_i - x_{i-1})^2 + 2{,}000(x_i - r_i)],$$

because $s_n = x_{n-1}$. Also,

$$f_n^*(s_n) = \min_{r_n \le x_n \le 255} f_n(s_n, x_n).$$

Hence $$f_n(s_n, x_n) = 200(x_n - s_n)^2 + 2{,}000(x_n - r_n) + f_{n+1}^*(x_n)$$

(with f_5^* defined to be *zero* because costs after stage 4 are irrelevant to the analysis). A summary of these basic relationships is given in Fig. 8.5.

Consequently, the *recursive relationship* relating the f_n^* functions is

$$f_n^*(s_n) = \min_{r_n \le x_n \le 255} \{200(x_n - s_n)^2 + 2{,}000(x_n - r_n) + f_{n+1}^*(x_n)\}.$$

The dynamic programming approach uses this relationship to identify successively these functions—$f_4^*(s_4)$, $f_3^*(s_3)$, $f_2^*(s_2)$, $f_1^*(255)$—and the corresponding minimizing x_n.

Solution Procedure

Stage 4: Beginning at the *last* stage ($n = 4$), we already know that $x_4^* = 255$, so the necessary results are

$n = 4$:

s_4	$f_4^*(s_4)$	x_4^*
$200 \le s_4 \le 255$	$200(255 - s_4)^2$	255

Stage 3: For the problem consisting of just the last *two* stages ($n = 3$), the recursive relationship reduces to

$$f_3^*(s_3) = \min_{200 \le x_3 \le 255} \{200(x_3 - s_3)^2 + 2{,}000(x_3 - 200) + f_4^*(x_3)\}$$

$$= \min_{200 \le x_3 \le 255} \{200(x_3 - s_3)^2 + 2{,}000(x_3 - 200) + 200(255 - x_3)^2\},$$

where the possible values of s_3 are $240 \le s_3 \le 255$.

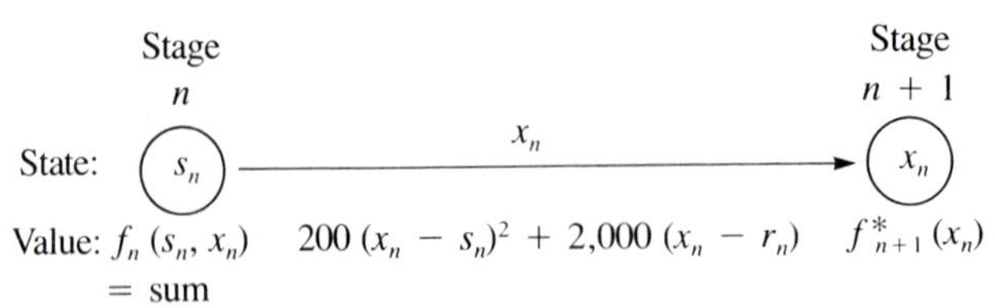

Figure 8.5 The basic structure for the Local Job Shop problem.

One way to solve for the value of x_3 that minimizes $f_3(s_3, x_3)$ for any particular value of s_3 is the graphical approach illustrated in Fig. 8.6.

However, a faster way is to use *calculus*. We want to solve for the minimizing x_3 in terms of s_3 by considering s_3 to have some fixed (but unknown) value. Therefore, equate to zero the first (partial) derivative of $f_3(s_3, x_3)$ with respect to x_3,

$$\begin{aligned}\frac{\partial}{\partial x_3} f_3(s_3, x_3) &= 400(x_3 - s_3) + 2{,}000 - 400(255 - x_3)\\ &= 400(2x_3 - s_3 - 250)\\ &= 0,\end{aligned}$$

which yields

$$x_3^* = \frac{s_3 + 250}{2}.$$

Because the second derivative is positive, and because this solution lies in the feasible interval for x_3 ($200 \le x_3 \le 255$) for all possible s_3 ($240 \le s_3 \le 255$), it is indeed the desired minimum.

Note a key difference between the nature of this solution and those obtained for the preceding examples where there were only a few possible states to consider. We now have an *infinite* number of possible states ($240 \le s_3 \le 255$), so it is no longer feasible to solve separately for x_3^* for each possible value of s_3. Therefore, we instead have solved for x_3^* as a *function* of the unknown s_3.

Using

$$\begin{aligned}f_3^*(s_3) = f_3(s_3, x_3^*) &= 200\left(\frac{s_3 + 250}{2} - s_3\right)^2 + 200\left(255 - \frac{s_3 + 250}{2}\right)^2\\ &\quad + 2{,}000\left(\frac{s_3 + 250}{2} - 200\right)\end{aligned}$$

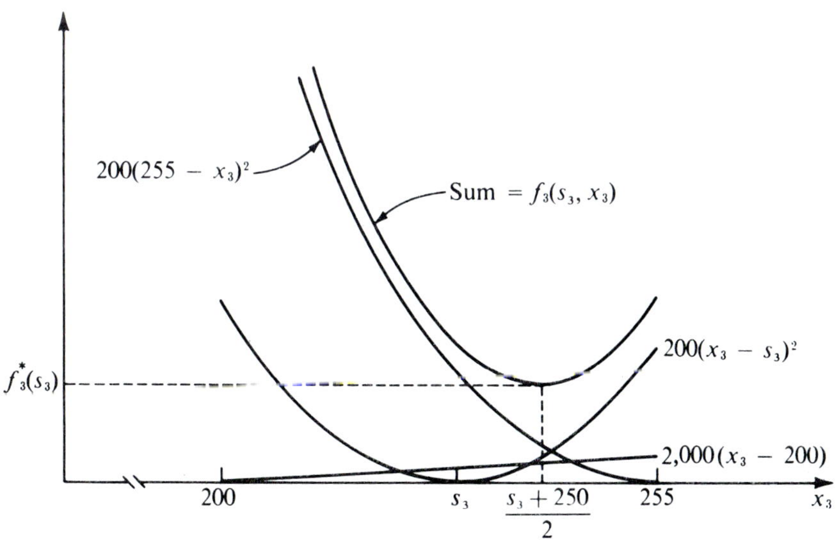

Figure 8.6 Graphical solution for $f_3^*(s_3)$ for the Local Job Shop problem.

and reducing this expression algebraically completes the required results for the two-stage problem summarized as follows.

$n = 3$:

s_3	$f_3^*(s_3)$	x_3^*
$240 \leq s_3 \leq 255$	$50(250 - s_3)^2 + 50(260 - s_3)^2 + 1{,}000(s_3 - 150)$	$\dfrac{s_3 + 250}{2}$

Stage 2: The three-stage ($n = 2$) and four-stage problems ($n = 1$) are solved in a similar fashion. Thus for $n = 2$,

$$\begin{aligned} f_2(s_2, x_2) &= 200(x_2 - s_2)^2 + 2{,}000(x_2 - r_2) + f_3^*(x_2) \\ &= 200(x_2 - s_2)^2 + 2{,}000(x_2 - 240) \\ &\quad + 50(250 - x_2)^2 + 50(260 - x_2)^2 + 1{,}000(x_2 - 150). \end{aligned}$$

The possible values of s_2 are $220 \leq s_2 \leq 255$ and the feasible region for x_2 is $240 \leq x_2 \leq 255$. The problem is to find the minimizing value of x_2 in this region, so that

$$f_2^*(s_2) = \min_{240 \leq x_2 \leq 255} f_2(s_2, x_2).$$

Setting to zero the partial derivative with respect to x_2,

$$\begin{aligned} \frac{\partial}{\partial x_2} f_2(s_2, x_2) &= 400(x_2 - s_2) + 2{,}000 - 100(250 - x_2) - 100(260 - x_2) + 1{,}000 \\ &= 200(3x_2 - 2s_2 - 240) \\ &= 0, \end{aligned}$$

yields

$$x_2 = \frac{2s_2 + 240}{3}.$$

Because

$$\frac{\partial^2}{\partial x_2^2} f_2(s_2, x_2) = 600 > 0,$$

this value of x_2 is the desired minimizing value *if* it is *feasible* ($240 \leq x_2 \leq 255$). Over the possible s_2 ($220 \leq s_2 \leq 255$), this solution actually is feasible *only* if $240 \leq s_2 \leq 255$.

Therefore, we still need to solve for the *feasible* value of x_2 that minimizes $f_2(s_2, x_2)$ when $220 \leq s_2 < 240$. The key to analyzing the behavior of $f_2(s_2, x_2)$ over the feasible region for x_2 again is the partial derivative of $f_2(s_2, x_2)$. When $s_2 < 240$,

$$\frac{\partial}{\partial x_2} f_2(s_2, x_2) > 0, \quad \text{for } 240 \leq x_2 \leq 255,$$

so that $x_2 = 240$ is the desired minimizing value.

The next step is to plug these values of x_2 into $f_2(s_2, x_2)$ to obtain $f_2^*(s_2)$ for $s_2 \geq 240$ and $s_2 < 240$. After considerable algebraic manipulation, the following results are obtained.

$n = 2$: s_2	$f_2^*(s_2)$	x_2^*
$220 \le s_2 \le 240$	$200(240 - s_2)^2 + 115{,}000$	240
$240 \le s_2 \le 255$	$\frac{200}{9}[2(250 - s_2)^2 + (265 - s_2)^2 + 30(3s_2 - 575)]$	$\frac{2s_2 + 240}{3}$

Stage 1: For the four-stage problem ($n = 1$),

$$f_1(s_1, x_1) = 200(x_1 - s_1)^2 + 2{,}000(x_1 - r_1) + f_2^*(x_1).$$

Because $r_1 = 220$, the feasible region for x_1 is $220 \le x_1 \le 255$. The expression for $f_2^*(x_1)$ will differ in the two portions, $220 \le x_1 \le 240$ and $240 \le x_1 \le 255$, of this region. Therefore,

$$f_1(s_1, x_1) = \begin{cases} 200(x_1 - s_1)^2 + 2{,}000(x_1 - 220) + 200(240 - x_1)^2 + 115{,}000, \\ \qquad\qquad\qquad\qquad \text{if } 220 \le x_1 \le 240 \\ 200(x_1 - s_1)^2 + 2{,}000(x_1 - 220) + \frac{200}{9}[2(250 - x_1)^2 \\ \qquad + (265 - x_1)^2 + 30(3x_1 - 575)], \text{ if } 240 \le x_1 \le 255. \end{cases}$$

Considering first the case where $220 \le x_1 \le 240$,

$$\frac{\partial}{\partial x_1} f_1(s_1, x_1) = 400(x_1 - s_1) + 2{,}000 - 400(240 - x_1)$$

$$= 400(2x_1 - s_1 - 235).$$

It is known that $s_1 = 255$ (spring employment), so that

$$\frac{\partial}{\partial x_1} f_1(s_1, x_1) = 800(x_1 - 245) < 0$$

for all $x_1 \le 240$. Therefore, $x_1 = 240$ is the minimizing value of $f_1(s_1, x_1)$ over the region, $220 \le x_1 \le 240$.

When $240 \le x_1 \le 255$,

$$\frac{\partial}{\partial x_1} f_1(s_1, x_1) = 400(x_1 - s_1) + 2{,}000 - \frac{200}{9}[4(250 - x_1) + 2(265 - x_1) - 90]$$

$$= \frac{400}{3}(4x_1 - 3s_1 - 225).$$

Because

$$\frac{\partial^2}{\partial x_1^2} f_1(s_1, x_1) > 0, \qquad \text{for all } x_1,$$

set

$$\frac{\partial}{\partial x_1} f_1(s_1, x_1) = 0,$$

which yields

$$x_1 = \frac{3s_1 + 225}{4}.$$

Because $s_1 = 255$, it follows that $x_1 = 247.5$ minimizes $f_1(s_1, x_1)$ over the region, $240 \le x_1 \le 255$.

Note that this region, $240 \le x_1 \le 255$, includes $x_1 = 240$, so that $f_1(s_1, 240) > f_1(s_1, 247.5)$. In the next-to-last paragraph, we found that $x_1 = 240$ minimizes $f_1(s_1, x_1)$ over the region $220 \le x_1 \le 240$. Consequently, we now can conclude that $x_1 = 247.5$ also minimizes $f_1(s_1, x_1)$ over the *entire* feasible region, $220 \le x_1 \le 255$.

Our final calculation is to find $f_1^*(s_1)$ for $s_1 = 255$ by plugging $x_1 = 247.5$ into the expression for $f_1(255, x_1)$ that holds for $240 \le x_1 \le 255$. Hence

$$\begin{aligned} f_1^*(255) &= 200(247.5 - 255)^2 + 2{,}000(247.5 - 220) \\ &\quad + \frac{200}{9}[2(250 - 247.5)^2 + (265 - 247.5)^2 + 30(742.5 - 575)] \\ &= 185{,}000. \end{aligned}$$

These results are summarized as follows:

$n = 1$:

s_1	$f_1^*(s_1)$	x_1^*
255	185,000	247.5

Therefore, tracing back through the tables for $n = 2$, $n = 3$, and $n = 4$, respectively, and setting $s_n = x_{n-1}^*$ each time, the resulting optimal solution is $x_1^* = 247.5$, $x_2^* = 245$, $x_3^* = 247.5$, $x_4^* = 255$, with a total estimated cost per cycle of \$185,000.

To conclude our illustrations of deterministic dynamic programming, we give one example that requires *more than one* variable to describe the state at each stage.

Example 5—A Linear Programming Problem

Consider the following linear programming problem:

$$\text{Maximize} \quad Z = 3x_1 + 5x_2,$$

subject to

$$\begin{aligned} x_1 \quad\quad &\le 4 \\ 2x_2 &\le 12 \\ 3x_1 + 2x_2 &\le 18 \end{aligned}$$

and

$$x_1 \ge 0, \quad x_2 \ge 0.$$

One way of solving small linear (or nonlinear) programming problems like this one is by dynamic programming, which is illustrated below.

FORMULATION: This problem requires making two interrelated decisions, namely, the level of activity 1, x_1, and the level of activity 2, x_2. Therefore, these two activities can be interpreted as the two *stages* in a dynamic programming formulation. Although they can be taken in either order, let stage n = activity n ($n = 1, 2$). Thus x_n is the decision variable at stage n.

What are the *states?* In other words, given that the decision had been made at prior stages (if any), what information is needed about the current state of affairs before the decision can be made at stage n? Reflection might suggest that the required information is the *amount of slack* left in the functional constraints. Interpret the right-hand side of these constraints (4, 12, and 18) as the total available amount of resources 1, 2, and 3, respectively. Then the state s_n can be defined as

State s_n = amount of the respective resources still available for allocation to the remaining activities ($n, \ldots, 2$).

(Note that the definition of the state is analogous to that for *distribution of effort* problems, including Examples 2 and 3, except that there are now three resources to be allocated instead of just one.) Thus

$$s_n = (R_1, R_2, R_3),$$

where R_i is the amount of resource i remaining to be allocated ($i = 1, 2, 3$). Therefore,

$$s_1 = (4, 12, 18),$$

$$s_2 = (4 - x_1, 12, 18 - 3x_1).$$

However, when we begin by solving for stage 2, we shall not yet know the value of x_1, and so use $s_2 = (R_1, R_2, R_3)$ at that point.

Therefore, in contrast to the preceding examples, this problem has *three* state variables (i.e., a *state vector* with three components) at each stage rather than one. From a theoretical standpoint, this difference is not particularly serious. It only means that, instead of considering all possible values of the one state variable, we must consider all possible *combinations* of values of the several state variables. However, from the standpoint of computational efficiency, this difference tends to be a very serious complication. Because the number of combinations, in general, can be as large as the *product* of the number of possible values of the respective variables, the number of required calculations tends to "blow up" rapidly when additional state variables are introduced. This phenomenon has been given the apt name of the **curse of dimensionality**.

Each of the three state variables is *continuous*. Therefore, rather than consider each possible combination of values separately, we must use the approach introduced in Example 4 of solving for the required information as a *function* of the state of the system.

Despite these complications, this problem is small enough that it can still be solved without great difficulty. To solve it, we need to introduce the usual dynamic programming notation. Thus,

$f_2(R_1, R_2, R_3, x_2)$ = contribution of activity 2 to Z if the system starts in state (R_1, R_2, R_3) at stage 2 and the decision is x_2

$= 5x_2,$

$$f_1(4, 12, 18, x_1) = \text{contribution of activities 1 and 2 to } Z \text{ if the system starts in state (4, 12, 18) at stage 1, the immediate decision is } x_1\text{, and then an optimal decision is made at stage 2.}$$

$$= 3x_1 + \underset{\substack{x_2 \le 12 \\ 2x_2 \le 18 - 3x_1 \\ x_2 \ge 0}}{\text{maximum}} \{5x_2\}.$$

Similarly, for $n = 1, 2,$

$$f_n^*(R_1, R_2, R_3) = \max_{x_n} f_n(R_1, R_2, R_3, x_n),$$

where this maximum is taken over the *feasible* values of x_n. Consequently, using the relevant portions of the constraints of the problem,

(1) $$f_2^*(R_1, R_2, R_3) = \underset{\substack{2x_2 \le R_2 \\ 2x_2 \le R_3 \\ x_2 \ge 0}}{\text{maximum}} \{5x_2\},$$

(2) $$f_1(4, 12, 18, x_1) = 3x_1 + f_2^*(4 - x_1, 12, 18 - 3x_1),$$

(3) $$f_1^*(4, 12, 18) = \underset{\substack{x_1 \le 4 \\ 3x_1 \le 18 \\ x_1 \ge 0}}{\text{maximum}} \{3x_1 + f_2^*(4 - x_1, 12, 18 - 3x_1)\}.$$

Equation (1) will be used to solve the stage 2 problem. Equation (2) shows the basic dynamic programming structure for this problem, as also depicted in Fig. 8.7. Equation (3) gives the *recursive relationship* between f_1^* and f_2^* that will be used to solve the stage 1 problem.

Solution Procedure

Stage 2: To solve at the last stage ($n = 2$), Eq. (1) indicates that x_2^* must be the *largest* value of x_2 that *simultaneously* satisfies $2x_2 \le R_2$, $2x_2 \le R_3$, and $x_2 \ge 0$. Assuming that $R_2 \ge 0$ and $R_3 \ge 0$, so that feasible solutions exist, this largest value is the smaller of $R_2/2$ and $R_3/2$. Thus the solution is

$n = 2$:

(R_1, R_2, R_3)	$f_2^*(R_1, R_2, R_3)$	x_2^*
$R_2 \ge 0, R_3 \ge 0$	$5 \min \left\{\frac{R_2}{2}, \frac{R_3}{2}\right\}$	$\min \left\{\frac{R_2}{2}, \frac{R_3}{2}\right\}$

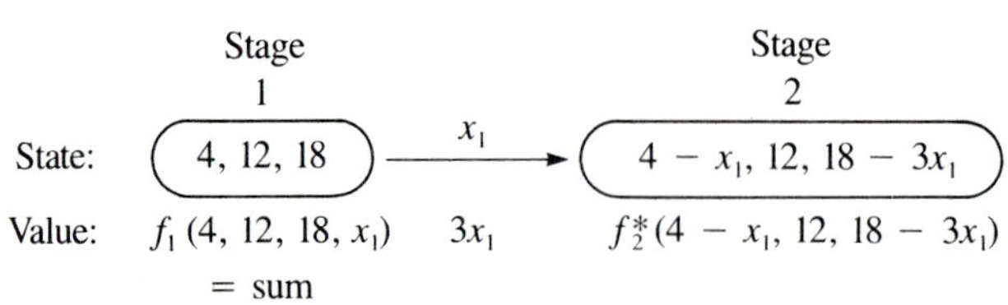

Figure 8.7 The basic structure for the linear programming problem.

Stage 1: To solve the two-stage problem ($n = 1$), we plug the solution just obtained for $f_2^*(R_1, R_2, R_3)$ into Eq. (3). For stage 2,

$$(R_1, R_2, R_3) = (4 - x_1, 12, 18 - 3x_1),$$

so that $f_2^*(4 - x_1, 12, 18 - 3x_1) = 5 \min \left\{ \frac{R_2}{2}, \frac{R_3}{2} \right\} = 5 \min \left\{ \frac{12}{2}, \frac{18 - 3x_1}{2} \right\}$

is the specific solution plugged into Eq. (3). After combining its constraints on x_1, Eq. (3) then becomes

$$f_1^*(4, 12, 18) = \max_{0 \le x_1 \le 4} \left\{ 3x_1 + 5 \min \left\{ \frac{12}{2}, \frac{18 - 3x_1}{2} \right\} \right\}.$$

Over the feasible interval, $0 \le x_1 \le 4$, notice that

$$\min \left\{ \frac{12}{2}, \frac{18 - 3x_1}{2} \right\} = \begin{cases} 6, & \text{if } 0 \le x_1 \le 2 \\ 9 - \frac{3}{2} x_1, & \text{if } 2 \le x_1 \le 4, \end{cases}$$

so that $3x_1 + 5 \min \left\{ \frac{12}{2}, \frac{18 - 3x_1}{2} \right\} = \begin{cases} 3x_1 + 30, & \text{if } 0 \le x_1 \le 2 \\ 45 - \frac{9}{2} x_1, & \text{if } 2 \le x_1 \le 4. \end{cases}$

Because both

$$\max_{0 \le x_1 \le 2} \{3x_1 + 30\} \quad \text{and} \quad \max_{2 \le x_1 \le 4} \left\{ 45 - \frac{9}{2} x_1 \right\}$$

achieve their maximum at $x_1 = 2$, it follows that $x_1^* = 2$, and that this maximum is 36, as given in the following table.

$n = 1$:

(R_1, R_2, R_3)	$f_1^*(R_1, R_2, R_3)$	x_1^*
(4, 12, 18)	36	2

Because $x_1^* = 2$ leads to

$$R_1 = 4 - 2 = 2, \qquad R_2 = 12, \qquad R_3 = 18 - 3(2) = 12$$

for stage 2, the $n = 2$ table yields $x_2^* = 6$. Consequently, $x_1^* = 2$, $x_2^* = 6$ is the optimal solution for this problem, and the $n = 1$ table shows that the resulting value of Z is 36.

8.4 Probabilistic Dynamic Programming

Probabilistic dynamic programming differs from deterministic dynamic programming in that the state at the next stage is *not* completely determined by the state and policy decision at the current stage. Rather, there is a *probability distribution* for what the next state will be. However, this probability distribution still is completely determined by the state and policy decision at the current stage. The resulting basic structure for probabilistic dynamic programming is described diagrammatically in Fig. 8.8.

For purposes of this diagram, we let S denote the number of possible states at stage $n + 1$ and label these states on the right-side as $1, 2, \ldots, S$. The system goes to state i with probability p_i ($i = 1, 2, \ldots, S$) given the state s_n and decision x_n at stage n. If the system goes to state i, C_i is the contribution of stage n to the objective function.

When Fig. 8.8 is expanded to include all the possible states and decisions at all the stages, it is sometimes referred to as a **decision tree**. If the decision tree is not too large, it provides a useful way of summarizing the various possibilities that may occur.

Because of the probabilistic structure, the relationship between $f_n(s_n, x_n)$ and the $f^*_{n+1}(s_{n+1})$ necessarily is somewhat more complicated than for deterministic dynamic programming. The precise form of this relationship will depend upon the form of the overall objective function.

To illustrate, suppose that the objective is to *minimize* the *expected sum* of the contributions from the individual stages. In this case, $f_n(s_n, x_n)$ would represent the minimum expected sum from stage n onward, *given* that the state and policy decision at stage n are s_n and x_n, respectively. Consequently,

$$f_n(s_n, x_n) = \sum_{i=1}^{S} p_i[C_i + f^*_{n+1}(i)],$$

with

$$f^*_{n+1}(i) = \min_{x_{n+1}} f_{n+1}(i, x_{n+1}),$$

where this minimization is taken over the *feasible* values of x_{n+1}.

Example 6 has this same form. Example 7 will illustrate another form.

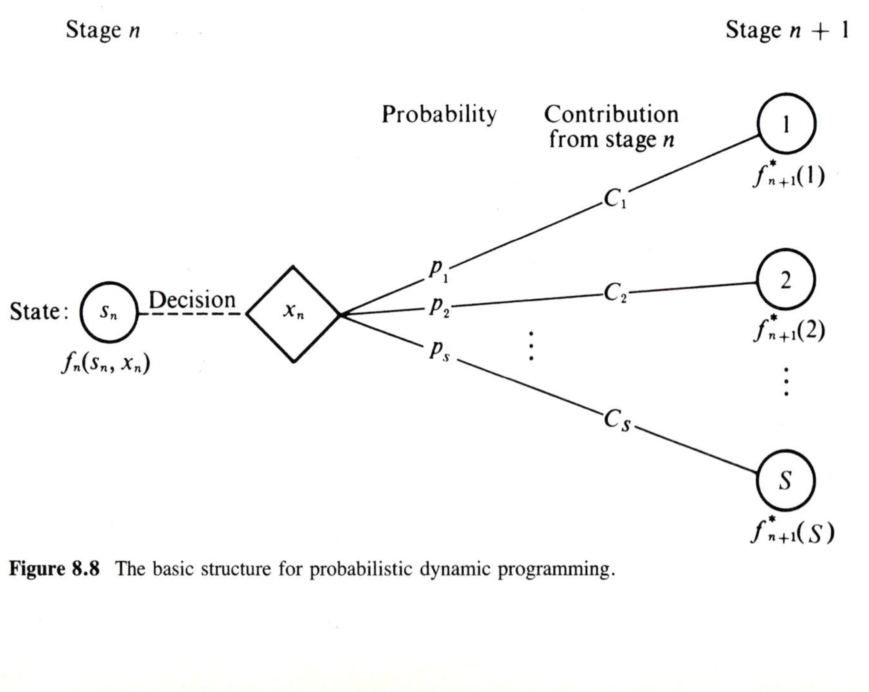

Figure 8.8 The basic structure for probabilistic dynamic programming.

Example 6—Determining Reject Allowances

The HIT-AND-MISS MANUFACTURING COMPANY has received an order to supply one item of a particular type. However, the customer has specified such stringent quality requirements that the manufacturer may have to produce more than one item to obtain an item that is acceptable. The number of *extra* items produced in a production run is called the *reject allowance*. Including a reject allowance is common practice when producing for a custom order, and it seems advisable in this case.

The manufacturer estimates that each item of this type that is produced will be *acceptable* with probability $\frac{1}{2}$ and *defective* (without possibility for rework) with probability $\frac{1}{2}$. Thus the number of acceptable items produced in a lot of size L will have a *binomial distribution;* that is, the probability of producing *zero* acceptable items in such a lot is $(\frac{1}{2})^L$.

Marginal production costs for this product are estimated to be \$100 per item (even if defective), and excess items are worthless. In addition, a setup cost of \$300 must be incurred whenever the production process is set up for this product, and a completely new setup at this same cost is required for each subsequent production run if a lengthy inspection procedure reveals that a completed lot has not yielded an acceptable item. The manufacturer has time to make no more than three production runs. If an acceptable item has not been obtained by the end of the third production run, the cost to the manufacturer in lost sales income and in penalty costs would be \$1,600.

The objective is to determine the policy regarding the lot size (1 + reject allowance) for the required production run(s) that minimizes total expected cost for the manufacturer.

FORMULATION: A dynamic programming formulation for this problem is

Stage n = production run n ($n = 1, 2, 3$),
x_n = lot size for stage n,
State s_n = number of acceptable items still needed (one or zero) when beginning stage n.

Thus, at stage 1, the state $s_1 = 1$. If at least one acceptable item is obtained subsequently, the state changes to $s_n = 0$, after which no additional costs need to be incurred.

Because of the stated objective for the problem,

$f_n(s_n, x_n)$ = total expected cost for stages $n, \ldots, 3$ if the system starts in state s_n at stage n, the immediate decision is x_n, and optimal decisions are made thereafter,

$$f_n^*(s_n) = \min_{x_n=0,1,\ldots} f_n(s_n, x_n),$$

where $f_n^*(0) = 0$. Using \$100 as the unit of money, the contribution to cost from stage n is $(K + x_n)$ regardless of the next state, where

$$K = \begin{cases} 0, & \text{if } x_n = 0 \\ 3, & \text{if } x_n > 0. \end{cases}$$

Therefore, for $s_n = 1$,

$$f_n(1, x_n) = K + x_n + (\tfrac{1}{2})^{x_n} f^*_{n+1}(1) + [1 - (\tfrac{1}{2})^{x_n}] f^*_{n+1}(0)$$

$$= K + x_n + (\tfrac{1}{2})^{x_n} f^*_{n+1}(1)$$

[where $f^*_4(1)$ is defined to be 16, the terminal cost if no acceptable items have been obtained]. A summary of these basic relationships is given in Fig. 8.9.

Consequently, the *recursive relationship* for the dynamic programming calculations is

$$f^*_n(1) = \min_{x_n = 0, 1, \ldots} \{K + x_n + (\tfrac{1}{2})^{x_n} f^*_{n+1}(1)\}$$

for $n = 1, 2, 3$.

Solution Procedure: The calculations using this recursive relationship are summarized as follows.

$n = 3$:

s_3 \ x_3	$f_3(1, x_3) = K + x_3 + 16(\frac{1}{2})^{x_3}$							
	0	1	2	3	4	5	$f^*_3(s_3)$	x^*_3
0	0						0	0
1	16	12	9	8	8	$8\frac{1}{2}$	8	3 or 4

$n = 2$:

s_2 \ x_2	$f_2(1, x_2) = K + x_2 + (\frac{1}{2})^{x_2} f^*_3(1)$						
	0	1	2	3	4	$f^*_2(s_2)$	x^*_2
0	0					0	0
1	8	8	7	7	$7\frac{1}{2}$	7	2 or 3

$n = 1$:

s_1 \ x_1	$f_1(1, x_1) = K + x_1 + (\frac{1}{2})^{x_1} f^*_2(1)$						
	0	1	2	3	4	$f^*_1(s_1)$	x^*_1
1	7	$7\frac{1}{2}$	$6\frac{3}{4}$	$6\frac{7}{8}$	$7\frac{7}{16}$	$6\frac{3}{4}$	2

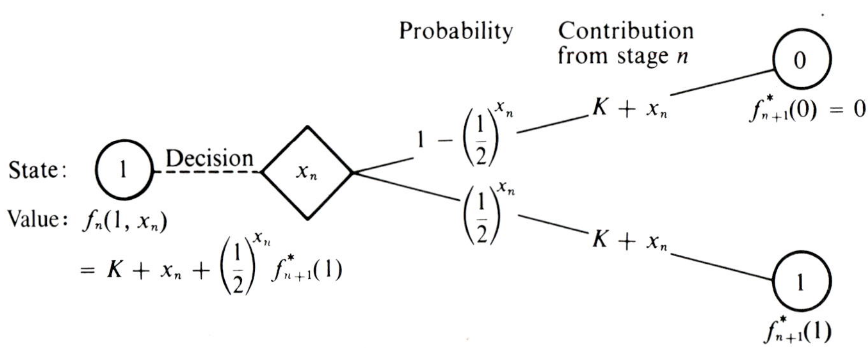

Figure 8.9 The basic structure for the Hit-and-Miss Manufacturing Co. problem.

Thus the optimal policy is to produce *two* items on the first production run; if none is acceptable, then produce either *two* or *three* items on the second production run; if none is acceptable, then produce either *three* or *four* items on the third production run. The *total expected cost* for this policy is \$675.

Example 7—Winning in Las Vegas

An enterprising young statistician believes that she has developed a system for winning a popular Las Vegas game. Her colleagues do not believe that her system works, so they have made a large bet with her that, starting with three chips, she will not have five chips after three plays of the game. Each play of the game involves betting any desired number of available chips and then either winning or losing this number of chips. The statistician believes that her system will give her a probability of $\frac{2}{3}$ of winning a given play of the game.

Assuming the statistician is correct, we now will use dynamic programming to determine her optimal policy regarding how many chips to bet (if any) at each of the three plays of the game. The decision at each play should take into account the results of earlier plays. The objective is to maximize the probability of winning her bet with her colleagues.

FORMULATION: The dynamic programming formulation for this problem is

$$\begin{aligned} \text{Stage } n &= n\text{th play of the game } (n = 1, 2, 3), \\ x_n &= \text{number of chips to bet at stage } n, \\ \text{State } s_n &= \text{number of chips in hand to begin stage } n. \end{aligned}$$

This definition of the state is chosen because it provides the needed information about the current situation for making an optimal decision on how many chips to bet next.

Because the objective is to maximize the probability that the statistician will win her bet, the objective function to be maximized at each stage must be the probability of finishing the three plays with at least five chips. Therefore,

$f_n(s_n, x_n) =$ probability of finishing the three plays with at least five chips, given that the statistician starts stage n in state s_n, makes the immediate decision x_n, and makes optimal decisions thereafter,

$$f_n^*(s_n) = \max_{x_n=0,1,\ldots,s_n} f_n(s_n, x_n).$$

The expression for $f_n(s_n, x_n)$ must reflect the fact that it may still be possible to accumulate five chips eventually even if the statistician should lose the next play. If she should lose, the state at the next stage would be $(s_n - x_n)$, and the probability of finishing with at least five chips would then be $f_{n+1}^*(s_n - x_n)$. If she should win the next play instead, the state would become $(s_n + x_n)$, and the corresponding probability would be $f_{n+1}^*(s_n + x_n)$. Because the assumed probability of winning a given play is $\frac{2}{3}$, it now follows that

$$f_n(s_n, x_n) = \frac{1}{3}f_{n+1}^*(s_n - x_n) + \frac{2}{3}f_{n+1}^*(s_n + x_n)$$

[where $f_4^*(s_4)$ is defined to be *zero* for $s_4 < 5$ and 1 for $s_4 \geq 5$]. Thus there is no direct contribution to the objective function from stage n in addition to the effect of then being in the next state. These basic relationships are summarized in Fig. 8.10.

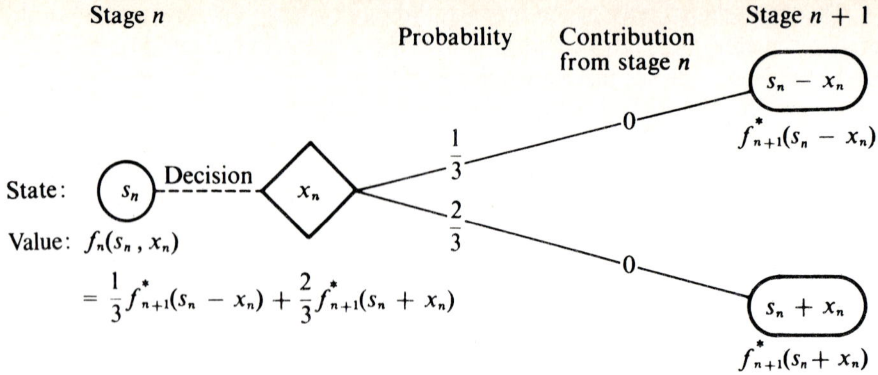

Figure 8.10 The basic structure for the Las Vegas problem.

Therefore, the *recursive relationship* for this problem is

$$f^*_n(s_n) = \max_{x_n=0,1,\ldots,s_n} \{\tfrac{1}{3} f^*_{n+1}(s_n - x_n) + \tfrac{2}{3} f^*_{n+1}(s_n + x_n)\},$$

for $n = 1, 2, 3$, with $f^*_4(s_4)$ as just defined.

Solution Procedure: This recursive relationship leads to the following computational results.

$n = 3$:

s_3	$f^*_3(s_3)$	x^*_3
0	0	—
1	0	—
2	0	—
3	$\frac{2}{3}$	2 (or more)
4	$\frac{2}{3}$	1 (or more)
≥ 5	1	0 (or $\leq s_3 - 5$)

$n = 2$:

s_2 \ x_2	$f_2(s_2, x_2) = \frac{1}{3} f^*_3(s_2 - x_2) + \frac{2}{3} f^*_3(s_2 + x_2)$						
	0	1	2	3	4	$f^*_2(s_2)$	x^*_2
0	0					0	—
1	0	0				0	—
2	0	$\frac{4}{9}$	$\frac{4}{9}$			$\frac{4}{9}$	1 or 2
3	$\frac{2}{3}$	$\frac{4}{9}$	$\frac{2}{3}$	$\frac{2}{3}$		$\frac{2}{3}$	0, 2, or 3
4	$\frac{2}{3}$	$\frac{8}{9}$	$\frac{2}{3}$	$\frac{2}{3}$	$\frac{2}{3}$	$\frac{8}{9}$	1
≥ 5	1					1	0 (or $\leq s_2 - 5$)

$n = 1$:

s_2 \ x_1	$f_1(s_1, x_1) = \frac{1}{3} f^*_2(s_1 - x_1) + \frac{2}{3} f^*_2(s_1 + x_1)$					
	0	1	2	3	$f^*_1(s_1)$	x^*_1
3	$\frac{2}{3}$	$\frac{20}{27}$	$\frac{2}{3}$	$\frac{2}{3}$	$\frac{20}{27}$	1

Therefore, the optimal policy is

$$x_1^* = 1 \begin{cases} \text{if win, } x_2^* = 1 \begin{cases} \text{if win, } x_3^* = 0 \\ \text{if lose, } x_3^* = 2 \text{ or } 3. \end{cases} \\ \\ \text{if lose, } x_2^* = 1 \text{ or } 2 \begin{cases} \text{if win, } x_3^* = \begin{cases} 2 \text{ or } 3 \text{ (for } x_2^* = 1) \\ 1, 2, 3, \text{ or } 4 \text{ (for } x_2^* = 2) \end{cases} \\ \text{if lose, bet is lost.} \end{cases} \end{cases}$$

This policy gives the statistician a probability of $\frac{20}{27}$ of winning her bet with her colleagues.

8.5 Conclusions

Dynamic programming is a very useful technique for making a *sequence of interrelated decisions*. It requires formulating an appropriate *recursive relationship* for each individual problem. However, it provides a great computational savings over using exhaustive enumeration to find the best combination of decisions, especially for large problems. For example, if a problem has 10 stages with 10 states and 10 possible decisions at each stage, then exhaustive enumeration must consider up to 10^{10} combinations, whereas dynamic programming need make no more than 10^3 calculations (10 for each state at each stage).

This chapter has considered only dynamic programming with a *finite* number of stages. Chapter 11 is devoted to a general kind of model for probabilistic dynamic programming where the stages continue to recur indefinitely—namely, *Markovian decision processes*.

SELECTED REFERENCES

1. Bellman, Richard, and Stuart Dreyfus: *Applied Dynamic Programming*, Princeton University Press, Princeton, N.J., 1962.
2. Cooper, Leon L., and Mary W. Cooper: *Introduction to Dynamic Programming*, Pergamon Press, Elmsford, N.Y., 1981.
3. Denardo, Eric V.: *Dynamic Programming Theory and Applications*, Prentice-Hall, Englewood Cliffs, N.J., 1982.
4. Dreyfus, Stuart E., and Averill M. Law: *The Art and Theory of Dynamic Programming*, Academic Press, New York, 1977.
5. Howard, Ronald A.: "Dynamic Programming," *Management Science*, **12**:317–345, 1966.

PROBLEMS

1. Consider the following network, where each number along a link represents the actual distance between the pair of nodes connected by that link.

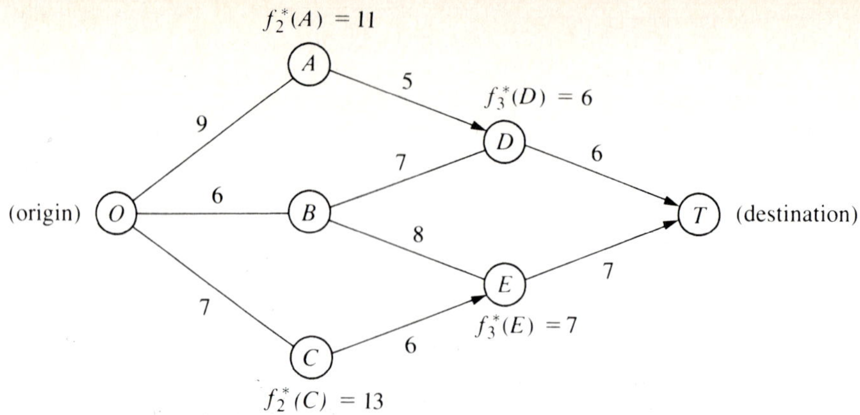

The objective is to find the shortest route from the origin to the destination.

(*a*) What are the stages and states for the dynamic programming formulation of this problem?

(*b*) Use dynamic programming to solve this problem. However, instead of using the usual tables, show your work graphically. In particular, redraw the preceding network, where the answers already are given for $f_n^*(s_n)$ for four of the nodes; then solve for and fill in $f_2^*(B)$ and $f_1^*(O)$. Draw an arrowhead that shows the optimal link to take out of each of the latter two nodes. Finally, identify the optimal route by following the arrows from node O onward to node T.

(*c*) Use dynamic programming to solve this problem by constructing the usual tables for $n = 3$, $n = 2$, and $n = 1$.

2. The sales manager for a publisher of college textbooks has six traveling salespeople to assign to three different regions of the country. She has decided that each region should be assigned at least one salesperson and that each individual salesperson should be restricted to one of the regions, but now she wants to determine how many salespeople should be assigned to the respective regions in order to maximize sales.

The following table gives the estimated increase in sales (in appropriate units) in each region if it were allocated various numbers of salespeople:

Number of Salespeople	*Region* 1	2	3
1	35	21	28
2	48	42	41
3	70	56	63
4	89	70	75

(*a*) Use dynamic programming to solve this problem. Instead of using the usual tables, show your work graphically by constructing and filling in a network such as the one shown for Prob. 1. Proceed as in part (*b*) of Prob. 1 by solving for $f_n^*(s_n)$ for each node (except the terminal node) and writing its value by the node. Draw an arrowhead to show the optimal link (or links in the case of a tie) to take out of each node. Finally, identify the resulting optimal route (or routes) through the network and the corresponding optimal solution (or solutions).

(*b*) Use dynamic programming to solve this problem by constructing the usual tables for $n = 3$, $n = 2$, and $n = 1$.

3.* The owner of a chain of three grocery stores has purchased five crates of fresh strawberries. The estimated probability distribution of potential sales of the strawberries before spoilage differs among the three stores. Therefore, the owner wants to know how he should allocate the five crates to the three stores to maximize expected profit.

For administrative reasons, the owner does not wish to split crates between stores. However, he is willing to distribute zero crates to any of his stores.

The following table gives the estimated expected profit at each store when it is allocated various numbers of crates:

Number of Crates	*Stores* 1	2	3
0	0	0	0
1	5	6	4
2	9	11	9
3	14	15	13
4	17	19	18
5	21	22	20

Use dynamic programming to determine how many of the five crates should be assigned to each of the three stores to maximize the total expected profit.

4. A college student has 7 days remaining before final examinations begin in her four courses, and she wants to allocate this study time as effectively as possible. She needs at least 1 day on each course, and she likes to concentrate on just one course each day, so she wants to allocate 1, 2, 3, or 4 days to each course. Having recently taken an operations research course, she decides to use dynamic programming to make these allocations to maximize the total grade points to be obtained from the four courses. She estimates that the alternative allocations for each course would yield the number of grade points shown in the following table:

Estimated Grade Points

Number of Study Days	*Course* 1	2	3	4
1	3	5	2	6
2	5	5	4	7
3	6	6	7	9
4	7	9	8	9

Solve this problem by dynamic programming.

5. A company is planning its advertising strategy for next year for its three major products. Since the three products are quite different, each advertising effort will focus on a single product. In units of millions of dollars, a total of 6 is available for advertising next year, where the advertising expenditure for each product must be an integer greater than or equal to 1. The vice-president for marketing has established the objective, namely, determine how much to spend on each product in order to maximize total sales. The following table gives the estimated increase in sales (in appropriate units) for the different advertising expenditures:

Advertising Expenditure	*Product* 1	2	3
1	7	4	6
2	10	8	9
3	14	11	13
4	17	14	15

Use dynamic programming to solve this problem.

6. A political campaign is entering its final stage, and polls indicate a very close election. One of the candidates has enough funds left to purchase TV time for a total of five prime-time commercials on TV stations located in four different areas. Based on polling information, an estimate has been made of the number of additional votes that can be won in the different broadcasting areas depending upon the number of commercials run. These estimates are given in the following table in units of thousands of votes:

Number of Commercials	*Area* 1	2	3	4
0	0	0	0	0
1	4	6	5	3
2	7	8	9	7
3	9	10	11	12
4	12	11	10	14
5	15	12	9	16

Use dynamic programming to determine how the five commercials should be distributed among the four areas in order to maximize the estimated number of votes won.

7. A county chairwoman of a certain political party is making plans for an upcoming presidential election. She has received the services of six volunteer workers for precinct work, and she wants to assign them to four precincts in such a way as to maximize their effectiveness. She feels that it would be inefficient to assign a worker to more than one precinct, but she is willing to assign no workers to any one of the precincts if they can accomplish more in other precincts.

The following table gives the estimated increase in the number of votes for the party's candidate in each precinct if it were allocated various numbers of workers:

Number of Workers	*Precinct* 1	2	3	4
0	0	0	0	0
1	4	7	5	6
2	9	11	10	11
3	15	16	15	14
4	18	18	18	16
5	22	20	21	17
6	24	21	22	18

This problem has several optimal solutions for how many of the six workers should be assigned to each of the four precincts to maximize the total estimated increase in the plurality of the party's candidate. Use dynamic programming to find all of them so the chairwoman can make the final selection based on other factors.

8. The *Build-Em-Fast Company* has agreed to supply its best customer with three widgits during *each* of the next three weeks, even though producing them will require some overtime work. The relevant production data are as follows:

Week	Maximum Production, Regular Time	Maximum Production, Overtime	Production Cost Per Unit, Regular Time
1	2	2	\$300
2	2	1	\$500
3	1	2	\$400

The cost per unit produced with overtime for each week is $100 more than for regular time. The cost of storage is $50 per unit for each week it is stored. There is already an inventory of two widgits on hand currently, but the company does not want to retain any widgits in inventory after the three weeks.

Management wants to know how many units should be produced in each week in order to maximize profit. Use dynamic programming to solve this problem.

9.* A company will soon be introducing a new product into a very competitive market and is currently planning its marketing strategy. The decision has been made to introduce the product in three phases. Phase 1 will feature making a special introductory offer of the product to the public at a greatly reduced price to attract first-time buyers. Phase 2 will involve an intensive advertising campaign to persuade these first-time buyers to continue purchasing the product at a regular price. It is known that another company will be introducing a new competitive product at about the time phase 2 will end. Therefore, phase 3 will involve a follow-up advertising and promotion campaign to try to keep the regular purchasers from switching to the competitive product.

A total of $4 million has been budgeted for this marketing campaign. The problem now is to determine how to allocate this money most effectively to the three phases. Let m denote the initial share of the market (expressed as a percentage) attained in phase 1, f_2 the fraction of this market share that is retained in phase 2, and f_3 the fraction of the remaining market share that is retained in phase 3. Given the following data, use dynamic programming to determine how to allocate the $4 million to maximize the *final share* of the market for the new product, i.e., to maximize mf_2f_3.

(*a*) Assume that the money must be spent in integer multiples of $1 million in each phase, where the minimum permissible multiple is 1 for phase 1 and 0 for phases 2 and 3. The following table gives the estimated effect of expenditures in each phase:

Millions of Dollars Expended	*Effect on Market Share*		
	m	f_2	f_3
0	—	0.2	0.3
1	20	0.4	0.5
2	30	0.5	0.6
3	40	0.6	0.7
4	50	—	—

(*b*) Now assume that *any* amount within the total budget can be spent in each phase, where the estimated effect of spending an amount x_i (in units of *millions* of dollars) in phase i ($i = 1, 2, 3$) is

$$m = 10x_1 - x_1^2$$

$$f_2 = 0.40 + 0.10x_2$$

$$f_3 = 0.60 + 0.07x_3.$$

[*Hint:* After solving for the $f_2^*(s)$ and $f_3^*(s)$ functions analytically, solve for x_1^* graphically.]

10. Consider an electronic system consisting of four components, each of which must function for the system to function. The reliability of the system can be improved by installing several parallel units in one or more of the components. The following table gives the probability that the respective components will function if they consist of one, two, or three parallel units:

Number of Parallel Units	*Probability of Functioning* Component 1	Component 2	Component 3	Component 4
1	0.5	0.6	0.7	0.5
2	0.6	0.7	0.8	0.7
3	0.8	0.8	0.9	0.9

The probability that the system will function is the product of the probabilities that the respective components will function.

The cost (in hundreds of dollars) of installing one, two, or three parallel units in the respective components is given by the following table:

Number of Parallel Units	*Cost* Component 1	Component 2	Component 3	Component 4
1	1	2	1	2
2	2	4	3	3
3	3	5	4	4

Because of budget limitations, a maximum of \$1,000 can be expended.

Use dynamic programming to determine how many parallel units should be installed in each of the four components to maximize the probability that the system will function.

11. Consider the following integer nonlinear programming problem.

$$\text{Maximize} \quad Z = x_1 x_2^2 x_3^3,$$

subject to

$$x_1 + 2x_2 + 3x_3 \leq 10,$$

$$x_1 \geq 1, \quad x_2 \geq 1, \quad x_3 \geq 1,$$

and

$$x_1, x_2, x_3 \text{ are integers.}$$

Use dynamic programming to solve this problem.

12.* Consider the following nonlinear programming problem.

$$\text{Maximize} \quad Z = 36x_1 + 9x_1^2 - 6x_1^3 + 36x_2 - 3x_2^3,$$

subject to

$$x_1 + x_2 \leq 3$$

and

$$x_1 \geq 0, \quad x_2 \geq 0.$$

Use dynamic programming to solve this problem.

13. Resolve the *Local Job Shop* employment scheduling problem (Example 4) when the total cost of changing the level of employment from one season to the next is changed to \$100 times the square of the difference in employment levels.

14. Consider the following nonlinear programming problem.

$$\text{Maximize} \quad Z = x_1^2 x_2,$$

subject to

$$x_1^2 + x_2 \leq 2.$$

(Note that there are no nonnegativity constraints.) Use dynamic programming to solve this problem.

15. Consider the following nonlinear programming problem.

$$\text{Maximize} \quad Z = x_1^3 + 4x_2^2 + 16x_3,$$

subject to

$$x_1 x_2 x_3 = 4$$

and

$$x_1 \geq 1, \quad x_2 \geq 1, \quad x_3 \geq 1.$$

(*a*) Solve by dynamic programming when, in addition to the given constraints, all three variables also are required to be *integer*.
(*b*) Use dynamic programming to solve the problem as given (continuous variables).

16. Consider the following nonlinear programming problem.

$$\text{Maximize} \quad Z = x_1(1 - x_2)x_3,$$

subject to

$$x_1 - x_2 + x_3 \leq 1$$

and

$$x_1 \geq 0, \quad x_2 \geq 0, \quad x_3 \geq 0.$$

Use dynamic programming to solve this problem.

17. Consider the following linear programming problem.

$$\text{Maximize} \quad Z = 15x_1 + 10x_2,$$

subject to

$$x_1 + 2x_2 \leq 6$$

$$3x_1 + x_2 \leq 8$$

and

$$x_1 \geq 0, \quad x_2 \geq 0.$$

Use dynamic programming to solve this problem.

18. Consider the following nonlinear programming problem.

$$\text{Maximize} \quad Z = 5x_1 + x_2,$$

subject to

$$2x_1^2 + x_2 \leq 13$$

$$x_1^2 + x_2 \leq 9$$

and

$$x_1 \geq 0, \quad x_2 \geq 0.$$

Use dynamic programming to solve this problem.

19. Consider the following "fixed-charge" problem.

$$\text{Maximize} \quad Z = 3x_1 + 7x_2 + 6f(x_3),$$

subject to

$$x_1 + 3x_2 + 2x_3 \leq 6$$

$$x_1 + x_2 \leq 5$$

and

$$x_1 \geq 0, \quad x_2 \geq 0, \quad x_3 \geq 0,$$

where

$$f(x_3) = \begin{cases} 0, & \text{if } x_3 = 0 \\ -1 + x_3, & \text{if } x_3 > 0. \end{cases}$$

Use dynamic programming to solve this problem.

20. A backgammon player will be playing three consecutive matches with friends tonight. For each match, he will have the opportunity to place an even bet that he will win; the amount bet can be *any* quantity of his choice between zero and the amount of money he still has left after the bets on the preceding matches. For each match, the probability is $\frac{1}{2}$ that he will win the match and thus win the amount bet, whereas the probability is $\frac{1}{2}$ that he will lose the match and thus lose the amount bet. He will begin with \$75, and his goal is to have \$100 at the end. (Because these are friendly matches, he does not want to end up with more than \$100.) Therefore, he wants to find the optimal betting policy (including all ties) that maximizes the probability that he will have exactly \$100 after the three matches.

Use dynamic programming to solve this problem.

21. Imagine that you have \$5,000 to invest and that you will have an opportunity to invest that amount in either of two investments (*A* or *B*) at the beginning of each of the next 3 years. Both investments have uncertain returns. For investment *A* you will either lose your money entirely or (with higher probability) get back \$10,000 (a profit of \$5,000) at the end of the year. For investment *B* you will either get back just your \$5,000 or (with low probability) \$10,000 at the end of the year. The probabilities for these events are:

Investment	Amount Returned (\$)	Probability
A	0	0.3
	10,000	0.7
B	5,000	0.9
	10,000	0.1

You are allowed to make only (at most) *one* investment each year, and can invest only \$5,000 each time. (Any additional money accumulated is left idle.)

(*a*) Use dynamic programming to find the investment policy that maximizes the *expected amount of money* you will have after the 3 years.

(*b*) Use dynamic programming to find the investment policy that maximizes the *probability* that you will have at least \$10,000 after the 3 years.

22.* Suppose that the situation for the Hit-and-Miss Manufacturing Co. problem (Example 6) has changed somewhat. After a more careful analysis, you now estimate that each item produced will be *acceptable* with probability $\frac{2}{3}$, rather than $\frac{1}{2}$, so that the probability of producing *zero* acceptable items in a lot of size L is $(\frac{1}{3})^L$. Furthermore, there now is only enough time available to make *two* production runs. Use dynamic programming to determine the new optimal policy for this problem.

23. Reconsider Example 7. Suppose that the bet is changed to: "Starting with two chips, she will not have five chips after *five* plays of the game." By referring to the previous computational results, make additional calculations to determine what the new optimal policy is for the enterprising young statistician.

24. The Profit & Gambit Co. has a major product that has been losing money recently because of declining sales. In fact, during the current quarter of the year, sales will be 4 million units below the *break-even point*. Because the *marginal* revenue for each unit sold exceeds the *marginal* cost by \$5, this amounts to a loss of \$20 million for the quarter. Therefore, management must take action quickly to rectify this situation. Two alternative courses of action are being considered. One is to abandon the product immediately, incurring a cost of \$20 million for shutting down. The other alternative is to undertake an intensive advertising campaign to increase sales and then abandon the product (at the cost of \$20 million) only if the campaign is not sufficiently successful. Tentative plans for this advertising campaign have been developed and analyzed. It would extend over the next three quarters (subject to early cancellation), and the cost would be \$30 million in each of the three quarters. It is estimated that the increase in sales would be approximately 3 million units in the first quarter, another 2 million units in the second quarter, and another 1 million units in the third quarter. However, because of a number of unpredictable market variables, there is considerable uncertainty as to what impact the advertising actually would have, and careful analysis indicates that the estimates for each quarter could turn out to be off by as much as 2 million units in either direction. (To quantify this uncertainty, assume that the additional increases in sales in the three quarters are independent random variables having a uniform distribution with a range from 1 to 5 million, from 0 to 4 million, and from -1 to 3 million, respectively.) If the actual increases are too small, the advertising campaign can be discontinued and the product abandoned at the end of either of the next two quarters.

If the intensive advertising campaign were to be initiated and continued to its completion, it is estimated that the sales for some time thereafter would continue to be at about the same level as in the third (last) quarter of the campaign. Therefore, if the sales in that quarter still are below the break-even point, the product would be abandoned. Otherwise, it is estimated that the expected discounted profit thereafter would be $40 for each unit sold over the break-even point in the third quarter.

Use dynamic programming to determine the optimal policy maximizing expected profit.

9 Inventory Theory

9.1 Introduction

Keeping an inventory (stock of goods) for future sale or use is very common in business. Retail firms, wholesalers, manufacturing companies—and even blood banks—generally have a stock of goods on hand. How does such a facility decide upon its "inventory policy"; i.e., when and how much does it replenish? In a small firm the manager may keep track of inventory and make these decisions. However, since this may not be feasible even in small firms, many companies have saved large sums of money by using "scientific inventory management." In particular, they:

1. Formulate a mathematical model describing the behavior of the inventory system.
2. Derive an optimal inventory policy with respect to this model.
3. Frequently use a computer to maintain a record of the inventory levels and to signal when and how much to replenish.

There are several basic considerations involved in determining an inventory policy that must be reflected in the mathematical inventory model; these are illustrated in the following examples.

Example 1: A television manufacturing company produces its own speakers, which are used in the production of its television sets. The television sets are assembled on a continuous production line at a rate of 8,000 per month. The speakers are produced in batches because they do not warrant setting up a continuous production line, and relatively large quantities can be produced in a short time. The company is interested in determining when and how many to produce. Several costs must be considered:

1. Each time a batch is produced, a setup cost of $12,000 is incurred. This cost includes the cost of "tooling up," administrative costs, record keeping, and so forth. Note that the existence of this cost argues for producing speakers in large batches.
2. The production of speakers in large batches leads to a large inventory. The estimated cost of keeping a speaker in stock is 30 cents/month. This cost includes the cost of capital tied up, storage space, insurance, taxes, protection, and so on. The existence of a storage or holding cost argues for producing small batches.
3. The production cost of a single speaker (excluding the setup cost) is $10 and can be assumed to be a unit cost independent of the batch size produced. (In general, however, the unit production cost need not be constant and may decrease with batch size.)
4. Company policy prohibits deliberately planning for shortages of any of its components. However, a shortage of speakers occasionally crops up, and it has been estimated that each speaker that is not available when required costs $1.10/month. This cost includes the cost of installing speakers after the television set is fully assembled, storage space, delayed revenue, record keeping, and so forth.

Example 2: A wholesale distributor of bicycles is having trouble with shortages of the most popular inexpensive 10-speed model and is currently reviewing the inventory policy for this model. The distributor purchases this model bicycle from the manufacturer monthly and then supplies them to various bicycle shops in the western United States. Upon request from shops, the distributor wholesales bicycles to the individual shops in its region. The distributor has analyzed his costs and has determined that the following are important:

1. The shortage cost, i.e., the cost of not having a bicycle on hand when needed. Most models are easily reordered from the manufacturer, and stores usually accept a delay in delivery. Still, although shortages are permissible, the distributor feels that he incurs a loss, which he estimates to be $15 per bicycle. This cost represents an evaluation of the cost of the loss of goodwill, additional clerical costs incurred, and the cost of the delay in revenue received. On a very few competitive (in price) models, stores do not accept a delay, which results in lost sales. In this case, the cost of lost revenue must be included in the shortage cost.
2. The holding cost, i.e., the cost of maintaining an inventory, is $1 per bicycle remaining at the end of the month. This cost represents the costs of capital tied up, warehouse space, insurance, taxes, and so on.

3. The ordering cost, i.e., the cost of placing an order plus the cost of the bicycle, consists of two components: The paperwork involved in placing an order is estimated as \$200, and the actual cost of a bicycle is \$35.

These two examples indicate that there exists a trade-off between the costs involved; the next section discusses the basic cost components of inventory models.

9.2 Components of Inventory Models

Because inventory policies obviously affect profitability, the choice among policies depends upon their relative profitability. Some of the costs that determine this profitability are (1) the costs of ordering or manufacturing, (2) holding or storage costs, (3) unsatisfied demand or shortage penalty costs, (4) revenues, (5) salvage costs, and (6) discount rates. [Costs (1) to (3) were encountered in Examples 1 and 2.]

The cost of ordering or manufacturing an amount z can be represented by a function $c(z)$. The simplest form of this function is one that is directly proportional to the amount ordered, that is, $c \cdot z$, where c represents the unit price paid. Another common assumption is that $c(z)$ is composed of two parts: a term that is directly proportional to the amount ordered and a term that is constant K for z positive and zero for $z = 0$. For this case, if z is positive, the ordering, or production cost, is given by $K + c \cdot z$. The constant K is often referred to as the setup cost and generally includes the administrative cost of ordering, the preliminary labor, and other expenses of starting a production run. There are other assumptions that can be made about this ordering function, but this chapter is restricted to the two cases just described. In Example 1, the speakers are manufactured, and the setup cost for the production run is \$12,000. Furthermore, each speaker costs \$10, so that the *production* cost is given by

$$c(z) = 12{,}000 + 10z, \qquad \text{for } z > 0.$$

In Example 2, the distributor orders bicycles from the manufacturer, and the *ordering* cost is given by

$$c(z) = 200 + 35z, \qquad \text{for } z > 0.$$

The holding or storage costs represent the costs associated with the storage of the inventory until it is sold or used. They may include the cost of capital tied up, space, insurance, protection, and taxes attributed to storage. These costs may be a function of the maximum quantity held during a period, the average amount held, or the cumulated excess of supply over the amount required (demand). The latter viewpoint is usually taken in this chapter. In the bicycle example, the holding cost was \$1 per bicycle remaining at the end of the month. This cost can be interpreted as the interest lost in keeping capital tied up in an ''unnecessary'' bicycle for a month, cost of extra storage space, insurance, and so forth.

The unsatisfied demand or shortage penalty cost is incurred when the amount of the commodity required (demand) exceeds the available stock. This cost depends upon the structure of the model. One such case occurs when the demand exceeds the available inventory, and (1) it is met by a priority shipment, or (2) it is not met at all. In (1) the penalty cost can be viewed as the entire cost of the priority shipment that is used to meet the excess demand. In (2), the situation where the unsatisfied demand is lost, the penalty cost can be viewed as the loss in revenue. Either situation is known as "no backlogging of unsatisfied demand." The scenario of the bicycle example implies that there exist a few competitive (in price) bicycle models where unsatisfied demand is lost, thereby resulting in lost revenue, and hence the example is one where unsatisfied demand is not backlogged. The second case of demand not being fulfilled out of stock assumes that it is satisfied when the commodity next becomes available. The penalty cost can be interpreted as the loss of customers' goodwill, their subsequent reluctance to do business with the firm, the cost of delayed revenue, and extra record keeping. This case is known as "backlogging of unsatisfied demand." The speaker example calls for backlogging of unsatisfied demand. If a shortage occurs, the final assembly of the television set awaits the production of the next batch of speakers. Usually the unsatisfied demand cost is a function of the excess of demand over supply.

The revenue cost may or may not be included in the model. If it is assumed that both the price and the demand for the product are not under the control of the company, the revenue from sales is independent of the firm's inventory policy and may be neglected. However, if revenue is neglected in the model, the *loss in revenue* must then be included in the unsatisfied demand penalty cost whenever the firm cannot meet the demand and the sale is lost. This point is discussed at length in the 10-speed bicycles example presented on page 294. Furthermore, even in the case where demand is backlogged, the cost of the delay in revenue must also be included in the unsatisfied demand cost. With these interpretations, revenue will not be considered as a separate cost in the remainder of this chapter.

The salvage value of an item is the value of a leftover item at the termination of the inventory period. If the inventory policy is carried on for an indefinite number of periods, and if there is no obsolescence, there are no leftover items. What is left over at the end of one period is the amount available at the beginning of the next period. On the other hand, if the policy is to be carried out for only one period, the salvage value represents the disposal value of the item to the firm, say, the selling price. The negative of the salvage value is called the salvage cost. If there is a cost associated with the disposal of an item, the salvage cost may be positive. Because the storage costs generally are assumed to be a function of excess of supply over demand, the salvage costs can be combined with this cost and hence they are usually neglected in this chapter.

Finally, the discount rate takes into account the time value of money. When a firm ties up capital in inventory, it is prevented from using this money for alternative purposes. For example, it could invest this money in secure investments, say, government bonds, and have a return on investment a year hence of, say, 7 percent. Thus a dollar invested today would be worth \$1.07 a year hence, or alternatively, a dollar profit a year hence is equivalent to $\alpha = 1/\$1.07$ today. The quantity α is known as the discount factor. Thus, in considering the profitability of an inventory policy, the profit or costs a year hence should be multiplied by α; 2 years hence, by α^2; and so on.

Of course, the convention of choosing a discount factor α that is based upon the current value of a dollar delivered *one year* hence is arbitrary, and any time period could have been used, for example, one month. It is also evident that in problems having short time horizons, α may be assumed to be 1 (and thereby neglected) because the current value of a dollar delivered during this short time horizon does not change very much. However, in problems having long time horizons, the discount factor must be included.

In using quantitative techniques to seek optimal inventory policies, we use the criterion of minimizing the total (expected) discounted cost. Under the assumptions that the price and demand for the product are not under the control of the company and that the lost or delayed revenue is included in the shortage penalty cost, minimizing cost is equivalent to maximizing net income. Another criterion to be considered, although it is nonquantitative but nevertheless important in practice, is that the resultant inventory policy be simple; i.e., the rule for indicating *when to order* and *how much to order* must be able to be easily described. Most of the policies considered possess this property.

Inventory models are usually classified according to whether the demand for a period is known (deterministic demand) or whether it is a random variable having a known probability distribution (nondeterministic or random demand). The production of batches of speakers is an example of deterministic demand because it is assumed that they are used in television assemblies at a rate of 8,000 per month. The bicycle shops' purchases of bicycles from the distributor is an example of random demand. This classification is frequently coupled with whether or not there exist time lags in the delivery of the items ordered or produced. In both the speaker and the bicycle examples, there was an implication that the items appeared immediately after an order was placed. In fact, the production of speakers may require some time, and similarly, the delivery of bicycles may not be instantaneous, so that time lags may have to be incorporated into the inventory model.

Another possible classification relates to the way the inventory is reviewed, either continuously or periodically. In continuous review, an order is placed as soon as the stock level falls below the prescribed reorder point, whereas in the periodic review case, the inventory level is checked at discrete intervals, e.g., at the end of each week, and ordering decisions are made only at these times even if the inventory level dips below the reorder point during the preceding period. The production of speakers is an example of a continuous review, whereas the bicycle problem is an example of periodic review. Incidentally, in practice, a periodic review policy can be used to approximate a continuous review policy by making the time interval sufficiently small.

In this chapter inventory policies are classified according to whether the demand is deterministic or random, and models are developed for continuous and periodic review policies. Instantaneous delivery will be assumed throughout.[1]

9.3 Deterministic Models

This section is concerned with inventory problems where the actual demand is assumed to be known. Several models are considered, including the well-known economic lot-size formulation.

Continuous Review—Uniform Demand

The most common inventory problem faced by manufacturers, retailers, and wholesalers is concerned with the case where stock levels are depleted with time and then are replenished by the arrival of new items. A simple model representing this situation

[1] Results for the delivery-lag case often can be obtained from the corresponding instantaneous delivery model by a simple modification in the calculation of some of the inventory costs.

is given by the economic lot-size model. Items are assumed to be withdrawn continuously at a known constant rate denoted by a; that is, a units are required per unit time, say, per month. It is further assumed that items are produced (or ordered) in equal numbers, Q at a time, and all Q items arrive simultaneously when desired (fixed delivery lags will be considered later). The only costs to be considered are the setup cost K, charged at the time of the production (or ordering), a production cost (or purchase cost) of c dollars per item, and an inventory holding cost of h dollars per item per unit of time. The inventory problem is to determine how often to make a production run and what size it should be so that the cost per unit of time is a minimum. This is a continuous review inventory policy. We shall first assume that shortages are not allowed, and then we shall relax this assumption. The example of the production of speakers in television sets satisfies this model.

Shortages Not Permitted: A cycle can be viewed as the time between production runs. Thus, if 24,000 speakers are produced at each production run and are used at the rate of 8,000 per month, then the cycle length is 24,000/8,000 = 3 months. In general, the cycle length is Q/a. Figure 9.1 illustrates how the inventory level varies over time.

The cost per unit time is obtained as follows: The production cost per cycle is given by

$$\begin{cases} 0, & \text{if } Q = 0 \\ K + cQ, & \text{if } Q > 0. \end{cases}$$

The holding cost per cycle is easily obtained. The average inventory level during a cycle is $(Q + 0)/2 = Q/2$ items per unit of time, and the corresponding cost is $hQ/2$ per unit of time. Because the cycle length is Q/a, the holding cost per cycle is given by

$$\frac{hQ^2}{2a}.$$

Therefore, the total cost per cycle is

$$K + cQ + \frac{hQ^2}{2a},$$

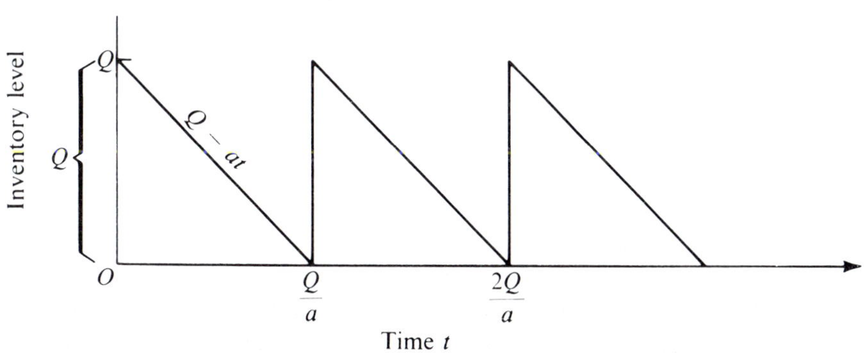

Figure 9.1 Diagram of inventory level as a function of time—no shortages permitted.

and the total cost per unit of time is

$$T = \frac{K + cQ + hQ^2/2a}{Q/a} = \frac{aK}{Q} + ac + \frac{hQ}{2}.$$

It is evident that the value of Q, say Q^*, that minimizes T is found from $dT/dQ = 0$. $dT/dQ = -aK/Q^2 + h/2 = 0$, so that

$$Q^* = \sqrt{\frac{2aK}{h}}$$

(because $d^2T/dQ^2 > 0$), which is the well-known economic lot-size result. Similarly, the time it takes to withdraw this optimum value of Q^*, say, t^*, is given by

$$t^* = \frac{Q^*}{a} = \sqrt{\frac{2K}{ah}}.$$

These results will now be applied to the speaker example. The appropriate parameters are

$$K = 12{,}000$$

$$h = 0.30$$

$$a = 8{,}000,$$

so that

$$Q^* = \sqrt{\frac{(2)(8{,}000)(12{,}000)}{0.30}} = 25{,}298$$

and

$$t^* = \frac{25{,}298}{8{,}000} = 3.2 \text{ months}.$$

Hence the production line is to be set up every 3.2 months and produce 25,298 speakers. Incidentally, the cost curve is rather flat near this optimal value, so that any production between 20,000 and 30,000 speakers is acceptable; this fact can be seen in Fig. 9.3.

Shortages Permitted: It may be profitable to permit shortages to occur because the cycle length can be increased with a resultant saving in setup costs. However, this benefit may be offset by the cost that is incurred when shortages occur, and hence a detailed analysis is required.

If shortages are allowed and are priced out at a cost of p dollars for each unit of demand unfilled for one unit of time, results similar to the no-shortage case can be obtained. Denote by S the stock on hand at the beginning of a cycle. The problem is summarized in Fig. 9.2.

The cost per unit time is obtained as follows: The production cost per cycle is given by

$$\begin{cases} 0, & \text{if } Q = 0 \\ K + cQ, & \text{if } Q > 0. \end{cases}$$

The holding cost per cycle is easily obtained. Note that the inventory level is positive for a time of S/a. The average inventory level *during this time* is $(S + 0)/2 = S/2$ items per unit of time, and the corresponding cost is $hS/2$ per unit of time. Hence the total holding cost incurred over the time the inventory level is positive is the

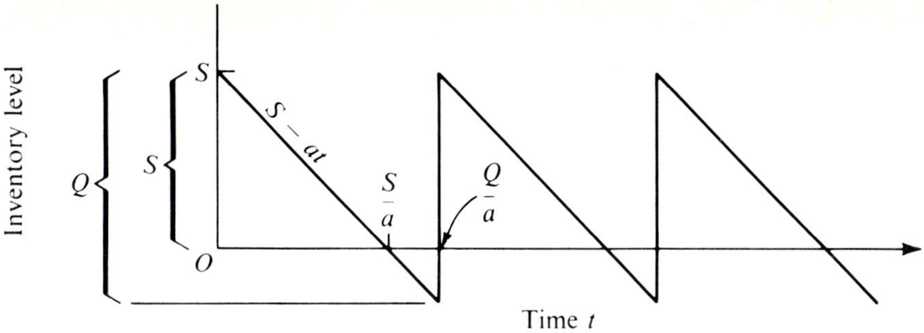

Figure 9.2 Diagram of inventory level as a function of time—shortages permitted.

holding cost per cycle, which is given by

$$\frac{hS}{2}\frac{S}{a} = \frac{hS^2}{2a}.$$

Similarly, shortages occur for a time $(Q - S)/a$. The average amount of shortages *during this time* is $[0 + (Q - S)]/2 = (Q - S)/2$ items per unit of time, and the corresponding cost is $p(Q - S)/2$ per unit of time. Hence the total shortage cost incurred over the time shortages exist is the shortage cost per cycle, which is given by

$$\frac{p(Q - S)}{2}\frac{(Q - S)}{a} = \frac{p(Q - S)^2}{2a}.$$

Therefore, the total cost per cycle is

$$K + cQ + \frac{hS^2}{2a} + \frac{p(Q - S)^2}{2a},$$

and the total cost per unit of time is

$$T = \frac{K + cQ + hS^2/2a + p(Q - S)^2/2a}{Q/a}$$

$$= \frac{aK}{Q} + ac + \frac{hS^2}{2Q} + \frac{p(Q - S)^2}{2Q}.$$

In this model there are two decision variables (S and Q), so the optimum values (S^* and Q^*) are found by setting the partial derivatives $\partial T/\partial S$ and $\partial T/\partial Q$ equal to zero. Thus

$$\frac{\partial T}{\partial S} = \frac{hS}{Q} - \frac{p(Q - S)}{Q} = 0.$$

$$\frac{\partial T}{\partial Q} = -\frac{aK}{Q^2} - \frac{hS^2}{2Q^2} + \frac{p(Q - S)}{Q} - \frac{p(Q - S)^2}{2Q^2} = 0.$$

Solving these equations simultaneously leads to

$$S^* = \sqrt{\frac{2aK}{h}}\sqrt{\frac{p}{p + h}}, \qquad Q^* = \sqrt{\frac{2aK}{h}}\sqrt{\frac{p + h}{p}}.$$

The optimal period length t^* is given by

$$t^* = \frac{Q^*}{a} = \sqrt{\frac{2K}{ah}}\sqrt{\frac{p+h}{p}}.$$

The maximum shortage is expressed as

$$Q^* - S^* = \sqrt{\frac{2aK}{p}}\sqrt{\frac{h}{p+h}}.$$

Further, from Fig. 9.2, the fraction of time that no shortage exists is given by

$$\frac{S^*/a}{Q^*/a} = \frac{p}{p+h},$$

which is independent of K.

If shortages are permitted in the speaker example, the cost is estimated as $p =$ \$1.10 per speaker. Again,

$$K = 12{,}000$$

$$h = 0.30$$

$$a = 8{,}000,$$

so

$$S^* = \sqrt{\frac{(2)(8{,}000)(12{,}000)}{0.30}}\sqrt{\frac{1.1}{1.1+0.3}} = 22{,}424,$$

$$Q^* = \sqrt{\frac{(2)(8{,}000)(12{,}000)}{0.30}}\sqrt{\frac{1.1+0.3}{1.1}} = 28{,}540,$$

and

$$t^* = \frac{28{,}540}{8{,}000} = 3.6 \text{ months.}$$

Hence, when shortages are permitted, the production line is to be set up every 3.6 months to produce 28,540 speakers. A shortage of 6,116 speakers is permitted. Note that Q^* and t^* are not very different from the no-shortage case.

Quantity Discounts, Shortages Not Permitted: The models considered have assumed that the unit cost of an item is the same, independent of the quantity produced. In fact, this assumption resulted in the optimal solutions being independent of this unit cost. Suppose, however, that there exist cost breaks; i.e., the unit cost varies with the quantity ordered. For example, suppose the unit cost of producing a speaker is $c_1 =$ \$11 if less than 10,000 speakers are produced, $c_2 =$ \$10 if production falls between 10,000 and 80,000 speakers, and $c_3 =$ \$9.50 if production exceeds 80,000 speakers. What is the optimal policy? The solution to this specific problem will reveal the general method.

From the results of the previously considered economic lot-size model (shortages not permitted), the total cost per unit time if the production cost is c_j is given by

$$T_j = \frac{aK}{Q} + ac_j + \frac{hQ}{2}, \qquad \text{for } j = 1, 2, 3.$$

A plot of T_j versus Q is shown in Fig. 9.3.

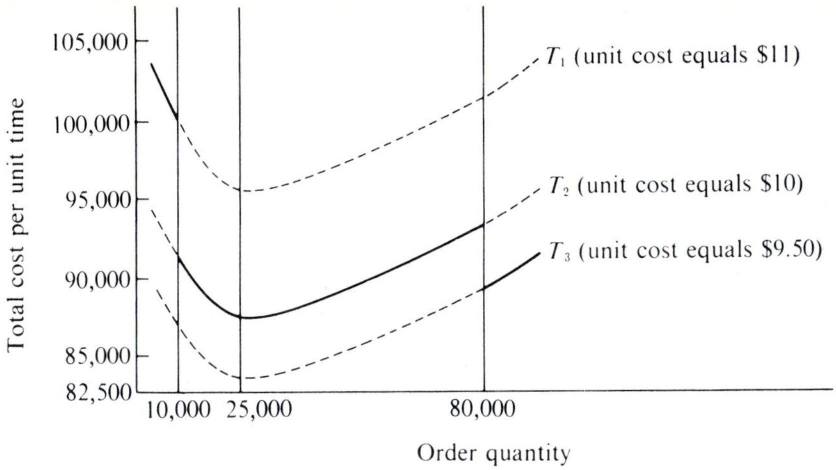

Figure 9.3 Total cost per unit time for speaker example with quantity discounts.

The feasible values of Q are shown by the solid lines, and it is only these regions that must be investigated. For each curve, the value of Q that minimizes T_j is easily found by the methods used in the previously considered economic lot-size model. For $K = 12{,}000$, $h = 0.30$, and $a = 8{,}000$, this value is

$$\sqrt{\frac{(2)(8{,}000)(12{,}000)}{0.30}} = 25{,}298.$$

This number is a feasible value for the cost function T_2. Because it is evident that for fixed Q, $T_j < T_{j-1}$ for all j, T_1 can be eliminated from further consideration. However, T_3 cannot be immediately discarded. Its minimum feasible value (which occurs at $Q = 80{,}000$) must be compared to T_2 evaluated at 25,298 (which is $87,589). Because T_3 evaluated at 80,000 equals $89,200, it is better to produce in quantities of 25,298, and thus this quantity is the optimal value for this set of quantity discounts. If the quantity discount led to a cost of $9 (instead of $9.50) when production exceeded 80,000, then T_3 evaluated at 80,000 would equal 85,200, and the optimal production quantity would become 80,000.

Although this analysis concerned a very specific problem, its extension to a general problem is evident. Furthermore, a similar analysis can be made for other types of quantity discounts, such as incremental quantity discounts, where a cost c_0 is incurred for the first q_0 items, c_1 for the next q_1 items, and so on.

Remarks: Several remarks can be made about economic lot-size models.

1. If it is assumed that the production (or purchase) cost of an item is constant throughout time, it does not appear in the optimal solution. This result is evident because no matter what policy is used, the same quantity is required, and hence this cost is fixed.

2. It was previously assumed that Q, the number of units produced, was constant from cycle to cycle. A little reflection will reveal that this assumption is really a result rather than an assumption.

3. These models can be viewed as a special case of an (s, S) policy. An (s, S) policy is usually used in the context of a periodic review policy, where at review time an order is placed to bring the inventory level up to S if the current inventory is less than or equal to s. Otherwise, no order is placed.[1] The symbol S then denotes the reorder level, and s denotes the reorder point. In the economic lot-size models, s denotes the inventory level when items are ordered, so that when shortages are not permitted the reorder point s is zero, and when shortages are permitted, s is equal to the negative of the maximum shortage; that is,

$$s = -\sqrt{\frac{2aK}{p}}\sqrt{\frac{h}{p+h}}.$$

Furthermore, because the economic lot-size models are continuously reviewed, when the inventory level equals s an order is placed, bringing the inventory up to the reorder level S. Hence, for economic lot-size models the (s, S) policy can be described as follows: When the inventory level reaches the reorder point s, place an order to bring the inventory level up to reorder level S; that is, order $Q = S - s$.

4. It is evident from the analysis presented that the reorder point will never be positive. A policy that calls for $s > 0$ cannot be optimal because it is dominated by a policy that calls for ordering the same Q, but only when the reorder point reaches zero. It is dominated in that the latter policy has the same setup and purchase costs but a uniformly smaller holding cost.

5. A known fixed delivery lag is easily accommodated. Denote by λ the lead time between the placing and receiving of the order. It is assumed that λ is constant over time and independent of the size of the order. It is evident that if it is desired to have the order arrive the moment the inventory level reaches s, then the order must be placed λ periods earlier. Thus the reorder point is simply

$$s + \lambda a,$$

where s is determined for the no-lag situation.[2]

Periodic Review—A General Model for Production Planning

The last section explored the economic lot-size model. The results were dependent upon the assumption of a constant demand rate. When this assumption is relaxed, i.e., when the amounts required from period to period are allowed to vary, the square-root formula no longer ensures a minimum cost solution.

Consider the following model, due to Wagner and Whitin.[3] As before, the only costs to be considered are the setup cost K, charged at the beginning of the period; a production cost (or purchase cost) of c dollars per item; and an inventory holding cost of h dollars per item, which is charged (arbitrarily) at the end of the time period. The choice of charging the inventory at the end of the period, and hence as a function of the excess of the supply over the requirement, is somewhat different from the holding charge incurred in the economic lot-size models. In the latter case, the average cost

[1] Usually (s, S) policies are described as ordering when the inventory level is less than s rather than when the inventory level is less than or equal to s. However, the costs are often the same if an optimal policy is followed.

[2] This result holds only when $\lambda a < Q$.

[3] Wagner, H. M., and T. M. Whitin: "Dynamic Version of the Economic Lot-Size Model," *Management Science*, **5**(1):89–96, 1958.

per unit of time was charged. Clearly, different policies can result from alternative ways of dealing with holding costs. In addition, $r_i = 1, 2, \ldots, n$ represents the requirements at time i, and it is assumed that these requirements must be met. Initially there is no stock on hand. For a horizon of n periods, the inventory problem is to determine how much should be produced at the beginning of each time period (assumed to be instantaneous) to minimize the total cost incurred over the n periods.

The model can be illustrated by the following variation of the speaker example.

EXAMPLE: A market survey conducted by the television manufacturer has indicated that the demand for television sets is seasonal rather than uniform. In particular, sales of 30,000 sets is forecast for the Christmas season (October to December), 20,000 for the winter slack season (January to March), 30,000 for the "new model" season (April to June), and 20,000 for the summer season (July to September). Because of the necessity of meeting the increased demand during the peak seasons, the television set production line was revamped. This revised production line enabled the company to introduce new equipment as well as to redesign some of its components, including the speakers. Hence the setup cost for speaker production is now \$20,000, but the unit cost is down to \$1. Furthermore, the holding cost of a speaker has been reduced to 20 cents per (3-month) period. Finally, the labor and equipment costs are such that the speakers must be produced in increments of 10,000. It is assumed that production of the television sets is completed and ready for shipment in the 3-month period prior to the season in which it is required. Thus the 30,000 sets required for the Christmas season are to be assembled during the July-to-September period. The speaker is the last component added to the television set, and it is easily installed. Furthermore, large quantities of speakers can be produced in a very short time, so their production and subsequent installation can be viewed as instantaneous. The problem is to determine how many to produce in each period while satisfying the requirements and minimizing the total cost. A solution, though not an optimal one, is given in Fig. 9.4; this policy calls for producing 30,000 speakers at the beginning of the first period (Christmas season), 60,000 speakers at the beginning of the second period, and 10,000 speakers at the beginning of the fourth period.

One approach to the solution of this model is to enumerate, for each of the 2^{n-1} combinations of either producing or not producing in each period, the possible

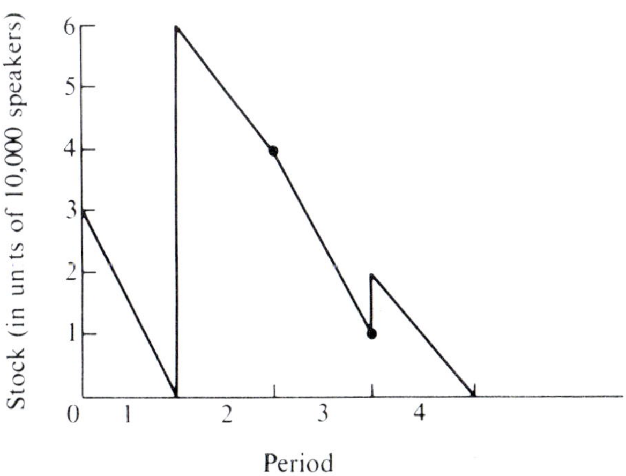

Figure 9.4 Production schedule that satisfies requirements.

quantities that can be produced. This approach is rather cumbersome, even for moderate-sized n. Hence a more efficient method is desirable. In particular, the method of dynamic programming introduced in Chap. 8 will be applied, followed by the introduction of an algorithm (in the ensuing section) that exploits the structure. Finally, a mathematical programming solution will be presented.

The Dynamic Programming Solution: Following the notation introduced for dynamic programming (Chap. 8) and interpreting the variables in the inventory context, the ith stage corresponds to the ith period; the state corresponds to the inventory entering period i and will be denoted by x_i; and the decision variable corresponds to the quantity produced (or ordered) at the beginning of period i and will be denoted by z_i.[1]

Let $B_i(x_i, z_i)$ denote the costs incurred in period i, given the entering inventory and the quantity produced. Recall that the only costs considered are the setup cost K, charged at the beginning of the period, a production cost (or purchase cost) of c dollars per item, and an inventory holding cost of h dollars per item charged at the end of the time period. The requirement for period i is r_i. Then $B_i(x_i, z_i)$ is given by

$$B_i(x_i, z_i) = \begin{cases} K + cz_i + h(x_i + z_i - r_i), & \text{if } z_i > 0 \\ h(x_i - r_i), & \text{if } z_i = 0. \end{cases}$$

Denote by $C_i(x_i, z_i)$ the total cost of the best overall policy from the beginning of period i to the end of the planning horizon, given that the inventory level entering period i is x_i and z_i is chosen to be produced; and let $C_i^*(x_i)$ denote the corresponding minimum value of $C_i(x_i, z_i)$, subject to the constraints that $z_i \geq 0$ and that the requirements for the periods are met. Therefore,

$$C_i^*(x_i) = \underset{\substack{z_i \geq 0 \\ z_i \geq r_i - x_i}}{\text{minimum}} [B_i(x_i, z_i) + C_{i+1}^*(x_i + z_i - r_i)],$$

for $i = 1, 2, \ldots, n$, and $C_{n+1}^*(\cdot)$ is defined to be zero. In the speaker example, there are four periods to be considered. (All calculations have been reduced in scale by a factor of 10,000.) The first iteration corresponds to $i = 4$, that is, a description of the optimal policy at the beginning of period 4. For this case,

$$C_4^*(x_4) = \underset{\substack{z_4 \geq 0 \\ z_4 \geq 2-x_4}}{\text{minimum}} [B_4(x_4, z_4)],$$

so that the immediate solution to the fourth-period problem is as follows:

x_4	z_4	$C_4^*(x_4)$	z_4^*
0	2	4	2
1	1	3	1
2	0	0	0

[1] For the notation in this chapter to be consistent with that used in inventory theory in general, it may differ somewhat from the notation introduced in Chap. 8.

The second iteration requires finding the optimal policy from the beginning of period 3 to the end of period 4. For this case,

$$C_3^*(x_3) = \underset{\substack{z_3 \geq 0 \\ z_3 \geq 3 - x_3}}{\text{minimum}} \, [B_3(x_3, z_3) + C_4^*(x_3 + z_3 - 3)],$$

so that the solution to the problem beginning at period 3 is as follows:

	$B_3(x_3, z_3) + C_4^*(x_3 + z_3 - 3)$							
x_3 \ z_3	0	1	2	3	4	5	$C_3^*(x_3)$	z_3^*
0	—	—	—	9.0	9.2	7.4	7.4	5
1	—	—	8	8.2	6.4	—	6.4	4
2	—	7	7.2	5.4	—	—	5.4	3
3	4.0	6.2	4.4	—	—	—	4.0	0
4	3.2	3.4	—	—	—	—	3.2	0
5	0.4	—	—	—	—	—	0.4	0

A typical entry can be verified. Let $x_3 = 2$ and $z_3 = 3$. Then 3 units are produced, given an initial inventory of 2. The cost of production is then $2 + 1(3) = 5$, and the inventory holding cost is $\frac{1}{5}(2 + 3 - 3) = 0.4$. The entering inventory in period 4 is then 2, so that $C_4^*(2) = 0$. Hence the total cost is given by $5 + 0.4 = 5.4$, which is the value given.

The third iteration requires finding the optimal policy from the beginning of period 2 to the end of period 4. For this case,

$$C_2^*(x_2) = \underset{\substack{z_2 \geq 0 \\ z_2 \geq 2 - x_2}}{\text{minimum}} \, [B_2(x_2, z_2) + C_3^*(x_2 + z_2 - 2)],$$

so that the solution to the problem beginning at period 2 is as follows:

	$B_2(x_2, z_2) + C_3^*(x_2 + z_2 - 2)$									
x_2 \ z_2	0	1	2	3	4	5	6	7	$C_2^*(x_2)$	z_2^*
0	—	—	11.4	11.6	11.8	11.6	12.0	10.4	10.4	7
1	—	10.4	10.6	10.8	10.6	11.0	9.4	—	9.4	6
2	7.4	9.6	9.8	9.6	10.0	8.4	—	—	7.4	0
3	6.6	8.8	8.6	9.0	7.4	—	—	—	6.6	0
4	5.8	7.6	8.0	6.4	—	—	—	—	5.8	0
5	4.6	7.0	5.4	—	—	—	—	—	4.6	0
6	4.0	4.4	—	—	—	—	—	—	4.0	0
7	1.4	—	—	—	—	—	—	—	1.4	0

Finally, the last iteration requires finding the optimal policy from the beginning of period 1 to the end of period 4. For this case,

$$C_1^*(0) = \underset{z_1 \geq 3}{\text{minimum}} \, [B_1(0, z_1) + C_2^*(z_1 - 3)],$$

so that the solution to the production planning problem is as follows:

	$B_1(0, z_1) + C_2^*(z_1 - 3)$												
x_1 \ z_1	0	1	2	3	4	5	6	7	8	9	10	$C_1^*(x_1)$	z_1^*
0	—	—	—	15.4	15.6	14.8	15.2	15.6	15.6	16.2	14.8	14.8	5 or 10

Therefore, the optimal production schedule is to produce all the speakers at the beginning of the first period, or produce 50,000 speakers (5 units) at the beginning of the first period and 50,000 speakers (5 units) at the beginning of the third period. The minimum cost is \$148,000.

It should be noted that the unit cost c is irrelevant to the problem because over all the time periods, all policies use the same number of items at the same total cost. Hence this cost could have been neglected, and the same optimal policies would have been obtained. Different costs in different periods are easily handled by this method of solution. For example, during the peak production periods (when 30,000 speakers are produced), the workers are fully engaged in the assembly of television sets, so that the speakers must be produced during overtime hours. Hence the unit cost is increased. This increase would be reflected in the calculations of $C_3^*(x_3)$ and $C_1^*(x_1)$.

Periodic Review—Production Planning: An Algorithm

In the previous section, dynamic programming was applied to solve the production planning model. Alternatively, by streamlining the dynamic programming approach, we shall develop an algorithm that exploits the structure of the model. Initially the production planning model we consider will be the same as that presented in the previous section, i.e., arbitrary demand requirement, a fixed setup cost, and linear production and holding costs.

The following result characterizes an optimal policy:

For an arbitrary demand requirement, a fixed setup cost, and linear production and holding costs, there is an optimum policy that produces only when the inventory level is zero.

To show why this result is true, choose any policy. Consider the time period that begins when production is made with zero stock level and ends with the first time production is made when the stock is not at zero level. For the policy given in Fig. 9.4, this interval starts at the beginning of period 2 and ends at the beginning of period 4, when one unit is produced. This time period is also shown by the solid lines in Fig. 9.5.

Consider the alternative policy, which implies production of 50,000 speakers at the beginning of period 2 and production of 20,000 speakers at the beginning of period 4. This policy is shown by the dashed lines in Fig. 9.5. This policy, B, dominates policy A in that the total cost is smaller. The setup and the production costs for both policies are the same. It is evident that the holding cost for B is smaller than that for A because there is always less stock on hand at the end of a period. Therefore, B is better than A, so that A cannot be optimal.

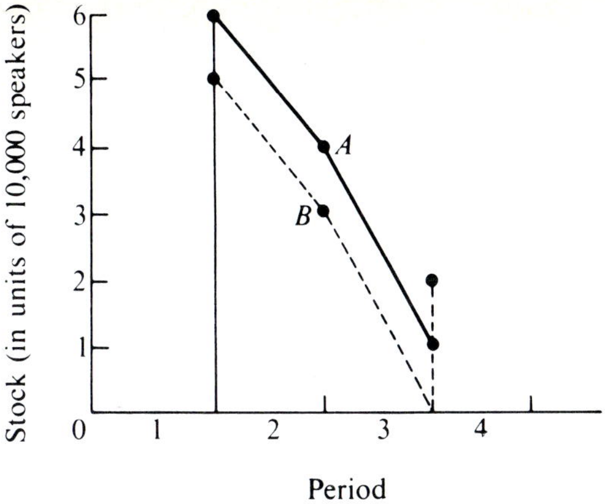

Figure 9.5 Production schedules.

This characterization of optimal policies can be used to determine which policies are not optimal. In addition, because it implies that the amount produced at the beginning of the *i*th period must be either 0, r_i, $r_i + r_{i+1}, \ldots, r_i + r_{i+1} + \cdots + r_n$, it can be exploited to obtain an efficient algorithm.

Suppose an optimal policy is presented. Consider the time from the initial production at the beginning of the first period to the first time the inventory level is again zero. The total cost of the subsequent periods must be a minimum for this reduced problem because the overall policy is optimal. Therefore, in the context of the speaker example, if the inventory level is zero for the first time (after the initial production) at the end of the second period, and if an optimal policy is being followed, all that remains to be done is to determine the optimal policy for the last two periods (periods 3 and 4) that have a requirement of 30,000 speakers for period 3 and 20,000 speakers for period 4.

Let C_i denote the total cost of the best overall policy from the beginning of period i, when no stock is available, to the end of the planning horizon, $i = 1, 2, \ldots, n$. A recursive relationship for C_i is given by

$$C_i = \underset{j=i,i+1,\ldots,n}{\text{minimum}} [C_{j+1} + K + c(r_i + r_{i+1} + \cdots + r_j)$$
$$+ h(r_{i+1} + 2r_{i+2} + 3r_{i+3} + \cdots + (j - i)r_j)],$$

where j can be viewed as an index that denotes the (end of the) period when the inventory reaches a zero level for the first time after the production at the beginning of period i. The cost C_{n+1} is 0, $c(r_i + r_{i+1} + \cdots + r_j)$ is the cost of the production from period i until the inventory level next reaches zero, and the quantity $h(\ )$ is the total holding cost of the inventory that results from the production from period i and remaining until the inventory level next reaches zero. This latter cost is charged at the end of every period as a function of the excess, if any, over the requirement.

The solution of this algorithm is much simpler than the dynamic programming approach. As in dynamic programming, $C_n, C_{n-1}, \ldots, C_2$ must be found before C_1 is obtained. However, the number of calculations is much smaller, and the number of possible production values is greatly reduced.

EXAMPLE: Returning to the speaker example, first consider the case of finding C_4, the cost of the optimal policy from the beginning of period 4 to the end of the planning horizon:

$$C_4 = C_5 + 2 + 1(2) = 0 + 2 + 2 = 4.0.$$

To find C_3 we must consider two cases; i.e., the first time after period 3 when the inventory reaches a zero level can occur at (1) the end of the third period or (2) the end of the fourth period. In the recursive relationship j may range over period 3 or 4, resulting in the costs $C_3^{(3)}$ or $C_3^{(4)}$, respectively. The policy associated with $C_3^{(3)}$ calls for producing only for period 3 and then following the optimal policy for period 4, whereas the policy associated with $C_3^{(4)}$ calls for producing for periods 3 and 4. The cost C_3 is then the minimum of $C_3^{(3)}$ and $C_3^{(4)}$. These cases are reflected by the policies given in Fig. 9.6.

$$C_3^{(3)} = C_4 + 2 + 1(3) = 4 + 2 + 3 = 9$$

and $$C_3^{(4)} = C_5 + 2 + 1(3 + 2) + \tfrac{1}{5}(2) = 0 + 2 + 5 + 0.4 = 7.4.$$

Hence $$C_3 = \min(7.4, 9.0) = 7.4.$$

To find C_2 we must consider three cases; i.e., the first time after period 2 when the inventory reaches a zero level can occur at (1) the end of the second period, (2) the end of the third period, or (3) the end of the fourth period. In the recursive relationship j may range over period 2, 3, or 4, resulting in costs $C_2^{(2)}$, $C_2^{(3)}$, or $C_2^{(4)}$, respectively. The cost C_2 is then the minimum of $C_2^{(2)}$, $C_2^{(3)}$, and $C_2^{(4)}$.

$$C_2^{(2)} = C_3 + 2 + 1(2) = 7.4 + 2 + 2 = 11.4,$$

$$C_2^{(3)} = C_4 + 2 + 1(2 + 3) + \tfrac{1}{5}(3) = 4 + 2 + 5 + 0.6 = 11.6,$$

and

$$C_2^{(4)} = C_5 + 2 + 1(2 + 3 + 2) + \tfrac{1}{5}[3 + 2(2)] = 0 + 2 + 7 + 1.4 = 10.4.$$

Hence $$C_2 = \min(11.4, 11.6, 10.4) = 10.4.$$

Finally, to find C_1 we must consider four cases; i.e., the first time after period 1 when the inventory reaches zero can occur at (1) the end of the first period, (2) the end of the second period, (3) the end of the third period, or (4) the end of the fourth period. In the recursive relationship j may range over period 1, 2, 3, or 4, resulting

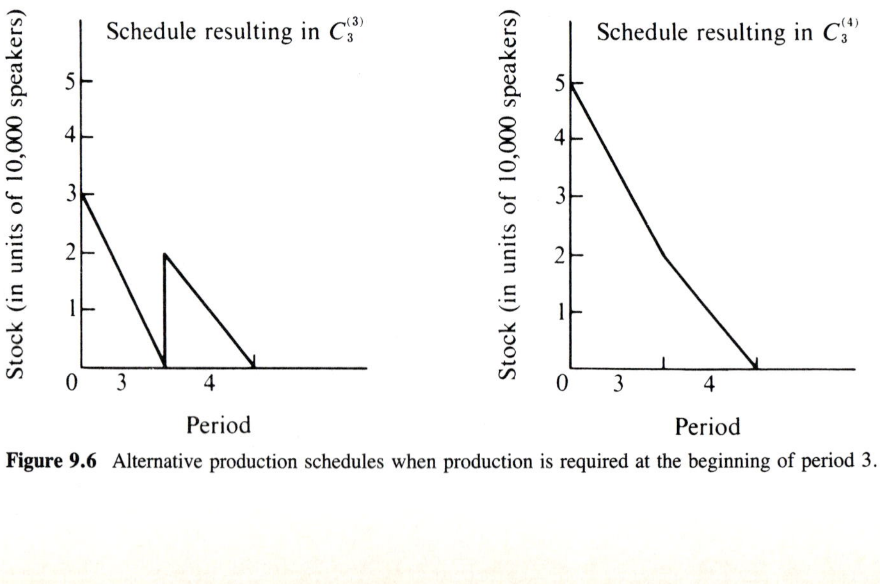

Figure 9.6 Alternative production schedules when production is required at the beginning of period 3.

in the costs $C_1^{(1)}$, $C_1^{(2)}$, $C_1^{(3)}$, $C_1^{(4)}$. The cost C_1 is then the minimum of $C_1^{(1)}$, $C_1^{(2)}$, $C_1^{(3)}$, and $C_1^{(4)}$.

$$C_1^{(1)} = C_2 + 2 + 1(3) = 10.4 + 2 + 3 = 15.4,$$

$$C_1^{(2)} = C_3 + 2 + 1(3 + 2) + \tfrac{1}{5}(2) = 7.4 + 2 + 5 + 0.4 = 14.8,$$

$$C_1^{(3)} = C_4 + 2 + 1(3 + 2 + 3) + \tfrac{1}{5}[2 + 2(3)] = 4 + 2 + 8 + 1.6 = 15.6,$$

and $$C_1^{(4)} = C_5 + 2 + 1(3 + 2 + 3 + 2) + \tfrac{1}{5}[2 + 2(3) + 3(2)]$$

$$= 0 + 2 + 10 + 2.8 = 14.8.$$

Hence, $$C_1 = \min(15.4, 14.8, 15.6, 14.8) = 14.8,$$

so that the optimal production schedule is to produce all the speakers at the beginning of the first period or to produce 50,000 speakers at the beginning of the first period and 50,000 speakers at the beginning of the third period (the same solution as obtained previously). Note that the 14.8 comes from $C_1^{(4)}$ and $C_1^{(2)}$. The policy associated with $C_1^{(4)}$ calls for producing items for periods 1, 2, 3, 4 at period 1 whereas $C_1^{(2)}$ calls for producing items for periods 1 and 2 at period 1, and then following an optimal policy from period 3 on. Such a policy has already been shown to call for production for periods 3 and 4 together.

Again it should be noted that the unit cost c is irrelevant to the problem because over all the time periods, all policies use the same number of items at the same total cost. Hence this cost could have been neglected, and the same optimal policies would have been obtained.

The characterization of the optimal policy, and the subsequent algorithm for finding an optimal policy, depended upon the assumption that the holding and production costs were linear. This constraint can be relaxed to include *concave production and holding costs*. In fact, any increasing function of the holding cost will serve as an alternative condition if the production cost is linear. However, these alternative conditions require a modification in the algorithm for finding an optimal policy. If the production cost function is denoted by $c[\cdot]$ and holding cost function $h[\cdot]$, the recursive relationship for C_i becomes

$$C_i = \underset{j=i,i+1,\ldots,n}{\text{minimum}} \{C_{j+1} + K + c[r_i + r_{i+1} + \cdots + r_j] + h[r_{i+1} + r_{i+2} + r_{i+3} + \cdots + r_j] + h[r_{i+2} + r_{i+3} + \cdots + r_j] + \cdots + h[r_j]\},$$

where $C_{n+1} = 0$.[1]

A natural extension of the production planning model permits shortages to occur. This extension has been studied by Zangwill[2] and differs from the production planning model studied in that shortage costs are incurred for each unit of demand unfilled for one unit of time. Zangwill characterizes the form of the optimal policy and gives an efficient recursive relationship for finding the optimal policy. These results apply when the production, holding, and shortage costs, per period, are concave functions (thereby including the case of linear costs).

[1] In the expression for C_i there is no holding cost when $j = i$.

[2] Zangwill, W. I.: "A Deterministic Multi-Period Production Scheduling Model with Backlogging," *Management Science*, **13**(1):105–119, 1966.

Other results have been obtained for this production planning model under the assumption that the production, holding, and shortage costs are convex. Convex production costs arise, for example, when there are several sources of limited production at different unit costs in a period. If one uses these sources up to capacity in order of ascending unit cost (as is optimal), the resulting production cost is convex in the total amount produced. Such an assumption about the production cost precludes the use of the setup charge.

Periodic Review—Production Planning: Integer Programming Formulation

The final technique for solving the production planning model is to formulate it as an integer programming problem. Instead of presenting the general solution, we shall again solve the speaker production example.

Again, let z_i denote the quantity produced at the beginning of period i.[1] The costs to be considered are

$$\text{Production costs} = \text{setup costs} + c(z_1 + z_2 + z_3 + z_4)$$

and

$$\begin{aligned}\text{Holding costs} &= h(z_1 - r_1) + h(z_1 + z_2 - r_1 - r_2) \\ &\quad + h(z_1 + z_2 + z_3 - r_1 - r_2 - r_3) \\ &= 3h(z_1 - r_1) + 2h(z_2 - r_2) + h(z_3 - r_3) \\ &= 0.20[3(z_1 - 3) + 2(z_2 - 2) + (z_3 - 3)].\end{aligned}$$

Hence the problem to be solved is to minimize the sum of the production and holding costs; that is,

$$\text{Minimize} \quad W = \text{setup cost} + 1(z_1 + z_2 + z_3 + z_4) + 0.20[3(z_1 - 3) + 2(z_2 - 2) + (z_3 - 3)],$$

subject to

$$\begin{aligned} z_1 &\le r_1 + r_2 + r_3 + r_4 = 10 \\ z_2 &\le r_2 + r_3 + r_4 = 7 \\ z_3 &\le r_3 + r_4 = 5 \\ z_4 &\le r_4 = 2 \\ z_1 &\ge r_1 = 3 \\ z_1 + z_2 &\ge r_1 + r_2 = 5 \\ z_1 + z_2 + z_3 &\ge r_1 + r_2 + r_3 = 8 \\ z_1 + z_2 + z_3 + z_4 &= r_1 + r_2 + r_3 + r_4 = 10 \end{aligned}$$

and $z_1 \ge 0, \quad z_2 \ge 0, \quad z_3 \ge 0, \quad z_4 \ge 0,$ and integer valued.

Unfortunately, this model is not an integer linear programming model because the objective function is not linear as a result of the "setup costs" term that appears

[1] Again, all values are expressed in units of 10,000 speakers.

in the objective function. However, this problem can be resolved as follows:
Define variables v_1, v_2, v_3, and v_4 such that

$$v_1 \le 1, \quad v_2 \le 1, \quad v_3 \le 1, \quad v_4 \le 1,$$

$$v_1 \ge 0, \quad v_2 \ge 0, \quad v_3 \ge 0, \quad v_4 \ge 0,$$

and v_1, v_2, v_3, and v_4 are integers (v_i is then 0 or 1).

Add the following constraints to the problem:

$$z_1 \le \left(\sum_{i=1}^{4} r_i\right) v_1 = 10v_1$$

$$z_2 \le \left(\sum_{i=2}^{4} r_i\right) v_2 = 7v_2$$

$$z_3 \le \left(\sum_{i=3}^{4} r_i\right) v_3 = 5v_3$$

$$z_4 \le r_4 v_4 = 2v_4.$$

Thus if $v_i = 0$, then $z_i = 0$, and if $v_i = 1$, then z_i is less than or equal to the maximum production in period i. The objective function can now be written as a linear function subject to linear constraints; that is,

$$\text{Minimize} \quad W = K(v_1 + v_2 + v_3 + v_4) + 1(z_1 + z_2 + z_3 + z_4) + 0.20[3(z_1 - 3) + 2(z_2 - 2) + (z_3 - 3)],$$

subject to

$$z_1 \le 10v_1$$
$$z_2 \le 7v_2$$
$$z_3 \le 5v_3$$
$$z_4 \le 2v_4$$
$$z_1 \ge 3$$
$$z_1 + z_2 \ge 5$$
$$z_1 + z_2 + z_3 \ge 8$$
$$z_1 + z_2 + z_3 + z_4 = 10$$

and
$$\begin{cases} z_1 \ge 0, \quad z_2 \ge 0, \quad z_3 \ge 0, \quad z_4 \ge 0, & \text{and integer valued,} \\ v_1 \ge 0, \quad v_2 \ge 0, \quad v_3 \ge 0, \quad v_4 \ge 0, & \\ v_1 \le 1, \quad v_2 \le 1, \quad v_3 \le 1, \quad v_4 \le 1, & \text{and integer valued.} \end{cases}$$

The solution to this problem may be obtained by using an integer programming algorithm; it yields the same solution as obtained previously, i.e., produce 100,000 speakers at the beginning of the first period ($z_1 = 10$, $v_1 = 1$, and all other z_i and v_i equal zero), or produce 50,000 speakers at the beginning of the first period and 50,000 speakers at the beginning of the third period ($z_1 = z_3 = 5$, $v_1 = v_3 = 1$, and all other z_i and v_i equal zero).

9.4 Stochastic Models

This section is concerned with inventory problems where the demand for a period is a random variable having a known probability distribution. Both single-period and multiperiod models are analyzed.

A Single-Period Model with No Setup Cost

The second example discussed in Sec. 9.1 is concerned with a wholesale distributor of 10-speed bicycles. Suppose that this distributor is offered very favorable terms on the purchase of a model of a name-brand bicycle whose production is to be discontinued. This opportunity appears to be ideal for the forthcoming Christmas season, where, because production has been discontinued, the stores have been informed that no reorders are possible. The cost of each bicycle is \$20, and it will be assumed that there is no setup cost incurred. The cost of maintaining an inventory is −\$9 per bicycle. This cost includes \$1, which represents the cost of capital tied up, warehouse space, and so on, and −\$10, which is what the distributor can get for each bicycle remaining in the inventory after the Christmas season (the salvage value). Note that this cost of maintaining an inventory is obtained by combining the bicycle storage cost with the bicycle salvage value and results in a negative holding cost. Each bicycle is sold for \$45 for a profit of \$25.

Two remaining cost components still require discussion, i.e., the unsatisfied demand cost and the revenue. If the demand exceeds the supply, those customers who fail to purchase a bicycle may bear some ill will, thereby resulting in a "cost" to the distributor. This unsatisfied demand cost is simply the per-item quantification of the loss of goodwill times the unsatisfied demand whenever a shortage occurs. In the bicycle example, this cost is considered to be negligible.

If we adopt the criterion of maximizing net income, we must include revenue in the model. Indeed, net income is equal to total revenue minus the costs incurred (ordering, inventory holding, and unsatisfied demand). The total revenue component is simply the sales price of a bicycle (\$45) times the demand *minus* the sales price times the unsatisfied demand whenever a shortage occurs. The former is independent of the inventory policy, and, hence, can be neglected, whereas the latter is just the lost revenue when a shortage exists. This latter term behaves just like the aforementioned description of the unsatisfied demand cost, and, hence, the two (the loss of goodwill and the lost revenue) can be combined (added) and considered to be the resultant *unsatisfied demand cost*. The unsatisfied demand cost will be so interpreted throughout this chapter. Thus, in the bicycle example, the unsatisfied demand is simply \$45[1] times the unsatisfied demand whenever a shortage exists.

Because the revenue times the demand term is independent of the inventory policy chosen, its omission from the net income expression results in terms that are

[1] In general, the cost of the loss of goodwill, assumed to be negligible in the bicycle example, must be added to the revenue to determine the complete *unsatisfied demand cost*.

the negative of total cost. Hence maximizing net income in this model is equivalent to minimizing total cost.

The discussion about the interpretation of the unsatisfied demand cost assumed that the unsatisfied demand was lost. For the case where this unsatisfied demand is met by a priority shipment, the same principles prevail. The total revenue component becomes the sales price of a bicycle ($45) times the demand *minus* the unit cost of the priority shipment times the unsatisfied demand whenever a shortage occurs. If the West Coast distributor is forced to meet the unsatisfied demand by purchasing bicycles from the Midwest distributor at the same cost that bicycles are sold to the retail outlets ($45) plus an air freight charge of, say, $2, then the appropriate unsatisfied demand cost is $47 per bicycle. Of course, any costs associated with loss of goodwill, if present, would be added to this amount.

The previous discussion concerned the costs involved in the model, and little attention was paid to the concept of the demand for the bicycles. Unfortunately, the distributor does not know what the demand for these bicycles will be from the stores in the western United States; i.e., the demand is a random variable, and hence he does not know how many bicycles will be required. However, an optimal inventory policy can be obtained if information about the probability distribution of demand is available. Let D represent the random variable demand, and denote by $P_D(d)$ the probability that the demand equals d; that is,

$$P_D(d) = P\{D = d\}.$$

It will be assumed that $P_D(d)$ is known for all values of d; that is, the probability distribution is specified.

In general, we consider the following inventory model. Items are purchased (or produced) for a *single period* at a cost of c dollars per item. The holding cost, the net unit cost of storing leftover items minus their salvage value, is given by h dollars per item and charged as a function of excess stock over the amount required. The cost of unsatisfied demand, e.g., lost revenue from sales or the cost of supplying a required unit, is given by p dollars per unit ($p > c$). It is assumed that there is no initial inventory on hand. Denote by y the quantity purchased (or produced) at the beginning of the period,[1] and let D be a random variable that denotes the demand during the period.

This single-period model may represent the inventory of an item that (1) becomes obsolete quickly, such as the bicycle in the example or a daily newspaper; (2) spoils quickly, such as vegetables; (3) is stocked only once, such as spare parts for a single production run of a new model airplane; or (4) has a future that is uncertain beyond a single period.

With the aforementioned structure, the question of how much inventory to have becomes relevant. More than the expected demand is probably desirable, but certainly less than the maximum demand is required. A trade-off is needed between (1) the risk of being short and thereby incurring shortage costs and (2) the risk of having an excess and thereby incurring wasted costs of ordering and holding excess units. One

[1] In the previous models the symbol z was used to denote the quantity produced, and x represented the inventory level at the beginning of the period. The symbol y is introduced here to denote the inventory level to be achieved after the ordering decision is made; that is, y is the amount ordered up to, so that $y = x + z$. Because the initial inventory level is assumed to be zero (that is, $x = 0$), y and z are the same.

reasonable criterion is to choose the inventory level that minimizes the expected value (in the statistical sense) of the sums of these costs.

The amount sold is given by

$$\begin{cases} D, & \text{if } D < y \\ y, & \text{if } D \geq y \end{cases} = \min(D, y).$$

Hence the cost incurred if the demand is D and y is stocked is given by

$$C(D, y) = cy + p \max(0, D - y) + h \max(0, y - D).$$

Because the demand is a random variable [with probability distribution $P_D(d)$], this cost is also a random variable. The expected cost is then given by $C(y)$, where

$$\begin{aligned} C(y) = E[C(D, y)] &= \sum_{d=0}^{\infty} [cy + p \max(0, d - y) + h \max(0, y - d)]P_D(d) \\ &= cy + \sum_{d=y}^{\infty} p(d - y)P_D(d) + \sum_{d=0}^{y-1} h(y - d)P_D(d). \end{aligned}$$

It is evident that $C(y)$ depends upon the probability distribution $P_D(d)$. Frequently a representation of this probability distribution is difficult to find, particularly when the demand ranges over a large number of possible values. Hence this discrete random variable is often approximated by a continuous random variable. Furthermore, when demand ranges over a large number of possible values, this approximation will generally yield small differences in numerical values in the optimal amounts of inventory to stock. In addition, when discrete demand is used the resulting expressions may become slightly more difficult to solve analytically. Hence, unless otherwise stated, continuous demand is assumed throughout the remainder of this chapter. The probability *density* function of this *continuous* random variable will be denoted by $\varphi_D(\xi)$. The expected cost $C(y)$ is then expressed as[1]

$$\begin{aligned} C(y) = E[C(D, y)] &= \int_0^{\infty} [cy + p \max(0, \xi - y) + h \max(0, y - \xi)]\varphi_D(\xi)\, d\xi \\ &= cy + \int_y^{\infty} p(\xi - y)\varphi_D(\xi)\, d\xi + \int_0^{y} h(y - \xi)\varphi_D(\xi)\, d\xi \\ &= cy + L(y), \end{aligned}$$

where $L(y)$ is often called the expected shortage plus holding cost. It then becomes necessary to find the value of y, say y^0, which minimizes $C(y)$. *The optimal quantity to order,* y^0, *is that value which satisfies*

$$\Phi(y^0) = \frac{p - c}{p + h},$$

[1] The standard notation of probability theory is being followed. If X is a random variable having density function $f_X(y)$, and $g(X)$ is a function of X, then

$$E[g(X)] = \int_{-\infty}^{\infty} g(y)f_X(y)\, dy.$$

Thus the expected cost is given by $C(y)$.

where the function $\Phi(a)$ is the cumulative distribution function of the demand random variable; that is,

$$\Phi(a) = \int_0^a \varphi_D(\xi)\, d\xi.$$

The derivation of this solution is given on page 299.

If D is assumed to be a discrete random variable having the cumulative distribution function

$$F_D(b) = \sum_{d=0}^{b} P_D(d),$$

a similar result for the optimal order quantity is obtained. In particular, *the optimal quantity to order, y^0, is the smallest integer such that*

$$F_D(y^0) \geq \frac{p - c}{p + h}.$$

Example: Returning to the bicycle example described in this section, assume that the demand has an exponential distribution given by

$$\varphi_D(\xi) = \begin{cases} \dfrac{1}{10{,}000} e^{-\xi/10{,}000}, & \xi \geq 0 \\ 0, & \text{otherwise.} \end{cases}$$

From the data given,

$$c = 20$$
$$p = 45$$
$$h = -9.$$

Because the demand density is exponential,

$$\Phi(a) = \int_0^a \frac{1}{10{,}000} e^{-\xi/10{,}000}\, d\xi = 1 - e^{-a/10{,}000}.$$

The optimum quantity to order, y^0, is that value which satisfies

$$1 - e^{-y^0/10{,}000} = \frac{45 - 20}{45 - 9} = 0.6944,$$

$$y^0 = 11{,}856.$$

Therefore, the distributor should stock 11,856 bicycles in the Christmas season. Note that this number is slightly more than the expected demand.

Whenever the demand is exponential with expectation λ, y^0 can easily be obtained from the relation

$$y^0 = -\lambda \ln \left(\frac{c + h}{p + h}\right).$$

Model with Initial Stock Level: As a slight variation of the previous model, suppose that the distributor has 500 bicycles of the aforementioned type on hand.

How does this stock influence the optimal inventory policy? In particular, suppose that the initial stock level is given by x, and the problem is to determine how much is to be made available, y, at the beginning of the period. Thus $(y - x)$ is to be ordered so that:

$$\text{Amount available } (y) = \text{initial stock } (x) + \text{amount ordered } (y - x).$$

The cost equation presented earlier remains identical except for the term that was previously cy. This term now becomes $c(y - x)$, so that minimizing the expected cost is given by:

$$\underset{y \geq x}{\text{Minimum}} \left[c(y - x) + \int_y^\infty p(\xi - y)\varphi_D(\xi)\, d\xi + \int_0^y h(y - \xi)\varphi_D(\xi)\, d\xi \right].$$

The constraint $y \geq x$ must be added because it is assumed that items on hand at the beginning of the period cannot be depleted or returned. The optimum policy is described as follows.

The inventory policy which satisfies, for $p > c$,

$$\underset{y \geq x}{\text{Minimum}} \left[-cx + \left\{ \int_y^\infty p(\xi - y)\varphi_D(\xi)\, d\xi + \int_0^y h(y - \xi)\varphi_D(\xi)\, d\xi + cy \right\} \right]$$

is given by $y = \{^{y^0}_{x}$, *so that*

If $x < y^0$, *order up to* y^0 *(order* $y^0 - x$*),*

If $x \geq y^0$, *do not order,*

where y^0 *satisfies*

$$\Phi(y^0) = \frac{p - c}{p + h}.$$

Thus, in the bicycle example, if there are 500 bicycles on hand, the optimal policy is to order up to 11,856 bicycles (which implies ordering 11,356 additional bicycles). On the other hand, if there are 12,000 bicycles already on hand, the optimal policy is not to order.

Model with Nonlinear Penalty Costs: Similar results for these models can be obtained for other than linear holding and shortage penalty costs. Denote the holding cost by

$$\begin{cases} h[y - D], & \text{if } y \geq D \\ 0, & \text{if } y < D, \end{cases}$$

where $h[\cdot]$ is a mathematical function, not necessarily linear.

Similarly, the shortage penalty cost can be denoted by

$$\begin{cases} p[D - y], & \text{if } D \geq y \\ 0, & \text{if } D < y, \end{cases}$$

where $p[\cdot]$ is also a function, not necessarily linear.

Thus the total expected cost is given by

$$c(y - x) + \int_y^\infty p[\xi - y]\varphi_D(\xi)\, d\xi + \int_0^y h[y - \xi]\varphi_D(\xi)\, d\xi,$$

where x is the amount on hand.

If $L(y)$ is defined as the expected shortage plus holding cost, that is,

$$L(y) = \int_y^\infty p[\xi - y]\varphi_D(\xi)\,d\xi + \int_0^y h[y - \xi]\varphi_D(\xi)\,d\xi,$$

then the total expected cost can be written as

$$c(y - x) + L(y).$$

The optimal policy is obtained by minimizing this expression, subject to the constraint that $y \geq x$; that is:

$$\underset{y \geq x}{\text{Minimum}}\ [c(y - x) + L(y)].$$

If $L(y)$ is strictly convex[1] [*a sufficient condition being that the shortage and holding costs each are convex and $\varphi_D(\xi) > 0$*], *then the optimal policy is given by*

$$\begin{cases} \text{if } x < y^0, & \text{order up to } y^0 \\ \text{if } x \geq y^0, & \text{do not order,} \end{cases}$$

where y^0 is the value of y that satisfies the expression

$$\frac{dL(y)}{dy} + c = 0.$$

Derivation of Results for the Single-Period Model with No Setup Cost and Linear Shortage and Holding Costs[2]: A useful result for finding policies that minimize the expected cost is as follows.

Let D be a random variable having a density function

$$\begin{cases} \varphi_D(\xi), & \text{if } \xi \geq 0 \\ 0, & \text{otherwise.} \end{cases}$$

Denote by $\Phi(a)$ the cumulative distribution function; that is,

$$\Phi(a) = \int_0^a \varphi_D(\xi)\,d\xi.$$

Let $g(\xi, y)$ be defined as

$$g(\xi, y) = \begin{cases} c_1(y - \xi), & \text{if } y > \xi,\ c_1 > 0 \\ c_2(\xi - y), & \text{if } y \leq \xi,\ c_2 > 0, \end{cases}$$

and

$$G(y) = \int_0^\infty g(\xi, y)\varphi_D(\xi)\,d\xi + cy,$$

where $c > 0$. Then $G(y)$ is minimized at $y = y^0$, where y^0 is the solution to

$$\Phi(y^0) = \frac{c_2 - c}{c_2 + c_1}.$$

To see why this value of y^0 minimizes $G(y)$, note that, by definition,

$$G(y) = c_1 \int_0^y (y - \xi)\varphi_D(\xi)\,d\xi + c_2 \int_y^\infty (\xi - y)\varphi_D(\xi)\,d\xi + cy.$$

[1] See Appendix 1 for the definition of a convex function.

[2] This section may be omitted by the less mathematically inclined reader.

Taking the derivative (see Appendix 2) and setting it equal to zero leads to

$$\frac{dG(y)}{dy} = c_1 \int_0^y \varphi_D(\xi)\, d\xi - c_2 \int_y^\infty \varphi_D(\xi)\, d\xi + c = 0.$$

This expression implies that

$$c_1\, \Phi(y^0) - c_2[1 - \Phi(y^0)] + c = 0,$$

because

$$\int_0^\infty \varphi_D(\xi)\, d\xi = 1.$$

Solving this expression results in

$$\Phi(y^0) = \frac{c_2 - c}{c_2 + c_1}.$$

Checking the second derivative indicates that

$$\frac{d^2G(y)}{dy^2} = (c_1 + c_2)\varphi_D(y) \geq 0$$

for all y, so that the result is obtained.

To apply this result, it is sufficient to show that

$$C(y) = cy + \int_y^\infty p(\xi - y)\varphi_D(\xi)\, d\xi + \int_0^y h(y - \xi)\varphi_D(\xi)\, d\xi$$

has the form of $G(y)$.

Clearly, $c_1 = h$, $c_2 = p$, and $c = c$, so that the optimal quantity to order, y^0, is that value which satisfies

$$\Phi(y^0) = \frac{p - c}{p + h}.$$

To derive the results for the case where the initial stock level is x, recall that it is necessary to solve the relationship

$$\min_{y \geq x}\left[-cx + \left\{\int_y^\infty p(\xi - y)\varphi_D(\xi)\, d\xi + \int_0^y h(y - \xi)\varphi_D(\xi)\, d\xi + cy\right\}\right].$$

Note that the expression in braces has the form of $G(y)$, with $c_1 = h$, $c_2 = p$, and $c = c$. Hence the cost function to be minimized can be written as

$$\min_{y \geq x}[-cx + G(y)].$$

It is clear that $-cx$ is a constant, so that it is sufficient to find the y that satisfies the expression

$$\min_{y \geq x} G(y).$$

Hence the value of y^0 that minimizes $G(y)$ satisfies

$$\Phi(y^0) = \frac{p - c}{p + h}.$$

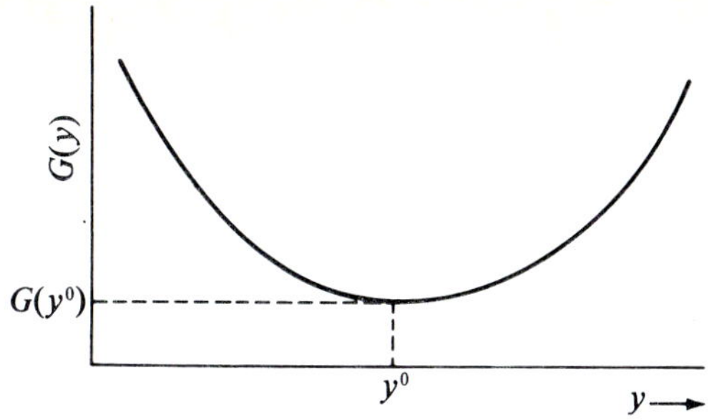

Figure 9.7 Graph of $G(y)$.

Furthermore, $G(y)$ must be a convex function, because

$$\frac{d^2G(y)}{dy^2} \geq 0.$$

Also,

$$\lim_{y \to 0} \frac{dG(y)}{dy} = c - p,$$

which is negative,[1] and

$$\lim_{y \to \infty} \frac{dG(y)}{dy} = h + c,$$

which is positive. Hence $G(y)$ must be as shown in Fig. 9.7. Thus the optimal policy must be given by the following:

If $x < y^0$, order up to y^0 because y^0 can be achieved together with the minimum value $G(y^0)$,
If $x \geq y^0$, do not order because any $G(y)$, with $y > x$, must exceed $G(x)$.

A similar argument can be constructed for obtaining optimal policies with nonlinear penalty costs when $L(y)$ is strictly convex.

A Single-Period Model with a Setup Cost

In discussing the bicycle example in this section, it was assumed that there was no fixed cost incurred in ordering the bicycles for the Christmas season. In actual fact, the cost of placing this special order is \$800, and this cost should be included in the analysis of the model. In fact, inclusion of the setup cost generally causes major changes in the results.

In general, the setup cost will be denoted by K. To begin with, the shortage and holding costs will each be assumed to be linear. Their resultant effect is then given by $L(y)$, where

$$L(y) = p \int_y^{\infty} (\xi - y)\varphi_D(\xi)\, d\xi + h \int_0^y (y - \xi)\varphi_D(\xi)\, d\xi.$$

[1] If $c - p$ is nonnegative, $G(y)$ will be a monotone increasing function. This implies that the item should not be stocked; that is, $y^0 = 0$.

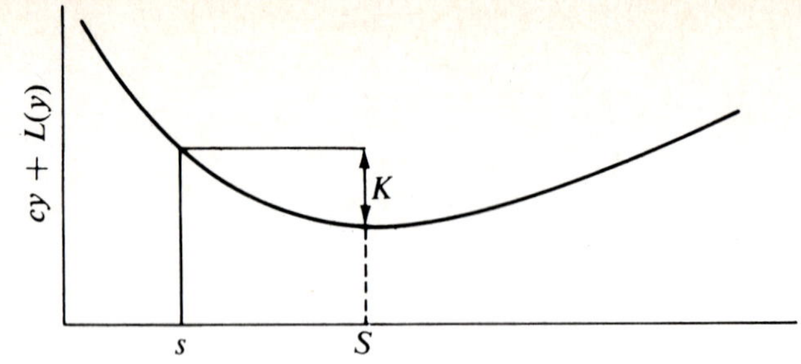

Figure 9.8 Graph of $cy + L(y)$.

Thus the total expected cost incurred if one orders up to y is given by

$$\begin{cases} K + c(y - x) + L(y), & \text{if } y > x \\ L(x), & \text{if } y = x. \end{cases}$$

Note that $cy + L(y)$ is the same expected cost considered in the previous section when the setup cost was omitted. If $cy + L(y)$ is drawn as a function of y, it will appear as shown in Fig. 9.8.[1] Define S as the value of y that minimizes $cy + L(y)$, and define s as the smallest value of y for which $cs + L(s) = K + cS + L(S)$. From Fig. 9.8 it is evident that if $x > S$, then $K + cy + L(y) > cx + L(x)$, for all $y > x$. Hence $K + c(y - x) + L(y) > L(x)$, where the left-hand side of the inequality represents the expected total cost if one orders up to y, and the right-hand side of the inequality represents the expected total cost if no ordering occurs. Hence the optimum policy indicates that if $x > S$, do not order. If $s \le x \le S$, it is again evident from Fig. 9.8 that

$$K + cy + L(y) \ge cx + L(x), \qquad \text{for all } y > x,$$

so that

$$K + c(y - x) + L(y) \ge L(x).$$

Again, no ordering is less expensive than ordering. Finally, if $x < s$, it follows from Fig. 9.8 that

$$\min_{y \ge x} [K + cy + L(y)] = K + cS + L(S) < cx + L(x),$$

or

$$\min_{y \ge x} [K + c(y - x) + L(y)] = K + c(S - x) + L(S) < L(x),$$

so that it pays to order. The minimum cost is incurred if one orders up to S.

Thus the optimum ordering policy can be summarized as follows:

$$\begin{cases} \textit{if } x < s, & \textit{order up to } S \\ \textit{if } x \ge s, & \textit{do not order.} \end{cases}$$

The value of S is obtained from

$$\Phi(S) = \frac{p - c}{p + h},$$

[1] In the derivation of results for the single-period model with no setup cost and linear shortage and holding costs, $cy + L(y)$ was denoted by $G(y)$ and rigorously shown to be a convex function of the form plotted in Fig. 9.8.

and s is the smallest value that satisfies the expression

$$cs + L(s) = K + cS + L(S).$$

This policy is an (s, S) policy mentioned previously, and it has had extensive use in industry.

Example: Referring to the bicycle example of the previous section,

$$y^0 = S = 11{,}856.$$

If $K = 800$, $c = 20$, $p = 45$, and $h = -9$, s is obtained from

$$20s + 45\int_s^\infty (\xi - s)\frac{1}{10{,}000}e^{-\xi/10{,}000}\,d\xi - 9\int_0^s (s - \xi)\frac{1}{10{,}000}e^{-\xi/10{,}000}\,d\xi$$

$$= 800 + 20(11{,}856) + 45\int_{11{,}856}^\infty (\xi - 11{,}856)\frac{1}{10{,}000}e^{-\xi/10{,}000}\,d\xi$$

$$-9\int_0^{11{,}856}(11{,}856 - \xi)\frac{1}{10{,}000}e^{-\xi/10{,}000}\,d\xi,$$

so that

$$s = 10{,}674.$$

Hence the optimal policy calls for ordering up to $S = 11{,}856$ bicycles if the amount on hand is less than $s = 10{,}674$. Otherwise, no order is placed.

Solution When the Demand Distribution Is Exponential: It may be of interest to solve this model, in general, when the distribution of demand, D, is exponential; that is,

$$\varphi_D(\xi) = \frac{1}{\lambda}e^{-\xi/\lambda}, \qquad \text{for } \xi > 0.$$

If Δ is defined as $S - s$, then Δ is the solution to the equation

$$e^{\Delta/\lambda} = \frac{K}{\lambda(c + h)} + \frac{\Delta}{\lambda} + 1.$$

Furthermore, a good approximation for Δ is given by

$$\Delta \simeq \sqrt{\frac{2\lambda K}{c + h}}.$$

Note that

$$s = S - \Delta.$$

These results are easily obtained. From the no-setup cost results,

$$1 - e^{-S/\lambda} = \frac{p - c}{p + h}$$

or

$$S = \lambda \ln\left(\frac{h + p}{h + c}\right).$$

For any y,

$$cy + L(y) = cy + h\int_0^y (y - \xi)\frac{1}{\lambda}e^{-\xi/\lambda}\,d\xi + p\int_y^\infty (\xi - y)\frac{1}{\lambda}e^{-\xi/\lambda}\,d\xi$$

$$= (c + h)y + \lambda(h + p)e^{-y/\lambda} - \lambda h.$$

Evaluating $cy + L(y)$ at the point $y = s$ and $y = S$ leads to

$$(c + h)s + \lambda(h + p)e^{-s/\lambda} - \lambda h = K + (c + h)S + \lambda(h + p)e^{-S/\lambda} - \lambda h,$$

or

$$(c + h)s + \lambda(h + p)e^{-s/\lambda} = K + (c + h)S + \lambda(c + h).$$

Although this last equation does not have a closed-form solution, it can be solved numerically quite easily. An approximate analytical solution can be obtained as follows. By letting $\Delta = S - s$, the last equation yields

$$e^{\Delta/\lambda} = \frac{K}{\lambda(c + h)} + \frac{\Delta}{\lambda} + 1.$$

If Δ/λ is close to zero, $e^{\Delta/\lambda}$ can be expanded into a Taylor series around zero. If the terms beyond the quadratic term are neglected, the result becomes

$$1 + \frac{\Delta}{\lambda} + \frac{\Delta^2}{2\lambda^2} \simeq \frac{K}{\lambda(c + h)} + \frac{\Delta}{\lambda} + 1,$$

so that

$$\Delta \simeq \sqrt{\frac{2\lambda K}{c + h}}.$$

Using the approximation in the bicycle example results in

$$\Delta = \sqrt{\frac{(2)(10{,}000)(800)}{20 - 9}} = 1{,}206,$$

which is quite close to the exact value of $\Delta = 1{,}182$.

Model with Nonlinear Penalty Costs: Again it is evident that these results can be extended easily to any strictly convex expected shortage plus holding cost, $L(y)$. This extension results in a strictly convex $cy + L(y)$, similar to Fig. 9.8. Hence the optimal ordering policy is of the form

$$\begin{cases} \text{if } x < s, & \text{order up to } S \\ \text{if } x \geq s, & \text{do not order,} \end{cases}$$

where S is the value of y that satisfies

$$c + \frac{dL(y)}{dy} = 0,$$

and s is the smallest value that satisfies the expression

$$cs + L(s) = K + cS + L(S).$$

A Two-Period Inventory Model with No Setup Cost

The single-period model was illustrated with a bicycle example where the distributor had only one opportunity to place an order. In many situations an opportunity to place an order occurs periodically, e.g., monthly, and the inventory manager must make a decision about whether and how much to stock. The manager may be interested in these decisions for a horizon of the next 2 months, 12 months, 18 months, or even possibly forever. Even for a horizon of 2 months, the periodic review policy of using the optimal one-period solution twice is not generally the optimal policy for the two-period problem. Smaller costs can usually be achieved by viewing the problem from a two-period viewpoint and then using the methods of dynamic programming introduced in Chap. 8 to obtain the best inventory policy. In fact, the only difference in concept in the dynamic programming approach for solving the inventory problem compared to the material in Chap. 8 is the probabilistic aspects that are introduced by the random demand.

Two-Period Model—No Setup Cost: Suppose that the bicycle distributor is permitted to make at most one reorder, to occur on November 15, after placing an initial order for the special type 10-speed bicycle on October 15. The assumptions about the costs are similar to those presented before. Assume that the purchase leads to immediate delivery; shortages at the end of the first period are to be made up, if they exist (backlogging of orders is possible at the end of the first period, but not at the end of the second period); no disposal of stock is permitted. Furthermore, the demands D_1, D_2 for the two periods are independent, identically distributed random variables having density $\varphi_D(\xi)$. The purchase cost is linear, that is, cz, where z is the amount ordered (no setup cost), and the shortage and holding costs are also linear, with respective unit costs denoted by p and h.

As indicated earlier, the solution to this problem is not to use the optimal one-period solution twice. Smaller costs can be achieved by viewing the problem from a two-period dynamic programming viewpoint. Order the time periods so that the beginning of time period 1 implies that there are two periods left in the horizon. Similarly, the beginning of time period 2 implies that there is one period left in the horizon (the last period is beginning). The problem is to find critical numbers that describe the optimum ordering policy. It will be shown that these numbers are single critical numbers for each period. These numbers will be denoted by y_1^0 and y_2^0. Furthermore, y_1 and y_2 (without the superscript) represent any amount of stock ordered up to at the beginning of the respective period.

Denote by $C_1(x_1)$ the expected cost of following an optimum policy (minimum cost) from the beginning of period 1 to the end of period 2, given that there are x_1 units on hand. Similarly, denote by $C_2(x_2)$ the expected cost of following an optimum policy (minimum cost) from the beginning of period 2, given that there are x_2 units on hand. $C_1(x_1)$ is the expression sought because this expression is obtained by following the optimum policy for the entire (two-period) horizon. To obtain $C_1(x_1)$ it is necessary to first find $C_2(x_2)$. From the results for the single-period model, the optimal policy for a one-period problem is given by a single critical number found from

$$\Phi(y_2^0) = \frac{p - c}{p + h};$$

that is, if x_2 is the amount available at the beginning of the last period, then

$$\begin{cases} \text{order } (y_2^0 - x_2), & \text{if } x_2 < y_2^0 \\ \text{do not order}, & \text{if } x_2 \geq y_2^0. \end{cases}$$

The cost of this optimum policy can be expressed as

$$C_2(x_2) = \begin{cases} L(x_2), & \text{if } x_2 \geq y_2^0 \\ c(y_2^0 - x_2) + L(y_2^0), & \text{if } x_2 < y_2^0, \end{cases}$$

where y_2^0 is the single critical number just determined, and $L(z)$ is the expected shortage plus holding cost for a single period when there are z units available. $L(z)$ can be expressed as

$$L(z) = \int_z^\infty p(\xi - z)\varphi_D(\xi)\,d\xi + \int_0^z h(z - \xi)\varphi_D(\xi)\,d\xi.$$

At the beginning of period 1, the costs incurred consist of the purchase cost $c(y_1 - x_1)$, the expected shortage plus holding cost $L(y_1)$, and the costs associated with following an optimal policy during the second period. Thus the expected cost of following the optimal policy for two periods is given by

$$C_1(x_1) = \underset{y_1 \geq x_1}{\text{minimum}} \{c(y_1 - x_1) + L(y_1) + E[C_2(x_2)]\},$$

where $E[C_2(x_2)]$ is obtained as follows: Note that x_2 is a random variable that depends upon the amount of stock on hand at the beginning of period 2; that is, $x_2 = y_1 - D_1$. Thus,

$$C_2(x_2) = C_2(y_1 - D_1) = \begin{cases} L(y_1 - D_1), & \text{if } y_1 - D_1 \geq y_2^0 \\ c(y_2^0 - y_1 + D_1) + L(y_2^0), & \text{if } y_1 - D_1 < y_2^0. \end{cases}$$

Hence $C_2(x_2)$ is a random variable, and its expected value is given by

$$E[C_2(x_2)] = \int_0^\infty C_2(y_1 - \xi)\varphi_D(\xi)\,d\xi = \int_0^{y_1 - y_2^0} L(y_1 - \xi)\varphi_D(\xi)\,d\xi$$
$$+ \int_{y_1 - y_2^0}^\infty [c(y_2^0 - y_1 + \xi) + L(y_2^0)]\varphi_D(\xi)\,d\xi.$$

Note that because shortages are permitted, $(y_1 - \xi)$ can be negative; further note that $E[C_2(x_2)]$ is just a function of y_1 and y_2^0, with y_2^0 obtained from the solution to the single-period problem. Thus $C_1(x_1)$ can now be expressed as

$$C_1(x_1) = \underset{y_1 \geq x_1}{\text{minimum}} \left\{ c(y_1 - x_1) + L(y_1) + \int_0^{y_1 - y_2^0} L(y_1 - \xi)\varphi_D(\xi)\,d\xi \right.$$
$$\left. + \int_{y_1 - y_2^0}^\infty [(y_2^0 - y_1 + \xi) + L(y_2^0)]\varphi_D(\xi)\,d\xi \right\}.$$

It can easily be shown that $C_1(x_1)$ has a unique minimum so that the optimal value of y_1, denoted by y_1^0, satisfies the equation

$$-p + (p + h)\Phi(y_1^0) + (c - p)\Phi(y_1^0 - y_2^0) + (p + h)\int_0^{y_1^0 - y_2^0} \Phi(y_1^0 - \xi)\varphi_D(\xi)\,d\xi = 0,$$

where $\Phi(a)$ denotes the cumulative distribution function; that is,

$$\Phi(a) = \int_0^a \varphi_D(\xi)\,d\xi.$$

Thus, if x_1 is the amount available at the beginning of period 1, then the optimal policy is to

$$\begin{cases} \text{order } (y_1^0 - x_1), & \text{if } x_1 < y_1^0 \\ \text{not order}, & \text{if } x_1 \geq y_1^0. \end{cases}$$

If the density function of demand is uniform over the range 0 to t, that is,

$$\varphi_D(\xi) = \begin{cases} \dfrac{1}{t}, & \text{if } 0 \le \xi \le t \\ 0, & \text{otherwise,} \end{cases}$$

then y_1^0 can be obtained from the expression

$$y_1^0 = \sqrt{(y_2^0)^2 + \left[\frac{2t(c-p)}{p+h}\right] y_2^0 + \frac{t^2[2p(p+h)+(h+c)^2]}{(p+h)^2}} - \frac{t(h+c)}{p+h}.$$

Finally, if the density function of demand is exponential, that is,

$$\varphi_D(\xi) = \frac{1}{\lambda} e^{-\xi/\lambda}, \qquad \text{for } \xi > 0,$$

the optimal value of y_1^0 satisfies the relationship

$$(h+c)e^{-(y_1^0-y_2^0)/\lambda} + (p+h)e^{-y_1^0/\lambda} + \frac{(p+h)(y_1^0-y_2^0)}{\lambda} e^{-y_1^0/\lambda} = 2h + c.$$

An alternative way of finding y_1^0 is to let z^0 denote $(y_1^0 - y_2^0)/\lambda$. Then z^0 satisfies the relationship

$$e^{-z^0}[(h+c) + (p+h)e^{-y_2^0/\lambda} + z^0(p+h)e^{-y_2^0/\lambda}] = 2h + c,$$

and

$$y_1^0 = \lambda z^0 + y_2^0.$$

Example: Consider the following example. The cost of producing an item is \$10 per item ($c = 10$). If any excess inventory remains at the end of a period, it is charged at \$10 per item ($h = 10$); if no excess appears, there is no holding charge. If a shortage occurs within a period, there is a penalty cost of \$15 per item ($p = 15$). The density function of demand is given by

$$\varphi_D(\xi) = \begin{cases} \dfrac{1}{10}, & \text{if } 0 \le \xi \le 10 \\ 0, & \text{otherwise.} \end{cases}$$

It is necessary to find the optimum two-period policy. For linear costs, the equation for the optimal single-period model becomes

$$\Phi(y_2^0) = \frac{p-c}{p+h} = \frac{15-10}{15+10} = \frac{1}{5}.$$

Hence, because $\Phi(y_2^0) = y_2^0/10$,

$$y_2^0 = 2.$$

Because the distribution of demand is uniform (with parameter $t = 10$), the optimal value of y_1^0 is found from

$$y_1^0 = \sqrt{(2)^2 + \left[\frac{2(10)(10 - 15)}{(15 + 10)}\right] 2 + (10)^2 \left[\frac{2(15)(15 + 10) + (10 + 10)^2}{(15 + 10)^2}\right]} - \frac{10(10 + 10)}{15 + 10}$$

$$= \sqrt{4 - 8 + 184} - 8 = 13.42 - 8 = 5.42.$$

Substituting $y_1^0 = 5$ and $y_1^0 = 6$ into $C_1(x_1)$ leads to a smaller value with $y_1^0 = 5$. Thus the optimal policy can be described as follows: If the initial amount of stock on hand is less than 5, order up to 5 units (order $5 - x_1$ unit). Otherwise, do not order. After a period has elapsed, and at the beginning of the last period, if the amount on hand is less than 2 units, order up to 2 units (order $2 - x_2$ units, where x_2 may be negative).

This two-period model has been solved by using the same penalty shortage cost for each of the two periods, even though unsatisfied demand is backlogged at the end of the first period and lost at the end of the second period. This is a deficiency of the model as presented, but it can be remedied by using different penalty shortage costs in each period. The computational procedure of the modified model is similar, and no additional difficulties are added.

Multiperiod Models—An Overview

The two-period model can be extended to several periods or to an infinite number of periods. This section presents a summary of multiperiod results that have practical importance.

MULTIPERIOD MODEL—NO SETUP COST: This model is an extension of the two-period problem. Suppose there exists a horizon of n periods for which we need to determine the optimal inventory policy. As before, assume that production leads to immediate delivery; shortages are to be backlogged except at the final period when they are lost; no disposal of stock is permitted. Furthermore, the demands for the n periods are independent, identically distributed random variables having density $\varphi_D(\xi)$. The purchase cost is linear, that is, cz, where z is the amount ordered (no setup cost), and the expected (one-period) shortage plus holding penalty cost $L(y)$ is strictly convex [which is the case if each cost is linear and $\varphi_D(\xi) > 0$]. A cost discounting factor is included; it is denoted by α, $0 < \alpha < 1$.

Again the time periods are ordered so that the beginning of time period 1 implies that there are n periods left in the horizon. Similarly, the beginning of time period n implies there is one period left in the horizon (the last period is beginning). The problem is to find critical numbers that describe the optimal ordering policy. As in the two-period model, the numbers $y_1^0, y_2^0, \ldots, y_n^0$ are difficult to obtain numerically, but it can be shown that the optimal policy has the form

At the beginning of period i, $i = 1, 2, \ldots, n$,

$$\begin{cases} \textit{order up to } y_i^0 \ (\textit{order } y_i^0 - x_i), & \textit{if } x_i < y_i^0 \\ \textit{do not order}, & \textit{if } x_i \geq y_i^0. \end{cases}$$

Furthermore, $\quad y_n^0 \leq y_{n-1}^0 \leq \cdots \leq y_2^0 \leq y_1^0.$

For the infinite-period model (where the inventory decisions are made indefinitely), there exists a single critical period number, y^0, so that the optimal policy now has the form

At the beginning of period i, i = 1, 2, . . . ,

$$\begin{cases} \text{order up to } y^0 \text{ (order } y^0 - x_i), & \text{if } x_i < y^0 \\ \text{do not order}, & \text{if } x_i \geq y^0. \end{cases}$$

Furthermore, y^0 is easily obtained because y^0 is the value of y that satisfies the expression

$$\frac{dL(y)}{dy} + c(1 - \alpha) = 0.$$

For the case of linear shortage and holding costs, unit costs of p and h, respectively, y^0 simply satisfies

$$\Phi(y^0) = \frac{p - c(1 - \alpha)}{p + h}.$$

A Variation of the Multiperiod Inventory Model—No Setup Cost: A slight modification of the aforementioned model leads to some simple, but interesting, results.[1] Consider the n-period model, but now assume that stock left over at the end of the final period can be salvaged with a return of the initial purchase cost c. Similarly, if there is a shortage at this time, the items are supplied also at the purchase price c. These two changes may lead to more realism in the model, but, what is equally important, they lead to rather simple optimal policies. In particular, the same single critical number y^0 is used for *all* periods, and, furthermore, the optimal policy is the same as that just presented in the infinite-period model; i.e.,

At the beginning of period i, i = 1, 2, . . . , n,

$$\begin{cases} \text{order up to } y^0 \text{ (order } y^0 - x_i), & x_i < y^0 \\ \text{do not order}, & x_i \geq y^0, \end{cases}$$

where y^0 satisfies the expression

$$\frac{dL(y)}{dy} + c(1 - \alpha) = 0.$$

Of course, this same result also holds for the infinite-period model. For the case of linear shortage and holding costs, unit costs of p and h, respectively, y^0 simply satisfies

$$\Phi(y^0) = \frac{p - c(1 - \alpha)}{p + h}.$$

Example of a Multiperiod Model: Example 2, Sec. 9.1, concerned the western distributor of bicycles who is having trouble with shortages of the most popular 10-speed model. The unit shortage cost p was estimated to be \$15. The holding cost h was determined to be \$1 per bicycle remaining at the end of the month. The actual

[1] This formulation is due to A. F. Veinott, Jr.: "The Optimal Inventory Policy for Batch Orderings," *Operations Research,* **13**(3):424–432, 1965.

cost of a bicycle is \$35. Orders may be placed on the first working day of each month. The distributor always places an order for some model bicycles each month, so he is willing to assume that the marginal setup cost is zero for this most popular model. The discount factor may be assumed to be $\alpha = 0.995$. From past history, the distribution of demand can be approximated by a uniform distribution given by

$$\varphi_D(\xi) = \begin{cases} \dfrac{1}{800}, & \text{if } 0 \le \xi \le 800 \\ 0, & \text{otherwise.} \end{cases}$$

The distributor expects to stock this model indefinitely, so that the infinite-period model is appropriate.

Because the shortage and holding costs are linear, y^0 satisfies

$$\Phi(y^0) = \frac{p - c(1 - \alpha)}{p + h}.$$

Because the demand distribution is uniform,

$$\Phi(y^0) = \frac{y^0}{800},$$

and

$$\frac{p - c(1 - \alpha)}{p + h} = \frac{15 - 35(1 - 0.995)}{15 + 1} = 0.927,$$

so $y_0 = 741$. Thus, if the number of bicycles on hand, x, at the first of each month is fewer than 741, the optimal policy calls for ordering up to 741 (ordering $741 - x$ bicycles). Otherwise, no order is placed. Note that if this policy is to be in effect for only a finite number of months, say, 24, and if (1) bicycles remaining in stock after 24 months can be salvaged at \$35 per bicycle and (2) unsatisfied demand remaining at the end of 24 months can be supplied to the western distributor from, say, the eastern distributor, at \$35 per bicycle, then the policy of ordering up to 741 bicycles each month is still optimal.

MULTIPERIOD MODEL WITH SETUP COST: The introduction of a fixed setup cost K that is incurred when ordering often adds more realism to the model. Unfortunately, however, the mathematics becomes very cumbersome, and the results available are primarily those that characterize the form of the optimal policy. In particular, if the ordering cost is assumed to be $K + cz$ for $z > 0$ and zero for $z = 0$, and if $L(y)$ is strictly convex, then the optimal policy has the form

At the beginning of period i, $i = 1, 2, \ldots, n$,

$$\begin{cases} \textit{order up to } S_i \textit{ (order } S_i - x_i), & \textit{if } x_i < s_i \\ \textit{do not order,} & \textit{if } x_i \ge s_i. \end{cases}$$

This policy is the familiar (s, S) policy alluded to earlier in the discussion of one-period models. As previously mentioned, unfortunately exact computations of s_i and S_i for the finite- or infinite-horizon model are extremely difficult. However, the importance of this result cannot be minimized. Even if the exact s_i and S_i are unknown, it is important to know that one should consider using policies of this form rather than a policy from another class.

(k, Q) Policies for a Multiperiod Model without Setup Cost: In the discussion of the previous models, any quantity could be ordered at the time of inventory review. Suppose that an additional constraint is placed on the ordering policy; i.e., each order for stock must be some nonnegative integral multiple of Q, a fixed positive constraint.

As before, the demands for the n periods are assumed to be independent, identically distributed random variables having density $\varphi_D(\xi)$. The purchase cost is linear, and the expected (one-period) shortage plus holding penalty cost $L(y)$ is strictly convex. α is the cost discounting factor, $0 < \alpha < 1$. At the beginning of each period the system is reviewed. An order may be placed for any nonnegative integral multiple of Q, a fixed positive number. Thus orders must be placed in multiples of some standard batch size, e.g., a case or a truckload. When the demand exceeds the inventory on hand, the excess demand is backlogged until it is subsequently filled by a delivery. In addition, it is assumed that stock left over at the end of the final period n can be salvaged with a return of the initial purchase cost c. Similarly, if there is a shortage at this time, the items are backlogged also at the purchase price c. This model was introduced by Veinott.[1] He shows that a (k, Q) policy is optimal. A (k, Q) policy is described as follows:

> *If at the beginning of a period the stock on hand is less than k, an order should be placed for the smallest multiple of Q that will bring the stock level to at least k (and probably higher); otherwise, an order should not be placed. The same parameter k is used in each period.*

The parameter k is chosen as follows: Let y^0 be the minimizer of $G(y) = (1 - \alpha)cy + L(y)$. $G(y)$ must be of the shape (convex) shown in Fig. 9.9. Then k is any number for which $k \le y^0 \le k + Q$ and $G(k) = G(k + Q)$. Refer to Fig. 9.9. If a "ruler" of length Q is placed horizontally into the "valley," k is found to be that value of the abscissa to the left of y^0 where the ruler intersects the valley. Note that if the initial inventory on hand lies in R_1, then Q is ordered; if it lies in R_2, then $2Q$ is ordered; and so on.

It should be noted that the same value of parameter k is used for each period of a finite-horizon model as well as for the infinite-horizon model. In the latter situation, this policy is the same optimal policy that would have been obtained if salvage costs and backlogging *at the last period* were omitted.

Advanced Mathematical Topics in Multiperiod Stochastic Models: This section will consider some of the models presented in the discussion of multiperiod stochastic models and derive the results presented. The reader who is not interested in mathematical derivations may skip this section.

Multiperiod Inventory Model with No Setup Cost: This model is an extension of the two-period model. Suppose there exists a horizon of n periods, for which it is required to determine the optimum inventory policy. As before, assume that the production leads to immediate delivery; shortages must be backlogged, except at the final period when they are lost; no disposal of stock is permitted. Furthermore, the demands for the n periods are independent, identically distributed random variables having density $\varphi_D(\xi)$. The purchase cost is linear, that is, cz, where z is the amount ordered, and the

[1] This formulation is due to A. F. Veinott, Jr.: "The Optimal Inventory Policy for Batch Orderings," *Operations Research*, **13**(3):424–432, 1965.

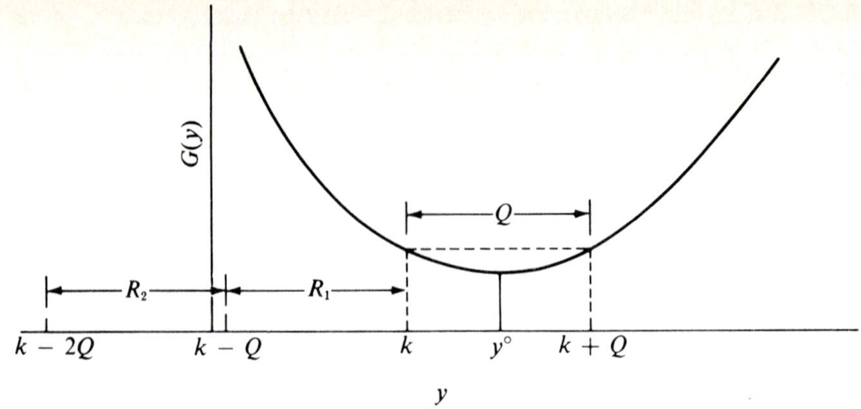

Figure 9.9 Plot of the $G(y)$ function.

expected (one period) shortage plus holding penalty cost $L(y)$ is strictly convex. α is the cost discounting factor, $0 < \alpha < 1$.

$C_1(x_1)$ is defined as the minimum expected cost of following an optimum policy from the beginning of the first period to the end of the horizon (period n), given that there are x_1 units on hand now. Similarly, define $C_2(x_2)$, $C_3(x_3)$, . . . , $C_{n-1}(x_{n-1})$, $C_n(x_n)$.

At the beginning of period 1, a decision about the inventory is made; that is, $y_1 - x_1$ is ordered ($y_1 \geq x_1$), and a cost incurred. This is given by

$$c(y_1 - x_1) + L(y_1).$$

During this first period, a demand D_1 occurs, so that at the beginning of the second period, the amount on hand, x_2, is a random variable equal to $(y_1 - D_1)$. If an optimum policy is carried on from the beginning of the second period to the end of the horizon, the discounted (to the beginning of period 1) expected minimum cost of the future $(n - 1)$ periods is given by

$$\alpha E[C_2(x_2)] = \alpha E[C_2(y_1 - D_1)] = \alpha \int_0^\infty C_2(y_1 - \xi)\varphi_D(\xi)\, d\xi.$$

Thus, the expected cost of following the optimal policy for the n-period horizon is obtained and is given by

$$C_1(x_1) = \underset{y_1 \geq x_1}{\text{minimum}}\{c(y_1 - x_1) + L(y_1) + \alpha E[C_2(y_1 - D_1)]\}.$$

This dynamic programming approach can be used to obtain the expected cost of following an optimal policy from the beginning of any period i, $i = 1, 2, \ldots, n$, to the end of period n, given that there are x_i units on hand; that is,

$$C_i(x_i) = \underset{y_i \geq x_i}{\text{minimum}}\{c(y_i - x_i) + L(y_i) + \alpha E[C_{i+1}(y_i - D_i)]\},$$

where $C_{n+1}(x_{n+1}) = 0$. In principle then it is possible to reach a solution by first finding $C_n(x_n)$, given that $C_{n+1}(x_{n+1}) = 0$, then finding $C_{n-1}(x_{n-1})$ from a knowledge of $C_n(x_n)$, and so on until $C_1(x_1)$ is obtained. This is rather cumbersome, and some qualitative results would be useful.

This model has the following properties, which are stated without proof but can be found in the work of Arrow, Karlin, and Scarf (see Selected References at the end of the chapter):

1. The optimal ordering policy in each period is a single critical number $y_{1,n}^0$, $y_{2,n}^0, \ldots, y_{n-1,n}^0, y_{n,n}^0$, where the second subscript is used to indicate that an n-period horizon is being considered.
2. The critical numbers are monotonic, that is,
$$y_{n,n}^0 \le y_{n-1,n}^0 \le \cdots \le y_{2,n}^0 \le y_{1,n}^0 \le \cdots \le y^0,$$
where y^0 satisfies
$$L'(y^0) + c(1 - \alpha) = 0.$$
3. To indicate that an n-period horizon is being considered, let the minimum expected cost of following an optimal policy from the beginning of the first period to the end period n, given that there are x units on hand, now be denoted by $C_{1,n}(x)$. Then for every x the sequence $C_{1,n}(x)$ converges. Hence, one may define a function
$$C(x) = \lim_{n\to\infty} C_{1,n}(x).$$
4. The limit function satisfies the equation
$$C(x) = \min_{y\ge x}\{c(y - x) + L(y) + \alpha \int_0^\infty C(y - \xi)\varphi_D(\xi)\, d\xi\}.$$
5. The value of y that satisfies (4) is the maximum of y^0 and x, where y^0 is obtained from the equation in (2).
6. $\lim_{n\to\infty} y_{1,n}^0 = y^0$.
7. $\lim_{n\to\infty} |(y^0 - y_{1,n}^0)/\alpha^n| \le 2c/L''(y^0)$. This indicates how fast the $y_{1,n}^0$ converge to y^0.

The solution referred to in (5) for the infinite-stage model, which is given in (2), is readily obtained if one is willing to assume that the form of the solution is a single critical number and $x \le y$.[1] The total cost can be written as

$$c(y - x) + L(y) + \alpha[cD_1 + L(y)] + \alpha^2[cD_2 + L(y)] + \cdots + \alpha^i[cD_i + L(y)] + \cdots,$$

where D_i is the demand during the ith period. Since $D_i \ge 0$, one always orders up to y, and, hence the expected holding and shortage cost for each period is $L(y)$. The total expected cost is then obtained as

$$cy - cx + L(y)(1 + \alpha + \alpha^2 + \cdots) + cE(D)\alpha[1 + \alpha + \alpha^2 + \cdots]$$

or

$$cy - cx + \frac{L(y)}{1-\alpha} + \frac{cE(D)\alpha}{1-\alpha}.$$

Taking the derivative with respect to y and setting it equal to zero lead to

$$L'(y^0) + c(1 - \alpha) = 0.$$

[1] If $x > y$ and $E(D) > 0$, the law of large numbers ensures that eventually $x \le y$.

For the example of the two-period model, it is of interest to find y^0 for the infinite-period model. α is now included and is chosen equal to 0.75:

$$L'(y^0) = h\Phi(y^0) - p[1 - \Phi(y^0)] = \frac{10}{10}y^0 - 15\left[1 - \frac{y^0}{10}\right] = \frac{5}{2}y^0 - 15.$$

Therefore y^0 satisfies

$$\tfrac{5}{2}y^0 - 15 = -10(0.25)$$

or

$$y^0 = 5.$$

Hence, the infinite-period model says that if the amount on hand does not exceed 5, order up to 5 units. Furthermore, from property (2), for any finite-stage model the critical numbers are always less than or equal to 5.

Just as in the case of the two-period model, the multiperiod model has been solved by using the same penalty shortage cost for each period, even though unsatisfied demand is backlogged at the end of each period, except at the end of the nth period, where it is assumed to be lost. Again, this is a deficiency of the model as presented, but it can be remedied by either using a different penalty shortage cost in the last period or by letting $C_{n+1}(x_{n+1})$ denote the additional cost of the lost demand. In either case the dynamic programming model can be solved, at least in principle. Of course the qualitative result presented above about the form of the optimal policy may no longer be valid, but the optimal policy will still be represented by single critical numbers. Furthermore, if the original formulation of the multiperiod model is retained, and if the number of periods is large, then the effect of the handling of the shortage cost incurred in the last period may be negligible because of discounting.

A Variation of the Multiperiod Inventory Model with No Setup Cost: There is another interesting formulation due to Veinott, Jr.,[1] which avoids the aforementioned problem with the shortage cost and also leads to the same optimal policy as obtained before. Consider the n-period model presented earlier, but now assume that stock left over at the end of the final period can be salvaged with a return of the initial purchase cost c. Similarly, if there is a shortage at this time, the items are supplied also at the purchase price c. These two changes could lead to more realism in the model, but, as important, they lead to rather simple optimal policies. In particular, the optimal ordering policy for the n- and infinite-period models is a single critical number policy of the form

$$\begin{cases} \text{order up to } y^0, \text{ if } x_i < y^0 \\ \text{do not order, } \quad \text{if } x_i \geq y^0 \end{cases} \qquad \text{for all } i,$$

where y^0 satisfies the equation

$$L'(y^0) + c(1 - \alpha) = 0.$$

These results are rather striking because a slight modification in the model leads to the same single critical number for *all* periods, both finite and infinite. It should also be noted that for the infinite-stage model the two models, with or without including salvage value, are equivalent because of the discount factor.

[1] A. F. Veinott, Jr., "The Optimal Inventory Policy for Batch Orderings," *Operations Research*, **13**(3):424–432, 1965.

In deriving the results for this model, recall that stock left over at the end of period n can be salvaged with a return of the initial purchase cost c. Similarly, if there is a shortage at this time, the items are supplied also at the purchase price c. It will again be assumed that the expected shortage plus holding penalty cost $L(y)$ is strictly convex. If y_i and x_i are the amount of stock ordered up to and the amount of stock available, respectively, at the beginning of the ith period, the total cost of producing items (including salvage or backlogging at the end of the horizon but not shortage and holding costs) can be written as

$$c(y_1 - x_1) + \alpha c(y_2 - x_2) + \cdots + \alpha^{n-1}c(y_n - x_n) - \alpha^n c(y_n - D_n)$$
$$= \sum_{i=1}^{n} \alpha^{i-1}c(y_i - x_i) - \alpha^n c(y_n - D_n),$$

where D_n represents the demand during the last period. In general let D_i be the demand during the ith period, so that

$$x_i = y_{i-1} - D_{i-1}, \qquad \text{for } i = 2, 3, \ldots, n.$$

The total cost of producing the items may be written as

$$c(y_1 - x_1) + \sum_{i=2}^{n} \alpha^{i-1}c(y_i - y_{i-1} + D_{i-1}) - \alpha^n \cdot c(y_n - D_n)$$
$$= cy_1 - cx_1 + \sum_{i=2}^{n} \alpha^{i-1}cy_i - \sum_{i=2}^{n} \alpha^{i-1}cy_{i-1}$$
$$+ \sum_{i=2}^{n} \alpha^{i-1}cD_{i-1} - \alpha^n cy_n + \alpha^n cD_n$$
$$= -cx_1 + \sum_{i=1}^{n} \alpha^{i-1}cy_i - \sum_{i=2}^{n+1} \alpha^{i-1}cy_{i-1} + \sum_{i=2}^{n+1} \alpha^{i-1}cD_{i-1}$$
$$= -cx_1 + (1 - \alpha)\sum_{i=1}^{n} \alpha^{i-1}cy_i + \sum_{i=1}^{n} \alpha^i cD_i.$$

The expected total cost of production is then given by

$$c\left\{-x_1 + (1 - \alpha)\sum_{i=1}^{n} \alpha^{i-1}y_i + E\left[\sum_{i=1}^{n} \alpha^i D_i\right]\right\}$$
$$= c(1 - \alpha)\sum_{i=1}^{n} \alpha^{i-1}y_i + \left\{-cx_1 + cE(D)\sum_{i=1}^{n} \alpha^i\right\}.$$

Notice that the bracketed quantity on the right, in the expression for the expected total cost of production, in no way depends upon the decision variables $y_1, \ldots, y_n$ and hence may be ignored in the minimization that follows.

Since the expected shortage plus holding cost in period i is given by $L(y_i)$, the total expected cost is given by

$$\sum_{i=1}^{n} \alpha^{i-1}G(y_i) + \left[-cx_1 + cE(D)\sum_{i=1}^{n} \alpha^i\right],$$

where $$G(y_i) = (1 - \alpha)cy_i + L(y_i).$$

To minimize this total expected cost, one need only minimize

$$\sum_{i=1}^{n} \alpha^{i-1} G(y_i).$$

If it can be shown that this sum is minimized term by term, that is,

$$\min_{y \geq x} G(y) = \begin{cases} G(y^0), & \text{if } y^0 > x \\ G(x), & \text{if } y^0 \leq x, \end{cases}$$

where y^0 is defined through

$$G'(y^0) = (1 - \alpha)c + L'(y^0) = 0,$$

the following result holds:

The optimal ordering policy for the n- and infinite-period models is a single critical number policy of the form

$$\begin{cases} \textit{order up to } y^0, & \textit{if } x_k < y^0 \\ \textit{do not order}, & \textit{if } x_k \geq y^0 \end{cases} \qquad \textit{for all } k,$$

where y^0 satisfies the equation

$$L'(y^0) + c(1 - \alpha) = 0.$$

To show that y^0 does indeed have this property, note that under the assumption that $L(y)$ is strictly convex $G(y)$ must also be strictly convex and appear as shown in Fig. 9.10. Now, consider any policy other than the single critical number policy described above. Let $y_n, y_{n-1}, \ldots, y_1$ be the inventory levels associated with it and $y'_n, y'_{n-1}, \ldots, y'_1$ the inventory levels associated with the single critical number policy.

Case 1: Suppose $y'_i \leq y^0$.
The single critical number policy calls for ordering up to y^0, so that $G(y'_i) = G(y^0)$.
Hence

$$G(y'_i) \leq G(y_i).$$

Case 2: Suppose $y'_i > y^0$.
If $y'_i > y^0$, so is every other earlier y'_i. In fact,

$$y'_1 \geq y'_2 \geq \cdots \geq y'_i.$$

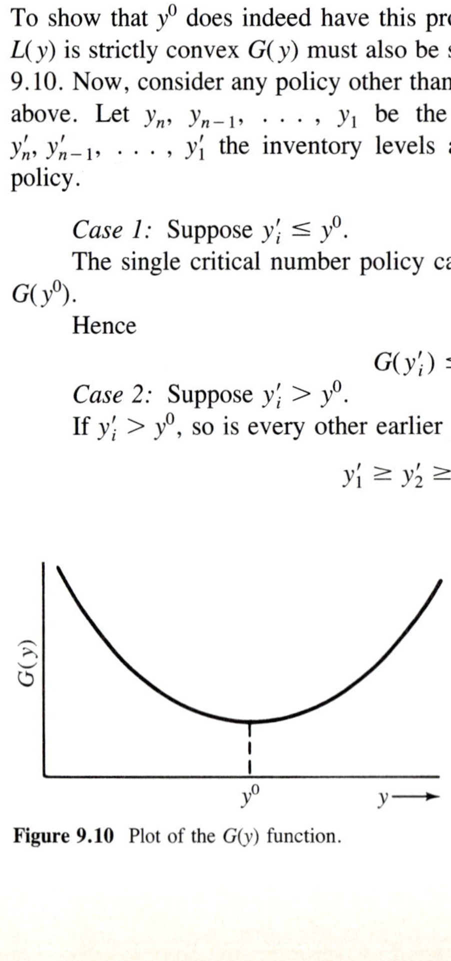

Figure 9.10 Plot of the $G(y)$ function.

Hence, no ordering ever took place under the single critical number policy, so that $x_i' = x_1 - D_1 - D_2 - \cdots - D_{i-1}$.

Furthermore,

$$x_i = x_1 - D_1 - D_2 - \cdots - D_{i-1} + \text{any amount ordered,}$$

so that

$$y_i \geq x_i \geq x_i' = y_i'$$

or

$$y_i \geq y_i'.$$

Hence

$$G(y_i) \geq G(y_i').$$

In both cases the inventory levels associated with following the single critical number policy lead to total expected costs which are no greater (and possibly smaller) than that obtained by following a different policy. Hence, the single critical number policy must be optimal.

The results easily extend to the infinite-case model.

Dynamic Inventory Models with Lead Times: Up to now it has been assumed that stock is obtained instantaneously after ordering. This assumption is now relaxed, and a lead time, λ, between the placing and receiving of an order is considered. λ is assumed constant over time and independent of the size of the order. Again, backlogging is permitted. One-period costs are a function of the stock on hand at the beginning of the period *after* delivery of old orders but before the placing of a new one. Under these assumptions, it can be shown that the optimal policy can be obtained from the dynamic programming expressions

$$C_i(u_i) = \underset{y_i \geq u_i}{\text{minimum}} \left\{ K\delta(y_i - u_i) + c(y_i - u_i) + \alpha^\lambda \int_0^\infty L(y_i - \eta)\varphi_D^{(\lambda)}(\eta)\,d\eta \right.$$
$$\left. + \alpha \int_0^\infty C_{i+1}(y_i - \xi)\varphi_D(\xi)\,d\xi \right\},$$

where

$$K\delta(y_i - u_i) = \begin{cases} K, & \text{if } y_i > u_i \\ 0, & \text{if } y_i = u_i, \end{cases}$$

and u_i represents the amount of stock on hand at the beginning of period i, plus the amount on order in the next $\lambda - 1$ periods. $\varphi_D^{(\lambda)}(\eta)$ is the density of the sum of λ demands, where demands are independent, identically distributed random variables each having density $\varphi_D(\xi)$.

It is evident that $C_i(u_i)$ is the usual zero lead time equation with a modified expected shortage plus holding cost $\mathcal{L}(y)$, where

$$\alpha^\lambda \mathcal{L}(y) = \alpha^\lambda \int_0^\infty L(y - \eta)\varphi_D^{(\lambda)}(\eta)\,d\eta.$$

The optimal polices for this model are exactly the same as the zero lead time models given previously. For example, for the infinite-state dynamic inventory model with no setup cost, the critical number is obtained from the solution of

$$L'(y^0) + c(1 - \alpha) = 0.$$

The solution to this problem with a λ-period lead time is found from

$$\alpha^\lambda \mathcal{L}'(y^0) + c(1 - \alpha) = 0,$$

where $\alpha^{(\lambda)}\mathcal{L}(y)$ is obtained as indicated above. If $\varphi_D(\xi)$ is exponential with parameter μ, then $\varphi_D^{(\lambda)}(\eta)$ has a gamma distribution with parameters μ and λ.

Continuous Review Model with Fixed Delivery Lag—Backlogging: In Sec. 9.3, a deterministic continuous review model, i.e., the economic lot-size model, was considered. The demand was assumed to be continuous, and required at a known constant rate. This model was classified as *continuous review* in that the inventory was continuously monitored, and orders were placed at any time; i.e., orders were placed whenever the inventory level reached the order point (or the trigger point). This ordering procedure is in contrast to the models considered in this section, which assume stochastic demand and periodic review; i.e., the inventory is monitored at certain times, and orders are placed at only these times. The model considered now is analogous to the economic lot-size model (continuous review), but where the demand for the item is stochastic and there is a fixed delivery lead time before an order is received. Only (s, S)-type policies[1] are considered; i.e., when the inventory falls to a level s, an order is placed to bring the inventory up to S (a quantity $Q = S - s$ is ordered). This model is often called a lot-size reorder-point model; a quantity Q is ordered whenever the inventory level reaches the reorder level s. Unsatisfied demand will be assumed to be filled immediately upon replenishment of the inventory; i.e., unsatisfied demand will be backlogged. In such models it is useful to consider the inventory level as the *inventory position*. The **inventory position** is defined as the amount on hand plus the amount ordered less the back-orders. Using the inventory position as the measure of inventory avoids the problem that may occur when the received order quantity fails to bring the on-hand inventory above the reorder point.

The model can be described in detail as follows. Inventory is stockpiled and used as demand dictates. When the *inventory position* reaches s, an order is placed for Q units to bring the *inventory position* up to level S. There is a fixed delivery lead time of length λ before the order is received, and the demand for items from inventory during the time λ is assumed to be a continuous random variable, D, having probability density function denoted by $\varphi_D(\xi)$. The mean demand is assumed equal to $a\lambda$; that is,

$$E(D) = a\lambda,$$

where a is the expected number of items demanded per unit of time.

Figure 9.11 illustrates how the inventory level varies over time. Note that this diagram can be viewed as a series of cycles, with a cycle defined as the time between receipt (or placement) of consecutive orders. Also, note that if the demand during the period λ is large, it is possible to have a negative inventory level. It is assumed that this unsatisfied demand will be backlogged. In Fig. 9.11, the inventory position is represented by the dashed lines during the period of the delivery lead time, and by the solid lines at other times.

The costs to be considered are the ordering cost K, the cost of the units ordered, cQ (both charged at the time of the ordering), an inventory holding cost of h dollars per item per unit of time, and a shortage cost of p dollars for each unit of demand unfilled and independent of the duration of the shortage. The inventory policy is to track the inventory position so that when the inventory position reaches s, an order of size Q is placed; this order will be delivered after a period of length λ. The problem,

[1] Under *Remark 3* after the economic lot-size model, it was pointed out that these models can be viewed as a special case of an (s, S) policy.

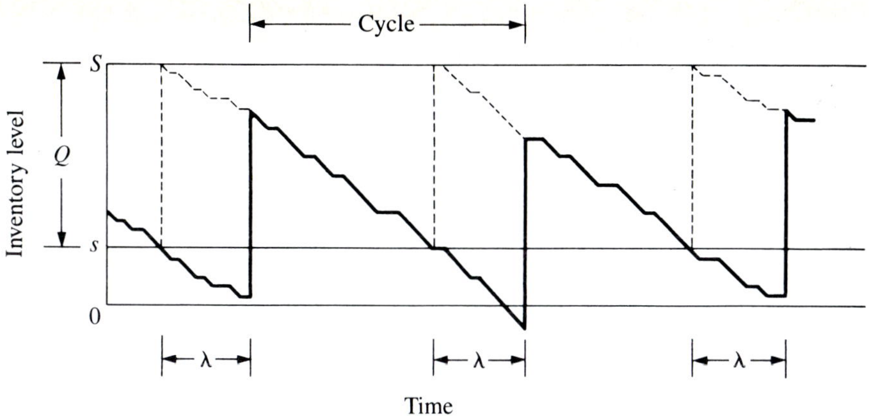

Figure 9.11 Diagram of inventory level as a function of time.

then, is to determine when to place an order (find the order point or trigger point, s) and to determine what size it should be (find the order quantity, Q), so that the expected total cost per unit time is a minimum.

The expected total cost per unit time [$C(Q, s)$] consists of the sum of the following three components: the expected ordering cost per unit of time [$E(OC)$], the expected holding cost per unit of time [$E(HC)$], and the expected shortage cost per unit of time [$E(SC)$]; that is,

$$C(Q, s) = E(OC) + E(HC) + E(SC).$$

In order to evaluate the terms of this expression, we assume initially that λ is sufficiently small that there is never more than a single order outstanding and that the reorder point, s (based on the inventory position), is always nonnegative. The first assumption guarantees that the inventory on hand when an order is received will always fall above the reorder point because, otherwise, more than one order would be outstanding. If p/h is sufficiently large, as is usually the case in practice, these assumptions are generally satisfied.

The expected ordering cost per unit time, $E(OC)$, is simply the ordering cost incurred per cycle times the expected number of cycles per unit of time. The ordering cost incurred per cycle is

$$K + cQ.$$

In order to find the expected number of cycles per unit of time, assume that the unit of time is, say, a year. Suppose that Q is 500 units, the lag in delivery time is 1 month ($\frac{1}{12}$ year), and the expected demand in 1 month is 100 units. These suppositions imply that $a = 1{,}200$ since a is defined as the expected demand per year (the unit of time chosen). It is evident that the expected time to deplete 500 units of inventory is $\frac{5}{12}$ year (500/1,200). Because a cycle is defined as the time between the receipt of orders of 500 units, the expected cycle length is equal to $\frac{5}{12}$ year, so that in 1 year the expected number of cycles is $\frac{12}{5} = 2.4$. In general, the expected number of cycles per unit time is given by a/Q. The expression for the expected ordering cost, $E(OC)$, is then given by

$$E(OC) = \left(\frac{a}{Q}\right)(K + cQ).$$

The expected holding cost per unit time is equal to the expected inventory holding cost per cycle times the expected number of cycles per unit time (already obtained as a/Q). The expected holding cost per cycle is the expected holding cost per unit of inventory for a cycle times the average amount of inventory held during the cycle.

The expected holding cost per unit of inventory for a cycle is the unit holding cost, h, times the expected length of a cycle Q/a, that is, hQ/a. The average amount of inventory held during a cycle can be obtained by averaging the on-hand inventory at the beginning and end of a cycle. From Fig. 9.10, the expected inventory level at the beginning of the cycle is given by $S - a\lambda$, and the expected inventory level at the end of the cycle is $s - a\lambda$ [assuming the expected number of stockouts (negative inventory) can be neglected, which is reasonable if the back-order state during a cycle is small compared to the length of the cycle]. Hence the average amount of inventory during a cycle can be approximated by

$$\frac{(S - a\lambda) + (s - a\lambda)}{2} = \frac{Q + s - a\lambda + s - a\lambda}{2} = \frac{Q}{2} + s - a\lambda,$$

so that $$E(HC) = \left(\frac{a}{Q}\right)\left(\frac{hQ}{a}\right)\left(\frac{Q}{2} + s - a\lambda\right) = h\left(\frac{Q}{2} + s - a\lambda\right).$$

The expected shortage cost per unit time, $E(SC)$, is simply the expected shortage cost incurred per cycle times the expected number of cycles per unit time (already obtained as a/Q). The expected shortage cost per cycle is just p times the expected number of shortages that occur during the lag period λ (a shortage can occur only when the demand during the delivery lead time exceeds s); i.e.,

$$p \int_s^\infty (\xi - s)\phi_D(\xi)\, d\xi,$$

$$E(SC) = \left[\left(\frac{a}{Q}\right)\left(p \int_s^\infty (\xi - s)\phi_D(\xi)\, d\xi\right)\right].$$

Adding the expressions for $E(OC)$, $E(HC)$, and $E(SC)$ leads to

$$C(Q, s) = \frac{aK}{Q} + ac + h\left[\left(\frac{Q}{2}\right) + s - a\lambda\right] + \left(\frac{pa}{Q}\right)\int_s^\infty (\xi - s)\phi_D(\xi)\, d\xi.$$

Because there are two decision variables (Q and s), the optimum values (Q^* and s^*) are found by minimizing $C(Q, s)$ with respect to the variables Q and s, that is, setting the partial derivatives $\partial C(Q, s)/\partial Q$ and $\partial C(Q, s)/\partial s$ equal to zero. Thus

$$\frac{\partial C(Q, s)}{\partial Q} = \frac{-aK}{Q^2} + \frac{h}{2} - \frac{pa\left[\int_s^\infty (\xi - s)\varphi_D(\xi)\, d\xi\right]}{Q^2} = 0$$

$$\frac{\partial C(Q, s)}{\partial s} = h - \frac{pa\left[\int_s^\infty \varphi_D(\xi)\, d\xi\right]}{Q} = 0.$$

Solving these equations simultaneously[1] leads to

$$(1) \qquad Q^* = \sqrt{\frac{\left\{2a\left[K + p\int_{s^*}^{\infty}(\xi - s^*)\varphi_D(\xi)\,d\xi\right]\right\}}{h}},$$

$$(2) \qquad \int_{s^*}^{\infty}\varphi_D(\xi)\,d\xi = \frac{hQ^*}{pa}.$$

Unfortunately, solving these equations simultaneously and obtaining a general closed-form expression for Q^* and s^* is not possible, so an iterative procedure is desirable. Such a procedure is as follows.

1. As an initial step, assume that p equals zero and obtain a value of Q from Eq. (1). Note that this equation is just the expression for Q in the deterministic economic lot-size model when shortages are not permitted.
2. Solve for s in Eq. (2) using the value of Q found in step 1.
3. Using the value of s found in step 2, solve for a new Q using Eq. (1).
4. Repeat steps 2 and 3 until successive values of Q and s are sufficiently close.

In practice, this procedure will generally converge in just a few iterations.

Remarks: Several remarks can be made about this continuous review model.

1. Note that in Eq. (2), the integral $\int_{s^*}^{\infty}\varphi_D(\xi)\,d\xi$ is just the probability that the random variable demand, D, during the lead time exceeds s^*, that is, $P\{D > s^*\}$. Hence, (hQ^*/pa) must fall between 0 and 1. If the algorithm ever leads to a value of $(hQ/pa) > 1$, this is an indication that the shortage cost (relative to the holding cost) is too small, with the result that the number of back-orders during a cycle will tend to be large. This will contradict the approximation of neglecting the number of stockouts made in the calculation of the expected holding cost, so that the derived formulas become inappropriate.

2. If the lead time is close to the average cycle length, Q/a, more than one order may be outstanding. Using the inventory position as the measure of inventory still leads to an operational rule, i.e., order when the inventory position reaches s. However, the number of stockouts may become too large to be neglected in the expected holding cost calculations.

3. The quantity $(s - a\lambda)$ is known as the safety stock, and represents "protection" against a stockout during the lead time. The probability of a stockout is simply $P\{D > s\} = \int_{s}^{\infty}\varphi_D(\xi)\,d\xi$ and can be obtained from hQ^*/pa when the optimal reorder point and lot size are used.

4. Because there is no closed-form solution for Eqs. (1) and (2), it is worth considering some special cases of distribution of demand. If the density function of

[1] Note that the optimum values of Q and s are independent of c, the unit cost of the items ordered. A little reflection indicates the total number of units ordered is independent of the values of Q and s, and, therefore, can be neglected in determining the optimum values of these parameters.

demand is uniform over the range from 0 to t, that is,

$$\varphi_D(\xi) = \begin{cases} \dfrac{1}{t}, & \text{if } 0 \le \xi \le t \\ 0, & \text{otherwise,} \end{cases}$$

then Eq. (2) leads to

$$\int_s^\infty \varphi_D(\xi)\, d\xi = 1 - \frac{s}{t},$$

so that
$$s^* = \frac{t(pa - hQ^*)}{pa} = t\left(1 - \frac{hQ^*}{pa}\right).$$

Furthermore, from Eq. (1),

$$\int_s^\infty (\xi - s)\varphi_D(\xi)\, d\xi = \frac{t}{2} + \frac{s^2}{2t} - s$$

and
$$Q^* = \sqrt{\frac{(2aK + apt + aps^{*2}/t - 2aps^*)}{h}}.$$

Finally, if the density function of demand is exponential, i.e.,

$$\varphi_D(\xi) = \left(\frac{1}{a\lambda}\right) e^{-\xi/a\lambda}, \qquad \text{for } \xi > 0$$

then Eq. (2) leads to

$$\int_s^\infty \varphi_D(\xi)\, d\xi = e^{-s/a\lambda},$$

so that
$$s^* = -a\lambda \ln\left(\frac{hQ^*}{pa}\right).$$

Furthermore, from Eq. (1),

$$\int_s^\infty (\xi - s)\varphi_D(\xi)\, d\xi = a\lambda e^{-s/a\lambda},$$

and
$$Q^* = \sqrt{\frac{2aK + 2a^2\lambda pe^{-s^*/a\lambda}}{h}} = \sqrt{\frac{2a}{h}}(K + a\lambda pe^{-s^*/a\lambda}).$$

Example: Consider the speaker example presented in Sec. 9.1. It was assumed that $K = 12{,}000$, $h = 0.30$ per speaker per month, and $a = 8{,}000$ per month. In this example, the demand was assumed to be fixed at the rate of 8,000 speakers per month. Now we assume that the demand is random and has a uniform density over the range from 0 to 16,000, with the unit of time considered to be 1 month and the delivery lead time also 1 month. Note that the expected demand is 8,000 speakers, so that a is again 8,000 and λ is 1. The penalty for unsatisfied demand will be \$5 per speaker. Using the results just obtained for this distribution, with the iterative procedure step 1 leads to (setting p equal to 0)

$$Q = \sqrt{\frac{2(8{,}000)(12{,}000)}{0.30}} = 25{,}298.$$

Using this value of Q, solve for s; that is,

$$s = 16{,}000\left[1 - \frac{(0.30)(25{,}298)}{(5)(8{,}000)}\right] = 12{,}964.$$

Using this value of s, solve for Q; that is,

$$Q = \sqrt{\frac{(8{,}000)\left\{2(12{,}000) + (5)(16{,}000) + \left[\frac{5(12{,}964)^2}{16{,}000}\right] - 2(5)(12{,}964)\right\}}{0.30}}$$

$$= 26{,}773.$$

Using this value of Q, solve for s; that is,

$$s = 16{,}000\left[1 - \frac{(0.30)(26{,}773)}{(5)(8{,}000)}\right] = 12{,}787.$$

Three more iterations lead to (Q = 26,945, s = 12,767), (Q = 26,965, s = 12,764), and finally, to (Q^* = 26,968, s^* = 12,764). The probability of a stockout is given by

$$P\{D > s^*\} = \frac{hQ^*}{pa} = 0.20.$$

Consider this same example, but now assume that the distribution of demand is exponential with mean 8,000. Step 1 in the iterative procedure again leads to Q = 25,298. Using this value of Q, solve for s; that is,

$$s = -8{,}000 \ln \frac{(0.30)(25{,}298)}{(5)(8{,}000)} = 13{,}297.$$

Using this value of s, solve for Q; that is,

$$Q = \sqrt{\frac{2(8{,}000)}{(0.30)}[12{,}000 + (8{,}000)(5)e^{-(13{,}297)/8{,}000}]} = 32{,}323.$$

Several further iterations lead to the final solution of (Q^* = 34,532, s^* = 10,808). The probability of a stockout is given by

$$P\{D > s^*\} = \frac{hQ^*}{pa} = 0.26.$$

Continuous Review Model with Fixed Delivery Lag—No Backlogging: This model is similar to the previous model except that unsatisfied demand will be assumed to be lost, i.e., unsatisfied demand will not be backlogged, and now lost revenue will be included in the unsatisfied demand penalty cost. The derivation of the costs contains the same assumptions that were made in the backlogging case, so the subsequent expressions will lead to approximate results. The expected ordering cost per unit of time, $E(OC)$, and the expected shortage cost per unit of time, $E(SC)$, are the same for both models. The only cost that differs is the expected holding cost per unit of time, $E(HC)$, and the key to its determination is to note that the average amount of inventory held during the cycle differs by the expected number of shortages that occur during the lag period λ; i.e., the no-backlogging case has an expected inventory level per cycle greater than the backlogging cost by an amount equal to the expected number of shortages, which is given by $\int_s^\infty (\xi - s)\varphi_D(\xi)\, d\xi$. Therefore, the expected holding cost per unit of time, $E(HC)$, can be written as

$$E(HC) = h\left[\left(\frac{Q}{2}\right) + s - a\lambda + \int_s^\infty (\xi - s)\varphi_D(\xi)\, d\xi\right].$$

The expected total cost per unit of time, $C(Q, s)$, can be expressed as

$$C(Q, s) = \frac{aK}{Q} + ac + h\left[\left(\frac{Q}{2}\right) + s - a\lambda + \int_s^\infty (\xi - s)\varphi_D(\xi)\, d\xi\right] + \left(\frac{pa}{Q}\right)\int_s^\infty (\xi - s)\varphi_D(\xi)\, d\xi.$$

Because there are two decision variables (Q and s), the optimum values (Q^* and s^*) are found by minimizing $C(Q, s)$ with respect to the variables Q and s, that is, setting the partial derivatives $\partial C(Q, s)/\partial Q$ and $\partial C(Q, s)/\partial s$ equal to zero. Thus,

$$\frac{\partial C(Q, s)}{\partial Q} = \frac{-aK}{Q^2} + \frac{h}{2} - \frac{pa\left[\int_s^\infty (\xi - s)\varphi_D(\xi)\, d\xi\right]}{Q^2} = 0$$

$$\frac{\partial C(Q, s)}{\partial s} = h\int_0^s \varphi_D(\xi)\, d\xi - \frac{pa\left[\int_s^\infty \varphi_D(\xi)\, d\xi\right]}{Q} = 0.$$

Solving these equations simultaneously[1] leads to

(3) $$Q^* = \sqrt{\frac{\left\{2a\left[K + p\int_{s^*}^\infty (\xi - s^*)\varphi_D(\xi)\, d\xi\right]\right\}}{h}},$$

(4) $$\int_{s^*}^\infty \varphi_D(\xi)\, d\xi = \frac{hQ^*}{hQ^* + pa}.$$

Unfortunately, solving these equations simultaneously and obtaining a general closed-form expression for Q^* and s^* is not possible, so an iterative procedure is desirable. Such a procedure is as follows.

[1] Note that the optimum values of Q and s are independent of c, the unit cost of the items ordered. A little reflection indicates the total number of units ordered is independent of the values of Q and s, and, therefore, can be neglected in determining the optimum values of these parameters.

1. As an initial step, assume that p equals zero and obtain a value of Q from Eq. (3). Note that this equation is just the expression for Q in the deterministic economic lot-size model when shortages are not permitted.
2. Solve for s in Eq. (4) using the value of Q found in step 1.
3. Using the value of s found in step 2, solve for a new Q using Eq. (3).
4. Repeat steps 2 and 3 until successive values of Q and s are sufficiently close.

In practice, this procedure will generally converge in just a few iterations.

MULTIPRODUCT INVENTORY MODELS: The previous sections dealt with inventory systems for single-product models, but most real inventory systems involve many products with various types of interactions, such as joint storage and budget limitations and product substitutability. One important reason for studying single-product models first is that they provide insight into solving multiproduct problems, and furthermore, it is often possible to "factor" an N-product problem into N one-product problems without loss of optimality. (This factoring can be done if the demand and cost for each product can be treated independently of the other products.) There has been some work on multiproduct models in which such factorization is not possible. In these models, stocks of a single product at different locations or echelons of a supply system can also be conveniently viewed as stocks of different products.[1] A multiproduct model proposed by A. F. Veinott, Jr.,[2] is a direct analog of the single-product multiperiod inventory model with no setup cost as presented earlier. This multiproduct model considers N products and m different classes of demands for these products. The demand classes in a period may be classified by such characteristics as time of occurrence, essentiality, products desired, and acceptable substitutes. Under the usual restrictions on the form of the costs, the optimal policy is a single critical number policy for each product; i.e., if x_{ik} is the amount on hand of product k at period i, order up to y_{ik} if $x_{ik} < y_{ik}$; otherwise, do not order (assuming initially that $x_{1k} \leq y_{1k}$ for all k).

This model has many useful applications. For example, suppose there are two products, with product 1 serving as a substitute for product 2, and all unsatisfied demands are lost. If we choose for this model a stocking policy that supplies unsatisfied demand for product 2 with excess stock from product 1 (if available), we can obtain the optimal critical numbers. This example can be interpreted as a two-echelon inventory model with demands for a single product that cannot be satisfied at echelon 2 being transmitted up to echelon 1.

A second application concerns two products that serve as substitutes for each other. Again, all unsatisfied demands are assumed to be lost. The stocking policy supplies the unsatisfied demand for one product with the excess stock from the other (if available). This example can be interpreted as a two-location inventory model with an end-of-period redistribution of excess stock (if any) at one location to satisfy a shortage (if one exists) at the other location.

[1] An excellent summary is given in a paper by A. F. Veinott, Jr.: "The Status of Mathematical Inventory Theory," *Management Science,* **12**(11):745–777, 1966.

[2] Veinott, A. F., Jr.: "Optimal Policy for a Multi-Product, Dynamic Non-Stationary Inventory Problem," *Management Science,* **12**(3):206–222, 1965.

A different variation of the two-product inventory models was presented by D. Iglehart.[1] Inventories of product 2 are maintained to provide capability for production of product 1. For example, if product 1 is a car, product 2 might be machinery or labor. An optimal policy describes the amount of product 1 and product 2 that must be produced in period i to minimize the total cost, subject to certain constraints.

9.5 Conclusions

The inventory models presented here are rather simplified, but they serve the purpose of introducing the general nature of inventory models. Furthermore, they are sufficiently accurate representations of many actual inventory situations so that they frequently are useful in practice. For example, the economic lot-size formulas have been particularly widely used, although they are sometimes modified to include some type of stochastic demand. The multiperiod models with stochastic demand have been important in characterizing the types of policies to follow, for example, (s, S) policies, even though the optimal values of s and S are difficult to obtain. Nevertheless, many inventory situations possess complications that must still be taken into account, e.g., interaction between products. Several complex models have been formulated in an attempt to fit such situations, but they still leave a wide gap between practice and theory. Continued growth is occurring in the computerization of inventory data processing, along with accompanying growth in scientific inventory management.

SELECTED REFERENCES

1. Arrow, K. J., S. Karlin, and H. Scarf: *Studies in the Mathematical Theory of Inventory and Production,* Stanford Univ. Press. Stanford, Calif., 1958.
2. Brown, R. G.: *Decision Rules for Inventory Management,* Holt, Rinehart and Winston, New York, 1967.
3. Buchan, J., and E. Koenigsberg: *Scientific Inventory Management,* Prentice-Hall, Englewood Cliffs, N.J., 1963.
4. Buffa, E. S.: *Modern Production/Operations Management,* 6th ed., Wiley, New York, 1980.
5. Buffa, E. S., and J. G. Miller: *Production-Inventory Systems: Planning and Control,* 3d ed., Richard D. Irwin, Homewood, Ill., 1979.
6. Hadley, G., and T. Whitin: *Analysis of Inventory Systems,* Prentice-Hall, Englewood Cliffs, N.J., 1963.
7. Hax, A., and D. Candea: *Production and Inventory Management,* Prentice-Hall, Englewood Cliffs, N.J., 1984.
8. Johnson, L. A., and D. C. Montgomery: *Operations Research in Production Planning, Scheduling and Inventory Control,* Wiley, New York, 1974.
9. Peterson, R., and E. A. Silver: *Decision Systems for Inventory Management and Production Planning,* Wiley, New York, 1979.
10. Starr, M., and D. Miller: *Inventory Control: Theory and Practice,* Prentice-Hall, Englewood Cliffs, N.J., 1962.

[1] Iglehart, D.: "Capital Accumulation and Production for the Firm: Optimal Dynamic Policies," *Management Science,* **12**(3):193–205, 1965.

1.* Suppose that the demand for a product is 30 units per month, and the items are withdrawn uniformly. The setup cost each time a production run is made is \$15. The production cost is \$1 per item, and the inventory holding cost is \$0.30 per item per month.

(*a*) Assuming shortages are not allowed, determine how often to make a production run and what size it should be.

(*b*) If shortages cost \$3 per item per month, determine how often to make a production run and what size it should be.

2. The demand for a product is 600 units per week, and the items are withdrawn uniformly. The items are ordered, and the setup cost is \$25. The unit cost of each item is \$3, and the inventory holding cost is \$0.05 per item per week.

(*a*) Assuming shortages are not allowed, determine how often to order and what size the order should be.

(*b*) If shortages cost \$2 per item per week, determine how often to order and what size the order should be.

3. Solve Prob. 2 with shortages permitted and assume a delivery lag of 1 week.

4.* Solve the economic lot-size model problem presented in Sec. 9.3 when shortages are permitted but when the cost is \$5 per speaker.

5. A taxi company uses gasoline at the rate of 8,500 gallons/month. The gasoline costs \$1.05/gallon, with a setup cost of \$1,000. The inventory holding cost is 1 cent/gallon/month.

(*a*) Assuming shortages are not allowed, determine how often and how much to order.

(*b*) If shortages cost 50 cents/gallon/month, determine how often and how much to order.

6. Solve Prob. (5*a*) by assuming that the cost of gasoline drops to \$1.00/gallon if at least 50,000 gallons are purchased.

7. Solve Prob. (5*a*) if the cost of gasoline is \$1.20/gallon for the first 20,000 gallons purchased, \$1.10 for the next 20,000 gallons, and \$1.00/gallon thereafter.

8. Consider the economic lot-size problem with shortages permitted, as presented in Sec. 9.3. Suppose, however, that it has been determined that $S/Q = 0.8$. Derive the expression for the optimal value of Q.

9. In the economic lot-size model, suppose the stock is replenished uniformly (rather than instantaneously) at the rate of b items per unit time until the lot size Q is fulfilled. Withdrawals from the inventory are made at the rate of a items per unit time, where $a < b$. Replenishments and withdrawals of the inventory are made simultaneously. For example, if Q is 60, b is 3 per day, and a is 2 per day, then 3 units of stock arrive each day until day 20. Units are withdrawn at the rate of 2 per day for 30 days, at which time 3 units of stock arrive each day for 20 days, etc. The diagram of inventory versus time is given below.

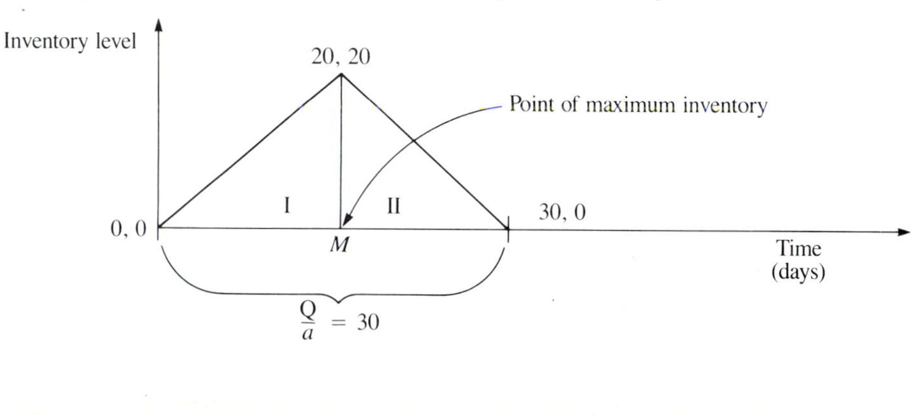

(*a*) Find the total cost per unit time in terms of the setup cost K, production number Q, unit cost c, holding cost h, withdrawal rate a, and replenishment rate b. (*Hint:* In determining the average inventory level, divide the cycle into the two intervals I and II shown in the diagram. Then find the average inventory level in each interval.)

(*b*) Determine the economic lot size.

10. Suppose the requirement for the next 5 months is given by $r_1 = 2$, $r_2 = 4$, $r_3 = 2$, $r_4 = 2$, and $r_5 = 3$. Items are ordered; the setup cost is \$4, the purchase cost is \$1, and the holding cost is \$0.30. Determine the optimal production schedule that satisfies the monthly requirements. Use dynamic programming.

11. Solve Prob. 10 by assuming that the production costs are given by \$3 $(1 + \log_e X)$, where X is the amount produced in a month.

12. Solve Prob. 10 by using the algorithm presented in Sec. 9.3.

13. Solve Prob. 11 by using the algorithm presented in Sec. 9.3.

14. Formulate Prob. 10 as an integer programming problem.

15.* Solve the production planning model for the production of speakers when the requirements are increased by 1 unit in each period.

16. Solve the production planning model for the production of speakers when the unit costs during the first and third periods are increased to \$1.40. Use dynamic programming.

17. Develop an algorithm to solve the production planning model that uses forward induction. (*Hint*: If an optimal policy is being followed and if the time of the *last* production is known for this optimal policy, the total cost of the previous periods must be a minimum for this reduced problem because the overall policy is optimal.)

18.* Consider a situation where a particular product is produced and placed in in-process inventory until it is needed in a subsequent production process. The number of units required in each of the next 3 months, as well as the setup cost and the regular-time unit production cost that would be incurred in each month, are

Month	Requirements	Setup Cost ($)	Regular-Time Unit Cost ($)
1	1	5	8
2	3	10	10
3	2	5	9

There currently is 1 unit in inventory, and we want to have 2 units in inventory at the end of the 3 months. A maximum of 3 units can be produced on regular-time production in each month, although one additional unit can be produced on overtime at a cost that is \$2 larger than the regular-time unit production cost. The cost of storage is \$2 per unit for each extra month that it is stored.

Use dynamic programming to determine how many units should be produced in each month to minimize total cost.

19. Consider a situation where a particular product is produced and placed in in-process inventory until it is needed in a subsequent production process. The number of units required in each of the next two months, as well as the setup cost, holding cost (charged as a function of excess of supply over requirement and charged at the end of the period), and regular-time unit production cost, are as follows:

Month	Requirement	Setup Cost ($)	Holding Cost ($)	Unit Cost ($)
1	3	5	0.30	9
2	4	5	0.30	9

Determine the optimal production schedule that satisfies the monthly requirements. Use the algorithm presented in Sec. 9.3.

20. A newspaper stand purchases newspapers for 18 cents and sells them for 25 cents. The shortage cost is 25 cents per newspaper (because the dealer buys papers at retail price to satisfy shortages). The holding cost is 0.1 cent. The demand distribution is a uniform distribution between 200 and 300. Find the optimal number of papers to buy.

21. Suppose the demand D for a spare airplane part has an exponential distribution with parameter $\frac{1}{50}$; that is,

$$\varphi_D(\xi) = \begin{cases} \frac{1}{50}e^{-\xi/50}, & \xi \geq 0 \\ 0, & \text{otherwise.} \end{cases}$$

This airplane will be obsolete in 1 year, hence all production is to take place at the present time. The production costs now are $1,000 per item—that is, $c = 1{,}000$—but they become $10,000 per item if they must be supplied at later dates—that is, $p = 10{,}000$. The holding costs, charged on the excess after the end of the period, are $300 per item.

(*a*) Determine the required number of spare parts.

(*b*) Suppose that the manufacturer has 23 parts already in inventory (from a similar, but now obsolete airplane). Determine the optimal inventory policy.

(*c*) Suppose that p cannot be determined now, but the manufacturer wishes to order a quantity so that the probability of a shortage equals 0.05. How many units should be ordered?

(*d*) If the manufacturer were following an optimal policy, but ordered the quantity in part (*b*), what is the implied value of p?

22.* A bread manufacturer distributes bread to grocery stores daily. The cost of the bread is 40 cents per loaf. The company sells the bread to the stores for 60 cents per loaf sold, provided that it is disposed of as fresh bread (sold on the day it is baked). Bread not sold is returned to the company. The company has a store outlet that sells bread that is a day or more old for 30 cents per loaf. This salvage cost represents the holding cost. The unsatisfied demand cost is estimated to be 40 cents per loaf. If the demand has a uniform distribution between 1,000 and 2,000 loaves, find the optimal daily number of loaves that the manufacturer should produce.

23. A student majoring in operations research enjoys optimizing his personal decisions. He is analyzing one such decision currently, namely, how much money to take out of his savings account (if any) to buy travelers' checks before leaving on a summer vacation trip to Europe.

He already has used the money he had in his checking account to buy travelers' checks worth $1,200, but this may not be enough. In fact, he has estimated the probability distribution of what he will need as shown in the following table:

Amount needed ($)	1,000	1,100	1,200	1,300	1,400	1,500	1,600	1,700
Probability	0.05	0.10	0.15	0.25	0.20	0.10	0.10	0.05

If he turns out to have less than he needs, then he will have to leave Europe 1 week early for every \$100 short. Because he places a value of \$150 on each week in Europe, each week lost would thereby represent a net imputed loss of \$50 to him. However, every \$100 travelers' check costs an extra \$1. Furthermore, *each* such check left over at the end of the trip (which would be redeposited in the savings account) represents a loss of \$2 in interest that could have been earned in the savings account during the trip, so he does not want to purchase too many.

Using these data, determine the optimal decision on how many additional \$100 travelers' checks (if any) the student should purchase from his savings account money.

24.* Find the optimal ordering policy for a one-period model, where the demand has a probability density

$$\varphi_D(\xi) = \begin{cases} \frac{1}{20}, & \text{if } 0 \le \xi \le 20 \\ 0, & \text{otherwise,} \end{cases}$$

and the costs are

$$\begin{aligned} \text{Holding} &= \$1 \text{ per item,} \\ \text{Shortage} &= \$3 \text{ per item,} \\ \text{Setup} &= \$1.50, \\ \text{Production} &= \$2 \text{ per item.} \end{aligned}$$

25. The campus bookstore must decide how many textbooks to order for a course that will be offered only once. The number of students who will take the course is a random variable D, whose distribution can be approximated by a (continuous) uniform distribution on the interval [40, 60]. After the quarter starts, the value of D becomes known. If D exceeds the number of books available, the known shortfall is made up by placing a rush order at a cost of \$14 plus \$2 per book over the normal ordering cost. If D is less than the stock on hand, the extra books are returned for their original ordering cost less \$1 each. What is the order quantity that minimizes the expected cost?

26. Consider the following inventory model, which is a single-period model with known density of demand $\varphi_D(\xi) = e^{-\xi}$, for $\xi > 0$ and zero elsewhere. There are two costs connected with the model: The first is the purchase cost, given by $c \cdot (y - x)$; the second is the unsatisfied demand cost, which is just a constant, p (independent of the amount of unsatisfied demand).

(*a*) If x units are available and goods are ordered up to y, write the expression for the expected loss, and describe completely the optimal policy.

(*b*) If a fixed cost K is also incurred whenever an order is placed, describe the optimal policy.

27. Using the approximation for finding the optimal policy for a single-period model when the density of demand has an exponential distribution, find this policy when

$$\varphi_D(\xi) = \begin{cases} \frac{1}{25} e^{-\xi/25}, & \text{if } \xi \ge 0 \\ 0, & \text{otherwise,} \end{cases}$$

and the costs are

$$\begin{aligned} \text{Holding} &= 40 \text{ cents per item,} \\ \text{Shortage} &= \$1.50 \text{ per item,} \\ \text{Purchase price} &= \$1 \text{ per item,} \\ \text{Setup} &= \$10. \end{aligned}$$

28.* There are production processes for which the difference between the cost of producing the maximum number of units allowed by some capacity restriction and the cost of producing any number of units less than this maximum is negligible; i.e., ordering is by batches. Consider a one-stage model, where the only two costs are holding costs given by

$$h(y - D) = \frac{3}{10}(y - D),$$

and the penalty cost of unsatisfied demand given by

$$p(D - y) = 2.5(D - y).$$

The density function for demand is given by

$$\varphi_D(\xi) = \begin{cases} \dfrac{e^{-\xi/25}}{25}, & \text{for } \xi \geq 0 \\ 0, & \text{otherwise.} \end{cases}$$

If you order, you must order in batches of 100 units, and this quantity is delivered instantaneously. Thus, if x denotes the quantity on hand, and if you do not order, then $y = x$. If you order one batch, then $y = x + 100$. Let $G(y)$ denote the total expected cost of this inventory problem when there are y units available for the period (after you have ordered).

(a) Write the expression for $G(y)$.

(b) What is the optimal ordering policy?

29. Consider the following inventory situation. Demands are independent with common density given by the following:

$$\varphi_D(\xi) = \begin{cases} \dfrac{e^{-\xi/25}}{25}, & \text{for } \xi \geq 0 \\ 0, & \text{otherwise.} \end{cases}$$

Orders may be placed at the start of each period without setup cost at a price of $c = 10$. There are a holding cost of 6 per unit remaining in stock at the end of each period and a penalty cost of 15 per unit quantity backlogged.

(a) Find the optimal one-period policy.

(b) Find the optimal two-period policy.

30. Consider the following inventory situation. Demands are independent with common density $\varphi_D(\xi) = \frac{1}{50}$, $0 < \xi < 50$. Orders may be placed at the start of each period without setup cost at a price of $c = 10$. There are a holding cost of 8 per unit remaining in stock at the end of each period and a penalty cost of 15 per unit quantity backlogged.

(a) Find the optimal one-period policy.

(b) Find the optimal two-period policy.

31.* Find the optimal inventory policy for the following two-period model by using a discount factor of $\alpha = 0.9$. Let the density of the demand D be given by

$$\varphi_D(\xi) = \begin{cases} \dfrac{1}{25} e^{-\xi/25}, & \text{if } \xi \geq 0 \\ 0, & \text{otherwise,} \end{cases}$$

and the costs are

Holding = 25 cents per item,
Shortage = \$2 per item,
Purchase price = \$1 per item.

Stock left over at the end of the final period is salvaged for \$1 per item, and shortages remaining at this time are also available at \$1 per item.

32. Solve Prob. 31 for a two-period model assuming no salvage value, no backlogging at the end of the second period, and no discounting.

33. Solve Prob. 31 for an infinite-period model by using a discount factor of $\alpha = 0.90$.

34.* Determine the optimum inventory policy when the goods are to be ordered at the end of every month from now on. The cost of ordering up to y when x is available is given by $2(y - x)$. Similarly, the cost of not satisfying a consumer demand of D is given by $5(D - y)$. The density function for the random variable, demand, is given by $\varphi_D(\xi) = e^{-\xi}$. The storage costs are given by $(y - D)$ and represent the expense of storing unsold stock. The losses at each succeeding stage are equivalent to a loss of 95 percent of that at the previous stage.

35. Solve the inventory problem given in Prob. 34 but assuming that the policy is to be used for only one year (a 12-stage model). Shortages are backlogged each month, except that any shortages remaining at the end of the year are made up by purchasing similar items at a unit cost of \$2. Any remaining inventory at the end of the year can be sold at a unit price of \$2.

36. A supplier of high-fidelity receiver kits is interested in using an optimal inventory policy. The distribution of demand per month is uniform between 2,000 and 3,000 kits. The cost of each kit is \$150. The holding cost is estimated to be \$2 per kit per month, and the unsatisfied demand cost is \$30 per kit per month. Using a discount factor of $\alpha = 0.90$, find the optimal inventory policy for this "infinite"-horizon problem.

37. The weekly demand for a certain type of electronic calculator is estimated to be given by

$$\varphi_D(\xi) = \begin{cases} \dfrac{1}{1{,}000} e^{-\xi/1{,}000}, & \xi \ge 0 \\ 0, & \text{otherwise.} \end{cases}$$

The unit cost of these calculators is \$8. The holding cost is 7 cents per calculator per week. The unsatisfied demand cost is \$2 per calculator per week. Using a discount factor of $\alpha = 0.95$, find the optimal inventory policy for this infinite-horizon problem.

38. Consider an infinite-period inventory model with no setup cost and nonlinear penalty and holding costs. The demands in each period are independent and are exponentially distributed with mean 1. The penalty cost is exponential; if demand exceeds supply by z units, then the penalty cost is $\exp(\frac{3}{4}z)$. There is a fixed holding cost of 3; that is, if supply exceeds demand by *any* amount, the cost is 3. The ordering cost is 1 per unit, and the discount factor is 0.95. Find the optimal ordering policy.

39.* Consider an infinite-period inventory model in which the demands are independent, identically distributed random variables. Denote the expected demand in a period by μ. Assume the cost of ordering z units ($z \ge 0$) is $c \cdot z$ ($c > 0$). Let α ($0 < \alpha < 1$) be the discount factor. Assume that all unsatisfied demand is backlogged. Finally, suppose that when y is the inventory on hand after ordering but before the occurrence of a demand of size D in a period, a cost $(y - D)^2$ is incurred. When $y > D$, the cost is a charge for carrying the inventory; when $y < D$, the cost is a charge for backlogging demand. Describe the optimal ordering policy, and give simple formulas for its parameters in terms of c, α, and μ.

40. Find the optimal (k, Q) policy for Prob. 28 for an infinite-period model by using a discount factor of $\alpha = 0.90$.

41. For the case of linear shortage and holding costs, unit costs of p and h, respectively, show that the value of y^0 that satisfies

$$\Phi(y^0) = \frac{p - c(1 - \alpha)}{p + h}$$

is equivalent to the value of y that satisfies

$$\frac{dL(y)}{dy} + c(1 - \alpha) = 0$$

where $L(y)$, the expected shortage plus holding cost, is given by

$$L(y) = \int_y^\infty p(\xi - y)\varphi_D(\xi)\, d\xi + \int_0^y h(y - \xi)\varphi_D(\xi)\, d\xi.$$

42.* Solve Prob. (1*b*) assuming that the demand is random with uniform distribution over the interval (0, 30), unsatisfied demand is backlogged, and the delivery lead time is $\frac{1}{2}$ month (2 weeks).

43. Solve Prob. (2*b*) assuming that the demand is random with an exponential distribution with mean 600, unsatisfied demand is backlogged, and the delivery lead time is 1 week.

44. Solve Prob. (5*b*) assuming that the demand is random with uniform distribution over the interval (0, 4,250), unsatisfied demand is backlogged, and the delivery lead time is $\frac{1}{4}$ month (1 week).

45. Solve Prob. (5*b*) assuming that the demand is random with an exponential distribution with mean 2,125, unsatisfied demand is backlogged, and the delivery lead time is $\frac{1}{4}$ month (1 week).

46. Consider the continuous review model with fixed delivery lag—backlogging, given on p. 318 of the text. Supposing that the order quantity Q is fixed in advance, derive the expression for the optimal reorder point, s^*, when the distribution of demand is exponential with mean $a\lambda$.

47. Suppose that the criterion of minimizing the expected total inventory cost per unit time subject to meeting a service level is chosen for the continuous review model with fixed delivery lag—backlogging. That is, the decision maker wishes to choose an "optimal" policy subject to being assured that the probability of a shortage occurring during a cycle will not exceed V, or

$$P\{D > s\} \le V.$$

(*a*) Assuming that the density function of demand is uniform over the range from 0 to t, and that the inventory policy is to place an order of size Q whenever the inventory position reaches s, find the relation between s and V.
(*b*) Using the smallest value of s in part (*a*), find the expression for $C(Q, s)$, the expected total cost per unit of time, in terms of the only remaining variable Q. Call this cost $C(Q)$.
(*c*) Show that optimal order size Q is given by

$$Q = \sqrt{\frac{2a}{h}\left(K + \frac{ptV^2}{2}\right)}.$$

48. Solve Prob. 47 when the density function of demand is exponential, or

$$\varrho_D(\xi) = \left(\frac{1}{a\lambda}\right) e^{-\xi/a\lambda}, \qquad \text{for } \xi > 0.$$

For part (*c*), show that

$$Q = \sqrt{\frac{2a}{h}(K + a\lambda p V)}.$$

49. Consider the continuous review model with fixed delivery lag—no backlogging. Suppose the density function of demand is uniform over the range from 0 to t. Show that Eq. (4) becomes

$$s^* = t\left(\frac{pa}{hQ^* + pa}\right),$$

and Eq. (3) becomes

$$Q^* = \sqrt{\frac{(2aK + apt + aps^{*2}/t - 2aps^*)}{h}}.$$

50. Consider the continuous review model with fixed delivery lag—no backlogging. Suppose the density function of demand is exponential. Show that Eq. (4) becomes

$$s^* = -a\lambda \ln\left(\frac{hQ^*}{hQ^* + pa}\right),$$

and Eq. (3) becomes

$$Q^* = \sqrt{\frac{2a}{h}}\,(K + a\lambda p e^{-s^*/a\lambda}).$$

10

Forecasting

10.1 Introduction

Chapter 9, ''Inventory Theory,'' was concerned with finding optimal inventory policies. These policies were derived from inventory models and are, in part, dependent upon some forecast of sales or use of the items of interest. Forecasting is an essential component of any successful inventory system. However, forecasting need not be associated solely with problems of inventory control. Other examples where forecasting plays an important role in industry include marketing, financial planning, and production. Indeed, it is extremely rare to think of examples where managerial decisions are made in the absence of some form of forecasting. We should emphasize that a forecast is not the final product itself; it is to be used as a tool in making a managerial decision.

Forecasts can be obtained by using qualitative or quantitative techniques. In the former case, a forecast is usually the result of an expression of one or more experts' personal judgment or opinion, and it is often called a judgmental technique. For example, a major research university calls in its leading economists every September to obtain their judgment on what to expect as an inflation rate for the next academic year—a number crucial to the budgeting process. This number is generally arrived at by consensus after prolonged discussion by the economists.

Two distinct quantitative techniques are used in forecasting, both of which are conventional statistical techniques, i.e., time series analysis and regression analysis. A statistical time series is simply a series of numerical values that a random variable takes on over a period of time. For example, the daily market closing prices of a particular stock over the period of a year constitute a time series. Time series analysis exploits techniques that utilize these data for forecasting the values that the variable of interest will take on in a future period. For example, the West Coast distributor of 10-speed bicycles (Chap. 9) wants to make quarterly sales forecasts for planning purposes. He has data on sales during previous quarters; i.e., he has the values that the random variable quarterly sales take on, and he wants to forecast the sales that will occur during the next quarter (or subsequent quarters). Forecasting, in general, is concerned with an analysis of past time series data in order to estimate one or more future values of the time series. The forecast depends upon a model of behavior of the time series.

In regression analysis,[1] the variable to be forecast (the dependent variable) is expressed as a mathematical function of other (independent) variables. For example, forecasting the total sales of a textbook in a given period may be functionally related to the mail order sales during the same period. Data on mail order sales and total sales over previous periods may be used to forecast total sales in a future period given the mail order sales for that period.

The three types of forecasts alluded to may be used in conjunction with one another. Indeed, the judgmental technique is often used together with an appropriate time series analysis.

10.2 Judgmental Techniques

Judgmental techniques are, by their very nature, subjective, and they may involve such qualities as intuition, expert opinion, and experience. They generally lead to forecasts that are based upon qualitative criteria. One commonly used technique is to bring together a group of experts who interact with each other and produce a consensus forecast. The example of the group of university economists who were asked to forecast the inflation rate is an example of the **expert group technique**.

Perhaps the most important judgmental technique is called the Delphi method. Like the expert group technique, the **Delphi method** utilizes a group of experts (not in a meeting setting). In addition to this group, there are one or more decision makers who ultimately are responsible for making the forecast. Finally, there is a staff who perform the duties associated with the method. These duties include the preparation of questionnaires and the analysis of their results.

The Delphi method first utilizes a questionnaire that is sent to the panel of experts and then analyzed. Based upon the results of the first questionnaire, a second questionnaire is developed and sent to the same panel of experts, together with the results of the first questionnaire. This second questionnaire is then completed by the

[1] Regression is simply an expression of the form of the function of the expected value of the dependent variable, given the values of the independent variables.

panel of experts and returned for analysis. Based upon the results of the two questionnaires, and using their own expertise, the decision makers ultimately come forth with a forecast. The key to the Delphi method is the feedback of the information contained in the first questionnaire to the panel of experts. Thus each member of the panel has access to information that he or she may have lacked originally, so that each member of the panel has all the same information when completing the second questionnaire.

Of course, the success of the Delphi method hinges on the quality of the design of the questionnaires. Occasionally, more than two iterations may be used if they are deemed desirable. This situation occurs when there appears to be sufficient divergence in the first two questionnaires to warrant a third round in the hope that the feedback from the results of the second round will lead to more convergence in the third.

10.3 Time Series

A **time series** can be viewed as the representation of the outcomes of a random variable of concern over a fixed period of time, usually taken at equally spaced intervals. The daily closing prices of AT&T stock taken over the last year comprise a time series. The quarterly unemployment rate from January 1980 to July 1984 comprises a time series. This time series is shown in Fig. 10.1. The quarterly sales of the West Coast distributor of 10-speed bicycles over the last three quarters also comprise a time series. The behavior of a time series can be displayed in graphical form, bar graphical form, or tabular form, where the first method is generally most descriptive of the pattern of behavior of the series.

Because a time series is a description of the past, a logical procedure for forecasting the future is to make use of this historical data. If history is to repeat itself—i.e., if the past data are indicative of what we can expect in the future—we can postulate an underlying mathematical model that is representative of the process. Indeed, if this model is known, except possibly for certain parameters, we can generate forecasts. Alternatively, if the model is not known, the past data may be suggestive of its form.

In most realistic situations, knowledge of the exact form of the model that generates the time series is unknown. Frequently, a model is chosen by observing the outcomes of the time series over a period of time. Several typical time series patterns are shown in Fig. 10.2. Figure 10.2*a* shows a time series that might be observed if the generating process were represented by a constant level superimposed with random fluctuations. Figure 10.2*b* shows a time series that might be observed if the generating process were represented by a linear trend superimposed with random fluctuations. Finally, Fig. 10.2*c* shows a time series that might be observed if the generating process were represented by a constant level superimposed with a seasonal effect together with random fluctuations. There are many other plausible representations, but these three are very useful in practice and will therefore be considered in this chapter.

Once the form of the model is chosen, an appropriate mathematical model of the generating process of the time series is identified, except possibly for the unknown values of parameters. For example, suppose that the generating process of the time series is identified as a constant level model superimposed with random fluctuations, that is, Fig. 10.2*a*. Such a representation can be given by

$$X_t = A + e_t,$$

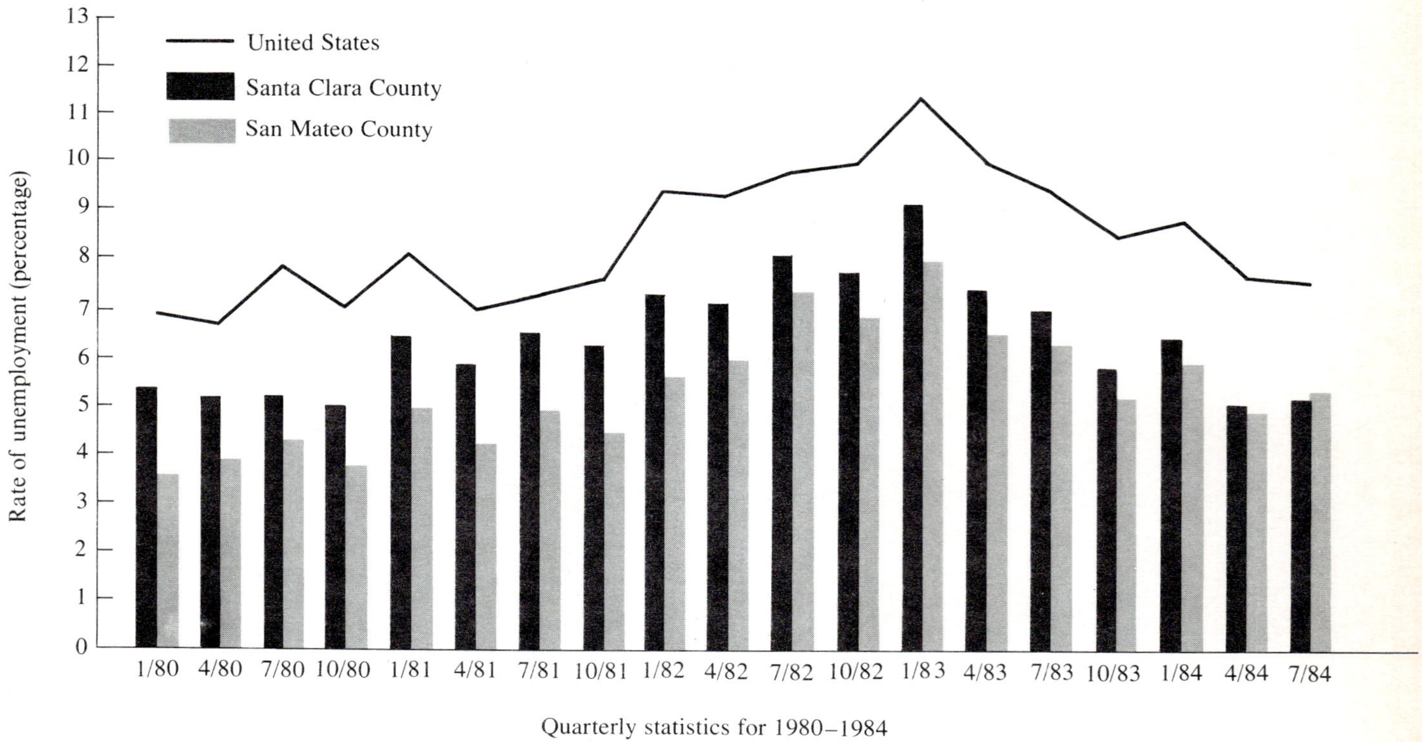

Figure 10.1 Unemployment rate (not seasonally adjusted).

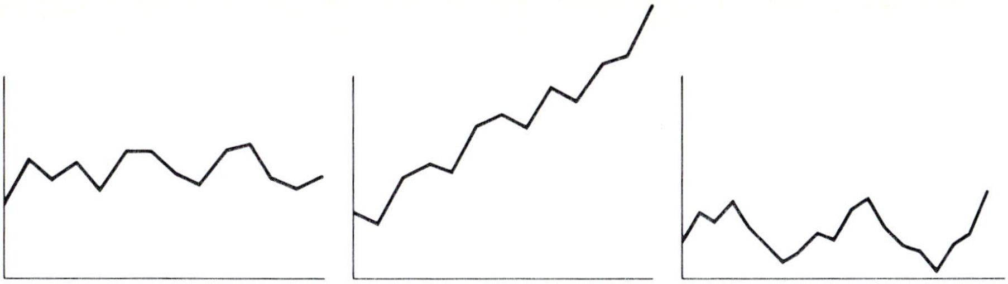

Figure 10.2 Typical time series patterns.

where X_t is the random variable that is observed at time t, A is the "constant value" of the model, and e_t is the random error occurring at time t (frequently assumed to have expected value equal to zero and constant variance). Let F_{t+1} denote the forecast of the value of the time series at time $t + 1$. It is reasonable to expect that F_{t+1} will be a function of some, or all, of the observed values of the time series prior to time $t + 1$.

10.4 Forecasting Techniques for Constant Level Models

An appropriate model is that given at the end of Sec. 10.3, i.e.,

$$X_t = A + e_t,$$

where X_t is the random variable that is observed at time t, A is the "constant value" of the model, and e_t is the random error occurring at time t (frequently assumed to have expected value equal to zero and constant variance). The following are four techniques that are often used in practice.

Last Value Forecasting Procedure

Denote by x_t the value the random variable X_t takes on. The last value forecasting procedure is to assume that the forecast at time $t + 1$, that is, F_{t+1}, equals the value of the time series observed at time t, that is, x_t. Therefore,

$$F_{t+1} = x_t$$

For example, the West Coast distributor of 10-speed bicycles may make quarterly sales forecasts for planning purposes. He has data on sales during previous quarters; i.e., he has the values the random variable quarterly sales take on, and he would like to forecast the sales that will occur in the forthcoming quarter or in subsequent quarters.

Using the last value forecasting procedure, the bicycle distributor uses the sales from last quarter as the forecast of the sales for future quarters. This forecasting procedure has the disadvantage of being imprecise; i.e., its variance is large because it is based upon a sample of size 1. It is worth considering only if the underlying assumption about the "constant level" model is "shaky," and the process is changing so rapidly that anything before time t is almost irrelevant or misleading; and/or the assumption that constant variance is unreasonable, and the conditional variance at time t is very small. This simple technique using past data is frequently compared with the results of using a more sophisticated technique to determine whether or not the sophisticated technique, which is generally more cumbersome, is of any value.

Average Forecasting Procedure

The bicycle distributor may use all his past quarterly data to forecast the sales for future quarters; i.e., he may choose

$$F_{t+1} = \sum_{i=1}^{t} \frac{x_i}{t}.$$

This estimate is an excellent one if the process is entirely stable, i.e., assumptions about the underlying model are correct. However, besides being too cumbersome when using large masses of data, one does not want to use data that are too old because there always exists skepticism about the persistance of the underlying model for too long a period.

Moving Average Forecasting Procedure

In using a moving average estimate, the bicycle distributor uses only the last n periods; that is,

$$F_{t+1} = \sum_{i=t-n+1}^{t} \frac{x_i}{n}.$$

Not only does this forecasting technique use all the relevant history in the last n periods, but it is easily updated from period to period; i.e., the first observation is lopped off and the last one is added. The moving average estimator combines the advantages of the previous estimators in that it uses only recent history and represents multiple observations. A disadvantage of this procedure is that it places as much weight on x_{t-n+1} as on x_t, and intuitively one would expect a good procedure to place more weight on the most recent observation.

Exponential Smoothing Forecasting Procedure

If the bicycle distributor uses exponential smoothing, then

$$F_{t+1} = \alpha x_t + (1 - \alpha)F_t,$$

where $0 < \alpha < 1$ is called the *smoothing constant*. Thus the forecast is just a weighted sum of the last observation and the previous forecast. The choice of α is discussed later. Note that the exponential smoothing technique represents a recursive relationship and can be expressed alternatively as

$$F_{t+1} = \alpha x_t + \alpha(1 - \alpha)x_{t-1} + \alpha(1 - \alpha)^2 x_{t-2} + \cdots.$$

In this form, it becomes evident that exponential smoothing gives the most weight to x_t and decreasing weights to earlier observations. Furthermore, the first form reveals that the forecast is simple to calculate because the data prior to period t need not be retained; all that is required is x_t and the previous forecast F_t. Another alternative form for the exponential smoothing technique is given by

$$F_{t+1} = F_t + \alpha(x_t - F_t),$$

which gives a heuristic justification for this procedure. Thus, the forecast of the time series at time $t + 1$ is just the previous forecast at time t plus the product of the forecasting error at time t and the discount factor. This alternative form is often simpler to use. Finally, a measure of effectiveness of exponential smoothing can be obtained under the assumption that the process is completely stable; that is, $X_1, X_2, \ldots$ are independent, identically distributed random variables with variance σ^2. It then follows that (for large t)

$$\text{var}[F_{t+1}] \approx \frac{\alpha\sigma^2}{2 - \alpha} = \frac{\sigma^2}{(2 - \alpha)/\alpha},$$

so that the variance is statistically equivalent to a moving average with $(2 - \alpha)/\alpha$ observations. If α is chosen equal to 0.1, then $(2 - \alpha)/\alpha = 19$. Thus the exponential smoothing technique is "equivalent" to a moving average procedure that uses 19 observations. However, it must be noted that when the aforementioned underlying assumptions are violated, exponential smoothing will react more quickly with superior "tracking."

An important drawback of exponential smoothing is that it lags behind a continuing trend; i.e., if the "constant" level model is incorrect and the mean is increasing steadily, then the forecast will be several periods behind. However, the procedure can be easily adjusted for trend (and even seasonally adjusted). Another disadvantage of exponential smoothing is that it is difficult to choose an appropriate smoothing constant α. Exponential smoothing can be viewed as a statistical filter that inputs raw data from a stochastic process and outputs smoothed estimates of a mean that varies with time. If α is chosen to be small, response to change is slow, with resultant smooth estimators. Similarly, if α is chosen to be large, response to change is fast, with resultant large variability in the output. Hence there is a need to compromise, depending upon the stability of the process. Furthermore, a "good" value of the smoothing constant depends upon the underlying stochastic process and the choice of a criterion to use in comparing constants. It has been suggested that α should not exceed 0.3 and that a reasonable choice for α is approximately 0.1. Of course it can be increased, perhaps temporarily, if an unusual change is expected or when starting. When starting, a reasonable approach is to choose the forecast for period 2 according to

$$F_2 = \alpha x_1 + (1 - \alpha)(\text{initial estimate}),$$

where some initial estimate of the constant level A must be chosen. If past data are available, such an estimate may be the "average" of these data. In the case of past history, another procedure for choosing α is to run a retrospective simulation of the process; i.e., for a fixed value of α and using past history, compare the forecasted quantity with the actual outcome and choose that value of α which is in some sense optimal. It is hoped that the process will behave in the future in the same manner as it has in the past.

10.5 A Forecasting Technique for Linear Trend Models

As indicated earlier, the procedure outlined in the previous section lags behind a continuing trend. Suppose that the generating process of the observed time series can be represented by a linear trend superimposed with random fluctuations, and the linear trend has slope B. The slope is called the *trend factor,* and in the bicycle example, it represents the units per period that the expected sales rate increases or decreases. The model can be represented by

$$X_t = A + Bt + e_t,$$

where X_t is the random variable that is observed at time t, A is a constant, B is the trend factor, and e_t is the random error occurring at time t (frequently assumed to have expected value equal to zero and constant variance). In the previous model (constant level), the forecast for period $t + 1$, based upon data from the previous periods, is the same as forecasts for periods $t + 1 + m$, for $m = 1, 2, \ldots$. For linear trend models, such a statement no longer holds. Hence, rather than referring to forecasts immediately, the concept of a "smoothed" level will be introduced. If x_t is the observed value of the time series at time t, then a "smoothed" level at time t, S_t, will be a linear combination of x_t and the "smoothed" value at the preceding time period $t - 1$ corrected by adding the trend (slope) to indicate the passage of a unit of time; i.e.,

$$S_t = \alpha x_t + (1 - \alpha)(S_{t-1} + B).$$

The forecast for time $t + 1$ can now be obtained; i.e.,

$$F_{t+1} = S_t + B.$$

Unfortunately, the trend (slope) B is unknown so it must be estimated, and exponential smoothing can again be used for this purpose; i.e.,

$$B_t = \beta(S_t - S_{t-1}) + (1 - \beta)B_{t-1},$$

where B_t is the "smoothed" value of the trend at the end of period t, and $0 < \beta < 1$ is another (possibly different from α) smoothing constant.[1] Hence, the smoothed level at time S_t can now be expressed as

$$S_t = \alpha x_t + (1 - \alpha)(S_{t-1} + B_{t-1}),$$

and the forecast for m periods ahead, $m = 1, 2, \ldots$, is given by

$$F_{t+m} = S_t + mB_t.$$

The forecasting procedures can now be summarized as follows.

1. Using the observed value of the time series at the end of the tth period, x_t, the smoothed level of the time series at time $t - 1$, S_{t-1}, and the smoothed value of the trend at the end of period $t - 1$, B_{t-1}, the smoothed level of the time series at time t is given by

$$S_t = \alpha x_t + (1 - \alpha)(S_{t-1} + B_{t-1}).$$

2. From the smoothed level of the time series at time t, S_t (calculated in step 1), the smoothed level of the time series at time $t - 1$, S_{t-1}, and the

[1] The previous discussion concerning the choice of α is relevant to the choice of β.

smoothed value of the trend at the end of period $t - 1$, B_{t-1}, the smoothed value of the trend at the end of period t is given by

$$B_t = \beta(S_t - S_{t-1}) + (1 - \beta)B_{t-1}.$$

3. Forecast of the time series for m periods ahead, $m = 1, 2, \ldots,$ is given by

$$F_{t+m} = S_t + mB_t.$$

Thus, in the context of forecasting sales of bicycles, the actual sales of bicycles, x_t, is obtained at the end of period t; the smoothed level of the sales at the end of period $t - 1$, S_{t-1}, and the smoothed level of the trend at the end of period $t - 1$, B_{t-1}, are known so that S_t can be obtained; the smoothed level of the trend at the end of period $t - 1$, B_{t-1}, and the smoothed level of sales at times t and $t - 1$, S_t and S_{t-1}, respectively, are known so that B_t can be obtained; and finally, the forecast of sales m periods ahead is then given by F_{t+m}.

As in the case of exponential smoothing for a constant level model, an initial value is required to start the smoothing process for the linear trend model. This initialization is frequently obtained by fitting a straight line to some past data (using methods shown in Sec. 10.8). The fitted line can be used to obtain an initial value of the smoothed level of the time series, S_0, and an initial value of the smoothed trend level, B_0. Thus,

$$S_1 = \alpha x_1 + (1 - \alpha)(S_0 + B_0),$$

and

$$B_1 = \beta(S_1 - S_0) + (1 - \beta)B_0.$$

If a forecast for period 2 is desired, it can be obtained from

$$F_{1+1} = F_2 = S_1 + B_1$$

Example: Suppose that quarterly sales for bicycles were 2,800, 2,925, and 3,040, respectively. Use exponential smoothing based upon the first three observations to forecast sales for the fifth period, using $\alpha = \beta = 0.1$. From past data (prior to the three data points), a straight line was fit. The value on the line corresponding to the last observed time is 2,750, and the slope is 100. Thus $S_0 = 2{,}750$ and $B_0 = 100$. The smoothed value of sales at the end of period 1 is given by

$$S_1 = 0.1(2{,}800) + 0.9(2{,}750 + 100) = 2{,}845.$$

The smoothed value of the trend at the end of period 1 is given by

$$B_1 = 0.1(2{,}845 - 2{,}750) + 0.9(100) = 99.5.$$

Repeating this procedure for the second observation leads to

$$S_2 = 0.1(2{,}925) + 0.9(2{,}845 + 99.5) = 2{,}943$$

and

$$B_2 = 0.1(2{,}943 - 2{,}845) + 0.9(99.5) = 99.4.$$

Finally, the third observation results in

$$S_3 = 0.1(3{,}040) + 0.9(2{,}943 + 99.4) = 3{,}042$$

and

$$B_3 = 0.1(3{,}042 - 2{,}943) + 0.9(99.4) = 99.4.$$

Therefore, the forecast of sales for the fifth period is

$$F_{3+2} = F_5 = 3{,}042 + 2(99.4) = 3{,}241.$$

If a forecast for the second period were desired (made at the end of the first period), it would result in

$$F_2 = S_1 + B_1 = 2{,}845 + 99.5 = 2{,}945.$$

Similarly, if a forecast for the third period were desired (made at the end of the second period), it would result in

$$F_3 = S_2 + B_2 = 2{,}943 + 99.4 = 3{,}042.$$

10.6 A Forecasting Technique for Constant Level with Seasonal Effects Models

In many important forecasting problems, there exist seasonal effects that must be accounted for in the model. Suppose that the generating process of the observed time series can be represented by a constant level superimposed with seasonal effects and random fluctuations. Such a model can be represented by

$$X_t = AI_t^* + e_t,$$

where X_t is the random variable that is observed at time t, A is a constant, I_t^* is the seasonal index or factor for period t, and e_t is the random error occurring at time t (frequently assumed to have expected value equal to zero and constant variance). Unfortunately, both A and I_t^* are unknown, and "smoothed" levels at time t for both of these factors are useful prior to making a forecast. Exponential smoothing can again be used for this purpose; that is,

$$S_t = \alpha\left(\frac{x_t}{I_{t-p}}\right) + (1 - \alpha)S_{t-1},$$

$$I_t = \gamma\left(\frac{x_t}{S_t}\right) + (1 - \gamma)I_{t-p},$$

and the forecast for the next period is given by

$$F_{t+1} = S_t I_{t-p+1},$$

where p is the number of periods in the seasonal cycle—e.g., if the seasonal periods are autumn, winter, spring, summer, then $p = 4$, and $0 < \gamma < 1$ is another smoothing constant. Note that I_{t-p} represents the "smoothed" value of the seasonal index for period t computed for the same season p periods ago; e.g., the seasonal index for the autumn of 1990 is based upon 1989 autumn data.

The forecast for m periods ahead, $m = 1, 2, \ldots,$ is given by

$$F_{t+m} = S_t I_{t-p+m}.$$

If m is greater than p, I_{t-p+m} has not, as yet, been calculated. For this case, I_{t-p+m} is to be interpreted as the last *computed* value of the corresponding seasonal index; e.g., if a forecast for the autumn of 1993 is desired, and no data later than the winter of 1990 are available, the seasonal index factor for autumn of 1990 is used in place of $I_{\text{aut 1992}}$.

The technique for forecasting can be summarized as follows.

1. Using the observed value of the time series at the end of the tth period, x_t, and the smoothed value of the seasonal index for period t computed for the same season p periods ago, I_{t-p}, the smoothed level of the time series at time t is given by

$$S_t = \alpha\left(\frac{x_t}{I_{t-p}}\right) + (1 - \alpha)S_{t-1}.$$

2. From the smoothed level of the time series at time period t, S_t (calculated in step 1), the smoothed level of the time series at time period $t - 1$, S_{t-1}, and the smoothed value of the seasonal index for period t computed for the same season p periods ago, I_{t-p}, the updated smoothed level of the seasonal index for period t is given by

$$I_t = \gamma\left(\frac{x_t}{S_t}\right) + (1 - \gamma)I_{t-p}.$$

3. Forecast of the time series for m periods ahead, $m = 1, 2, \ldots,$ is given by

$$F_{t+m} = S_t I_{t-p+m}.$$

As in the other exponential smoothing procedures, initial values are required to start the smoothing process. For models containing seasonal factors, a full cycle of seasonal data is required, as is suggested in the discussion about I_{t-p}. A description of the initialization procedure is best given in the context of an example. Consider the bicycle example of Sec. 10.5, and suppose that bicycle sales are seasonal (autumn, winter, spring, summer). Upon examination of past data, the trend model has been found to be inappropriate and has been replaced by a seasonal model as described above. The data are given in columns 1 and 2 of Table 10.1.

A reasonable way to initially estimate the seasonal factors is to divide last year's quarterly sales by the quarterly average sales over the last year, that is, $(2{,}786 + 2{,}928 + 3{,}025 + 3{,}061)/4 = 2{,}950$. These values are shown in column 3 of Table 10.1. The initial estimate of the constant level A is chosen to be the average of the four quarters over the past year, that is, $S_0 = 2{,}950$. Exponential smoothing is to be used to forecast the bicycle sales for autumn, winter, and spring of this year and, ultimately, for the summer quarter of this year. Use $\alpha = 0.1$ and $\gamma = 0.2$.

The forecast for autumn is given by

$$F_{\text{aut}} = S_0(I_{\text{old aut}}) = 2{,}950(0.944) = 2{,}785.$$

Table 10.1 **Bicycle Sales**

	1 Old (Last) Year	2 New (This) Year	3 Initial Seasonal Factor
Autumn	2,786	2,800	0.944
Winter	2,928	2,925	0.993
Spring	3,025	3,040	1.025
Summer	3,061		1.038
Total	11,800		4.000

In order to obtain the forecast for the winter quarter, the smoothed level of the time series, S_{aut}, must be obtained. The smoothed level of the seasonal factor for next year's autumn quarter forecast, $I_{\text{new aut}}$, is also obtained.

$$S_{\text{aut}} = \alpha\left(\frac{x_{\text{aut}}}{I_{\text{old aut}}}\right) + (1 - \alpha)S_0$$

$$= 0.1\left(\frac{2{,}800}{0.944}\right) + 0.9(2{,}950) = 2{,}952,$$

and

$$I_{\text{new aut}} = \gamma\left(\frac{x_{\text{aut}}}{S_{\text{aut}}}\right) + (1 - \gamma)I_{\text{old aut}}$$

$$= 0.2\left(\frac{2{,}800}{2{,}952}\right) + 0.8(0.944) = 0.945.$$

The forecast for the winter quarter is then given by

$$F_{\text{aut}+1} = F_{\text{win}} = S_{\text{aut}}(I_{\text{old win}}) = 2{,}952(0.993) = 2{,}931.$$

In order to obtain the forecast for the spring quarter, the smoothed level of the time series, S_{win}, must be obtained. The smoothed level of the seasonal factor for next year's winter quarter forecast, $I_{\text{new win}}$, is also obtained.

$$S_{\text{win}} = \alpha\left(\frac{x_{\text{win}}}{I_{\text{old win}}}\right) + (1 - \alpha)S_{\text{aut}}$$

$$= 0.1\left(\frac{2{,}925}{0.993}\right) + 0.9(2{,}952) = 2{,}951,$$

and

$$I_{\text{new win}} = \gamma\left(\frac{x_{\text{win}}}{S_{\text{win}}}\right) + (1 - \gamma)I_{\text{old win}}$$

$$= 0.2\left(\frac{2{,}925}{2{,}951}\right) + 0.8(0.993) = 0.993.$$

The forecast for the spring quarter is then given by

$$F_{\text{win}+1} = F_{\text{spr}} = S_{\text{win}}(I_{\text{old spr}}) = 2{,}951(1.025) = 3{,}025.$$

Finally, in order to obtain the forecast for the summer quarter, the smoothed level of the time series, S_{spr}, must be obtained. The smoothed level of the seasonal factor for next year's spring quarter forecast, $I_{\text{new spr}}$, is also obtained.

$$S_{\text{spr}} = \alpha\left(\frac{x_{\text{spr}}}{I_{\text{old spr}}}\right) + (1 - \alpha)S_{\text{win}}$$

$$= 0.1\left(\frac{3{,}040}{1.025}\right) + 0.9(2{,}951) = 2{,}952,$$

and

$$I_{\text{new spr}} = \gamma\left(\frac{x_{\text{spr}}}{S_{\text{spr}}}\right) + (1 - \gamma)I_{\text{old spr}}$$

$$= 0.2\left(\frac{3{,}040}{2{,}952}\right) + 0.8(1.025) = 1.026.$$

The forecast for the summer quarter is then given by

$$S_{\text{spr}+1} = S_{\text{sum}} = S_{\text{spr}}(I_{\text{old sum}}) = 2{,}952(1.038) = 3{,}064.$$

Although these forecasts appear to yield good results, a forecasting technique based upon a model of a generating process of the time series that is represented by a linear trend superimposed with seasonal effects and random fluctuations (a model not presented in this text) may be more appropriate (based upon the limited amount of past data presented).

10.7 Forecasting Errors

Several forecasting techniques have been presented together with different underlying models of the time series. How does one compare these techniques, especially if the generating process is unknown, as is frequently the case in practice? Some measure of performance is called for.

Define the **forecast error**, E_t, as the difference between the observed value of the time series for period t and the forecast for period t, that is,

$$E_t = x_t - F_t.$$

The forecast error is also referred to as the **residual**. If the underlying model of the time series is appropriate for the forecasting technique chosen, the E_t should behave; in a probabilistic sense, like the random error associated with the stochastic process, and the technique that produces "small" values of E_t is desirable. A measure of "small" is the mean square error (MSE) associated with the forecasting technique. If there are n time periods, then

$$\text{MSE} = \frac{E_1^2 + E_2^2 + \cdots + E_n^2}{n}.$$

For the bicycle example using winter and spring values together with the appropriate forecasts, the following are mean square errors for the three forecasting techniques discussed earlier:

Last value forecasting $\quad \text{MSE} = \dfrac{(2{,}925 - 2{,}800)^2 + (3{,}040 - 2{,}925)^2}{2} = 14{,}425,$

Linear trend forecasting $\quad \text{MSE} = \dfrac{(2{,}925 - 2{,}945)^2 + (3{,}040 - 3{,}042)^2}{2} = 202,$

Seasonal forecasting $\quad \text{MSE} = \dfrac{(2{,}925 - 2{,}931)^2 + (3{,}040 - 3{,}025)^2}{2} = 131.$

The best technique of the three, based upon the MSE measure of effectiveness, is the seasonal forecasting technique. Last value forecasting performs poorest. Of course, these results are based upon past data (and very few points at that), and there is no assurance that, say, the seasonal forecasting technique will perform well with future data, unless the underlying model is appropriate for that forecasting technique.

Note that the simple bicycle example illustrates a useful concept. Before choosing any forecasting technique, it is worthwhile "trying it out" on some past data. This will reveal whether or not a forecasting technique is reasonable for the past data and, if the times series does not change, may give an indication of how well the forecasting technique will perform in the future.

10.8 The Box-Jenkins Method

It has already been pointed out that procedures and models frequently are not coordinated in practice. The beauty of the Box-Jenkins method, a sophisticated and complex technique that requires a great amount of past data (a minimum of 50 time periods), is that the model and the procedure are coordinated. There is a systematic approach to identifying an appropriate model, chosen from a rich class of models. The historical data are used to test the validity of the model. The model also generates an appropriate forecasting procedure.

The Box-Jenkins method is iterative in nature. First, a model is chosen. To choose this model, we must compute autocorrelations and partial autocorrelations and examine their patterns. An autocorrelation measures the correlation between time series values separated by a fixed number of periods. This fixed number of periods is called the *lag*. Therefore, the autocorrelation for a lag of two periods measures the correlation between every other observation; i.e., it is the correlation between the original time series and the same series moved forward two periods. The partial autocorrelation is a conditional autocorrelation between the original time series and the same series moved forward a fixed number of periods, holding the effect of the other lagged times fixed. We can compute both the autocorrelations and the partial autocorrelations for all lags; we can do it easily with a computer. From the autocorrelations and the partial autocorrelations, we can identify the form of one or more possible models because a rich class of models is characterized by these parameters. Actually, we compute the sample autocorrelations and the sample partial autocorrelations, but these computations are "good" estimates, because we assume large amounts of data. Now that we have identified the model—i.e., we have identified the functional form—we must estimate the parameters associated with the model. This estimate is made using the historical data. The functional form then becomes known (or approximately known), and we can compute the residuals and examine their behavior. Similarly, we can examine the behavior of the estimated parameters. If both the "residuals" and the "estimated parameters" behave as expected under the presumed model, the model appears to be validated. If they do not, then the model should be modified and the procedure repeated until a model is validated. At this point, we can obtain a forecast.

For example, suppose that the sample autocorrelations and the sample partial autocorrelations are calculated (usually by computer) as shown in Fig. 10.3. The sample autocorrelations appear to decrease exponentially as a function of the time lags, while the sample partial autocorrelations appear to have spikes at the first and second time lags of the observations followed by values that seem to be of negligible magnitude. This behavior of the sample autocorrelations and the sample partial autocorrelations is characteristic of the functional form,

$$X_t = B_0 + B_1 X_{t-1} + B_2 X_{t-2} + e_t.$$

Assuming this functional form, we use the time series data to estimate B_0, B_1, and B_2 (usually by means of a computer program). Denote these estimates by b_0, b_1, and b_2, respectively. Using these estimates, together with the time series data, we obtain

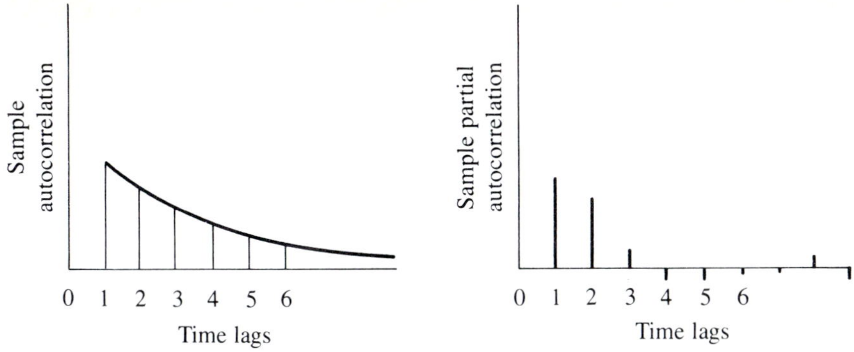

Figure 10.3 Plot of sample autocorrelation and partial autocorrelation versus time lags.

the residuals

$$x_t - (b_0 + b_1 x_{t-1} + b_2 x_{t-2}).$$

If the assumed functional form is adequate, the residuals and the estimated parameters should behave in a predictable manner. In particular, the sample residuals should behave as approximately independent, normally distributed random variables, each having mean zero and variance σ^2 (assuming that e_t, the random error at time period t, has mean zero and variance σ^2). The estimated parameters should be uncorrelated and significantly different from zero. Statistical tests are available for this diagnostic checking.

The Box-Jenkins procedure appears to be a complex one, and it is. Fortunately, computer software is available for the procedure. The programs calculate the sample autocorrelations and the sample partial autocorrelations necessary for identifying the form of the model. They also estimate the parameters of the model and do the diagnostic checking. These programs, however, cannot accurately identify one or more models that are compatible with the autocorrelations and the partial autocorrelations. This expertise can be acquired, but it is beyond the scope of this text. Although the Box-Jenkins method is complicated, the resultant forecasts are extremely accurate, and, when the time horizon is short, better than most other forecasting techniques. Further, the procedure produces a measure of the forecast error.

10.9 Forecasting with Linear Regression

Quantitative forecasting has been discussed in terms of a time series representation of a generating process of a model. An alternative technique that is sometimes appropriate is regression analysis. In regression analysis the variable to be forecast, denoted as the dependent variable, is expressed as a mathematical function of other variables, denoted as independent variables. In other words, the relationship between the de pendent variable and the independent variables are such that knowledge of values that the independent variables take on furnishes information about the corresponding value of the dependent variable. Hence, a forecast (of the dependent variable) can be made, provided that the values of the independent variables are known at the time such a forecast is desired. Regression analysis for a single independent variable is described in Section 4.6, and the reader is urged to review this material. In a forecasting context, the prototype example is appropriate. Recall that a publisher of textbooks is concerned about the initial press run for his books. He sells books both through bookstores and through mail orders. This latter method uses an extensive advertising campaign through publishing media and direct mail. The advertising campaign is conducted prior to the publication of the book. The sales manager has noted that there appears to exist a linear relationship between the number of mail orders and the number sold through bookstores during the first year. He suggests that this relationship be exploited to forecast the initial press run for subsequent books. Denote by X the mail order sales for a book, and denote by Y the number of bookstore sales of this book. An appropriate model for this relationship is then given by

$$E[Y|X = x] = A + Bx.$$

Using past data, estimate the parameters A and B (using the technique of least squares described in Section 4.6), and obtain a forecast of the bookstore sales of a forthcoming book from information on its mail order sales. Using the data given in Table 4.5, the least-squares estimate of the bookstore sales y, when the mail order sales is x, is given by

$$\tilde{y} = -19{,}041.9 + 17.930x.$$

If the mail order sales on a particular book is 1,400, the forecasted bookstore sales is given by

$$\tilde{y} = -19{,}041.9 + 17.930(1{,}400) = 6{,}060.$$

Another example is mentioned in Section 10.5 in the context of a generating process of a time series being represented by a linear trend superimposed with random fluctuations, that is,

$$X_t^{\dagger} = A + Bt + e_t,$$

where A is a constant, B is the slope, and e_t is the random error (assumed to have expected value equal to zero and constant variance). Again, it follows that

$$E[X_t] = A + Bt,$$

and A and B can be estimated by the methods given in Section 4.6. Thus, the value of X at time t can be forecasted for a given value of t. This second example has a special structure when t takes on integer values starting with 1 so that the least-squares estimates of A and B result in simplified expressions.

When using regression methods for forecasting, it is very important to plot the data. Visual inspection is helpful in determining the appropriateness of the assumed functional form.

10.10 Conclusions

It is important to consider the entire forecasting system carefully. The need to obtain a forecast has to be identified at the appropriate management level. The historical data required must be compiled. By studying these data, an appropriate model can be structured. A forecasting procedure that behaves well under the model should be selected. The forecasting procedure may require choosing one or more parameters—e.g., the smoothing constant α in exponential smoothing—and the historical data may

† The symbol X_t also can be read as X given t or as $X|t$. Also note that the variable to be forecast and the independent variable are X_t and t, respectively, rather than Y and X.

prove useful in making this choice. Finally, the forecasting process should be viewed as dynamic, with the accumulated new data compared with their associated forecasts. These data may also be used to update the parameters and the form of the underlying model, as well as the parameters of the forecasting procedure itself.

SELECTED REFERENCES

1. Box, G. E. P., and G. M. Jenkins: *Time Series Analysis, Forecasting and Control*, Holden-Day, San Francisco, 1976.
2. Brown, R. G.: *Statistical Forecasting for Inventory Control*, McGraw-Hill, New York, 1959.
3. Brown, R. G.: *Smoothing, Forecasting, and Prediction of Discrete Time Series*, Prentice-Hall, Englewood Cliffs, N.J., 1972.
4. Gardner, E. S., Jr.: "Exponential Smoothing: The State of the Art," *Journal of Forecasting*, **4**:1–38, 1985.
5. Gilchrist, W. G.: *Statistical Forecasting*, Wiley, New York, 1976.
6. Hax, A. C., and D. Candea: *Production and Inventory Management*, Prentice-Hall, Englewood Cliffs, N.J., 1984.
7. Hoff, J. C.: *A Practical Guide to Box-Jenkins Forecasting*, Lifetime Learning Publications, Belmont, Calif., 1983.
8. Johnson, L. A., and D. C. Montgomery: "Operations Research in Production," *Planning, Scheduling, and Inventory Control*, Wiley, New York, 1974.
9. Levin, R. I., D. S. Rubin, and J. P. Stinson: *Quantitative Approaches to Management*, 6th ed., McGraw-Hill, New York, 1986.
10. Montgomery, D. C., and L. A. Johnson: *Forecasting and Time Series Analysis*, McGraw-Hill, New York, 1976.

PROBLEMS

1.* Suppose that the previous forecast was 2,083, the actual value of the variable of interest for the last period was 1,975, and the oldest value of the variable of interest was 1,945. Using the moving average technique based upon the most recent four observations, what is the new forecast for the next period?

2. Suppose that the previous forecast was 2,083, the actual value of the variable of interest for the last period was 1,975, and $\alpha = 0.3$. Using exponential smoothing, what is the new forecast for the next period?

3.* Suppose that the previous forecast was 782, the actual value of the variable of interest for the last period was 794, and $\alpha = 0.1$. Using exponential smoothing, what is the new forecast for the next period?

4. Suppose that the previous forecast was 782, the actual value of the variable of interest for the last period was 794, and the oldest value of the variable of interest was 805. Using the moving average technique based upon the most recent observations, what is the new forecast for the next period?

5. You are the new person at a statistical forecasting service, and you have been asked to update a moving average forecast based upon the most recent 10 observations. You know that the previous forecast was 1,551, the actual value of the variable of interest for the latest period (i.e., zero periods ago) is 1,532, and the oldest value (i.e., 10 periods ago) of the variable of interest was 1,632. What is the new forecast for the next period?

6. If α is set equal to zero or 1 in the exponential smoothing expression, what happens to the forecast?

7. Use the bicycle sales example presented in the section on exponential smoothing adjusted for trend, and solve the problem in this example using simple exponential smoothing with $\alpha = 0.3$.

8. A company uses exponential smoothing with $\alpha = \frac{1}{2}$ to forecast demand for a product. For each month, the company keeps a record of the forecasted demand (made at the end of the previous month) and also the actual demand. Some of the records have been lost; the remaining data appear in the table below.

	January	February	March	April	May	June
Forecast			400	380	390	380
Actual	400		360	—	—	

(*a*) Using only data in the table for March, April, May, and June, determine the actual demands in April and May.

(*b*) Suppose now that a clerical error is discovered; the actual demand in January was 432, not 400 as shown in the table. Using only the actual demands going back to January (even though the February actual demand is unknown), give the corrected forecast for June.

9. Use the bicycle sales example presented in the section on exponential smoothing adjusted for trend, and solve the problem in this example using $\alpha = \beta = 0.3$.

10.* The U.S. unemployment rate time series shown in Fig. 10.1 can be represented in tabular form as follows:

Date	Unemployment Rate (%)	Date	Unemployment Rate (%)	Date	Unemployment Rate (%)
1/80	6.9	1/81	8.2	1/82	9.4
4/80	6.7	4/81	7.0	4/82	9.2
7/80	7.9	7/81	7.3	7/82	9.8
10/80	7.1	10/81	7.5	10/82	9.9

Date	Unemployment Rate (%)	Date	Unemployment Rate (%)
1/83	11.4	1/84	8.8
4/83	10.0	4/84	7.6
7/83	9.4	7/84	7.5
10/83	8.4	10/84	—

(*a*) Starting with a forecast for 10/80, use the moving average technique based upon the past three periods to forecast each unemployment rate through 10/84.

(*b*) Examine the residuals for 10/80 through 7/84, and compute the mean square error, i.e.,

$$\frac{\sum E_i^2}{16}.$$

11. Use the unemployment data of Prob. 10.

(*a*) Starting with a forecast for 10/80, use the exponential smoothing technique with an initial estimate of 7.2 percent and $\alpha = 0.1$, i.e.,

$$\text{Forecast for } 10/80 = (0.1)7.9 + (0.9)7.2,$$

to forecast each unemployment rate through 10/84.

(b) Examine the residuals for 10/80 through 7/84, and compute the mean square error, i.e.,

$$\frac{\sum E_i^2}{16}.$$

12. (a) Solve Prob. 11 using $\alpha = 0.3$.
(b) Which α ($\alpha = 0.1$ or 0.3) would you use to forecast the unemployment rate for 10/84?

13. Based upon the results of Probs. 10(b) and 11(b), which forecasting procedure would you choose, moving average or exponential smoothing?

14. Use the unemployment data of Prob. 10.
(a) Starting with a one-step forecast for 10/80, use the exponential smoothing technique adjusted for trend, with an initial estimate of 7.2 percent for unemployment and with an initial estimate of trend of 0.2 percent, to forecast each unemployment rate through 10/84. Use $\alpha = \beta = 0.1$.
(b) Examine the residuals for 10/80 through 7/84, and compute the mean square error, i.e.,

$$\frac{\sum E_i^2}{16}.$$

15. Based upon the results of Probs. 10(b), 11(b), and 14(b), which forecasting procedure would you choose, moving average, exponential smoothing, or exponential smoothing adjusted for trend?

16. In order to plan for a suitable labor force, you need to know the demand for a particular product. The following table presents demand over the past 11 quarters.

Quarter	Demand	Quarter	Demand	Quarter	Demand
1	546	5	647	9	736
2	528	6	594	10	724
3	530	7	665	11	813
4	508	8	630	12	—

(a) Starting with a forecast for quarter 5, use the moving average technique based upon the past 4 quarters to forecast each demand through quarter 12.
(b) Examine the residuals for quarter 5 through quarter 12, and compute the mean square error, i.e.,

$$\frac{\sum E_i^2}{7}.$$

17. Use the quarterly demand data of Prob. 16.
(a) Starting with a forecast for quarter 3, use the exponential smoothing technique with an initial estimate of 546 and $\alpha = 0.1$, i.e.,

$$\text{Forecast for quarter 3} = (0.1)528 + (0.9)546,$$

to forecast each demand through quarter 12.
(b) Examine the residuals for quarter 3 through quarter 11, and compute the mean square error, i.e.,

$$\frac{\sum E_i^2}{9}.$$

18. (*a*) Solve Prob. 17 using $\alpha = 0.3$.
(*b*) Which α ($\alpha = 0.1$ or 0.3) would you use to forecast the demand for quarter 12?

19. Based upon the results of Probs. 16(*b*) and 17(*b*), which forecasting procedure would you choose, moving average or exponential smoothing?

20. Use the quarterly demand data of Prob. 16.
(*a*) Starting with a one-step forecast for quarter 3, use the exponential smoothing technique adjusted for trend, with an initial estimate of 546 for demand and an initial estimate of trend of 18, to forecast each demand through quarter 12. Use $\alpha = \beta = 0.1$.
(*b*) Examine the residuals for quarter 3 through quarter 11, and compute the mean square error, i.e.,

$$\frac{\sum E_i^2}{9}.$$

21. Based upon the results of Probs. 16(*b*), 17(*b*), and 20(*b*), which forecasting procedure would you choose, moving average, exponential smoothing, or exponential smoothing adjusted for trend?

22. Use the unemployment data of Prob. 10.
(*a*) Assume a seasonal effects model (four periods per year) and use the data for 1980 to estimate the initial seasonal factors. Use exponential smoothing with seasonal effects to forecast the unemployment for each period of the year 1981. Use the final forecast (10/81) to forecast the unemployment rate for 10/84. Let $\alpha = \gamma = 0.1$.
(*b*) Examine the residuals for the four periods of 1981, and compute the mean square error, i.e.,

$$\frac{\sum E_i^2}{4}.$$

23. Solve Prob. 22 using 1982 data to estimate the initial seasonal factors, and forecast the unemployment rate for each period of the year 1983. Use the final forecast (10/83) to forecast the unemployment rate for 10/84.

24.* Suppose that a time series behaves as follows:

t	Y	t	Y
1	430	6	514
2	446	7	532
3	464	8	548
4	480	9	570
5	498	10	591

(*a*) Using the method of least squares, estimate the line $A + Bt$.
(*b*) From the line found in (*a*), forecast Y_{11}.
(*c*) Starting with a forecast in period 3, use the exponential smoothing technique with an initial estimate of 430 and $\alpha = 0.1$, i.e.,

$$\text{Forecast for period 3} = (0.1)446 + (0.9)430,$$

to forecast each demand through period 11.
(*d*) Starting with a one-step forecast for period 3, use the exponential smoothing technique adjusted for trend, with an initial estimate of 430 for the variable of interest and an initial estimate of trend of 18, to forecast each demand through period 11. Use $\alpha = \beta = 0.2$.

25. If a particle is dropped at time $t = 0$, physical theory indicates that the relationship between r, the distance traveled, and t, the time elapsed, is $r = gt^k$ for some positive constants g and k. A transformation to linearity can be obtained by taking logarithms:

$$\log r = \log g + k \log t.$$

Letting $y = \log r$, $A = \log g$, and $x = \log t$, this relation becomes $y = A + kx$. Due to random error in measurement, however, it can only be stated that $E(Y|x) = A + kx$. Assume Y is normally distributed with mean $A + kx$ and variance σ^2.

A physicist who wishes to estimate k and g performs the following experiment: At time 0 the particle is dropped. At time t the distance r is measured. He performs this experiment five times, obtaining the following data:

$y = \log r$	$x = \log t$
−3.95	−2.0
−2.12	−1.0
0.08	0.0
2.20	+1.0
3.87	+2.0

(*a*) Obtain least-squares estimates for k and $\log g$, and forecast the distance traveled when $\log t = +3.0$.

(*b*) Starting with a forecast for $\log r$ when $\log t = 0$, use the exponential smoothing technique with an initial estimate of $\log r = -3.95$ and $\alpha = 0.1$, i.e.,

Forecast of $\log r$ (when $\log t = 0$) $= (0.1)(-2.12) + (0.9)(-3.95)$,

to forecast each $\log r$ for all integer $\log t$ through $\log t = +3.0$.

(*c*) Repeat part (*b*) using the exponential smoothing technique adjusted for trend and a one-step forecast. Use an inital estimate of trend equal to the slope found in part (*a*). Let $\beta = 0.1$.

The following data have been calculated:

$$\bar{y} = 0.016$$

$$\sum(x_i - \bar{x})^2 = 10$$

$$\sum(x_i - \bar{x})(y_i - \bar{y}) = 19.96$$

$$\sum(y_i - \tilde{y}_i)^2 = 0.0858$$

$$\frac{s_{y|x}}{\sqrt{\sum(x_i - \bar{x})^2}} = 0.053$$

$$s_{y|x}\sqrt{\frac{1}{5}} = 0.076.$$

Note: All logarithms are to the base 10.

26. Assume that the independent variable takes on the values $1, 2, \ldots, n$. Derive the least-squares equations for estimating the slope and intercept.

11

Markovian Decision Processes and Applications

11.1 Introduction

Section 5.2 introduced the concept of a dynamic system evolving over time. The behavior of such a system resulted in an analysis of a particular type of stochastic process. The ideas presented can be illustrated by considering the following maintenance-model example. A production process contains a machine that deteriorates rapidly in both quality and output under heavy usage, so that it is inspected periodically, say, at the end of each day. Immediately after inspection, the condition of the machine is noted and classified into one of four possible states:

State	Condition
0	Good as new
1	Operable—minor deterioration
2	Operable—major deterioration
3	Inoperable—output of unacceptable quality

Let X_t denote the observed state of the machine after inspection at the end of the tth day. It is reasonable to assume that the state of the system evolves according to some probabilistic "laws of motion," so that the sequence of states $\{X_t\}$ can be viewed as a stochastic process. Furthermore, it will be assumed that the stochastic process is a finite-state Markov chain (see Sec. 5.3), with known transition matrix given by

State	0	1	2	3
0	0	$\frac{7}{8}$	$\frac{1}{16}$	$\frac{1}{16}$
1	0	$\frac{3}{4}$	$\frac{1}{8}$	$\frac{1}{8}$
2	0	0	$\frac{1}{2}$	$\frac{1}{2}$
3	0	0	0	1

From this transition matrix, it becomes evident that once the machine becomes inoperable (enters state 3), it remains inoperable. Therefore, the analysis of this stochastic process is probably uninteresting because state 3 is an absorbing state, and eventually the machine will enter this state and just remain there; i.e., after some time period, X_t will always equal 3. Clearly, from a practical point of view, this model is intolerable because a machine that is inoperable cannot continue to remain in the production process and must be replaced (or repaired). This action of replacement alters the behavior of the system, so that the system now evolves over time according to the joint effect of the probabilistic laws of motion and the action of replacing an inoperable machine. Note that the action of replacing an inoperable machine can be thought of as defining a maintenance policy.

When a machine becomes inoperable and is replaced, the replacement machine is as good as new; i.e., the machine is found to be in state 0 at the time of the regular inspection at the end of the next day. As a practical matter, the replacement process can be thought of as taking 1 day to complete so that production is lost for this period.

The costs incurred while this system evolves contains several components. When the system is in state 0, 1, or 2, defective items may be produced during the next day, and the expected costs are given by

State	Expected Cost Due to Producing Defective Items
0	0
1	\$1,000
2	\$3,000

If the machine is replaced, a replacement cost of \$4,000 is incurred, together with a cost of lost production (lost profit) of \$2,000. Hence the total cost incurred whenever the system is in state 3 is \$6,000.

The stochastic process resulting from the system with the aforementioned maintenance policy, i.e., replacing an inoperable machine, is still a finite-state Markov chain, but with the transition matrix now given by

State	0	1	2	3
0	0	$\frac{7}{8}$	$\frac{1}{16}$	$\frac{1}{16}$
1	0	$\frac{3}{4}$	$\frac{1}{8}$	$\frac{1}{8}$
2	0	0	$\frac{1}{2}$	$\frac{1}{2}$
3	1	0	0	0

It may be interesting to evaluate the cost of this maintenance policy. If the (long-run) expected average cost per day or the (long-run) actual average cost per day is an appropriate measure, the results appearing in Sec. 5.7 (under the subsections on expected average cost per unit time or expected average cost per unit time for complex cost functions) are appropriate.

By noting that $p_{ij}^{(4)} > 0$ for all i and j, it is evident that every state is positive recurrent and belongs to one class. The steady-state equations can be written as

$$\begin{aligned}
\pi_0 &= \pi_3, \\
\pi_1 &= \tfrac{7}{8}\pi_0 + \tfrac{3}{4}\pi_1, \\
\pi_2 &= \tfrac{1}{16}\pi_0 + \tfrac{1}{8}\pi_1 + \tfrac{1}{2}\pi_2, \\
\pi_3 &= \tfrac{1}{16}\pi_0 + \tfrac{1}{8}\pi_1 + \tfrac{1}{2}\pi_2, \\
1 &= \pi_0 + \pi_1 + \pi_2 + \pi_3.
\end{aligned}$$

The simultaneous solution is

$$\begin{aligned}
\pi_0 &= \tfrac{2}{13}, \\
\pi_1 &= \tfrac{7}{13}, \\
\pi_2 &= \tfrac{2}{13}, \\
\pi_3 &= \tfrac{2}{13}.
\end{aligned}$$

Hence the long-run expected average cost per day is given by

$$0\pi_0 + 1{,}000\pi_1 + 3{,}000\pi_2 + 6{,}000\pi_3 = \frac{25{,}000}{13} = \$1{,}923.08,$$

and this cost represents the cost of this maintenance policy.

11.2 Markovian Decision Models

The previous section introduced an example of a maintenance model for a machine and presented a maintenance policy; i.e., when a machine becomes inoperable, it is replaced; otherwise, the machine is left alone. In other words, a decision is made to take the action *replace the machine* when it is found to be in state 3, whereas a decision is made to take the action *leave the machine as is* when it is found to be in state 0, 1, or 2. Even when these two actions are the only permissible ones, there are still other policies that can be generated; e.g., when the machine becomes inoperable or is found to be operable but with major deterioration (machine is in state 2 or 3), replace it; otherwise, leave the machine as is. Note that this policy generates a different transition matrix, i.e.,

State	0	1	2	3
0	0	$\frac{7}{8}$	$\frac{1}{16}$	$\frac{1}{16}$
1	0	$\frac{3}{4}$	$\frac{1}{8}$	$\frac{1}{8}$
2	1	0	0	0
3	1	0	0	0

To make the machine-maintenance example more realistic, suppose that a third action is permitted: overhaul. When a machine is overhauled, the machine is returned to state 1 (operable—minor deterioration) at the time of the regular inspection at the end of the next day. As a practical matter, the overhaul process, like replacement, can be thought of as requiring a day to complete, so that production is lost for this period. Furthermore, overhauling the machine costs \$2,000 and will not be considered as a viable decision when the machine becomes inoperable.

In viewing this dynamic system, it is evident that the system evolves over time according to the joint effect of the probabilistic laws of motion and the sequence of decisions made (actions taken). In particular, the machine is inspected at the end of each day and its state is recorded. A decision as to which action to take must be made, i.e.,

Decision	Action
1	Do nothing
2	Overhaul (return system to state 1)
3	Replace (return system to state 0)

For the general model it will be assumed that a system is observed at time $t = 0, 1, \ldots,$ and classified into one of a finite number of states labeled $0, 1, \ldots, M$. Let $\{X_t, t = 0, 1, \ldots\}$ denote the sequence of observed states. After each observation, one of K (finite) possible decisions (actions), labeled $1, 2, \ldots, K$, is taken.[1] Let $\{\Delta_t, t = 0, 1, \ldots\}$ denote the sequence of actual decisions made.

A policy, denoted by R, is a rule for making decisions at each point in time. In principle, a policy could use all the previously observed information up to time t, that is, the entire history of the system consisting of $X_0, X_1, \ldots, X_t$ and $\Delta_0, \Delta_1, \Delta_2, \ldots, \Delta_{t-1}$. However, for most problems encountered in practice, it is sufficient to confine consideration to those policies that depend upon only the observed state of the system at time t, X_t, and the possible decisions available. Hence a policy R can be viewed as a rule that prescribes decision $d_i(R)$ when the system is in state i, $i = 0, 1, \ldots, M$. Thus R is completely characterized by the values

$$\{d_0(R), d_1(R), \ldots, d_M(R)\}.$$

[1] In general, the number of possible decisions may depend upon the state of the system. Such a case is considered in Sec. 11.7.

Note that this description assumes that whenever the system is in state i, the decision to be made is the same for all values of t. Policies possessing this property are called *stationary policies*.

In the example, the interesting policies are

Policy	Verbal Description	$d_0(R)$	$d_1(R)$	$d_2(R)$	$d_3(R)$
R_a	Replace in state 3	1	1	1	3
R_b	Replace in state 3, overhaul in state 2	1	1	2	3
R_c	Replace in states 2, 3	1	1	3	3
R_d	Replace in states 1, 2, 3	1	3	3	3

Note that policy R_a is the policy described in the previous section, and R_c is the policy alluded to earlier in this section. Furthermore, recall that each policy results in a different transition matrix.

It has been noted that a system evolves over time according to the joint effect of the probabilistic laws of motion and the sequence of decisions made; its path is dependent upon its initial state, X_0. It is assumed that whenever the system is in state i and decision $d_i(R) = k$ is made, the system moves to a new state j, with known transition probability $p_{ij}(k)$, for all $i, j = 0, 1, \ldots, M$ and $k = 1, 2, \ldots, K$. Thus, if a given policy R is followed, the resultant stochastic process is a Markov chain with a known transition matrix (dependent upon the policy chosen). Unless otherwise noted, throughout this chapter it is assumed for technical reasons that the *Markov chain associated with every transition matrix is irreducible.*

In the example, the following transition matrices are obtained:

	R_a			
State	0	1	2	3
0	0	$\frac{7}{8}$	$\frac{1}{16}$	$\frac{1}{16}$
1	0	$\frac{3}{4}$	$\frac{1}{8}$	$\frac{1}{8}$
2	0	0	$\frac{1}{2}$	$\frac{1}{2}$
3	1	0	0	0

	R_b			
State	0	1	2	3
0	0	$\frac{7}{8}$	$\frac{1}{16}$	$\frac{1}{16}$
1	0	$\frac{3}{4}$	$\frac{1}{8}$	$\frac{1}{8}$
2	0	1	0	0
3	1	0	0	0

	R_c			
State	0	1	2	3
0	0	$\frac{7}{8}$	$\frac{1}{16}$	$\frac{1}{16}$
1	0	$\frac{3}{4}$	$\frac{1}{8}$	$\frac{1}{8}$
2	1	0	0	0
3	1	0	0	0

	R_d			
State	0	1	2	3
0	0	$\frac{7}{8}$	$\frac{1}{16}$	$\frac{1}{16}$
1	1	0	0	0
2	1	0	0	0
3	1	0	0	0

To summarize, given a distribution $P\{X_0 = i\}$ over the initial states of the system and a policy R, a system evolves over time according to the joint effect of the probabilistic laws of motion and the sequence of decisions made (actions taken). In particular, when the system is in state i and decision $d_i(R) = k$ is made, then the probability that the system is in state j at the next observed time period is given by $p_{ij}(k)$. This situation results in a sequence of observed states $X_0, X_1, \ldots$ and a sequence of decisions made, $\Delta_0, \Delta_1, \ldots$. This sequence of observed states and sequence of decisions made is called a **Markovian decision process**. The term *Markovian* is used because of the underlying assumptions made about the probabilistic laws of motion.

Four maintenance policies have been described, but their properties have not been evaluated. Questions such as "Which one is 'best'?" remain to be answered. To pursue this avenue, it is necessary to introduce a cost structure. When the system is in state i and decision $d_i(R) = k$ is made following policy R, a known cost C_{ik} is incurred. This cost may represent an expected rather than an actual cost. For example, in the maintenance problem, the cost of leaving a machine as is depends upon the random variable, the number of defective items produced during the next time period. The expected value of this cost function taken with respect to the distribution of the number of defective items will result in the desired cost C_{ik}.[1] It is important to reiterate that this cost depends upon only the state the system is found in and the decision made; that is,

$$C_{ik} = \text{known (expected) cost incurred during next transition if system is in state } i \text{ and decision } k \text{ is made.}$$

For the four maintenance policies, the costs can be obtained from the following information:

Decision	State \ Cost	Expected Cost Due to Producing Defective Items	Maintenance Cost	Cost (Lost Profit) of Lost Production	Total Cost Per Day
1. Leave machine as is	0	0	0	0	0
	1	\$1,000	0	0	\$1,000
	2	\$3,000	0	0	\$3,000
	3	∞*	0	0	∞
2. Overhaul	0,1,2	0	\$2,000	\$2,000	\$4.000
	3	0	∞*	\$2,000	∞
3. Replace	0,1,2,3	0	\$4,000	\$2,000	\$6,000

* Because leaving the machine in an inoperable condition or overhauling it when it is inoperable is prohibited by assumption, a cost of infinity is assigned. An alternative approach would be to omit these decisions from the set of possible decisions when the machine is found to be in state 3.

Note that the costs incurred when the decision is made to replace the machine are independent of the state of the system. This fact is evident because no production takes place during the ensuing day when this action is taken. Finally, the total expected

[1] See Sec. 5.7 under the subsection on expected average cost per unit time for complex cost functions for an additional example.

costs incurred per day are summarized as follows:

State \ Decision	C_{ik} (In Thousands of Dollars) 1	2	3
0	0	4	6
1	1	4	6
2	3	4	6
3	∞	∞	6

To compare policies, we must settle on an appropriate cost measure. One such measure associated with a policy is the (long-run) expected average cost per unit time; this measure will be the one used.[1] The results appearing in Sec. 5.7 are appropriate; i.e., for any policy, the (long-run) expected average cost per unit time, $E(C)$, can be calculated from the expression

$$E(C) = \sum_{i=0}^{M} C_{ik}\pi_i,$$

where $k = d_i(R)$ for each i, and $(\pi_0, \pi_1, \ldots, \pi_M)$ represents the steady-state distribution of the state of the system under the policy R being evaluated. Thus the policy that minimizes $E(C)$ is sought. Using this criterion, it is evident that the distribution over the initial states of the system is not important, because the long-run effect of the cost of the initial decision is negligible. In the maintenance example, it is necessary to solve for $(\pi_0, \pi_1, \ldots, \pi_M)$ under each of the four policies of interest and then use these results to obtain $E(C)$. The necessary calculations for R_a are given in Sec. 11.1; all are now summarized:

Policy	$\pi_0, \pi_1, \pi_2, \pi_3$	$E(C)$
R_a	$(\frac{2}{13}, \frac{7}{13}, \frac{2}{13}, \frac{2}{13})$	$\frac{1}{13}[2(0) + 7(1) + 2(3) + 2(6)] = \frac{25}{13} = 1.923$
R_b	$(\frac{2}{21}, \frac{5}{7}, \frac{2}{21}, \frac{2}{21})$	$\frac{1}{21}[2(0) + 15(1) + 2(4) + 2(6)] = \frac{35}{21} = 1.667$ min.
R_c	$(\frac{2}{11}, \frac{7}{11}, \frac{1}{11}, \frac{1}{11})$	$\frac{1}{11}[2(0) + 7(1) + 1(6) + 1(6)] = \frac{19}{11} = 1.727$
R_d	$(\frac{1}{2}, \frac{7}{16}, \frac{1}{32}, \frac{1}{32})$	$\frac{1}{32}[16(0) + 14(6) + 1(6) + 1(6)] = \frac{96}{32} = 3$

It is evident that policy R_b is the best. Among the four policies considered, the policy that calls for replacing the machine when it is found to be in state 3 and overhauling it when it is found to be in state 2 is the best, and the (long-run) expected average cost per day is $1,667.

The technique described here is just an exhaustive enumeration of a given set of possible policies. It is evident that direct enumeration becomes cumbersome when the number of policies is large and algorithms are desirable for finding optimal policies. The next three sections consider such algorithms.

[1] The ensuing results are also valid for the (long-run) actual average cost per unit time measure as noted in Sec. 5.7.

11.3 Linear Programming and Optimal Policies

Section 11.2 defined a policy, and we saw that a policy R can be viewed as a rule that prescribes decision $d_i(R)$ when the system is in state i. Thus R is characterized by the values

$$\{d_0(R), d_1(R), \ldots, d_M(R)\}.$$

Alternatively, R can be characterized by assigning values $D_{ik} = 0$ or 1 in the matrix

$$\text{State}\quad \begin{array}{c} \\ 0 \\ 1 \\ \vdots \\ M \end{array}\!\!\begin{array}{c} \text{Decision, } k \\ \begin{array}{cccc} 1 & 2 & \cdots & K \end{array} \\ \begin{bmatrix} D_{01} & D_{02} & \cdots & D_{0K} \\ D_{11} & D_{12} & \cdots & D_{1K} \\ \vdots & & & \vdots \\ D_{M1} & D_{M2} & \cdots & D_{MK} \end{bmatrix} \end{array},$$

where each row must contain a single 1 with the rest of the elements zero (i.e., each row sums to 1). When an element $D_{ik} = 1$, it can be interpreted as calling for decision k when the system is in state i. In the maintenance-model example, policy R_b can be characterized by the matrix

$$\text{State}\quad \begin{array}{c} \\ 0 \\ 1 \\ 2 \\ 3 \end{array}\!\!\begin{array}{c} \text{Decision, } k \\ \begin{array}{ccc} 1 & 2 & 3 \end{array} \\ \begin{bmatrix} 1 & 0 & 0 \\ 1 & 0 & 0 \\ 0 & 1 & 0 \\ 0 & 0 & 1 \end{bmatrix} \end{array};$$

i.e., replace the machine when it is in state 3, overhaul the machine when it is in state 2, and leave the machine as is when it is in state 0 or 1. This interpretation of the D_{ik} provides motivation for a linear programming formulation. It is hoped that the expected cost of a policy can be expressed as a linear function of the D_{ik} or a related variable, subject to linear constraints. Unfortunately, the D_{ik} are integers (zero or 1), and continuous variables are required for a linear programming formulation. This requirement can be handled by expanding the interpretation of a policy. The previous definition calls for making the same decision every time the system is in state i. The new interpretation of a policy will call for determining a probability distribution for the decision to be made when the system is in state i. Thus D_{ik} can now be viewed as[1]

$$D_{ik} = P\{\text{decision} = k | \text{state} = i\}, \quad k = 1, 2, \ldots, K,$$
$$i = 0, 1, \ldots, M.$$

Such a policy is called a *randomized policy,* whereas the policy calling for $D_{ik} = 0$ or 1 can be called a *deterministic policy*. Randomized policies can again be

[1] The right-hand side of this equation is read as the conditional probability that the decision k is made, given the system is in state i.

characterized by the matrix

$$
\begin{array}{c c}
 & \text{Decision, } k \\
 & \begin{array}{cccc} 1 & 2 & \cdots & K \end{array} \\
\text{State} \begin{array}{c} 0 \\ 1 \\ \vdots \\ M \end{array} & \left[\begin{array}{cccc} D_{01} & D_{02} & \cdots & D_{0K} \\ D_{11} & D_{12} & \cdots & D_{1K} \\ & & \vdots & \\ D_{M1} & D_{M2} & \cdots & D_{MK} \end{array}\right],
\end{array}
$$

where each row sums to 1, and now

$$0 \le D_{ik} \le 1.$$

Note that each row $(D_{i1}, D_{i2}, \ldots, D_{ik})$ is the probability distribution for the decision to be made when the system is in state i. As an example, suppose that a new policy, R_e, is to be used in the maintenance model. This policy is a randomized policy and is given by the matrix

$$
\begin{array}{c c}
 & \text{Decision, } k \\
 & \begin{array}{ccc} 1 & 2 & 3 \end{array} \\
\text{State} \begin{array}{c} 0 \\ 1 \\ 2 \\ 3 \end{array} & \left[\begin{array}{ccc} 1 & 0 & 0 \\ 1 & 0 & 0 \\ \frac{1}{4} & \frac{1}{4} & \frac{1}{2} \\ 0 & \frac{1}{2} & \frac{1}{2} \end{array}\right].
\end{array}
$$

This policy calls for observing the state of the machine at the end of the day. If it is found to be in state 0 or 1, it is left as is. If it is found to be in state 2, it is left as is with probability $\frac{1}{4}$, overhauled with probability $\frac{1}{4}$, and replaced with probability $\frac{1}{2}$. Presumably, a random device with these probabilities (possibly a table of random numbers) can be used to make the actual decision. Finally, if the machine is found to be in state 3, it is overhauled with probability $\frac{1}{2}$ and replaced with probability $\frac{1}{2}$.

The *linear programming formulation* is best expressed in terms of a variable y_{ik}, which is related to D_{ik} as follows. Let y_{ik} be the steady-state unconditional probability that the system is in state i *and* decision k is made; that is,

$$y_{ik} = P\{\text{state} = i \text{ and decision} = k\}.$$

From the rules of conditional probability,

$$y_{ik} = \pi_i D_{ik}.$$

Furthermore,

$$\pi_i = \sum_{k=1}^{K} y_{ik},$$

so that

$$D_{ik} = \frac{y_{ik}}{\pi_i} = \frac{y_{ik}}{\sum_{k=1}^{K} y_{ik}}.$$

There exist several constraints on y_{ik}:

1. $\sum_{i=0}^{M} \pi_i = 1$, so that $\sum_{i=0}^{M} \sum_{k=1}^{K} y_{ik} = 1$.

2. From results on steady-state probabilities (see Sec. 5.7),[1]

$$\pi_j = \sum_{i=0}^{M} \pi_i p_{ij}, \qquad \text{so that } \sum_{k=1}^{K} y_{jk} = \sum_{i=0}^{M} \sum_{k=1}^{K} y_{ik} p_{ij}(k), \qquad \text{for } j = 0, 1, \ldots, M.$$

3. $y_{ik} \geq 0$, $i = 0, 1, \ldots, M$ and $k = 1, 2, \ldots, K$.

The long-run expected average cost per unit time is given by

$$E(C) = \sum_{i=0}^{M} \sum_{k=1}^{K} \pi_i C_{ik} D_{ik} = \sum_{i=0}^{M} \sum_{k=1}^{K} C_{ik} y_{ik}.$$

Hence the problem is to choose the y_{ik} that

$$\text{Minimizes} \quad \sum_{i=0}^{M} \sum_{k=1}^{K} C_{ik} y_{ik},$$

subject to the constraints

(1) $$\sum_{i=0}^{M} \sum_{k=1}^{K} y_{ik} = 1.$$

(2) $$\sum_{k=1}^{K} y_{jk} - \sum_{i=0}^{M} \sum_{k=1}^{K} y_{ik} p_{ij}(k) = 0, \quad \text{for } j = 0, 1, \ldots, M.$$

(3) $$y_{ik} \geq 0, \quad i = 0, 1, \ldots, M; \quad k = 1, 2, \ldots, K.$$

This formulation is clearly a *linear programming problem* that can be solved by the simplex method. Once the y_{ik} are obtained, D_{ik} is easily found from

$$D_{ik} = \frac{y_{ik}}{\sum_{k=1}^{K} y_{ik}}.$$

The solution has some interesting properties. It will contain $(M + 1)$ basic variables $y_{ik} \geq 0$ (there is one redundant constraint). It can be shown that $y_{ik} > 0$ for at least one $k = 1, 2, \ldots, K$, for each $i = 0, 1, \ldots, M$. Therefore, it follows that $y_{ik} > 0$ for only *one* k for each $i = 0, 1, \ldots, M$; that is, $D_{ik} = 0$ or 1. In other words, the optimal policy is *deterministic* rather than *randomized*. Finally, since there are $(M + 2)$ functional constraints and $K(M + 1)$ original variables, "practical" problems tend to be large under this formulation, so that solutions may not be obtainable even with the simplex method.

EXAMPLE: We can formulate the machine-maintenance problem as a *linear program*; i.e.:

$$\begin{aligned} \text{Minimize} \quad & 4{,}000y_{02} + 6{,}000y_{03} + 1{,}000y_{11} + 4{,}000y_{12} + 6{,}000y_{13} \\ & + 3{,}000y_{21} + 4{,}000y_{22} + 6{,}000y_{23} + M_1 y_{31} + M_2 y_{32} \\ & + 6{,}000y_{33}, \end{aligned}$$

[1] The k is introduced in $p_{ij}(k)$ to indicate that the appropriate transition probability depends upon the decision k.

where M_1 and M_2 are taken to be large numbers, subject to

$$\sum_{i=0}^{3}\sum_{k=1}^{3} y_{ik} = 1,$$

$$\sum_{k=1}^{3} y_{0k} - (y_{03} + y_{13} + y_{23} + y_{33}) = 0,$$

$$\sum_{k=1}^{3} y_{1k} - (\tfrac{7}{8}y_{01} + y_{02} + \tfrac{3}{4}y_{11} + y_{12} + y_{22} + y_{32}) = 0,$$

$$\sum_{k=1}^{3} y_{2k} - (\tfrac{1}{16}y_{01} + \tfrac{1}{8}y_{11} + \tfrac{1}{2}y_{21}) = 0,$$

$$\sum_{k=1}^{3} y_{3k} - (\tfrac{1}{16}y_{01} + \tfrac{1}{8}y_{11} + \tfrac{1}{2}y_{21} + y_{31}) = 0,$$

and $\quad y_{ik} \geq 0, \quad i = 0, 1, 2, 3 \quad$ and $\quad k = 1, 2, 3.$

This linear program can be solved by using the simplex method.

The results yield all y_{ik} equal to zero, except for $y_{01} = \frac{2}{21}$, $y_{11} = \frac{5}{7}$, $y_{22} = \frac{2}{21}$, and $y_{33} = \frac{2}{21}$. Note that these values are just the steady-state probabilities for policy R_b, which is now seen to be the optimal policy. The corresponding

$$D_{ik} = \frac{y_{ik}}{\sum_{k=1}^{3} y_{ik}}$$

are given by $\quad D_{01} = D_{11} = D_{22} = D_{33} = 1,$

and all the remaining $D_{ik} = 0$. This policy calls for leaving the machine as is when it is in state 0 or 1, overhauling it when it is in state 2, and replacing it when it is in state 3.

11.4 Policy Improvement Algorithms for Finding Optimal Policies

A second algorithm for finding optimal policies is given by a *policy improvement technique*. The algorithm to be presented is useful in that it often leads to finding the optimal policy quickly, and also it is applicable under more general conditions than previously specified; e.g., under certain assumptions, the number of states may be countably infinite rather than finite.

Following the model of Sec. 11.2, and as a joint result of the current state i of the system and the decision $d_i(R) = k$ when operating under policy R, two things occur. A (expected) cost C_{ik} that depends upon only the observed state of the system and the decision made is incurred. The system moves to a new state j at the next observed time period, with transition probability given by $p_{ij}(k)$. If, in fact, a cost that depends upon both the initial and transited states is incurred, it is treated as follows. Denote by $q_{ij}(k)$ the (expected) cost incurred when the system is in state i and decision k is made, and then evolves to state j at the next observed time period. Then

$$C_{ik} = \sum_{j=0}^{M} q_{ij}(k)p_{ij}(k).$$

When a system operates as just described under policy R, it can be shown that there exist values $g(R)$, $v_0(R)$, $v_1(R)$, . . . , $v_M(R)$ that satisfy

$$g(R) + v_i(R) = C_{ik} + \sum_{j=0}^{M} p_{ij}(k)v_j(R), \qquad \text{for } i = 0, 1, 2, \ldots, M.$$

A heuristic justification for these relationships and an interpretation for these values are desirable. Denote by $v_i^n(R)$ the total expected cost of a system starting in state i (at the first observed time period) and evolving for n time periods. Then $v_i^n(R)$ consists of two components, namely, (1) C_{ik}, the cost incurred at the first observed time period as a result of the current state i and the decision $d_i(R) = k$ when operating under policy R, and (2) $\sum_{j=0}^{M} p_{ij}(k)v_j^{n-1}(R)$, the total expected cost of the system evolving over the remaining $n - 1$ time periods. Thus the recursive equation,

$$v_i^n(R) = C_{ik} + \sum_{j=0}^{M} p_{ij}(k)v_j^{n-1}(R),$$

for $i = 0, 1, 2, \ldots, M$ and $v_i^1(R) = C_{ik}$ for all i is obtained. It is of interest to explore the behavior of the total expected cost $v_i^n(R)$ as n gets large. Now, it is known that the long-run expected average cost *per unit time* following any policy R can be expressed as

$$g(R) = \sum_{i=0}^{M} \pi_i C_{ik},$$

which is independent of the starting state i. Hence $v_i^n(R)$ behaves approximately as $ng(R)$ for large n, and, in fact, can be expressed (neglecting certain fluctuations) as the sum of two components, one of which is independent of the initial state and one of which is dependent upon it; that is,

$$v_i^n(R) \approx ng(R) + v_i(R),$$

where $v_i(R)$ can be interpreted as the effect on the total expected cost due to starting in state i. Thus

$$v_i^n(R) - v_j^n(R) \approx v_i(R) - v_j(R),$$

so that $v_i(R) - v_j(R)$ is a measure of the effect of starting in state i rather than state j.

Substituting this linear approximation for $v_i^n(R)$ (assumed to be valid for large n) into the recursive equation for $v_i^n(R)$ leads to

$$g(R) + v_i(R) = C_{ik} + \sum_{j=0}^{M} p_{ij}(k)v_j(R),$$

for $i = 0, 1, \ldots, M$, so that these values satisfy the expressed equations.

Note that there are $M + 1$ equations with $M + 2$ unknowns, so that one of these variables may be chosen arbitrarily. By convention, $v_M(R)$ will be chosen equal to zero. Therefore, by solving a system of linear equations, the (long-run) expected average cost per unit time following policy R, $g(R)$ can be obtained. In principle, all policies can be enumerated, and that policy which minimizes $g(R)$ can be found. However, even for a moderate number of states and decisions, this technique is cumbersome. Fortunately, there exists an algorithm that can be used to evaluate policies and find the optimum one without complete enumeration. The algorithm begins by choosing an arbitrary policy R_1 and calculates the values of $g(R_1)$, $v_0(R_1)$, $v_1(R_1)$, . . . , $v_{M-1}(R_1)$ [recall that $v_M(R_1)$ is chosen equal to zero]. This step is called *value determination*. A better policy, denoted by R_2, is then constructed. This step is called *policy improvement*. Using the new policy R_2, we repeat the value-determination step. These steps continue until two successive iterations lead to identical policies, which signifies that the optimal policy has been obtained. In particular, the following steps are to be followed:

Step 1: Value determination For an arbitrarily chosen policy R_1, use $p_{ij}(k_1)$, C_{ik_1}, and $v_M(R_1) = 0$ to solve the set of $(M + 1)$ equations,

$$g(R_1) = C_{ik_1} + \sum_{j=0}^{M} p_{ij}(k_1)v_j(R_1) - v_i(R_1), \qquad i = 0, 1, \ldots, M,$$

for all $(M + 1)$ unknown values of $g(R_1)$, $v_0(R_1)$, $v_1(R_1)$, . . . , $v_{M-1}(R_1)$.

Step 2: Policy improvement Using the current values of $v_i(R_1)$ computed for policy R_1, find the alternative policy R_2 such that, for each state i, $d_i(R_2) = k_2$ is the decision that makes

$$C_{ik_2} + \sum_{j=0}^{M} p_{ij}(k_2)v_j(R_1) - v_i(R_1)$$

a minimum; i.e., for *each* state i, find the appropriate value of k_2 that

$$\underset{k_2=1,2,\ldots,K}{\text{Minimizes}} \quad \left\{ C_{ik_2} + \sum_{j=0}^{M} p_{ij}(k_2)v_j(R_1) - v_i(R_1) \right\},$$

and then set $d_i(R_2)$ equal to the minimizing value of k_2. This procedure defines a new policy, R_2.

If R_2 does not equal R_1, then return to step 1, using R_2 instead of R_1, and solve for $g(R_2)$, $v_0(R_2)$, $v_1(R_2)$, . . . , $v_{n-1}(R_2)$. Using these values, go to step 2 and find R_3. Continue in this fashion until you find two successive R's to be equal. When you find them, the optimal policy is achieved, and the algorithm terminates. In fact, it can be shown that

1. $g(R_{j+1}) \leq g(R_j)$, for $j = 1, 2, \ldots$; and
2. The algorithm terminates with the optimal solution in a finite number of iterations.

EXAMPLE: We shall solve the maintenance model presented in Sec. 11.2 by the *policy improvement algorithm*. Recall that the machine can be in one of four states: state 0, signifying the machine is as good as new; state 1, signifying the machine is operable with minor deterioration; state 2, signifying the machine is operable with major deterioration; and state 3, signifying the machine is inoperable. There exist three possible decisions: decision 1 implies leaving the machine as is; decision 2 implies overhaul, which returns it to state 1; and decision 3 implies replacement, which returns it to state 0. Each decision necessitates an action that affects the transition matrix, and there are costs C_{ik} associated with making decision k when the system is in state i. We want to find the optimal policy, and step 1 of the algorithm calls for choosing a policy arbitrarily. Choose the policy that calls for replacement of the machine when it is found to be in state 3; otherwise, leave the machine as is. Denote this policy by R_1. The transition matrix for this policy is given by

State	0	1	2	3
0	0	$\frac{7}{8}$	$\frac{1}{16}$	$\frac{1}{16}$
1	0	$\frac{3}{4}$	$\frac{1}{8}$	$\frac{1}{8}$
2	0	0	$\frac{1}{2}$	$\frac{1}{2}$
3	1	0	0	0

The costs incurred following policy R_1 are given by

State	C_{ik_1}
0	0
1	1,000
2	3,000
3	6,000

With this policy, the value-determination step requires solving the following four equations simultaneously for $g(R_1)$, $v_0(R_1)$, $v_1(R_1)$, and $v_2(R_1)$ [recall that $v_3(R_1)$ is arbitrarily taken to be zero]:

$$\begin{aligned} g(R_1) &= C_{0k_1} + \sum_{j=0}^{3} p_{0j}(k_1)v_j(R_1) - v_0(R_1) \\ &= C_{1k_1} + \sum_{j=0}^{3} p_{1j}(k_1)v_j(R_1) - v_1(R_1) \\ &= C_{2k_1} + \sum_{j=0}^{3} p_{2j}(k_1)v_j(R_1) - v_2(R_1) \\ &= C_{3k_1} + \sum_{j=0}^{3} p_{3j}(k_1)v_j(R_1) - v_3(R_1), \end{aligned}$$

or alternatively [with $v_3(R_1) = 0$],

$$\begin{aligned} g(R_1) &= + \tfrac{7}{8}v_1(R_1) + \tfrac{1}{16}v_2(R_1) - v_0(R_1) \\ &= 1,000 + \tfrac{3}{4}v_1(R_1) + \tfrac{1}{8}v_2(R_1) - v_1(R_1) \\ &= 3,000 + \tfrac{1}{2}v_2(R_1) - v_2(R_1) \\ &= 6,000 + v_0(R_1). \end{aligned}$$

The simultaneous solution to this system of equations yields

$$g(R_1) = \frac{25,000}{13} = 1,923$$

$$v_0(R_1) = -\frac{53,000}{13} = -4,077$$

$$v_1(R_1) = -\frac{34,000}{13} = -2,615$$

$$v_2(R_1) = \frac{28,000}{13} = 2,154.$$

Step 2 can now be applied. It is necessary to find the improved policy R_2, which has the property that $d_0(R_2) = k_2^0$, $d_1(R_2) = k_2^1$, $d_2(R_2) = k_2^2$, and $d_3(R_2) = k_2^3$ minimize the following expressions:

(0) $C_{0k_2^0} - p_{00}(k_2^0)4,077 - p_{01}(k_2^0)2,615 + p_{02}(k_2^0)2,154 + 4,077$

(1) $C_{1k_2^1} - p_{10}(k_2^1)4,077 - p_{11}(k_2^1)2,615 + p_{12}(k_2^1)2,154 + 2,615$

(2) $C_{2k_2^2} - p_{20}(k_2^2)4,077 - p_{21}(k_2^2)2,615 + p_{22}(k_2^2)2,154 - 2,154$

(3) $C_{3k_2^3} - p_{30}(k_2^3)4,077 - p_{31}(k_2^3)2,615 + p_{32}(k_2^3)2,154.$

To find k_2^0, the "best" decision when the machine is in state 0, it is necessary to evaluate the first expression for all possible decisions. Note that the appropriate transition probabilities and the costs C_{0k} depend upon the decisions made. A summary of the necessary calculations follows:

State 0

Decision	$p_{00}(k_2)$	$p_{01}(k_2)$	$p_{02}(k_2)$	$p_{03}(k_2)$	C_{0k_2}	Value of Expression 0
1	0	$\frac{7}{8}$	$\frac{1}{16}$	$\frac{1}{16}$	0	1,923
2	0	1	0	0	4,000	5,462
3	1	0	0	0	6,000	6,000

It is clear that $d_0(R_2) = k_2^0 = 1$ minimizes this first expression, so that under R_2 the appropriate decision when the system is in state 0 is to leave the machine as is.

Similar calculations are required to find $d_1(R_2) = k_2^1$, $d_2(R_2) = k_2^2$, and $d_3(R_2) = k_2^3$; these calculations are summarized as follows:

State 1

Decision	$p_{10}(k_2)$	$p_{11}(k_2)$	$p_{12}(k_2)$	$p_{13}(k_2)$	C_{1k_2}	Value of Expression 1
1	0	$\frac{3}{4}$	$\frac{1}{8}$	$\frac{1}{8}$	1,000	1,923
2	0	1	0	0	4,000	4,000
3	1	0	0	0	6,000	4,538

State 2

Decision	$p_{20}(k_2)$	$p_{21}(k_2)$	$p_{22}(k_2)$	$p_{23}(k_2)$	C_{2k_2}	Value of Expression 2
1	0	0	$\frac{1}{2}$	$\frac{1}{2}$	3,000	1,923
2	0	1	0	0	4,000	−769
3	1	0	0	0	6,000	−231

State 3

Decision	$p_{30}(k_2)$	$p_{31}(k_2)$	$p_{32}(k_2)$	$p_{33}(k_2)$	C_{3k_2}	Value of Expression 3
1	0	0	0	1	∞	∞
2	0	1	0	0	∞	∞
3	1	0	0	0	6,000	1,923

Thus $d_1(R_2) = k_2^1 = 1$, $d_2(R_2) = k_2^2 = 2$, and $d_3(R_2) = k_2^3 = 3$. Hence policy R_2 calls for leaving the machine alone when it is in state 0 or 1, overhauling it when it is in state 2, and replacing it when it is in state 3. Furthermore, because R_2 differs from R_1, at least one more iteration is required. The equations that must now be solved are given by [again setting $v_3(R_2) = 0$]

$$\begin{aligned} g(R_2) &= \qquad\qquad + \tfrac{7}{8}v_1(R_2) + \tfrac{1}{16}v_2(R_2) - v_0(R_2) \\ &= 1{,}000 \quad + \tfrac{3}{4}v_1(R_2) + \tfrac{1}{8}v_2(R_2) - v_1(R_2) \\ &= 4{,}000 \quad + v_1(R_2) \qquad\qquad\quad - v_2(R_2) \\ &= 6{,}000 + v_0(R_2). \end{aligned}$$

The simultaneous solution to these equations yields

$$g(R_2) = \frac{5{,}000}{3} = 1{,}667$$

$$v_0(R_2) = -\frac{13{,}000}{3} = -4{,}333$$

$$v_1(R_2) = -3{,}000$$

$$v_2(R_2) = -\frac{2{,}000}{3} = -667.$$

Step 2 can now be applied. We seek an improved policy R_3 that has the property that $d_0(R_3) = k_3^0$, $d_1(R_3) = k_3^1$, $d_2(R_3) = k_3^2$, and $d_3(R_3) = k_3^3$ minimizes the following expressions:

(0) $\quad C_{0k_3^0} - p_{00}(k_3^0)4{,}333 - p_{01}(k_3^0)3{,}000 - p_{02}(k_3^0)667 + 4{,}333$

(1) $\quad C_{1k_3^1} - p_{10}(k_3^1)4{,}333 - p_{11}(k_3^1)3{,}000 - p_{12}(k_3^1)667 + 3{,}000$

(2) $\quad C_{2k_3^2} - p_{20}(k_3^2)4{,}333 - p_{21}(k_3^2)3{,}000 - p_{22}(k_3^2)667 + 667$

(3) $\quad C_{3k_3^3} - p_{30}(k_3^3)4{,}333 - p_{31}(k_3^3)3{,}000 - p_{32}(k_3^3)667.$

The first iteration provides most of the necessary data (the transition probabilities and C_{ik}) required for determining the new policy, except for the values of each of the four expressions. These values are found to be

Decision	Value of Expression 0	Value of Expression 1	Value of Expression 2	Value of Expression 3
1	1,667	1,667	3,333	∞
2	5,333	4,000	1,667	∞
3	6,000	4,667	2,334	1,667

Thus $d_0(R_3) = k_3^0 = 1$, $d_1(R_3) = k_3^1 = 1$, $d_2(R_3) = k_3^2 = 2$, and $d_3(R_3) = k_3^3 = 3$, so this policy is identical to R_2. Because the policies on two successive iterations are the same, the optimal policy has been obtained. This optimal policy calls for leaving the machine as is when it is in state 0 or 1, overhauling it when it is in state 2, and replacing it when it is in state 3. Of course, this result is the same one found in Sec. 11.3.

11.5 Criterion of Discounted Costs

Throughout this chapter, we measured policies on the basis of the (long-run) expected average cost per unit time or the (long-run) actual average cost per unit time. An alternative measure is to find the expected long-run total discounted cost. A discount factor $\alpha < 1$ is specified, so that the present value of 1 unit of cost m periods in the future is α^m. α can be interpreted as equal to $1/(1 + i)$, where i is the current interest rate. This measure was used extensively in Chap. 9. We are seeking a policy that minimizes the expected long-run total discounted cost.

Policy Improvement Algorithm

The description of the Markovian decision process is as described previously. Given a distribution $P\{X_0 = i\}$ over the initial states of the system and a policy R, a system evolves over time according to the joint effect of the probabilistic laws of motion and the sequence of decisions made (actions taken). In particular, when the system is in state i and decision $d_i(R) = k$ is made, then the probability that the system is in state j at the next observed time period is given by $p_{ij}(k)$. Furthermore, a known expected cost C_{ik} is incurred. Denote by $V_i^n(R)$ the expected total discounted cost of a system starting in state i (at the first observed time period) and evolving for n time periods. Then $V_i^n(R)$ consists of two components, namely, (1) C_{ik}, the cost incurred at the first observed time period as a result of the current state i and the decision $d_i(R) = k$ when operating under policy R; (2) $\alpha \sum_{j=0}^{M} p_{ij}(k)V_j^{n-1}(R)$, the expected total discounted cost of the system evolving over the remaining $n - 1$ time periods. Thus the recursive equation,

$$V_i^n(R) = C_{ik} + \alpha \sum_{j=0}^{M} p_{ij}(k)V_j^{n-1}(R),$$

for $i = 0, 1, 2, \ldots, M$ and $V_i^1(R) = C_{ik}$ for all i is obtained. This policy can be evaluated by using the techniques associated with dynamic programming. It can be shown that as n approaches infinity, this expression converges to

$$V_i(R) = C_{ik} + \alpha \sum_{j=0}^{M} p_{ij}(k)V_j(R), \qquad \text{for } i = 0, 1, \ldots, M,$$

where $V_i(R)$ can now be interpreted as the expected long-run total discounted cost for a system starting in state i and continuing indefinitely. There are $M + 1$ equations and $M + 1$ unknowns, and hence $V_i(R)$ may be obtained by standard methods. For example, the machine-maintenance model will be solved using the policy that calls for leaving the machine as is when in state 0 or 1, overhauling it when it is in state 2, and replacing it when it is in state 3—that is, policy R_b. The discount factor will be chosen to be $\alpha = 0.9$. The following set of equations is obtained:

$$\begin{aligned}
V_0(R) &= \phantom{6{,}000} + 0.9[\tfrac{7}{8}V_1(R) + \tfrac{1}{16}V_2(R) + \tfrac{1}{16}V_3(R)]\\
V_1(R) &= 1{,}000 + 0.9[\tfrac{3}{4}V_1(R) + \tfrac{1}{8}V_2(R) + \tfrac{1}{8}V_3(R)]\\
V_2(R) &= 4{,}000 + 0.9[V_1(R)]\\
V_3(R) &= 6{,}000 + 0.9[V_0(R)].
\end{aligned}$$

The simultaneous solution to this system of equations yields

$$\begin{aligned}
V_0(R) &= 14{,}949\\
V_1(R) &= 16{,}262\\
V_2(R) &= 18{,}636\\
V_3(R) &= 19{,}454.
\end{aligned}$$

Thus, assuming that the system started in state 0, the expected long-run total discounted cost is \$14,949.

The aforementioned procedure not only evaluates a given policy, but also is suggestive of an algorithm to determine the optimal policy. The calculations are similar to those required in the value-determination step (step 1) of the policy improvement technique presented in Sec. 11.4. Indeed, an algorithm very similar to that presented in Sec. 11.4 is available. In particular, these steps are to be followed:

Step 1: Value determination For an arbitrarily chosen policy R_1, use $p_{ij}(k_1)$ and C_{ik_1} to solve the set of $(M + 1)$ equations

$$V_i(R_1) = C_{ik_1} + \alpha \sum_{j=0}^{M} p_{ij}(k_1)V_j(R_1), \qquad i = 0, 1, \ldots, M,$$

for all $(M + 1)$ unknown values of $V_i(R_1)$.

Step 2: Policy improvement Using the current values of $V_i(R_1)$, find the alternative policy R_2 such that, for each state i, $d_i(R_2) = k_2$ is the decision that makes

$$C_{ik_2} + \alpha \sum_{j=0}^{M} p_{ij}(k_2)V_j(R_1)$$

a minimum; i.e., for *each* state i, find the appropriate value of k_2 that

$$\underset{k_2=1,2,\ldots,K}{\text{Minimizes}} \left\{ C_{ik_2} + \alpha \sum_{j=0}^{M} p_{ij}(k_2) V_j(R_1) \right\},$$

and then set $d_i(R_2)$ = minimizing the value of k_2. This procedure defines a new policy R_2.

If R_2 does not equal R_1, then return to step 1 by using R_2 instead of R_1, and solve for $V_i(R_2)$, $i = 0, 1, \ldots, M$. Using these values, go to step 2 and find R_3. Continue in this fashion until you find two successive R's to be equal. When you find them, the optimal policy is achieved, and the algorithm terminates. In fact, it can be shown that:

1. $V_i(R_{j+1}) \leq V_i(R_j)$, for $i = 0, 1, \ldots, M$ and $j = 1, 2, \ldots$;
2. The algorithm terminates with the optimal solution in a finite number of iterations; and
3. The algorithm is valid without the assumption that the Markov chain associated with every transition matrix is irreducible.

Example: We shall obtain the optimal policy for the machine-maintenance problem by the *policy improvement algorithm*. The discount factor is chosen to be 0.9. The first step, the value-determination step, has already been carried out earlier in this section if the arbitrary policy chosen, R_1, calls for leaving the machine as is when it is in state 0 or 1, overhauling it when it is in state 2, and replacing it when it is in state 3. The appropriate V's are

$$V_0(R_1) = 14{,}949$$

$$V_1(R_1) = 16{,}262$$

$$V_2(R_1) = 18{,}636$$

$$V_3(R_1) = 19{,}454.$$

Step 2 can now be applied. We are seeking an improved policy R_2 that has the property that $d_0(R_2) = k_2^0$, $d_1(R_2) = k_2^1$, $d_2(R_2) = k_2^2$, and $d_3(R_2) = k_2^3$ minimizes the following expressions:

(0) $C_{0k_2^0} + 0.9[p_{00}(k_2^0)14{,}949 + p_{01}(k_2^0)16{,}262 + p_{02}(k_2^0)18{,}636 + p_{03}(k_2^0)19{,}454]$

(1) $C_{1k_2^1} + 0.9[p_{10}(k_2^1)14{,}949 + p_{11}(k_2^1)16{,}262 + p_{12}(k_2^1)18{,}636 + p_{13}(k_2^1)19{,}454]$

(2) $C_{2k_2^2} + 0.9[p_{20}(k_2^2)14{,}949 + p_{21}(k_2^2)16{,}262 + p_{22}(k_2^2)18{,}636 + p_{23}(k_2^2)19{,}454]$

(3) $C_{3k_2^3} + 0.9[p_{30}(k_2^3)14{,}949 + p_{31}(k_2^3)16{,}262 + p_{32}(k_2^3)18{,}636 + p_{33}(k_2^3)19{,}454]$.

Most of the necessary data can be taken from the first iteration of the example in Sec. 11.4 (the transition probabilities and C_{ik}). Using these data, the values of each of the four expressions are obtained as follows:

Decision	Value of Expression 0	Value of Expression 1	Value of Expression 2	Value of Expression 3
1	14,949	16,262	20,140	∞
2	18,636	18,636	18,636	∞
3	19,454	19,454	19,454	19,454

Thus $d_0(R_2) = k_2^0 = 1$, $d_1(R_2) = k_2^1 = 1$, $d_2(R_2) = k_2^2 = 2$, and $d_3(R_2) = k_2^3 = 3$, so that this policy is identical to R_1. Because the policies obtained on two successive iterations are the same, the optimal policy has been obtained. Again, the optimal policy calls for leaving the machine as is when it is in state 0 or 1, overhauling it when it is in state 2, and replacing it when it is in state 3—the same policy that was obtained by using the long-run expected average cost per day criterion.

Linear Programming Formulation

Just as there is a policy improvement algorithm for the expected long-run total discounted cost criterion, there is also a *linear programming formulation*. It can be shown that the statement of the linear programming problem is to choose the y_{ik} that

$$\text{Minimizes} \quad \sum_{i=0}^{M} \sum_{k=1}^{K} C_{ik} y_{ik},$$

subject to the constraints

$$(1) \qquad \sum_{k=1}^{K} y_{jk} - \alpha \sum_{i=0}^{M} \sum_{k=1}^{K} y_{ik} p_{ij}(k) = \beta_j, \qquad \text{for } j = 0, 1, \ldots, M,$$

where β_j are given constants such that $\beta_j > 0$ and $\sum_{j=0}^{M} \beta_j = 1$,[1] and

$$(2) \qquad y_{ik} \geq 0, \qquad i = 0, 1, \ldots, M, \qquad k = 1, 2, \ldots, K.$$

If we consider the policy defined by

$$D_{ik} = P\{\text{decision} = k | \text{state} = i\}$$

$$= \frac{y_{ik}}{\sum_{k=1}^{K} y_{ik}},$$

then the y_{ik} can be interpreted as a weighted (in a discounted sense) expected time of being in state i and making decision k, when $P\{X_0 = j\} = \beta_j$; that is, if

$$z_{ik}^n = P\{\text{at time } n, \text{ state } = i \text{ and decision } = k\},$$

then
$$y_{ik} = z_{ik}^0 + \alpha z_{ik}^1 + \alpha^2 z_{ik}^2 + \alpha^3 z_{ik}^3 + \cdots.$$

Again, it can be shown that the optimal policy is deterministic; that is, $D_{ik} = 0$ or 1. Furthermore, the technique is valid without the assumption that the Markov chain associated with every transition matrix is irreducible.

[1] The optimal policy is independent of the particular values of the β_j's, but, of course, the expected long-run total discounted cost is a function of these β_j's.

Example: Returning to the machine-maintenance model (with $\alpha = 0.9$), we can formulate the *linear program* as:

Minimize $4{,}000y_{02} + 6{,}000y_{03} + 1{,}000y_{11} + 4{,}000y_{12} + 6{,}000y_{13} + 3{,}000y_{21} + 4{,}000y_{22} + 6{,}000y_{23} + M_1y_{31} + M_2y_{32} + 6{,}000y_{33}$,

where M_1 and M_2 are taken to be large numbers, subject to

$$\sum_{k=1}^{3} y_{0k} - 0.9(y_{03} + y_{13} + y_{23} + y_{33}) = \tfrac{1}{4}$$

$$\sum_{k=1}^{3} y_{1k} - 0.9(\tfrac{7}{8}y_{01} + y_{02} + \tfrac{3}{4}y_{11} + y_{12} + y_{22} + y_{32}) = \tfrac{1}{4}$$

$$\sum_{k=1}^{3} y_{2k} - 0.9(\tfrac{1}{16}y_{01} + \tfrac{1}{8}y_{11} + \tfrac{1}{2}y_{21}) = \tfrac{1}{4}$$

$$\sum_{k=1}^{3} y_{3k} - 0.9(\tfrac{1}{16}y_{01} + \tfrac{1}{8}y_{11} + \tfrac{1}{2}y_{21} + y_{31}) = \tfrac{1}{4}$$

and $\qquad y_{ik} \geq 0, \qquad i = 0, 1, 2, 3, \qquad k = 1, 2, 3,$

where β_0, β_1, β_2, and β_3 are arbitrarily chosen to be $\frac{1}{4}$.

The optimal solution yields all y_{ik} equal to zero, except for $y_{01} = 1.210$, $y_{11} = 6.656$, $y_{22} = 1.067$, and $y_{33} = 1.067$. The corresponding

$$D_{ik} = \frac{y_k}{\sum_{k=1}^{3} y_{ik}}$$

are given by $D_{01} = D_{11} = D_{22} = D_{33} = 1$, and all the remaining $D_{ik} = 0$. This solution is the same as that obtained earlier in this section and calls for leaving the machine as is when it is in state 0 or 1, overhauling it when it is in state 2, and replacing it when it is in state 3. The minimized value of the objective function is \$17,325, and it is seen to be related to the V's of the optimal policy found in the discussion of the discounted cost policy improvement algorithm. Because $P\{X_0 = j\}$ was chosen to equal $\frac{1}{4}$ for all j,

$$17{,}325 = \tfrac{1}{4}[V_0(R) + V_1(R) + V_2(R) + V_3(R)]$$

$$= \tfrac{1}{4}[14{,}949 + 16{,}262 + 18{,}636 + 19{,}454].$$

Finite-Period Markovian Decision Processes and the Method of Successive Approximations

Chapter 8 introduced the concept of dynamic programming and characterized deterministic dynamic programming problems and their solutions. Many of these concepts have analogous interpretations with Markovian decision processes. In particular, suppose we seek the expected total discounted cost of a system starting in state i and evolving for n time periods when an optimal policy is followed. Note that a finite number of time periods are now being considered. This problem is analogous to deterministic dynamic programming, except that the Markov system evolves according to some probabilistic laws of motion rather than evolving in a deterministic fashion. The deterministic dynamic programming solution is suggestive of the solution to this probabilistic dynamic programming problem. Denote by V_i^n the expected total discounted cost of a system starting in state i and evolving for n time periods when an optimal policy is followed.[1] Using the *principle of optimization,* it follows that this cost function satisfies the recursive relationship,

$$V_i^{n+1} = \min_k \left\{ C_{ik} + \alpha \sum_{j=0}^{M} p_{ij}(k) V_j^n \right\}, \qquad i = 0, 1, \ldots, M.$$

Using this recursive relationship, the solution procedure moves backward period by period—each time finding the optimal policy for that period model—until it finds the optimal policy for the original problem. In particular, it is usually assumed that $V_0^0, V_1^0, \ldots, V_M^0$ is zero, so that V_i^1 can be obtained from

$$V_i^1 = \min_k \{C_{ik}\}, \qquad i = 0, 1, \ldots, M,$$

with the corresponding optimal decision becoming known. If this optimal policy is followed, V_i^1 is the minimum expected total discounted cost of a system starting in state i and evolving for one time period.

The V_i^2 can now be obtained from

$$V_i^2 = \min_k \left\{ C_{ik} + \alpha \sum_{j=0}^{M} p_{ij}(k) V_j^1 \right\}, \qquad i = 0, 1, \ldots, M,$$

with the corresponding optimal decisions becoming known. If this optimal policy is followed, V_i^2 is the minimum expected total discounted cost of a system starting in state i and evolving for two time periods.

In a similar manner, the V_i^T can be obtained from

$$V_i^T = \min_k \left\{ C_{ik} + \alpha \sum_{j=0}^{M} p_{ij}(k) V_j^{T-1} \right\}, \qquad i = 0, 1, \ldots, M,$$

with the corresponding optimal decisions becoming known. If this optimal policy is followed, V_i^T is the minimum expected total discounted cost of a system starting in state i and evolving for T time periods. Thus, solving a five-period problem requires solving the four-, three-, two-, and one-period problems also. An example of a three-period version of the machine-maintenance model will be solved later in this section. It should be noted that α can be set equal to 1 (no discounting) for finite-period problems, in which case the cost criterion becomes the expected total cost.

Until now this section has dealt with a finite-period version of a Markov decision process. When the criterion of discounted costs is used, it can be shown that the V_i^n converges to V_i as n approaches infinity, where V_i is the expected (long-run) total discounted cost of a system starting in state i and continuing indefinitely when an *optimal* policy is followed and satisfies

$$V_i = \min_k \left\{ C_{ik} + \alpha \sum_{j=0}^{M} p_{ij}(k) V_j \right\}, \qquad i = 0, 1, \ldots, M.$$

[1] This notation V_i^n for cost is now being used instead of the notation introduced in Chap. 8 to be consistent with the material introduced in the current chapter. In accordance with the notation of Chap. 8, the subscript (i) is the state variable, and the superscript (n) is equivalent to the stage, except that the stage is now measured by the system having "n periods to go" rather than being in period n. This change is due to the need to treat the infinite-period problem also.

Furthermore, we obtain the optimal policy by making appropriate decisions to minimize the right-hand side of the preceding equation. This solution is, indeed, the solution to the Markov decision process considered throughout earlier sections of this chapter.

Finding V_i and the corresponding optimal decisions is generally difficult, but it is relatively simple to approximate V_i and obtain the corresponding policy. This is what the **method of successive approximations** does. It uses the recursive relationship of the finite-period problem presented earlier; that is,

$$V_i^{n+1} = \min_k \left\{ C_{ik} + \alpha \sum_{j=0}^{M} p_{ij}(k) V_j^n \right\}.$$

The first step is to choose arbitrarily a set of values, $V_0^0, V_1^0, V_2^0, \ldots, V_M^0$, usually taken to be zero (as will be assumed from here on). Using the expression for V_i^{n+1}, V_i^1 can be obtained from

$$V_i^1 = \min_k \{C_{ik}\}, \qquad i = 0, 1, \ldots, M,$$

with the corresponding decisions becoming known. This step can be viewed as the first approximation to the optimal policy, and as noted earlier the V_i^1 can be interpreted as the expected total discounted cost of a system starting in state i and evolving for one period when an optimal policy is followed.

The next iteration uses the $V_0^1, V_1^1, V_2^1, \ldots, V_M^1$ found from the previous step. From the recurrence relationship presented, V_i^2 can be obtained from

$$V_i^2 = \min_k \left\{ C_{ik} + \alpha \sum_{j=0}^{M} p_{ij}(k) V_j^1 \right\},$$

with the corresponding decisions becoming known. This policy can be viewed as the second approximation to the optimal policy, and, as noted earlier, the V_i^2 can be interpreted as the expected total discounted cost of a system starting in state i and evolving for two periods when an optimal policy is followed.

Further iterations can be obtained by using the recursive relationship. For the Tth iteration, V_i^T can be interpreted as the expected total discounted cost of a system starting in state i and evolving for T periods when an optimal policy is followed. The number T can be made large, and V_i^T will become "close" to the optimal expected long-run total discounted cost, and, for sufficiently large T, the optimal policy will be obtained. However, there is no procedure for deciding when to terminate the method of successive approximations. A check can be made at any time to see whether the current iteration satisfies the policy improvement equations; if it does, then an optimal policy has been obtained.

Although the *method of successive approximations* may not lead to an optimal policy (using a finite number of iterations), it has one distinct advantage over the policy improvement and linear programming techniques: It never requires the solution of a system of simultaneous equations, and hence each iteration can be performed simply and quickly.

EXAMPLE: We shall solve the machine-maintenance model ($\alpha = 0.9$) by the *method of successive approximations*. Let $V_0^0 = V_1^0 = V_2^0 = V_3^0 = 0$. Then

$$V_0^1 = \min_k \{C_{0k}\} = 0 \qquad (k = 1)$$

$$V_1^1 = \min_k \{C_{1k}\} = 1{,}000 \qquad (k = 1)$$

$$V_2^1 = \min_k \{C_{2k}\} = 3{,}000 \qquad (k = 1)$$

$$V_3^1 = \min_k \{C_{3k}\} = 6{,}000 \qquad (k = 3).$$

Thus the first approximation calls for making decision 1 (leave the machine alone) when the system is in state 0, 1, or 2. When the system is in state 3, decision 3 (replace) is made.

The second iteration leads to

$$\begin{aligned}
V_0^2 &= \min \{0 + 0.9[\tfrac{7}{8}(1{,}000) + \tfrac{1}{16}(3{,}000) + \tfrac{1}{16}(6{,}000)], \\
&\qquad 4{,}000 + 0.9[1(1{,}000)],\ 6{,}000 + 0.9[1(0)]\} = 1{,}294 \qquad (k = 1) \\
V_1^2 &= \min \{1{,}000 + 0.9[\tfrac{3}{4}(1{,}000) + \tfrac{1}{8}(3{,}000) + \tfrac{1}{8}(6{,}000)], \\
&\qquad 4{,}000 + 0.9[1(1{,}000)],\ 6{,}000 + 0.9[1(0)]\} = 2{,}688 \qquad (k = 1) \\
V_2^2 &= \min \{3{,}000 + 0.9[\tfrac{1}{2}(3{,}000) + \tfrac{1}{2}(6{,}000)], \\
&\qquad 4{,}000 + 0.9[1(1{,}000)],\ 6{,}000 + 0.9[1(0)]\} = 4{,}900 \qquad (k = 2) \\
V_3^2 &= \qquad 6{,}000 + 0.9[1(0)] = 6{,}000 \qquad (k = 3).
\end{aligned}$$

Thus the second approximation calls for leaving the machine as is when it is in state 0 or 1, overhauling it when it is in state 2, and replacing it when it is in state 3. Note that this policy is the optimal one, even though the optimal cost has not been obtained.

The third iteration leads to

$$\begin{aligned}
V_0^3 &= \min \{0 + 0.9[\tfrac{7}{8}(2{,}688) + \tfrac{1}{16}(4{,}900) + \tfrac{1}{16}(6{,}000)], \\
&\qquad 4{,}000 + 0.9[1(2{,}688)],\ 6{,}000 + 0.9[1(1{,}294)]\} = 2{,}730 \qquad (k = 1) \\
V_1^3 &= \min \{1{,}000 + 0.9[\tfrac{3}{4}(2{,}688) + \tfrac{1}{8}(4{,}900) + \tfrac{1}{8}(6{,}000)], \\
&\qquad 4{,}000 + 0.9[1(2{,}688)],\ 6{,}000 + 0.9[1(1{,}294)]\} = 4{,}041 \qquad (k = 1) \\
V_2^3 &= \min \{3{,}000 + 0.9[\tfrac{1}{2}(4{,}900) + \tfrac{1}{2}(6{,}000)], \\
&\qquad 4{,}000 + 0.9[1(2{,}688)],\ 6{,}000 + 0.9[1(1{,}294)]\} = 6{,}419 \qquad (k = 2) \\
V_3^3 &= \qquad 6{,}000 + 0.9[1(1{,}294)]\} = 7{,}165 \qquad (k = 3).
\end{aligned}$$

Again the optimal policy is achieved, and the costs are getting closer to those of the optimal policy. This procedure can be continued, and V_0^n, V_1^n, V_2^n, and V_3^n will converge to 14,949, 16,262, 18,636, and 19,454, respectively. It should be noted that termination of the method of successive approximations after the second iteration would have resulted in an optimal policy, although there is no way to know this fact without solving the problem by other methods.

As indicated earlier, the *method of successive approximations solves a finite-period Markovian decision problem*. In particular, the optimal solution to the one-period machine-maintenance model calls for leaving the machine alone when it is in state 0, 1, or 2, and replacing it when it is in state 3. The minimum expected total discounted cost of the system starting in state i, $i = 0, 1, 2, 3$, and evolving for one period is given by 0, 1,000, 3,000, and 6,000, respectively. The optimal solution to the two-period machine-maintenance model is

Period 1 Leave machine alone when it is in state 0 or 1.
Overhaul machine when it is in state 2.
Replace machine when it is in state 3.

Period 2 Leave machine alone when it is in state 0, 1, or 2.
Replace machine when it is in state 3.

The minimum expected total discounted cost of the system starting in state i, $i = 0, 1, 2, 3$, and evolving for two periods is given by 1,294, 2,688, 4,900, and 6,000, respectively. Finally, the optimal solution to the three-period model is

Period 1 and Period 2 { Leave machine alone when it is in state 0 or 1.
Overhaul machine when it is in state 2.
Replace machine when it is in state 3.

Period 3 Leave machine alone when it is in state 0, 1, or 2.
Replace machine when it is in state 3.

The minimum expected total discounted costs over three periods, if the system starts in state i, $i = 0, 1, 2, 3$, are given by 2,730, 4,041, 6,419, and 7,165, respectively.

11.6 A Water-Resource Model

A multipurpose dam is used for generating electric power as well as for flood control. The capacity of the dam is 3 units. The probability distribution of the quantity of water, W_t, that flows into the dam during month t (for $t = 0, 1, \ldots$) is given by $P_W(m)$, where

$$P_W(0) = P\{W = 0\} = \tfrac{1}{6}$$
$$P_W(1) = P\{W = 1\} = \tfrac{1}{3}$$
$$P_W(2) = P\{W = 2\} = \tfrac{1}{3}$$
$$P_W(3) = P\{W = 3\} = \tfrac{1}{6}.$$

For the purpose of generating electric power, 1 unit of water is required. At the beginning of each month, water is released from the dam. The first unit is used to generate electric power and then used for irrigation purposes, the latter function being worth \$100,000. If additional units are released, they can also be used for irrigation purposes, and each unit is worth \$100,000. If the dam contains less than 1 unit at the beginning of a month, additional power must be purchased at a cost of \$300,000. If at any time the water in the dam exceeds the capacity of 3 units, the excess water is released through the spillways at no cost or gain.

A release policy is sought. Policies are to be compared on the basis of expected discounted cost, with discount factor $\alpha = 0.99$. The *policy improvement algorithm* will be used.

Let X_t denote the amount of water in the dam at time t. Then $X_t = 0, 1, 2, 3$. The natural laws of motion for this system (no water released) are given by the transition matrix:

State	0	1	2	3
0	$\frac{1}{6}$	$\frac{1}{3}$	$\frac{1}{3}$	$\frac{1}{6}$
1	0	$\frac{1}{6}$	$\frac{1}{3}$	$\frac{1}{2}$
2	0	0	$\frac{1}{6}$	$\frac{5}{6}$
3	0	0	0	1

For example, the element in the second row and fourth column, p_{13}, is obtained as follows: If the dam contains 1 unit of water now, then for it to contain 3 units of water a month later, 2 or 3 units of water must flow into the dam during the month (recall that dam capacity is 3 units, so that a flow of 3 units will result in 1 unit being released through the spillways). This occurs with probability $\frac{1}{3} + \frac{1}{6} = \frac{1}{2}$.

There are three possible decisions that can be made at the beginning of each month:

Decision	Action
1	Release 1 unit
2	Release 2 units
3	Release 3 units

It is clear that releasing no units is not a sensible action, because 1 unit is needed for electric power generation anyway. Thus a policy calls for determining how many units to release as a function of the quantity of water found in the dam. A typical policy R_1 might call for releasing all the water in the dam if it contains 0, 1, or 2 units, and releasing 2 units if it contains 3 units. The resultant transition matrix is given by

State	0	1	2	3
0	$\frac{1}{6}$	$\frac{1}{3}$	$\frac{1}{3}$	$\frac{1}{6}$
1	$\frac{1}{6}$	$\frac{1}{3}$	$\frac{1}{3}$	$\frac{1}{6}$
2	$\frac{1}{6}$	$\frac{1}{3}$	$\frac{1}{3}$	$\frac{1}{6}$
3	0	$\frac{1}{6}$	$\frac{1}{3}$	$\frac{1}{2}$

Of course, a policy that calls for releasing 3 units when there is only 1 unit in the dam is to be interpreted as calling for releasing all the available water. Necessary cost information can be obtained from the following data:

State	Decision	Cost (In Hundred Thousands)
0	1	3
	2	3
	3	3
1	1	−1
	2	−1
	3	−1
2	1	−1
	2	−2
	3	−2
3	1	−1
	2	−2
	3	−3

The policy R_1 will be used in the value-determination step (step 1) of the policy improvement algorithm. Using the cost information just given, the values of C_{ik_1} are

$$C_{0k_1} = 3$$

$$C_{1k_1} = -1$$

$$C_{2k_1} = -2$$

$$C_{3k_1} = -2.$$

The following four equations must be solved:

$$V_0(R_1) = 3 + 0.99[\tfrac{1}{6}V_0(R_1) + \tfrac{1}{3}V_1(R_1) + \tfrac{1}{3}V_2(R_1) + \tfrac{1}{6}V_3(R_1)]$$

$$V_1(R_1) = -1 + 0.99[\tfrac{1}{6}V_0(R_1) + \tfrac{1}{3}V_1(R_1) + \tfrac{1}{3}V_2(R_1) + \tfrac{1}{6}V_3(R_1)]$$

$$V_2(R_1) = -2 + 0.99[\tfrac{1}{6}V_0(R_1) + \tfrac{1}{3}V_1(R_1) + \tfrac{1}{3}V_2(R_1) + \tfrac{1}{6}V_3(R_1)]$$

$$V_3(R_1) = -2 + 0.99[\qquad \tfrac{1}{6}V_1(R_1) + \tfrac{1}{3}V_2(R_1) + \tfrac{1}{2}V_3(R_1)].$$

The simultaneous solution of these equations results in the values

$$V_0(R_1) = -103.881,\ V_1(R_1) = -107.881,$$

and

$$V_2(R_1) = -108.881,\ V_3(R_1) = -110.358.$$

Step 2 can now be applied. We want to find an improved policy R_2 that has the property that $d_0(R_2) = k_2^0$, $d_1(R_2) = k_2^1$, $d_2(R_2) = k_2^2$, and $d_3(R_2) = k_2^3$ minimizes the following expressions:

(0) $$C_{0k_2^0} + 0.99[-103.881p_{00}(k_2^0) - 107.881p_{01}(k_2^0) - 108.881p_{02}(k_2^0) - 110.358p_{03}(k_2^0)]$$

(1) $$C_{1k_2^1} + 0.99[-103.881p_{10}(k_2^1) - 107.881p_{11}(k_2^1) - 108.881p_{12}(k_2^1) - 110.358p_{13}(k_2^1)]$$

(2) $C_{2k_2^2} + 0.99[-103.881p_{20}(k_2^2) - 107.881p_{21}(k_2^2) - 108.881p_{22}(k_2^2) - 110.358p_{23}(k_2^2)]$

(3) $C_{3k_2^3} + 0.99[-103.881p_{30}(k_2^3) - 107.881p_{31}(k_2^3) - 108.881p_{32}(k_2^3) - 110.358p_{33}(k_2^3)].$

To find k_2^0, the best decision when the system is in state 0, it is necessary to evaluate the first expression for all possible decisions. It is clear that when the system is in state 0 (dam empty), there is no choice among the decisions because they all are equivalent. The data for the necessary calculations for evaluating expression (0) follow:

State 0

Decision	$p_{00}(k_2)$	$p_{01}(k_2)$	$p_{02}(k_2)$	$p_{03}(k_2)$	C_{0k_2}	Total Value of Expression 0
1, 2, 3	$\frac{1}{6}$	$\frac{1}{3}$	$\frac{1}{3}$	$\frac{1}{6}$	3	−103.881

Similarly, for state 1 there is no choice among the decisions because they are all equivalent. The data for the necessary calculations for evaluating follow:

State 1

Decision	$p_{10}(k_2)$	$p_{11}(k_2)$	$p_{12}(k_2)$	$p_{13}(k_2)$	C_{1k_2}	Total Value of Expression 1
1, 2, 3	$\frac{1}{6}$	$\frac{1}{3}$	$\frac{1}{3}$	$\frac{1}{6}$	−1	−107.881

For the remaining two states, the appropriate transition probabilities and costs generally depend upon the decisions made. The data for the necessary calculations for finding the best decisions, given the dam is in state 2 or 3, follow:

State 2

Decision	$p_{20}(k_2)$	$p_{21}(k_2)$	$p_{22}(k_2)$	$p_{23}(k)_2$	C_{2k_2}	Total Value of Expression 2
1	0	$\frac{1}{6}$	$\frac{1}{3}$	$\frac{1}{2}$	−1	−109.358
2, 3	$\frac{1}{6}$	$\frac{1}{3}$	$\frac{1}{3}$	$\frac{1}{6}$	−2	−108.881

State 3

Decision	$p_{30}(k_2)$	$p_{31}(k_2)$	$p_{32}(k_2)$	$p_{33}(k_2)$	C_{3k_2}	Total Value of Expression 3
1	0	0	$\frac{1}{6}$	$\frac{5}{6}$	−1	−110.011
2	0	$\frac{1}{6}$	$\frac{1}{3}$	$\frac{1}{2}$	−2	−110.358
3	$\frac{1}{6}$	$\frac{1}{3}$	$\frac{1}{3}$	$\frac{1}{6}$	−3	−109.881

Thus $d_0(R_2) = k_2^0 = d_1(R_2) = k_2^1 = 1$, 2, or 3; $d_2(R_2) = k_2^2 = 1$; and $d_3(R_2) = k_2^3 = 2$. Hence policy R_2 calls for releasing all the water when there is 1 unit in the dam, 1 unit of water when there are 2 units available in the dam, and 2 units when there are 3 units available in the dam. This policy differs from R_1, so that another iteration is required. For the value-determination step, the equations that must now be solved are

$$\begin{aligned}
V_0(R_2) &= 3 + 0.99[\tfrac{1}{6}V_0(R_2) + \tfrac{1}{3}V_1(R_2) + \tfrac{1}{3}V_2(R_2) + \tfrac{1}{6}V_3(R_2)]\\
V_1(R_2) &= -1 + 0.99[\tfrac{1}{6}V_0(R_2) + \tfrac{1}{3}V_1(R_2) + \tfrac{1}{3}V_2(R_2) + \tfrac{1}{6}V_3(R_2)]\\
V_2(R_2) &= -1 + 0.99[\tfrac{1}{6}V_1(R_2) + \tfrac{1}{3}V_2(R_2) + \tfrac{1}{2}V_3(R_2)]\\
V_3(R_2) &= -2 + 0.99[\tfrac{1}{6}V_1(R_2) + \tfrac{1}{3}V_2(R_2) + \tfrac{1}{2}V_3(R_2)].
\end{aligned}$$

The simultaneous solution of these equations results in the values $V_0(R_2) = -119.642$, $V_1(R_2) = -123.642$, $V_2(R_2) = -125.119$, and $V_3(R_2) = -126.119$.

Step 2 can now be applied. We want to find an improved policy R_3 that has the property that $d_0(R_3) = k_3^0$, $d_1(R_3) = k_3^1$, $d_2(R_3) = k_3^2$, and $d_3(R_3) = k_3^3$ minimizes the following expressions:

$$\begin{aligned}
&(0)\quad C_{0k_3^0} + 0.99[-119.642p_{00}(k_3^0) - 123.642p_{01}(k_3^0) - 125.119p_{02}(k_3^0) - 126.119p_{03}(k_3^0)]\\
&(1)\quad C_{1k_3^1} + 0.99[-119.642p_{10}(k_3^1) - 123.642p_{11}(k_3^1) - 125.119p_{12}(k_3^1) - 126.119p_{13}(k_3^1)]\\
&(2)\quad C_{2k_3^2} + 0.99[-119.642p_{20}(k_3^2) - 123.642p_{21}(k_3^2) - 125.119p_{22}(k_3^2) - 126.119p_{23}(k_3^2)]\\
&(3)\quad C_{3k_3^3} + 0.99[-119.642p_{30}(k_3^3) - 123.642p_{31}(k_3^3) - 125.119p_{32}(k_3^3) - 126.119p_{33}(k_3^3)].
\end{aligned}$$

The data on the transition matrices and the costs from the previous iteration can again be used; the resulting values of the expression are

Decision	Value of Expression 0	Value of Expression 1	Value of Expression 2	Value of Expression 3
1	−119.642	−123.642	−125.119	−125.693
2	−119.642	−123.642	−124.642	−126.119
3	−119.642	−123.642	−124.642	−125.642

Thus $d_0(R_3) = k_3^0 = d_1(R_3) = k_3^1 = 1$, 2, or 3; $d_2(R_3) = k_3^2 = 1$; and $d_3(R_3) = k_3^3 = 2$. Hence policy R_3 and policy R_2 are identical, and the optimal release policy calls for releasing all the water when there is 1 unit in the dam, 1 unit of water when there are 2 units available in the dam, and 2 units when there are 3 units available in the dam.

Of course, direct enumeration would have been just as simple a technique to use in this situation, but the policy improvement algorithm was used for illustrative purposes.

11.7 Inventory Model

In Chap. 5 the following inventory problem was considered. A camera store stocks a particular model camera that can be ordered weekly. Let $D_1, D_2, \ldots,$ represent the demand for this camera during the first week, the second week, . . . , respectively. It is assumed that the D_i are independent, identically distributed random variables having a Poisson distribution with parameter λ equal to 1. Let X_0 represent the number of cameras on hand at the outset, X_1 the number of cameras on hand at the end of week one, X_2 the number of cameras on hand at the end of week two, and so forth. On Saturday night the store places an order that is delivered in time for the opening of the store on Monday. The store uses an (s, S) ordering policy. If the number of cameras on hand at the end of the week is less than $s = 1$ (no cameras in stock), the store orders up to $S = 3$. Otherwise, the store does not order (if there are any cameras in stock, no order is placed). It is assumed that sales are lost when demand exceeds the inventory on hand (no backlogging). The cost structure considered calls for incurring a penalty cost of \$50 per unit for each unit of unsatisfied demand (lost sales). If $z > 0$ cameras are ordered, the cost incurred is $10 + 25z$ dollars. If no cameras are ordered, no ordering cost is incurred. Holding costs are to be neglected. In Sec. 5.7, this policy was evaluated by using the (long-run) expected average cost per unit time as the criterion. It is not evident that this policy is optimal, and the purpose of this section is to find the optimal policy. Even though we know that the optimal policy must be of the (s, S) form, we shall consider all possible policies, although we shall assume that three cameras is the maximum number of cameras that the store will stock. The *policy improvement algorithm* will be used first, followed by the *linear programming formulation.*

Because X_t represents the state of the system, i.e., the number of cameras on hand at the end of week t (before ordering), then $X_t = 0, 1, 2, 3$. Similarly, there are four possible decisions:

Decision	Action
0	Do not order
1	Order 1 camera
2	Order 2 cameras
3	Order 3 cameras

The possible transitions are given by[1]

	Decision 0			
State	0	1	2	3
0	1	0	0	0
1	$P\{D \geq 1\}$	$P\{D = 0\}$	0	0
2	$P\{D \geq 2\}$	$P\{D = 1\}$	$P\{D = 0\}$	0
3	$P\{D \geq 3\}$	$P\{D = 2\}$	$P\{D = 1\}$	$P\{D = 0\}$

[1] Note that in this example the set of possible decisions varies with the states.

Decision 1

State	0	1	2	3
0	$P\{D \geq 1\}$	$P\{D = 0\}$	0	0
1	$P\{D \geq 2\}$	$P\{D = 1\}$	$P\{D = 0\}$	0
2	$P\{D \geq 3\}$	$P\{D = 2\}$	$P\{D = 1\}$	$P\{D = 0\}$
3	Decision 1 not permitted			

Decision 2

State	0	1	2	3
0	$P\{D \geq 2\}$	$P\{D = 1\}$	$P\{D = 0\}$	0
1	$P\{D \geq 3\}$	$P\{D = 2\}$	$P\{D = 1\}$	$P\{D = 0\}$
2, 3	Decision 2 not permitted			

Decision 3

State	0	1	2	3
0	$P\{D \geq 3\}$	$P\{D = 2\}$	$P\{D = 1\}$	$P\{D = 0\}$
1, 2, 3	Decision 3 not permitted			

Recalling that the demand D is a Poisson random variable with parameter $\lambda = 1$, and using appendix Table A.5.4, these transitions can now be expressed as

Decision 0

State	0	1	2	3
0	1	0	0	0
1	0.632	0.368	0	0
2	0.264	0.368	0.368	0
3	0.080	0.184	0.368	0.368

Decision 1

State	0	1	2	3
0	0.632	0.368	0	0
1	0.264	0.368	0.368	0
2	0.080	0.184	0.368	0.368
3	Decision 1 not permitted			

Decision 2

State	0	1	2	3
0	0.264	0.368	0.368	0
1	0.080	0.184	0.368	0.368
2, 3	Decision 2 not permitted			

Decision 3

State	0	1	2	3
0	0.080	0.184	0.368	0.368
1, 2, 3	Decision 3 not permitted			

The cost information required is similar to that given in Sec. 5.7, and you are urged to review this material. A summary is given by

State	Decision	Actual Cost Per Week	Expected Cost Per Week, C_{ik}
0	0	$50D$	$50E(D) = 50$
	1	$35 + 50 \max \{(D-1), 0\}$	$35 + 50[1P\{D=2\} + 2P\{D=3\} + \cdots] = 53.4$
	2	$60 + 50 \max \{(D-2), 0\}$	$60 + 50[1P\{D=3\} + 2P\{D=4\} + \cdots] = 65.2$
	3	$85 + 50 \max \{(D-3), 0\}$	$85 + 50[1P\{D=4\} + 2P\{D=5\} + \cdots] = 86.2$
1	0	$50 \max \{(D-1), 0\}$	$50[1P\{D=2\} + 2P\{D=3\} + \cdots] = 18.4$
	1	$35 + 50 \max \{(D-2), 0\}$	$35 + 50[1P\{D=3\} + 2P\{D=4\} + \cdots] = 40.2$
	2	$60 + 50 \max \{(D-3), 0\}$	$60 + 50[1P\{D=4\} + 2P\{D=5\} + \cdots] = 61.2$
	3	Decision 3 not permitted	
2	0	$50 \max \{(D-2), 0\}$	$50[1P\{D=3\} + 2P\{D=4\} + \cdots] = 5.2$
	1	$35 + 50 \max \{(D-3), 0\}$	$35 + 50[1P\{D=4\} + 2P\{D=5\} + \cdots] = 36.2$
	2, 3	Decisions 2, 3 not permitted	
3	0	$50 \max \{(D-3), 0\}$	$50[1P\{D=4\} + 2P\{D=5\} + \cdots] = 1.2$
	1, 2, 3	Decisions 1, 2, 3 not permitted	

Choose the (s, S) policy already introduced as the initial policy for carrying out the value-determination step (step 1) of the *policy improvement algorithm*. This policy, R_1, calls for ordering up to 3 units whenever the system is in state 0 (no cameras on hand); otherwise, no order is placed. With this policy, the following four equations must be solved simultaneously for $g(R_1)$, $v_0(R_1)$, $v_1(R_1)$, and $v_2(R_1)$ [recall that $v_3(R_1)$ is arbitrarily taken to be zero]:

$$\begin{aligned} g(R_1) &= C_{0k_1} + \sum_{j=0}^{3} p_{0j}(k_1)v_j(R_1) - v_0(R_1) \\ &= C_{1k_1} + \sum_{j=0}^{3} p_{1j}(k_1)v_j(R_1) - v_1(R_1) \\ &= C_{2k_1} + \sum_{j=0}^{3} p_{2j}(k_1)v_j(R_1) - v_2(R_1) \\ &= C_{3k_1} + \sum_{j=0}^{3} p_{3j}(k_1)v_j(R_1) - v_3(R_1), \end{aligned}$$

or, alternatively,

$$\begin{aligned} g(R_1) &= 86.2 + 0.080v_0(R_1) + 0.184v_1(R_1) + 0.368v_2(R_1) - v_0(R_1) \\ &= 18.4 + 0.632v_0(R_1) + 0.368v_1(R_1) - v_1(R_1) \\ &= 5.2 + 0.264v_0(R_1) + 0.368v_1(R_1) + 0.368v_2(R_1) - v_2(R_1) \\ &= 1.2 + 0.080v_0(R_1) + 0.184v_1(R_1) + 0.368v_2(R_1). \end{aligned}$$

The simultaneous solution of this system of equations yields

$$g_1(R_1) = 31.43$$

$$v_0(R_1) = 85.00$$

$$v_1(R_1) = 64.38$$

$$v_2(R_1) = 31.49.$$

Step 2 can now be applied. It is necessary to find the improved policy R_2, which has the property that $d_0(R_2) = k_2^0$, $d_1(R_2) = k_2^1$, $d_2(R_2) = k_2^2$, and $d_3(R_2) = k_2^3$ minimizes the following expressions:

(0) $C_{0k_2^0} + p_{00}(k_2^0)85 + p_{01}(k_2^0)64.38 + p_{02}(k_2^0)31.49 - 85$

(1) $C_{1k_2^1} + p_{10}(k_2^1)85 + p_{11}(k_2^1)64.38 + p_{12}(k_2^1)31.49 - 64.38$

(2) $C_{2k_2^2} + p_{20}(k_2^2)85 + p_{21}(k_2^2)64.38 + p_{22}(k_2^2)31.49 - 31.49$

(3) $C_{3k_2^3} + p_{30}(k_2^3)85 + p_{31}(k_2^3)64.38 + p_{32}(k_2^3)31.49.$

To find the optimal decisions, the following data are required:

State 0

Decision	$p_{00}(k_2)$	$p_{01}(k_2)$	$p_{02}(k_2)$	C_{0k_2}	Total Value of Expression 0
0	1	0	0	50	50
1	0.632	0.368	0	53.4	45.81
2	0.264	0.368	0.368	65.2	37.92
3	0.080	0.184	0.368	86.2	31.43

State 1

Decision	$p_{10}(k_2)$	$p_{11}(k_2)$	$p_{12}(k_2)$	C_{1k_2}	Total Value of Expression 1
0	0.632	0.368	0	18.4	31.43
1	0.264	0.368	0.368	40.2	33.54
2	0.080	0.184	0.368	61.2	27.05

State 2

Decision	$p_{20}(k_2)$	$p_{21}(k_2)$	$p_{22}(k_2)$	C_{2k_2}	Total Value of Expression 2
0	0.264	0.368	0.368	5.2	31.43
1	0.080	0.184	0.368	36.2	34.94

State 3

Decision	$p_{30}(k_2)$	$p_{31}(k_2)$	$p_{32}(k_2)$	C_{3k_2}	Total Value of Expression 3
0	0.080	0.184	0.368	1.2	31.43

Thus $d_0(R_2) = k_2^0 = 3$, $d_1(R_2) = k_2^1 = 2$, $d_2(R_2) = k_2^2 = d_3(R_2) = k_2^3 = 0$. Hence policy R_2 calls for ordering up to three cameras whenever there is 0 or 1 camera in stock; otherwise, no ordering is done; i.e., if the number of cameras on hand at the end of the week is less than $s = 2$ cameras, the store orders up to $S = 3$ cameras. Because policy R_2 differs from policy R_1, another iteration is required. The following four equations must be solved simultaneously for $g(R_2)$, $v_0(R_2)$, $v_1(R_2)$, and $v_2(R_2)$:

$$\begin{aligned} g(R_2) &= 86.2 + 0.080v_0(R_2) + 0.184v_1(R_2) + 0.368v_2(R_2) - v_0(R_2) \\ &= 61.2 + 0.080v_0(R_2) + 0.184v_1(R_2) + 0.368v_2(R_2) - v_1(R_2) \\ &= 5.2 + 0.264v_0(R_2) + 0.368v_1(R_2) + 0.368v_2(R_2) - v_2(R_2) \\ &= 1.2 + 0.080v_0(R_2) + 0.184v_1(R_2) + 0.368v_2(R_2). \end{aligned}$$

The simultaneous solution of this system of equations yields

$$\begin{aligned} g_1(R_2) &= 30.33 \\ v_0(R_2) &= 85.00 \\ v_1(R_2) &= 60.00 \\ v_2(R_2) &= 30.68. \end{aligned}$$

Step 2 can now be applied. It is necessary to find the improved policy R_3, which has the property that $d_0(R_3) = k_3^0$, $d_1(R_3) = k_3^1$, $d_2(R_3) = k_3^2$, and $d_3(R_3) = k_3^3$ minimizes the following expressions:

(0) $C_{0k_3^0} + p_{00}(k_3^0)85 + p_{01}(k_3^0)60 + p_{02}(k_3^0)30.68 - 85$

(1) $C_{1k_3^1} + p_{10}(k_3^1)85 + p_{11}(k_3^1)60 + p_{12}(k_3^1)30.68 - 60$

(2) $C_{2k_3^2} + p_{20}(k_3^2)85 + p_{21}(k_3^2)60 + p_{22}(k_3^2)30.68 - 30.68$

(3) $C_{3k_3^3} + p_{30}(k_3^3)85 + p_{31}(k_3^3)60 + p_{32}(k_3^3)30.68.$

Using the data from the previous iteration, the relevant calculations are

Decision	Total Value of Expression 0	Total Value of Expression 1	Total Value of Expression 2	Total Value of Expression 3
0	50	34.20	30.33	30.33
1	44.20	36.01	34.65	—
2	36.01	30.33	—	—
3	30.33	—	—	—

Thus $d_0(R_3) = k_3^0 = 3$, $d_1(R_3) = k_3^1 = 2$, $d_2(R_3) = k_3^2 = d_3(R_3) = k_3^3 = 0$. Hence policy R_3 and policy R_2 are identical, so that the optimal policy calls for ordering up to three cameras when there is 0 or 1 camera in stock; otherwise, no ordering is done.

The *linear programming formulation* calls for finding the y_{ik} that

$$\text{Minimize} \quad 50y_{00} + 53.4y_{01} + 65.2y_{02} + 86.2y_{03} + 18.4y_{10} + 40.2y_{11} + 61.2y_{12} + 5.2y_{20} + 36.2y_{21} + 1.2y_{30},$$

subject to

$$y_{00} + y_{01} + y_{02} + y_{03} + y_{10} + y_{11} + y_{12} + y_{20} + y_{21} + y_{30} = 1,$$

$$y_{00} + y_{01} + y_{02} + y_{03} - [y_{00} + y_{01}(0.632) + y_{02}(0.264) + y_{03}(0.080) + y_{10}(0.632) + y_{11}(0.264) + y_{12}(0.080) + y_{20}(0.264) + y_{21}(0.080) + y_{30}(0.080)] = 0,$$

$$y_{10} + y_{11} + y_{12} - [y_{01}(0.368) + y_{02}(0.368) + y_{03}(0.184) + y_{10}(0.368) + y_{11}(0.368) + y_{12}(0.184) + y_{20}(0.368) + y_{21}(0.184) + y_{30}(0.184)] = 0,$$

$$y_{20} + y_{21} - [y_{02}(0.368) + y_{03}(0.368) + y_{11}(0.368) + y_{12}(0.368) + y_{20}(0.368) + y_{21}(0.368) + y_{30}(0.368)] = 0,$$

$$y_{30} - [y_{03}(0.368) + y_{12}(0.368) + y_{21}(0.368) + y_{30}(0.368)] = 0,$$

and

$$y_{00}, y_{01}, y_{02}, y_{03}, y_{10}, y_{11}, y_{12}, y_{20}, y_{21}, y_{30} \geq 0.$$

This linear program can be solved by using the simplex method. The results yield all y_{ik} equal to zero, except for

$$y_{03} = 0.148, \qquad y_{12} = 0.252, \qquad y_{20} = 0.368, \qquad y_{30} = 0.233.$$

The corresponding D_{ik} are given by

$$D_{03} = D_{12} = D_{20} = D_{30} = 1,$$

and all the remaining $D_{ik} = 0$.

11.8 Conclusions

The material presented in this chapter represents a powerful tool for formulating models and finding the optimal policies for controlling a large *class* of systems—those that are *Markovian decision processes*. These techniques are applicable to the solution of problems in such areas as queueing theory, inventory, maintenance, and probabilistic dynamic programming, in general.

Two algorithms were presented, the *policy improvement algorithm* and the *linear programming formulation*, for finding optimal policies. It is evident from the examples that data-collection requirements are high. Even if the solution converges rapidly in the policy improvement algorithm, completing step 2 requires considerable calculation for systems with a large number of states. Using the linear programming formulation with, say, 50 states and 25 decisions leads to 1,250 variables and 51 constraints (excluding the nonnegativity constraints), which represents a large linear program. Nevertheless, these two solution methods are useful for solving real-world problems.

When the cost criterion is the expected discounted cost, the *method of successive approximations* provides a valuable tool for approximating the optimal solution. Much simpler calculations are required for this algorithm than for the policy improvement or linear programming methods.

Considerable research activities have been devoted to the field of Markovian decision processes in recent years. C. Derman[1] has shown that in the expected average cost per unit time case, the optimal policy is deterministic (calls for always taking a particular action when the system is in a given state). Similarly, he has shown[2] that in the expected discounted cost case, the optimal policy is also deterministic. The policy improvement algorithm is due to R. Howard.[3] For the expected average cost per unit time case, he presents not only the algorithm for the situation where all the states belong to one class but also an algorithm that is applicable when there is more than one (a finite number) class of states. He also considers the continuous time case. The linear programming formulation using the expected average cost per unit time was first given by A. S. Manne,[4] who treated the case where all the states belong to one class. The linear programming formulation using the expected discounted cost was first given by F. d'Epenoux.[5] Finally, although the results presented in this chapter assumed the state space to be finite, most of the results are applicable to the case of a countable state space.

SELECTED REFERENCES

1. Bertsekas, D. P.: *Dynamic Programming and Stochastic Control*, Academic Press, New York, 1976.
2. Derman, C.: *Finite State Markovian Decision Processes*, Academic Press, New York, 1970.
3. Dreyfus, S., and A. Law: *The Art and Theory of Dynamic Programming,* Academic Press, New York, 1976.
4. Dynkin, E. B., and A. Yushkevich: *Controlled Markov Processes*, Springer-Verlag, New York, 1979.
5. Heyman, D., and M. Sobel: *Stochastic Models in Operations Research, Vol. 2,* McGraw-Hill, New York, 1982.
6. Ross, S.: *Introduction to Stochastic Dynamic Programming*, Academic Press, New York, 1983.

PROBLEMS

1.* During any period, a potential customer arrives at a certain facility with probability $\frac{1}{2}$. If there are already two people at the facility (including the one being served), the potential customer leaves the facility immediately and never returns. However, if there is one or fewer

[1] Derman, C.: "On Sequential Decisions and Markov Chains," *Management Science,* **9**:16–24, 1962.

[2] Derman, C.: "Markovian Sequential Control Processes—Denumerable State Space," *Journal of Mathematical Analysis and Applications*, **10**:295–302, 1965.

[3] Howard, R.: *Dynamic Programming and Markov Processes*, Technology Press, Cambridge, Mass., and Wiley, New York, 1960.

[4] Manne, A. S.: "Linear Programming and Sequential Decisions," *Management Science*, **6**:259–267, 1960.

[5] d'Epenoux, F.: "Sur un Probleme de Production et de Stockage dans l'Aléatoire," *Rev. Française Information Recherche Opérationnelle,* **141**:3–16, 1960; English translation: *Management Science*, **10**:98–108, 1963.

people, he enters the facility and becomes an actual customer. The manager of the facility has two types of service rates available. If she uses her "slow" service rate at a cost of \$3 during a period, a customer will be served and leave the facility with probability $\frac{3}{5}$. If she uses her "fast" service rate at a cost of \$9 during a period, a customer will be served and leave the facility with probability $\frac{4}{5}$. Note that the probability of more than one customer arriving or more than one customer being served in a period is zero. A profit of \$50 is earned when a customer is served.

(*a*) Use exhaustive enumeration to identify all the stationary deterministic policies.

(*b*) Use the *policy improvement algorithm* to determine the policy the manager should follow to minimize the expected long-run average cost per period.

(*Hint*: In computing the costs for services when two customers are at a facility, do not forget the opportunity cost of losing a potential customer.)

2.* Formulate Prob. 1 as a *linear programming* problem.

3. A person often finds that she is up to 1 hour late for work. If she is from 1 to 30 minutes late, \$4 is deducted from her paycheck; if she is from 31 to 60 minutes late for work, \$8 is deducted from her paycheck. If she drives to work at her normal speed (which is well under the speed limit), she can arrive in 20 minutes. However, if she exceeds the speed limit a little here and there on her way to work, she can get there in 10 minutes, but she runs the risk of getting a speeding ticket. With probability $\frac{1}{8}$ she will get caught speeding and will not only be fined \$20 but will also be delayed 10 minutes, so that it takes 20 minutes to reach work.

Let s be the time she finds she has to reach work before being late; that is, $s = 10$ means she has 10 minutes to get to work and $s = -10$ means she is already 10 minutes late for work. For simplicity, she considers s to be in one of four intervals: $(20, \infty)$, $(10, 19)$, $(-10, 9)$, and $(-20, -11)$.

The transition probabilities for s tomorrow if she does not speed today are given by

	$(20, \infty)$	$(10, 19)$	$(-10, 9)$	$(-20, -11)$
$(20, \infty)$	$\frac{3}{8}$	$\frac{1}{4}$	$\frac{1}{4}$	$\frac{1}{8}$
$(10, 19)$	$\frac{1}{2}$	$\frac{1}{4}$	$\frac{1}{8}$	$\frac{1}{8}$
$(-10, 9)$	$\frac{5}{8}$	$\frac{1}{4}$	$\frac{1}{8}$	0
$(-20, -11)$	$\frac{3}{4}$	$\frac{1}{4}$	0	0

The transition probabilities for s tomorrow if she speeds to work today are given by

	$(20, \infty)$	$(10, 19)$	$(-10, 9)$	$(-20, -11)$
$(20, \infty)$				
$(10, 19)$	$\frac{3}{8}$	$\frac{1}{4}$	$\frac{1}{4}$	$\frac{1}{8}$
$(-10, 9)$				
$(-20, -11)$	$\frac{5}{8}$	$\frac{1}{4}$	$\frac{1}{8}$	0

Note that there are no transition probabilities for $(20, \infty)$ and $(-10, 9)$, because she will get to work on time and from 1 to 30 minutes late, respectively, regardless of whether she speeds or not. Hence speeding when in these states would not be a logical choice.

Also note that the transition probabilities imply that the later she is for work and the more she has to rush to get there, the likelier she is to leave for work earlier the next day.

Use the *policy improvement algorithm* to determine when she should speed and when she should take her time getting to work.

4. Formulate Prob. 3 as a *linear programming* problem.

5. Every Saturday night a man plays poker, much to the dismay of his wife. Regardless of the kind of mood his wife is in, if he takes her out to dinner (at an expected cost of $14) before going to play poker, she will be in a good mood, with probability $\frac{7}{8}$, and a bad mood, with probability $\frac{1}{8}$, next Saturday night. However, if he goes to play poker without taking her out to dinner first, she will be in a good mood next Saturday, with probability $\frac{1}{8}$, and a bad mood, with probability $\frac{7}{8}$, regardless of her mood this week. Furthermore, if she happens to be in a bad mood and he does not take her to dinner, she will go to an exclusive store and buy a new outfit (at an expected cost of $75). Use the *policy improvement algorithm* to find the policy that the man should follow to minimize his long-run expected average cost per week.

6. Formulate Prob. 5 as a *linear programming* problem.

7.* When a tennis player serves, he gets two chances to serve in bounds. If he fails to do so twice, he loses the point (1 unit). If he attempts to serve an ace, he serves in bounds, with probability $\frac{3}{8}$. If he serves a lob, he serves in bounds, with probability $\frac{7}{8}$. If he serves an ace in bounds, he wins the point (1 unit), with probability $\frac{2}{3}$. With an inbounds lob, he wins the point (1 unit), with probability $\frac{1}{3}$. Use the *policy improvement algorithm* to determine the optimal strategy. (*Hint*: Let state 0 denote point over, two serves to go on next point; and let state 1 denote one serve left.)

8.* Formulate Prob. 7 as a *linear programming* problem.

9. A student is concerned about her car and does not like to get it dented. When she drives to school she has a choice of parking it on the street in one space, parking it on the street and taking up two spaces, or parking in the lot. If she parks on the street in one space, her car gets dented, with probability $\frac{1}{10}$. If she parks on the street and takes two spaces, the probability of a dent is $\frac{1}{50}$ and the probability of a $15 ticket is $\frac{3}{10}$. Parking in a lot costs $5, but the car will not get dented. If her car gets dented, she can have it repaired at the dealer, in which case it is out of commission for a day and costs her $50 in fees and cab fares. She can also drive her car dented, but she feels that the loss of pride and shame is worth about $9 a day. Using the *policy improvement algorithm*, determine the optimal policy.

10. Formulate Prob. 9 as a *linear programming* problem.

11. Each year Mr. Merrill has the chance to invest in two different no-load mutual funds: the Go-Go Fund or the Go-Slow Mutual Fund. At the end of each year, Mr. Merrill liquidates his holdings, takes his profits, and then reinvests. The yearly profits of the mutual funds are dependent upon how the market reacts each year. Recently the market has been oscillating around the 2,500 mark, according to the probabilities given in the following matrix:

$$\begin{array}{c} \\ 2{,}400 \\ 2{,}500 \\ 2{,}600 \end{array} \begin{array}{c} \begin{array}{ccc} 2{,}400 & 2{,}500 & 2{,}600 \end{array} \\ \begin{bmatrix} 0.3 & 0.5 & 0.2 \\ 0.1 & 0.5 & 0.4 \\ 0.2 & 0.4 & 0.4 \end{bmatrix} \end{array}$$

Each year the market moves up (or down) 100 points, the Go-Go Fund has profits (or losses) of $20, while the Go-Slow Fund has profits (or losses) of $10. If the market moves up (or down) 200 points in a year, the Go-Go Fund has profits (or losses) of $50, while the Go-Slow Fund has profits (or losses) of only $20. If the market does not change, there is no profit or loss for either fund. Use the *policy improvement algorithm* to determine how Mr. Merrill should invest each year.

12. Formulate Prob. 11 as a *linear programming* problem.

13. Consider the blood-inventory problem presented in Chap. 5, Prob. 15. Suppose now that the number of pints of blood delivered (on a regular delivery) can be specified at the time of delivery (instead of using the old policy of receiving one pint at each delivery). Thus, the number of pints delivered can be 0, 1, 2, or 3 (more than 3 pints can never be used). The cost of regular delivery is \$50 per pint, while the cost of an emergency delivery is \$100 per pint. Starting with the proposed policy given in Chap. 5, Prob. 15, perform two iterations of the *policy improvement algorithm*.

14. Formulate Prob. 13 as a *linear programming* problem.

15. A soap company specializes in a luxury type of bath soap. The sales of this soap fluctuate between two levels—"low" and "high"—depending upon two factors: (1) whether they advertise or not, and (2) the advertising and marketing of new products by competitors. The second factor is out of the company's control, but they are trying to determine what their own advertising policy should be. For example, the *marketing manager's proposal* is to advertise when sales are low but not to advertise when sales are high (a particular policy). Advertising in any quarter of a year has its primary impact on sales in the *following* quarter. Therefore, at the beginning of each quarter, the needed information is available to forecast accurately whether sales will be low or high that quarter and to decide whether to advertise that quarter.

The cost of advertising is \$1,000,000 for each quarter of a year in which it is done. When advertising during a quarter, the probability of having high sales the next quarter is $\frac{1}{2}$ or $\frac{3}{4}$, depending upon whether the current quarter's sales are high or low. These probabilities go down to $\frac{1}{4}$ or $\frac{1}{2}$ when advertising is not done during the current quarter. The company's quarterly profits (excluding advertising costs) are \$4,000,000 when sales are high but only \$2,000,000 when sales are low.

(*a*) Use exhaustive enumeration to identify all the stationary deterministic policies. Evaluate the expected average cost per quarter for the *marketing manager's proposal*.
(*b*) Use the *policy improvement algorithm* to find the optimal policy.

16. Formulate Prob. 15 as a *linear programming* problem.

17. Consider an infinite-period inventory problem involving a single product where, at the beginning of each period, a decision must be made about how may items to produce during that period. The setup cost is \$10, and the unit production cost is \$5. The storage cost for each item not sold during the period is \$4 (a *maximum* of 2 items can be stored). The demand during each period has a known probability distribution, namely, a probability of $\frac{1}{3}$ of 0, 1, and 2 items, respectively. If the demand exceeds the supply available during the period, then those sales are lost and a penalty cost (including lost revenue) is incurred, namely, \$8 and \$32 for a shortage of 1 and 2 items, respectively.

(*a*) Consider the (s, S) policy, $(s, S) = (1, 2)$, where enough items are produced to raise the current inventory level to 2 items if, and only if, the inventory level at the beginning of a period is less than 1 item; i.e., no items are present. Determine the (long-run) expected average cost per period for this policy. In finding the transition matrix for the Markov chain for this policy, let the states represent the inventory levels at the beginning of the period.
(*b*) Use exhaustive enumeration to identify all the stationary deterministic inventory policies.

18. Use the *policy improvement algorithm* to find the optimal policy in Prob. 17.

19. Formulate Prob. 17 as a *linear programming* problem.

20. Buck and Bill Bogus are twin brothers who work at a gas station and have a counterfeiting business on the side. Each day a decision is made as to who will go to work at the gas station, and who will stay home and run the printing press in the basement. Each day that the machine works properly, it is estimated that 60 usable \$20 bills can be produced. However, the machine is somewhat unreliable and breaks down frequently. If the machine is not working at the beginning of the day, Buck can have it in working order by the beginning of the next day with probability 0.6. If Bill works on the machine, the probability decreases to 0.5. If Bill operates the machine when it is working, the probability is $\frac{3}{5}$ that it will still be working at the beginning of the next day. If Buck operates the machine, it breaks down with probability $\frac{3}{5}$. (Assume for simplicity that all breakdowns occur at the end of the day.)

(*a*) Formulate the problem of minimizing costs, i.e., maximizing the value of the Bogus brothers' labor, as a Markov decision process. (Specify states, decisions, transition matrices, and C_{ik}'s.)

(*b*) Starting with the policy of always having Buck stay home to work with/on the machine, use the *policy improvement algorithm* to determine the optimal policy.

21. Formulate Prob. 20 as a *linear programming* problem.

22.* Suppose a person wants to dispose of a car. He receives an offer each month and must decide immediately whether or not to accept the offer. Once rejected, the offer is lost. The possible offers are $600, $800, and $1,000, made with probabilities $\frac{5}{8}$, $\frac{1}{4}$, and $\frac{1}{8}$, respectively (it may be assumed that successive offers are independent of each other). Suppose that there is a maintenance cost of $60 per month and that a discount factor of $\alpha = 0.95$ is specified. Using the *policy improvement algorithm*, find a policy that minimizes the expected long-run total discounted cost. (*Hint:* There are two actions: Accept or reject the offer. Let the state space at time t denote the offer at time t, augmented by the state ∞. The process goes to state ∞ whenever an offer is accepted, and it remains there at a monthly cost of 0.) Find the optimal policy using the *policy improvement algorithm*.

23.* Formulate Prob. 22 as a *linear programming* problem, assuming that the initial states are equally likely.

24.* In Prob. 22, use three iterations of the *method of successive approximations* to approximate the optimal solution.

25. The price of a certain stock is fluctuating among the prices $10, $20, and $30 from month to month. Market analysts have predicted that if the stock is at $10 during any month, it will be at $10 or $20 next month, with probabilities $\frac{4}{5}$ and $\frac{1}{5}$, respectively; if the stock is at $20, it will be at $10, $20, or $30 next month, with probabilities $\frac{1}{4}$, $\frac{1}{4}$, and $\frac{1}{2}$, respectively; and if the stock is at $30, it will be at $20 or $30 next month, with probabilities $\frac{3}{4}$ and $\frac{1}{4}$, respectively. Given a discount factor of 0.9, use the *policy improvement algorithm* to determine when to sell and when to hold the stock to maximize the expected long-run total discounted profits. (*Hint*: Augment the state space with a state that is reached with probability 1 when stock is sold and with probability 0 when the stock is held.)

26. Formulate Prob. 25 as a *linear programming* problem.

27. In Prob. 25, use three iterations of the *method of successive approximations* to approximate the optimal solution.

28. A person is in the market for a house. Until she finds one she lives in a hotel for $70 a day. When she buys one, she pays immediately and moves in the next day. She can look at a house (if she chooses to) at most once a day, and when she does look at a house, she pays a broker's fee of $50. The houses can cost $140,000, $170,000, and $200,000, and these costs each occur with probability $\frac{1}{3}$ on any given day when she looks at a house. There is a daily discount factor of 0.999. Use the *policy improvement algorithm* to find the optimal policy.

29. Formulate Prob. 28 as a *linear programming* problem.

30. In Prob. 28, use three iterations of the *method of successive approximations* to approximate the optimal solution.

31.* A farmer raises corn. Each year that he has a successful crop he grosses $17,000 on expenses of $6,000 for seed and labor. Sometimes his crop fails and he grosses only $8,000. Each year the farmer has a chance of using two types of fertilizer: Type *A*, at a cost of $2,000, guarantees that there is a 60 percent chance of having a successful crop the next year; type *B*, at a cost of \$3,000, guarantees that there is an 80 percent chance of a good crop next year. Using the *policy improvement algorithm* with a discount factor of 0.5, determine when the farmer should use fertilizers *A* and *B*.

32.* Formulate Prob. 31 as a *linear programming* problem.

33.* In Prob. 31, use three iterations of the *method of successive approximations* to approximate the optimal solution.

34. A chemical company produces two chemicals, denoted by 0 and 1. Each month a decision is made as to which chemical to produce that month. Because the demand for each chemical is predictable, it is known that if 1 is produced this month, there is a 70 percent chance that it will also be produced again next month. Similarly, if 0 is produced this month, there is only a 20 percent chance that it will be produced again next month.

To combat the emissions of pollutants, the chemical company has two processes, process *A*, which is efficient in combating the pollution from the production of 1 but not from 0, and process *B*, which is efficient in combating the pollution from the production of 0 but not from 1. The amount of pollution from the production of each chemical under both processes is

	0	1
A	100	10
B	10	30

Unfortunately, there is a time delay in setting up the pollution-control processes, so that a decision as to which process to use must be made in the month prior to the production decision.

(*a*) Use exhaustive enumeration to identify all the stationary deterministic policies.

(*b*) Use the *policy improvement algorithm* to determine a pollution-control policy that will minimize the present value of all future pollution at a discount factor of $\alpha = 0.5$.

35. Formulate Prob. 34 as a *linear programming* problem.

36. In Prob. 34, use two iterations of the *method of successive approximations* to approximate the optimal solution.

37. A man is playing a slot machine at \$1 per play. Each time that he wins the jackpot of \$10, he finds that if he pulls the lever hard, he has a 10 percent chance of winning on the next round. But if he pulls the lever gently, he has a 20 percent chance of hitting the jackpot on the next round. If he loses and then pulls the lever hard, he has an 80 percent chance of losing on the next round also. But if he pulls the lever gently, he has a 95 percent chance of losing again. Use the *policy improvement algorithm* with a discount factor of 0.9 to determine whether the player should pull the lever hard or gently.

38. Formulate Prob. 37 as a *linear programming* problem.

39. In Prob. 37, use two iterations of the *method of successive approximations* to approximate the optimal solution.

40. Solve Prob. 28 as a *four-period model.*

41.* Solve Prob. 31 as a *four-period model.*

42. Solve Prob. 34 as a *three-period model.*

43. Solve Prob. 37 as a *three-period model.*

44. Formulate the water-resource model presented in Sec. 11.6 as a *linear programming* problem.

45. Use three iterations of the *method of successive approximations* to approximate the optimal solution to the water-resource model presented in Sec. 11.6.

46. Solve the inventory model presented in Sec. 11.7 using the expected discounted cost as the cost criterion, with a discount factor of $\alpha = 0.90$.

47. Use three iterations of the *method of successive approximations* to approximate the optimal solution to Prob. 46.

48.* Find the optimal solution to a *four-period* machine-maintenance model using the data from the example presented in this chapter. (Use a discount factor of $\alpha = 0.90$.)

49. Solve the inventory model presented in Sec. 11.7 as a *four-period* problem.

50. Solve the water-resource model presented in Sec. 11.6 as a *three-period* problem.

12

Reliability

12.1 Introduction

The many definitions of reliability that exist depend upon the viewpoint of the user. However, they all have a common core that contains the statement that reliability, $R(t)$, is the probability that a device performs adequately over the interval $[0, t]$. In general, it is assumed that unless repair or replacement occurs, adequate performance at time t implies adequate performance during the interval $[0, t]$. The device under consideration may be an entire system, a subsystem, or a component.[1] Although this definition is simple, the systems to which it is applied are generally very complex. In principle, it is possible to break down the system into black boxes, with each black box being in one of two states: good or bad. Mathematical models of the system can then be abstracted from the physical processes and the theory of combinatorial probability used to predict the reliability of the system. The black boxes may be inde-

[1] A subsystem can be viewed as containing one or more components.

pendent of, or be very dependent upon, each other. For any reasonable system, such a probability analysis generally becomes so cumbersome that it must be considered impractical. Hence we seek other methods that either simplify the calculations or provide bounds on the reliability of the entire complex system.

As an example, consider an automobile. There are a large number of functional parts, wiring, and joints. These may be broken into subsystems, with each subsystem having a reliability associated with it. Possible subsystems are the engine, transmission, exhaust, body, carburetor, and brakes. A mathematical model of the automobile system can be abstracted and the theory of combinatorial probability used to predict the reliability of the automobile.

12.2 Structural Function of a System

Suppose an automobile can be divided into n components (subsystems). The performance of each component can be denoted by a random variable, X_i, that takes on the value $x_i = 1$ if the component performs satisfactorily for the desired time and $x_i = 0$ if the component fails during this time. In general, then, X_i is a binary random variable defined by

$$X_i = \begin{cases} 1, & \text{if component } i \text{ performs satisfactorily during time } [0, t] \\ 0, & \text{if component } i \text{ fails during time } [0, t]. \end{cases}$$

The performance of the system is measured by the binary random variable[1] $\phi(X_1, X_2, \ldots, X_n)$, where

$$\phi(X_1, X_2, \ldots, X_n) = \begin{cases} 1, & \text{if system performs satisfactorily during time } [0, t] \\ 0, & \text{if system fails during time } [0, t]. \end{cases}$$

The function ϕ is called the *structure function* of the system and is just a function of the n-component random variables. Thus the performance of the automobile is a function of its n components and takes on the value 1 if the automobile functions properly for the desired time and 0 if it does not. Because the performance of each component in the automobile takes on the value 1 or 0, then the function ϕ is defined over 2^n points, with each point resulting in a 1 if the automobile performs satisfactorily and a 0 if the automobile fails.

There are several important structure functions to consider, depending upon how the components are assembled. Three structure functions will be discussed in detail.

Series System

The series system is the simplest and most common of all the configurations. For a **series system**, the system fails if any component of the system fails; i.e., it performs satisfactorily if and only if all the components perform satisfactorily. The structure function for a series system is given by

$$\phi(X_1, X_2, \ldots, X_n) = X_1 X_2 \cdots X_n = \min\{X_1, X_2, \ldots, X_n\}.$$

This equation holds because each X_i is either 1 or 0. Hence the structure function takes on the value 1 if each X_i equals 1 or, alternatively, if the minimum of the X_i

[1] Note that X_i and ϕ are functions of the time t, but t will be suppressed for ease of notation.

equals 1. For example, suppose the automobile is divided into only two components: the engine (X_1) and the transmission (X_2). Then it is reasonable to assume that the automobile will perform satisfactorily for the desired time period if and only if the engine and the transmission both perform satisfactorily. Hence

$$\phi(X_1, X_2) = X_1 X_2,$$

and

$$\phi(1, 1) = 1, \qquad \phi(1, 0) = \phi(0, 1) = \phi(0, 0) = 0.$$

Parallel System

A **parallel system** of n components is defined to be a *system that fails if all components fail*, or alternatively, a *system that performs satisfactorily if at least one of the n components performs satisfactorily* (with all n components operating simultaneously). This property of parallel systems is often called *redundancy* (i.e., there are alternative components, existing within the system, to help the system operate successfully in case of failure of one or more components). The structure function for a parallel system is given by

$$\phi(X_1, X_2, \ldots, X_n) = 1 - (1 - X_1)(1 - X_2) \cdots (1 - X_n)$$

$$= \max\{X_1, X_2, \ldots, X_n\}.$$

This equation again follows because each X_i is either 1 or 0. The structure function takes on the value 1 if at least one of the X_i equals 1 or, alternatively, if the largest X_i equals 1. In the automobile example, the car is equipped with front disk (X_1) and rear drum (X_2) brakes. The automobile will perform successfully if either the front or rear brakes operate properly.[1] If one is concerned with the structure function of the brake subsystem, then

$$\phi(X_1\ X_2) = 1 - (1 - X_1)(1 - X_2) = X_1 + X_2 - X_1 X_2,$$

and

$$\phi(1, 1) = \phi(1, 0) = \phi(0, 1) = 1, \qquad \phi(0, 0) = 0.$$

k Out of *n* System

Some systems are assembled such that the system operates if k out of n components function properly. Note that the series system is a k out of n system, with $k = n$, and the parallel system is a k out of n system, with $k = 1$. The structure function for a k out of n system is given by

$$\phi(X_1, X_2, \ldots, X_n) = \begin{cases} 1, & \text{if } \sum_{i=1}^{n} X_i \geq k \\ 0, & \text{if } \sum_{i=1}^{n} X_i < k. \end{cases}$$

In the automobile example, consider a large truck equipped with eight tires. The structure function for the tire system is an example of a four out of eight system. (Although the system's performance may be degraded if fewer than eight tires are

[1] It is evident that the loss of the front or rear brakes will affect the braking capability of the automobile, but the definition of "perform successfully" may allow for either set working.

operating, rearrangement of the tire configuration will result in adequate performance as long as at least four tires are usable.)

It is reasonable to expect the performance of an automobile to improve if the performance of one or more components is improved. This improvement can be reflected in the characterization of the structure function, where, for example, one would expect $\phi(1, 0, 0, 1)$ to be no less than $\phi(1, 0, 0, 0)$. Hence it will be assumed that if $x_i \leq y_i$, for $i = 1, 2, \ldots, n$, then

$$\phi(y_1, y_2, \ldots, y_n) \geq \phi(x_1, x_2, \ldots, x_n).$$

A system possessing this property (ϕ is an increasing function of x) is called a **coherent** (or **monotone**) **system**.

12.3 System Reliability

The structure function of a system containing n components is a binary random variable that takes on the value 1 or 0. Furthermore, the reliability of this system can be expressed as[1]

$$R = P\{\phi(X_1, X_2, \ldots, X_n) = 1\}.$$

Thus, for a series system, the reliability is given by

$$R = P\{X_1 X_2 \cdots X_n = 1\} = P\{X_1 = 1, X_2 = 1, \ldots, X_n = 1\}.$$

When the usual terms for conditional probability are employed,

$$R = P\{X_1 = 1\}P\{X_2 = 1|X_1 = 1\}P\{X_3 = 1|X_1 = 1, X_2 = 1\} \cdots P\{X_n = 1|X_1 = 1, \ldots, X_{n-1} = 1\}.$$

In general, such conditional probabilities require careful analysis. For example, $P\{X_2 = 1|X_1 = 1\}$ is the probability that component 2 will perform successfully, given that component 1 performs successfully. Consider a system where the heat from component 1 affects the temperature of component 2 and thereby its probability of success. The performance of these components is then dependent, and the evaluation of the conditional probability is extremely difficult. If, on the other hand, the performance characteristics of these components do not interact, e.g., the temperature of one component does not affect the performance of the other component, then the components can be said to be independent. The expression for the reliability then simplifies and becomes

$$R = P\{X_1 = 1\}P\{X_2 = 1\} \cdots P\{X_n = 1\}.$$

When the components of a series system are assumed to be independent, it should be noted that the reliability is a function of the probability distribution of the X_i. This phenomenon is true for any system structure.

Unless otherwise specified, it will be assumed throughout the remainder of this chapter that the component performances are independent. Hence the probability dis-

[1] The time t is now suppressed in the notation. Recall that the time is implicitly included in determining whether or not the ith component performs satisfactorily.

tribution of the binary random variables X_i can be expressed as

$$P\{X_i = 1\} = p_i,$$

and $$P\{X_i = 0\} = 1 - p_i.$$

Thus, for systems composed of independent components, the reliability becomes a function of the p_i; that is,

$$R = R(p_1, p_2, \ldots, p_n).$$

Reliability of Series Systems

As previously indicated, for a series structure,

$$\begin{aligned} R(p_1, p_2, \ldots, p_n) &= P\{\phi(X_1, X_2, \ldots, X_n) = 1\} \\ &= P\{X_1 X_2 \cdots X_n = 1\} \\ &= P\{X_1 = 1, X_2 = 1, \ldots, X_n = 1\} \\ &= P\{X_1 = 1\}P\{X_2 = 1\} \cdots P\{X_n = 1\} \\ &= p_1 p_2 \cdots p_n. \end{aligned}$$

Thus, returning to the automobile example, if the probability that the engine performs satisfactorily is 0.95 and the probability that the transmission performs satisfactorily is 0.99, then the reliability of this automobile series subsystem is given by $R = (0.95)(0.99) = 0.94$.

Reliability of Parallel Systems

The structure function for a parallel system is

$$\phi(X_1, X_2, \ldots, X_n) = \max(X_1, X_2, \ldots, X_n),$$

and the reliability is given by

$$\begin{aligned} R(p_1, p_2, \ldots, p_n) &= P\{\max(X_1, X_2, \ldots, X_n) = 1\} \\ &= 1 - P\{\text{all } X_i = 0\} \\ &= 1 - P\{X_1 = 0, X_2 = 0, \ldots, X_n = 0\} \\ &= 1 - (1 - p_1)(1 - p_2) \cdots (1 - p_n). \end{aligned}$$

Thus, if the probability that the front disk brakes and the rear drum brakes perform satisfactorily is 0.99 for each, the subsystem reliability is given by

$$R = 1 - (0.01)(0.01) = 0.9999.$$

Reliability of *k* Out of *n* Systems

The structure function for a k out of n system is

$$\phi(X_1, X_2, \ldots, X_n) = \begin{cases} 1, & \text{if } \sum_{i=1}^{n} X_i \geq k \\ 0, & \text{if } \sum_{i=1}^{n} X_i < k, \end{cases}$$

and the reliability is given by

$$R(p_1, p_2, \ldots, p_n) = P\left\{\sum_{i=1}^{n} X_i \geq k\right\}.$$

The evaluation of this expression is, in general, quite difficult except for the case of $p_1 = p_2 = \cdots = p_n = p$. Under this assumption, $\Sigma_{i=1}^{n} X_i$ has a binomial distribution with parameters n and p, so that

$$R(p, p, \ldots, p) = \sum_{i=k}^{n} \binom{n}{i} p^i(1 - p)^{n-i}.$$

For the truck tire example, if each tire has a probability of 0.95 of performing satisfactorily, then the reliability of a four out of eight system is given by

$$R = \sum_{i=4}^{8} \binom{8}{i} (0.95)^i(0.05)^{8-i} = 0.9999.$$

For general structures, the system reliability calculations can become quite tedious. A technique for computing reliabilities for this general case will be presented in the next section. However, the final result of this section is to indicate that the reliability function of a system of independent components can be shown to be an increasing function of the p_i; that is, if $p_i \leq q_i$ for $i = 1, 2, \ldots, n$, then

$$R(q_1, q_2, \ldots, q_n) \geq R(p_1, p_2, \ldots, p_n).$$

This result is analogous to, and dependent upon, the assumption that the structure function of the system is coherent. The implication of this intuitive result is that the reliability of the automobile will improve if the reliability of one or more components is improved.

12.4 Calculation of Exact System Reliability

A representation of the structure of a system can be expressed in terms of a network. For example, consider the system that can be represented by the network in Fig. 12.1. This system consists of five components, connected in a somewhat complex manner. According to the network diagram, the system will operate successfully if there exists a flow from A (source) to D (sink) through the directed graph, i.e., if components 1 and 4 operate successfully, or components 2 and 5 operate successfully, or components 1, 3, and 5 operate successfully. In fact, each arc can be viewed as having capacity 1 or 0, depending upon whether or not the component is operating. If an arc has a 0

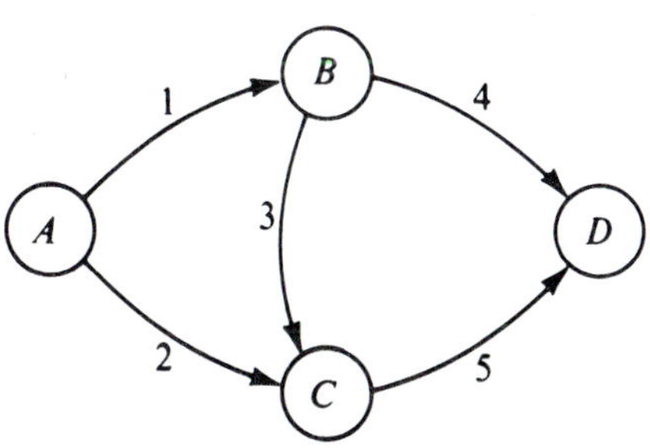

Figure 12.1 A five-component system.

attached to it (the component fails), then the network would lose that arc, and the system would operate successfully if and only if there was a path from the source to the sink in the resultant network. This situation is illustrated in Fig. 12.2, where the system still operates if components 3 and 4 fail but becomes inoperable if components 2, 3, and 4 fail. This suggests a possible method for computing the exact system reliability. Again, denote the performance of the ith component by the binary random variable X_i. Then X_i takes on the value 1 with probability p_i and 0 with probability $(1 - p_i)$. For each realization, $X_1 = x_1$, $X_2 = x_2$, $X_3 = x_3$, $X_4 = x_4$, and $X_5 = x_5$ (there are 2^5 such realizations), it is determined whether or not the system will operate, i.e., whether or not the structure function equals 1. The network consisting of those arcs with X_i equal to 1 contains at least one path if and only if the corresponding structure function equals 1. If a path is formed, the probability of obtaining this configuration is obtained. For the realization in Fig. 12.2*a*, a path is formed, and

$$P\{X_1 = 1, X_2 = 1, X_3 = 0, X_4 = 0, X_5 = 1\} = p_1 p_2 (1 - p_3)(1 - p_4) p_5.$$

Because each realization is disjoint, the system reliability is just the sum of the probabilities of those realizations that contain a path. Unfortunately, even for this simple system, 32 different realizations must be evaluated, and other techniques are desirable.

Another possible procedure for finding the exact reliability is to note that the reliability $R(p_1, p_2, \ldots, p_n)$ can be expressed as

$$R(p_1, p_2, \ldots, p_n) = P\{\text{maximum flow from source to sink} \geq 1\}.$$

This identity allows the concept of paths and cuts to be used. In reliability theory, the terminology of minimal paths and minimal cuts is introduced. A **minimal path** is a *minimal set of components that, by functioning, ensures the successful operation of the system*. For the example in Fig. 12.1, components 2 and 5 are a minimal path. A **minimal cut** is a *minimal set of components that, by failing, ensures the failure of the system*. In Fig. 12.1, components 1 and 2 are a minimal cut. For the system given in Fig. 12.1, the minimal paths and cuts are

Minimal Paths	Minimal Cuts
X_1X_4	X_1X_2
$X_1X_3X_5$	X_4X_5
X_2X_5	$X_2X_3X_4$
	X_1X_5

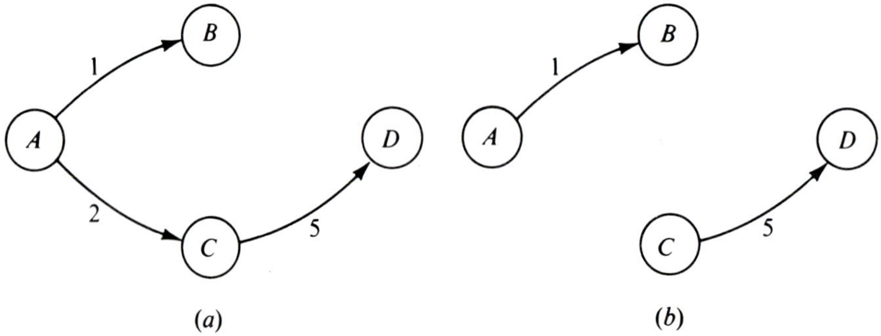

Figure 12.2 (*a*) System with components 3 and 4 failed; (*b*) system with components 2, 3, and 4 failed.

If we use all the *minimal paths,* there are *two ways* to obtain the *exact system reliability*. Because the system will operate if all the components in at least one of the minimal paths operate, the system reliability can be expressed as

$$R(p_1, p_2, p_3, p_4, p_5) = P\{\phi(X_1, X_2, X_3, X_4, X_5) = 1\}$$
$$= P\{(X_1X_4 = 1) \cup (X_1X_3X_5 = 1) \cup (X_2X_5 = 1)\}.$$

Using the algebra of sets,

$$\begin{aligned}R(p_1, p_2, p_3, p_4, p_5) &= P\{X_1X_4 = 1\} + P\{X_1X_3X_5 = 1\} \\ &\quad + P\{X_2X_5 = 1\} - P\{X_1X_3X_4X_5 = 1\} \\ &\quad - P\{X_1X_2X_4X_5 = 1\} - P\{X_1X_2X_3X_5 = 1\} \\ &\quad + P\{X_1X_2X_3X_4X_5 = 1\} \\ &= p_1p_4 + p_1p_3p_5 + p_2p_5 - p_1p_3p_4p_5 \\ &\quad - p_1p_2p_4p_5 - p_1p_2p_3p_5 + p_1p_2p_3p_4p_5 \\ &= 2p^2 + p^3 - 3p^4 + p^5, \qquad \text{when } p_i = p.\end{aligned}$$

Notice that there are $2^3 - 1 = 7$ terms in the expansion of the reliability function (in general, if there are r paths, then there are $2^r - 1$ terms in the expansion), so that this calculation is not simple.

The second method of determining the system reliability from paths is as follows: For the minimal path containing components 1 and 4, $X_1X_4 = 1$ if and only if both components function. This fact is similarly true for the other two minimal paths. However, the system will operate if all the components in at least one of the minimal paths operate. Hence paths operate as a parallel system, so that

$$\begin{aligned}\phi(X_1, X_2, X_3, X_4, X_5) &= \max[X_1X_4, X_1X_3X_5, X_2X_5] \\ &= 1 - (1 - X_1X_4)(1 - X_1X_3X_5)(1 - X_2X_5).\end{aligned}$$

Because $X_i^2 = X_i$, then

$$\begin{aligned}\phi(X_1, X_2, X_3, X_4, X_5) &= X_1X_4 + X_1X_3X_5 + X_2X_5 - X_1X_3X_4X_5 \\ &\quad - X_1X_2X_4X_5 - X_1X_2X_3X_5 + X_1X_2X_3X_4X_5.\end{aligned}$$

Noting that ϕ is a binary random variable taking on the values 1 and 0,

$$\begin{aligned}E[\phi(X_1, X_2, X_3, X_4, X_5)] &= P\{\phi(X_1, X_2, X_3, X_4, X_5) = 1\} \\ &= R(p_1, p_2, p_3, p_4, p_5).\end{aligned}$$

Therefore,

$$\begin{aligned}&R(p_1, p_2, p_3, p_4, p_5) \\ &\quad = E[X_1X_4 + X_1X_3X_5 + X_2X_5 - X_1X_3X_4X_5 - X_1X_2X_4X_5 \\ &\qquad - X_1X_2X_3X_5 + X_1X_2X_3X_4X_5] \\ &\quad = p_1p_4 + p_1p_3p_5 + p_2p_5 - p_1p_3p_4p_5 - p_1p_2p_4p_5 - p_1p_2p_3p_5 \\ &\qquad + p_1p_2p_3p_4p_5.\end{aligned}$$

This result is the same as the one obtained earlier and requires essentially the same amount of calculation.

If we use all the *minimal cuts,* there are also *two ways* to obtain the *exact system reliability*. Because the system will fail if and only if all the components in at least one of the minimal cuts fail, the system reliability can be expressed as

$$
\begin{aligned}
R(p_1, p_2, p_3, p_4, p_5) &= 1 - P\{\phi(X_1, X_2, X_3, X_4, X_5) = 0\} \\
&= 1 - P\{(X_1 = 0, X_2 = 0) \cup (X_4 = 0, X_5 = 0) \\
&\quad \cup (X_2 = 0, X_3 = 0, X_4 = 0) \cup (X_1 = 0, X_5 = 0)\} \\
&= 1 - P\{X_1 = 0, X_2 = 0\} - P\{X_4 = 0, X_5 = 0\} \\
&\quad - P\{X_2 = 0, X_3 = 0, X_4 = 0\} - P\{X_1 = 0, X_5 = 0\} \\
&\quad + P\{X_1 = 0, X_2 = 0, X_4 = 0, X_5 = 0\} \\
&\quad + P\{X_1 = 0, X_2 = 0, X_3 = 0, X_4 = 0\} \\
&\quad + P\{X_1 = 0, X_2 = 0, X_5 = 0\} \\
&\quad + P\{X_2 = 0, X_3 = 0, X_4 = 0, X_5 = 0\} \\
&\quad + P\{X_1 = 0, X_4 = 0, X_5 = 0\} \\
&\quad + P\{X_1 = 0, X_2 = 0, X_3 = 0, X_4 = 0, X_5 = 0\} \\
&\quad - P\{X_1 = 0, X_2 = 0, X_3 = 0, X_4 = 0, X_5 = 0\} \\
&\quad - P\{X_1 = 0, X_2 = 0, X_4 = 0, X_5 = 0\} \\
&\quad - P\{X_1 = 0, X_2 = 0, X_3 = 0, X_4 = 0, X_5 = 0\} \\
&\quad - P\{X_1 = 0, X_2 = 0, X_3 = 0, X_4 = 0, X_5 = 0\} \\
&\quad + P\{X_1 = 0, X_2 = 0, X_3 = 0, X_4 = 0, X_5 = 0\} \\
&= 1 - q_1q_2 - q_4q_5 - q_2q_3q_4 - q_1q_5 + q_1q_2q_3q_4 \\
&\quad + q_1q_2q_5 + q_2q_3q_4q_5 + q_1q_4q_5 - q_1q_2q_3q_4q_5,
\end{aligned}
$$

where $$q_i = 1 - p_i.$$

This result is, of course, algebraically equivalent to the one obtained previously, and it involves $2^4 - 1 = 15$ terms in the expansion of the reliability function. In general, if there are s cuts, there are $2^s - 1$ terms in the expansion.

The second method of determining the system reliability from cuts is: For the minimal cut containing components 1 and 2, $1 - (1 - X_1)(1 - X_2) = 0$ if and only if both components fail. This fact is similarly true for the other three cuts. However, the system will operate if at least one of the components in *each* cut operates. Hence cuts operate as a series system, so that

$$
\begin{aligned}
&\phi(X_1, X_2, X_3, X_4, X_5) \\
&\quad = \min[1 - (1 - X_1)(1 - X_2), 1 - (1 - X_4)(1 - X_5), \\
&\quad\quad 1 - (1 - X_2)(1 - X_3)(1 - X_4), 1 - (1 - X_1)(1 - X_5)] \\
&\quad = ([1 - (1 - X_1)(1 - X_2)][1 - (1 - X_4)(1 - X_5)] \\
&\quad\quad [1 - (1 - X_2)(1 - X_3)(1 - X_4)][1 - (1 - X_1)(1 - X_5)]) \\
&\quad = 1 - (1 - X_1)(1 - X_2) - (1 - X_4)(1 - X_5) \\
&\quad\quad - (1 - X_2)(1 - X_3)(1 - X_4) - (1 - X_1)(1 - X_5) \\
&\quad\quad + (1 - X_1)(1 - X_2)(1 - X_3)(1 - X_4) \\
&\quad\quad + (1 - X_1)(1 - X_2)(1 - X_5) \\
&\quad\quad + (1 - X_2)(1 - X_3)(1 - X_4)(1 - X_5) \\
&\quad\quad + (1 - X_1)(1 - X_4)(1 - X_5) \\
&\quad\quad - (1 - X_1)(1 - X_2)(1 - X_3)(1 - X_4)(1 - X_5).
\end{aligned}
$$

Taking expectations on both sides leads to the desired expression for the reliability. *Again,* this method requires essentially the same amount of calculation as required for the first procedure using cuts.

Although the results presented in this section were based upon the example, an extension to any system can be easily obtained. All minimal paths and/or cuts must be found and one of the four methods presented chosen.

As previously mentioned, if there are r paths and s cuts in the network, then calculating the exact reliability using paths will involve summing $2^r - 1$ terms, and using cuts will involve $2^s - 1$ terms. Hence the method using paths should be used if and only if $r \leq s$. Generally, however, it is simpler to find minimal paths rather than minimal cuts, so that the method using paths may have to be used because finding all cuts may be computationally infeasible. It is evident that finding the exact reliability of a system is quite difficult and that bounds are desirable, provided that the calculations are substantially reduced.

12.5 Bounds on System Reliability

It is evident that the calculations required to compute exact system reliability are numerous, and that other methods, such as obtaining upper and lower bounds, are desirable.

There exists a well-known result concerning binary random variables, i.e.:

If $X_1, X_2, \ldots, X_n$ are independent binary random variables that take on the value 1 or 0, and $Y_i = \Pi_{j \in J_i} X_j$, where the product ranges over all j that are elements in the set J_i, $i = 1, 2, \ldots, r$, then

$$P\{Y_1 = 0, Y_2 = 0, \ldots, Y_r = 0\} \geq P\{Y_1 = 0\}P\{Y_2 = 0\} \cdots P\{Y_r = 0\}.$$

Returning to the example of Sec. 12.4, it was pointed out that the system will operate if all the components in at least one of the minimal paths operate, so that

$$\begin{aligned} R(p_1, p_2, p_3, p_4, p_5) &= P\{\phi(X_1, X_2, X_3, X_4, X_5) = 1\} \\ &= 1 - P\{\text{all paths fail}\} \\ &= 1 - P\{X_1X_4 = 0, X_1X_3X_5 = 0, X_2X_5 = 0\}. \end{aligned}$$

From the aforementioned result on binary random variables,

$$\begin{aligned} R(p_1, p_2, p_3, p_4, p_5) &\leq 1 - P\{X_1X_4 = 0\}P\{X_1X_3X_5 = 0\}P\{X_2X_5 = 0\} \\ &= 1 - (1 - p_1p_4)(1 - p_1p_3p_5)(1 - p_2p_5) \\ &= 1 - (1 - p^2)^2(1 - p^3). \end{aligned}$$

when

$$p_i = p,$$

so that an upper bound is obtained. Similarly, in Sec. 12.4, it was pointed out that the system will operate if at least one of the components in *each* cut operates, so that

$R(p_1, p_2, p_3, p_4, p_5)$

$= P\{\phi(X_1, X_2, X_3, X_4, X_5) = 1\} = P\{$at least one of X_1, X_2 operates; at least one of X_4, X_5 operates; at least one of X_2, X_3, X_4 operates; at least one of X_1, X_5 operates$\}$

$$= P\{[1 - (1 - X_1)(1 - X_2)] = 1, [1 - (1 - X_4)(1 - X_5)] = 1, \\ [1 - (1 - X_2)(1 - X_3)(1 - X_4)] = 1, [1 - (1 - X_1)(1 - X_5)] = 1\}$$

$$= P\{(1 - X_1)(1 - X_2) = 0, (1 - X_4)(1 - X_5) = 0, \\ (1 - X_2)(1 - X_3)(1 - X_4) = 0, (1 - X_1)(1 - X_5) = 0\}.$$

Now $(1 - X_i)$ are independent binary random variables that take on the values 1 and 0, so that the result on binary random variables is again applicable; that is,

$$\begin{aligned}
&R(p_1, p_2, p_3, p_4, p_5) \\
&\quad \geq (P\{(1 - X_1)(1 - X_2) = 0\}P\{(1 - X_4)(1 - X_5) = 0\} \\
&\qquad\qquad P\{(1 - X_2)(1 - X_3)(1 - X_4) = 0\}P\{(1 - X_1)(1 - X_5) = 0\}) \\
&\quad = ([1 - (1 - p_1)(1 - p_2)][1 - (1 - p_4)(1 - p_5)] \\
&\qquad\qquad [1 - (1 - p_2)(1 - p_3)(1 - p_4)][1 - (1 - p_1)(1 - p_5)]) \\
&\quad = [1 - (1 - p)^2]^3[1 - (1 - p)^3],
\end{aligned}$$

when $$p_i = p,$$

so that a lower bound is obtained.

Thus we obtain an upper bound on the reliability based upon paths and a lower bound based upon cuts. For example, if $p_i = p = 0.9$, then

$$\begin{aligned}
0.9693 &= [1 - (0.1)^2]^3[1 - (0.1)^3] \leq R(0.9, 0.9, 0.9, 0.9, 0.9) \\
&\leq 1 - [1 - (0.9)^2]^2[1 - (0.9)^3] = 0.9902.
\end{aligned}$$

Furthermore, the exact reliability obtained from the expressions in Sec. 12.4 is given by

$$R(0.9, 0.9, 0.9, 0.9, 0.9) = (0.9)^2 + (0.9)^3 - 3(0.9)^4 + (0.9)^5 = 0.9712.$$

In general, this technique provides useful results in that the bounds are frequently quite narrow.

12.6 Bounds on Reliability Based upon Failure Times

The previous sections considered systems that performed successfully during a designated period or failed during this same period. An alternative way of viewing systems is to view their performance as a function of time.

Consider a component (or system) and its associated random variable, the time to failure, T. Denote the probability distribution of the time to failure of the component by F and its density function by f. In terms of the previous discussion, the random variables X and T are related in that X takes on the values

$$\begin{aligned}
&1, \quad \text{if } T \geq t \\
&0, \quad \text{if } T < t.
\end{aligned}$$

Then $$R(t) = P\{X = 1\} = 1 - F(t) = \int_t^\infty f(y)\, dy.$$

An appealing intuitive property in reliability is the failure rate. The **failure rate** $r(t)$ is defined for those values of t for which $F(t) < 1$ by

$$r(t) = \frac{f(t)}{R(t)}.$$

This function has a useful probabilistic interpretation; namely, $r(t)\, dt$ represents the conditional probability that an object surviving to age t will fail in the interval $[t, t + dt]$. This function is sometimes called the **hazard rate**.

In many applications, there is every reason to believe that the failure rate tends to increase because of the inevitable deterioration that occurs. Such a failure rate that remains constant or increases with age is said to have an **increasing failure rate** (IFR).

In some applications, the failure rate tends to decrease. It would be expected to decrease initially, for instance, for materials that exhibit the phenomenon of work hardening. Certain solid-state electronic devices are also believed to have a decreasing failure rate. Thus a failure rate that remains constant or decreases with age is said to have a **decreasing failure rate** (DFR).

The failure rate possesses some interesting properties. The time to failure distribution is completely determined by the failure rate. In particular, it is easily shown that

$$R(t) = 1 - F(t) = \exp\left[-\int_0^t r(\xi)\, d\xi\right].$$

Thus an assumption made about the failure rate has direct implications on the time to failure distribution. As an example, consider a component whose failure distribution is given by the exponential; i.e.,

$$F(t) = P\{T \le t\} = 1 - e^{-t/\theta}.$$

Thus $R(t)$ is given by $e^{-t/\theta}$, and the failure rate is given by

$$r(t) = \frac{(1/\theta)e^{-t/\theta}}{e^{-t/\theta}} = \frac{1}{\theta}.$$

Note that the exponential has a constant failure rate and hence has both IFR and DFR. In fact, using the expression relating the time to failure distribution and the failure rate, it is evident that a component having a constant failure rate must have a time to failure distribution that is exponential.

Bounds for IFR Distributions

Under either the IFR or DFR assumption, it is possible to obtain sharp bounds on the reliability in terms of moments and percentiles: In particular, such bounds can be derived from statements based upon the *mean time to failure*. This fact is particularly important because many design engineers present specifications in terms of mean time to failure.

Because the exponential distribution with constant failure rate is the boundary distribution between IFR and DFR distributions, it provides natural bounds on the survival probability of IFR and DFR distributions. In particular, it can be shown that if all that is known about the failure distribution is that it is IFR and has mean μ, then the greatest lower bound on the reliability that can be given is

$$R(t) \ge \begin{cases} e^{-t/\mu}, & \text{for } t < \mu \\ 0, & \text{for } t \ge \mu, \end{cases}$$

and the inequality is sharp; i.e., the exponential distribution with mean μ attains the lower bound for $t < \mu$, and the degenerate distribution concentrating at μ attains the lower bound for $t \ge \mu$. This situation can be represented graphically as shown in Fig. 12.3.

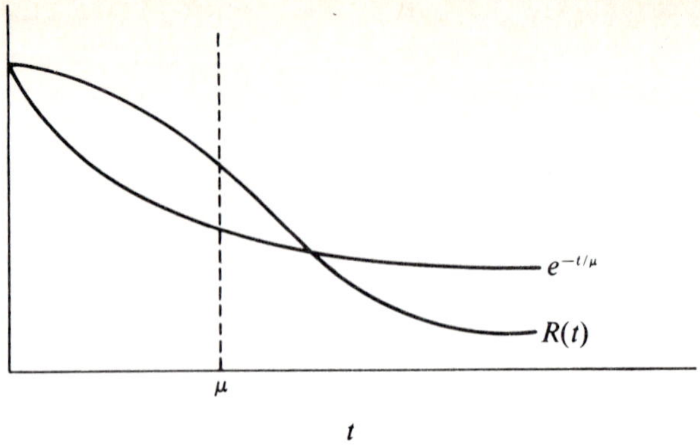

Figure 12.3 A lower bound on reliability for IFR distributions.

The least upper bound on $R(t)$ that can be obtained if we know only that F is IFR with mean μ is given by

$$R(t) \leq \begin{cases} 1, & \text{for } t \leq \mu \\ e^{-\omega t}, & \text{for } t > \mu, \end{cases}$$

where ω depends on t and satisfies $1 - \omega\mu = e^{-\omega t}$. It is important to note that the ω in the term $e^{-\omega t}$ is a function of t, so that a different ω must be found for each t. For fixed t and μ, this ω is obtained by finding the intersection of the linear function $(1 - \omega\mu)$ and the exponential function $e^{-\omega t}$. It can be shown that for $t > \mu$, such an intersection always exists.

Thus $R(t)$ for an IFR distribution with mean μ can be bounded above and below, as shown in Fig. 12.4. Note that the lower bound is the only one of consequence for $t < \mu$, and that the upper bound is the only one of consequence for $t > \mu$.

Increasing Failure Rate Average

Now that bounds on the reliability of a component have been obtained, what can be said about the preservation of *monotone failure rate*; i.e., what structures have the IFR property when their individual components have this property? Series structures of independent IFR (DFR) components are also IFR (DFR). k out of n structures consisting of n identical independent components, each having an IFR failure distribution, are also IFR; however, parallel structures of independent IFR components are not IFR unless they are composed of identical components. Thus it is evident that, even for some simple systems, there may not be a preservation of the monotone failure rate.

Instead of using the failure rate as a means for characterizing the reliability,

$$R(t) = \exp\left[-\int_0^t r(\xi)\, d\xi\right],$$

a somewhat less appealing characterization can be obtained from the failure-rate average function,

$$\int_0^t \frac{r(\xi)\, d\xi}{t} = -\frac{\log R(t)}{t}.$$

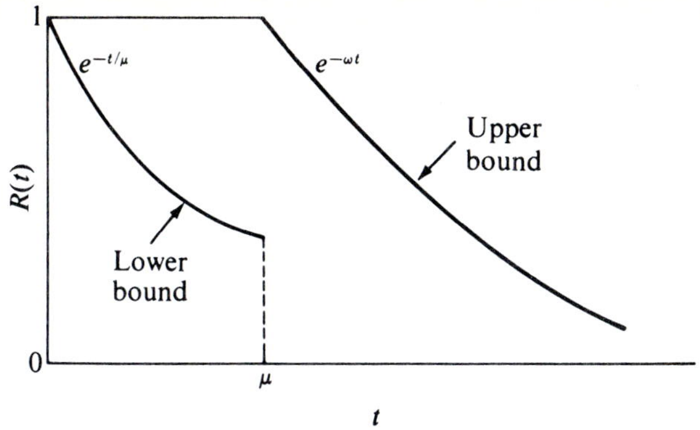

Figure 12.4 Upper and lower bounds on reliability for IFR distributions.

A distribution F such that $F(0) = 0$ is called **increasing failure rate average** (IFRA) if and only if

$$\frac{\int_0^t r(\xi)\, d\xi}{t}$$

is nondecreasing in $t \geq 0$. A similar definition is given for DFRA. It can be shown that a coherent system of independent components, each of which has an IFRA failure distribution, has a system failure distribution that is also IFRA.

As with IFR systems, there are bounds for IFRA systems. It can be easily shown that IFR distributions are also IFRA distributions (but not the reverse), and the same upper bound as given for IFR distributions is applicable here. A sharp lower bound for IFRA distributions with mean μ is given by

$$R(t) \geq \begin{cases} 0, & \text{for } t \geq \mu \\ e^{-bt}, & \text{for } t < \mu, \end{cases}$$

where b depends upon t and is defined by $e^{-bt} = b(\mu - t)$.

As an example, a monotone system containing only independent components, each of which is exponential (thereby IFRA), is itself IFRA, and the aforementioned bounds are applicable. Furthermore, these bounds are dependent only upon the system mean time to failure.

12.7 Conclusions

In recent years, the delivery of systems that perform adequately for a specified period of time in a given environment has become an important goal for both industry and government. In the space program, higher system reliability means the difference between life and death. In general, the cost of maintaining and/or repairing electronic equipment during the first year of operation often exceeds the purchase cost, giving impetus to the study and development of reliability techniques.

This chapter has been concerned with determining system reliability (or bounds) from a knowledge of component reliability or characteristics of components, such as failure rate or mean time to failure. Even the desirable state of knowing these values may lead to cumbersome and sometimes crude results. However, it must be emphasized that these values, e.g., component reliability or mean time to failure, are *not* known and are often just the design engineers' educated guesses. Furthermore, except in the case of the exponential distribution, knowledge of the mean time to failure leads to nothing but bounds. Also, it is evident that the reliability of components or systems depends heavily upon the failure rate, and the assumption of constant failure rate, which appears to be used frequently in practice, should not be made without careful analysis.

The contents of the chapter have not been concerned with the statistical aspects of reliability, i.e., estimating reliability from test data. This subject was omitted because our emphasis is on probability models, but this is not a reflection on its importance. The statistical aspects of reliability may very well be the important problem. Statistical estimation of component reliability is well in hand, but estimation of system reliability from component data is virtually an unsolved problem.

SELECTED REFERENCES

1. Barlow, R., J. Fussell, and N. Singpurwalla, eds.: *Reliability and Fault Tree Analysis,* SIAM Publications, Philadelphia, 1975.
2. Barlow, R. E., and F. Proschan: *Mathematical Theory of Reliability,* Wiley, New York, 1965.
3. Barlow, R. E., and F. Proschan: *Statistical Theory of Reliability and Life Testing,* Holt, Reinhart & Winston, New York, 1974.
4. Gnedenko, B. V., Yu. K. Belyayev, and A. D. Solovyev: *Mathematical Methods of Reliability Theory,* Academic Press, New York, 1969.
5. Kapur, K., and L. Lamberson: *Reliability in Engineering Design,* Wiley, New York, 1977.
6. Lieberman, G. J.: "The Status and Impact of Reliability Methodology," *Naval Research Logistics Quarterly,* **16**(1):17–35, 1969.
7. Ross, S.: *Introduction to Probability Models,* 3d ed., Academic Press, New York, 1985.
8. Shooman, M. L.: *Probabilistic Reliability—An Engineering Approach,* McGraw-Hill, New York, 1968.

PROBLEMS

1.* Show that the structure function for a three-component system that functions if and only if component 1 functions *and* at least one of components 2 or 3 functions, is given by

$$\begin{aligned}\phi(X_1X_2X_3) &= X_1 \max(X_2, X_3)\\ &= X_1[1 - (1 - X_2)(1 - X_3)].\end{aligned}$$

2. Show that the structure function for a four-component system that functions if and only if components 1 and 2 function *and* at least one of components 3 or 4 functions, is given by

$$\phi(X_1, X_2, X_3, X_4) = X_1X_2 \max(X_3, X_4).$$

3.* Find the reliability of the structure function given in Prob. 1 when each component has probability p_i of performing successfully.

4. Find the reliability of the structure function given in Prob. 2 when each component has probability p_i of performing successfully.

5. Consider a system consisting of three components (labeled 1, 2, 3) that operate simultaneously. The system is able to function satisfactorily as long as *any two* of the three components are still functioning satisfactorily. The goal is for the system to function satisfactorily for a length of time t, so the system's reliability, $R(t)$, is the probability that this will occur. The times until failure of the individual components are independently (but not identically) distributed, where p_i is the probability that the time until failure of component i exceeds t, for $i = 1, 2, 3$.

(*a*) Is this a k out of n system? If so, what are k and n?
(*b*) Draw a flow network representation of this system.
(*c*) Develop an explicit expression for the structure function of this system.
(*d*) Find $R(t)$ as a function of the p_i's.

6. Consider a system consisting of five components, labeled 1, 2, 3, 4, 5. The system is able to function satisfactorily as long as *at least one* of the following three combinations of components has *every* component in that combination functioning satisfactorily:

(1) Components 1 and 4;
(2) Components 2 and 5;
(3) Components 2, 3, and 4.

For a given amount of time t, let $R_i(t)$ be the known reliability of component i ($i = 1, 2, 3, 4, 5$), that is, the probability that this component will function satisfactorily for this length of time. Assume that the times until failure of the individual components are independently distributed. Let $R(t)$ be the unknown reliability of the overall system.

(*a*) Draw a flow network representation of this system.
(*b*) Develop an explicit expression for the structure function of this system.
(*c*) Find $R(t)$ as a function of the $R_i(t)$.

7. Suppose that there exist three different types of components, with two units of each type. Each unit operates independently, and each type has probability p_i of performing successfully. Either one or two systems can be built. One system can be assembled as follows: The two units of each type of component are put together in parallel, and the three types are then assembled to operate in series. Alternatively, two subsystems are assembled, each consisting of the three different types of components assembled in series. The final system is obtained by putting the two subsystems together in parallel. Which system has higher reliability?

8.* Consider the following network.

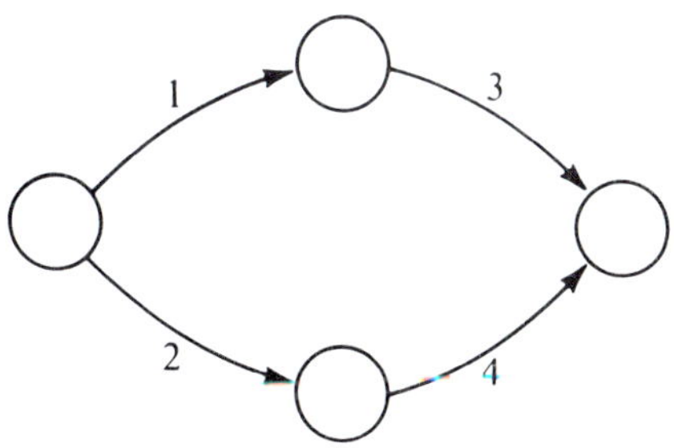

Assume that each component is independent with probability p_i of performing satisfactorily.

(*a*) Find all the minimal paths and cuts.
(*b*) Compute the exact system reliability, and evaluate it when $p_i = p = 0.90$.
(*c*) Find upper and lower bounds on the reliability, and evaluate them when $p_i = p = 0.90$.

9. Solve Prob. 8 by using the following network.

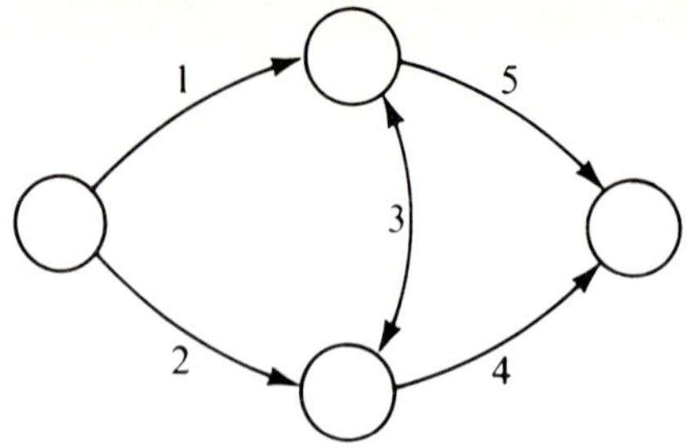

Note that component 3 flows in both directions.

10. Solve Prob. 8 by using the following network.

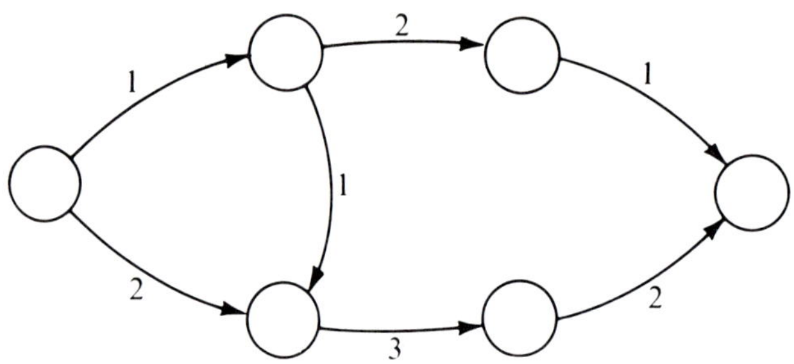

11. Solve Prob. 8 by using the following network.

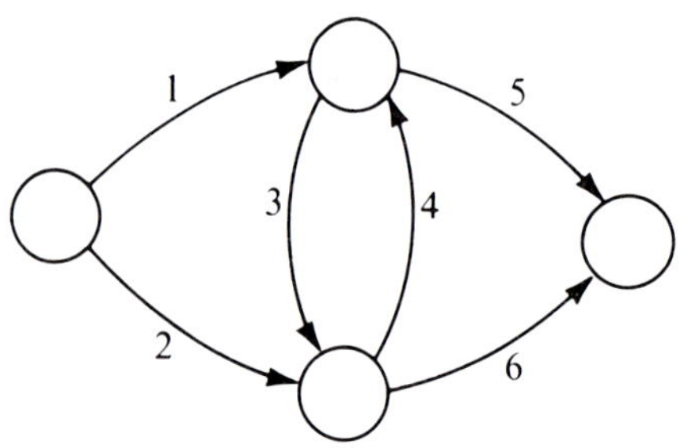

12.* Suppose F is IFR, with $\mu = 0.5$. Find upper and lower bounds on $R(t)$ for (*a*) $t = \frac{1}{4}$ and (*b*) $t = 1$.

13. A time to failure distribution is said to have a Weibull distribution if the cumulative distribution function is given by

$$F(t) = 1 - e^{-t^{\beta}/\eta}, \quad \eta, \beta > 0.$$

Find the failure rate, and show that the Weibull distribution is IFR when $\beta \geq 1$ and DFR when $0 < \beta \leq 1$.

14. Suppose that a system consists of two different, but independent, components, arranged into a series system. Further, assume that the time to failure for each component is exponential with parameter θ_i, $i = 1, 2$. Show that the distribution of the time to failure of the system is IFR.

15. Consider a parallel system consisting of two independent components whose time to failure distributions are exponential with parameters μ_1 and μ_2, respectively ($\mu_1 \neq \mu_2$). Show that the time to failure distribution of the system is not IFR.

Hint:

$$R(t) = P\{T_1 > t \text{ or } T_2 > t\} = 1 - P\{T_1 \leq t \text{ and } T_2 \leq t\}$$

$$= 1 - (1 - e^{-t/\mu_1})(1 - e^{-t/\mu_2}).$$

16. For Prob. 15, show that the time to failure distribution is IFRA.

13

Game Theory

Life is full of conflict and competition. Numerous examples involving adversaries in conflict include parlor games, military battles, political campaigns, advertising and marketing campaigns by competing business firms, and so forth. A basic feature in many of these situations is that the final outcome depends primarily upon the combination of strategies selected by the adversaries. *Game theory* is a mathematical theory that deals with the general features of competitive situations like these in a formal, abstract way. It places particular emphasis on the decision-making processes of the adversaries.

As briefly surveyed in Sec. 13.6, research on game theory continues to delve into rather complicated types of competitive situations. However, the focus in this chapter is on the simplest case, called **two-person, zero-sum games.** As the name implies, these games involve only two adversaries or *players* (who may be armies, teams, firms, and so on). They are called *zero-sum* games because one player wins whatever the other one loses, so that the sum of their net winnings is zero.

After introducing the basic model for two-person, zero-sum games in Sec. 13.1, the following four sections describe and illustrate different approaches to solving such games. We then conclude the chapter by mentioning some other kinds of competitive situations that are dealt with by other branches of game theory.

13.1 The Formulation of Two-Person, Zero-Sum Games

To illustrate the basic characteristics of two-person, zero-sum games, consider the game called *Odds and Evens*. This game consists simply of each player simultaneously showing either one finger or two fingers. If the number of fingers matches, so that the total number for both players is even, then the player taking Evens (say, player I) wins the bet (say $1) from the player taking Odds (player II). If the number does not match, player I would pay $1 to player II. Thus each player has two *strategies:* to show either one finger or two fingers. The resulting payoff to player I in dollars is shown in the *payoff table* given in Table 13.1.

In general, a two-person game is characterized by

1. The strategies of player I.
2. The strategies of player II.
3. The payoff table.

Before the game begins, each player knows the strategies he or she has available, the ones the opponent has available, and the payoff table. The actual play of the game consists of the players simultaneously choosing a strategy without knowing the opponent's choice.

A *strategy* may involve only a simple action, as when showing a certain number of fingers in the Odds and Evens game. On the other hand, in more complicated games involving a series of moves, a **strategy** is a *predetermined rule that specifies completely how one intends to respond to each possible circumstance at each stage of the game*. For example, a strategy for one side in *chess* would indicate how to make the next move for *every* possible position on the board, so the total number of possible strategies would be astronomical. Applications of game theory normally involve far less complicated competitive situations than chess, but the strategies involved can be fairly complex.

The **payoff table** shows the gain (positive or negative) for player I that would result from each combination of strategies for the two players. It is given only for player I because the table for player II is just the negative of this one, due to the zero-sum nature of the game.

The entries in the payoff table may be in any units desired, such as dollars, provided that they accurately represent the *utility* to player I of the corresponding outcome. However, utility is not necessarily proportional to the amount of money (or any other commodity) when large quantities are involved. For example, $2 million (after taxes) is probably worth much less than twice as much as $1 million to a poor person. In other words, given the choice between (1) a 50 percent chance of receiving $2 million rather than nothing and (2) being sure of getting $1 million, such an individual probably would much prefer the latter. On the other hand, the outcome

***Table 13.1* Payoff Table for the Odds and Evens Game**

		II	
		1	2
I	*1*	1	−1
	2	−1	1

corresponding to an entry of 2 in a payoff table should be "worth twice as much" to player I as the outcome corresponding to an entry of 1. Thus, given the choice, he or she should be indifferent between a 50 percent chance of receiving the former outcome (rather than nothing) and definitely receiving the latter outcome instead.[1]

A primary objective of game theory is the development of *rational criteria* for selecting a strategy. Two key assumptions are made:

1. *Both* players are *rational*.
2. *Both* players choose their strategies solely to *promote their own welfare* (no compassion for the opponent).

Game theory contrasts with *decision analysis* (see Chap. 14), where the assumption is that the decision maker is playing a game with a passive opponent, nature, which chooses its strategies in some random fashion.

We shall develop the standard game theory criteria for choosing strategies by means of illustrative examples. In particular, the next section presents a prototype example that illustrates the formulation of a two-person, zero-sum game and its solution in some simple situations. A more complicated variation of this game is then carried into Sec. 13.3 to develop a more general criterion. Sections 13.4 and 13.5 describe a *graphical procedure* and a *linear programming formulation* for solving such games.

13.2 Solving Simple Games—A Prototype Example

Two politicians are running against each other for the U.S. Senate. Campaign plans must now be made for the final 2 days, which are expected to be crucial because of the closeness of the race. Therefore, both politicians want to spend these days campaigning in two key cities: *Bigtown* and *Megalopolis*. To avoid wasting campaign time, they plan to travel at night and spend either one full day in each city or two full days in just one of the cities. However, since the necessary arrangements must be made in advance, neither politician will learn his (or her)[2] opponent's campaign schedule until after he has finalized his own. Therefore each politician has asked his campaign manager in each of these cities to assess what the impact would be (in terms of votes won or lost) from the various possible combinations of days spent there by himself and by his opponent. He then wishes to use this information to choose his best strategy on how to use these 2 days.

Formulation: To formulate this problem as a two-person, zero-sum game, we must identify the two *players* (obviously the two politicians), the *strategies* for each player, and the *payoff table*.

As the problem has been stated, each player has the following three strategies:

Strategy 1 = spend 1 day in each city.
Strategy 2 = spend both days in Bigtown.
Strategy 3 = spend both days in Megalopolis.

By contrast, the strategies would have been more complicated ones in a different situation where each politician would learn where his opponent will spend his first day before he finalizes his own plans for his second day. In that case, a typical strategy

[1] See Sec. 14.5 for a further discussion of the concept of *utility*.

[2] We use only *his* or *her* in our examples and problems for ease of reading; we do not mean to imply that only men or women are engaged in the various activities.

Table 13.2 **Formulation of Payoff Table for the Political Campaign Problem**

Strategy (Politician I) \ Strategy (Politician II)	Total Net Votes Won by Politician I (in Units of 1,000 Votes)		
	Politician II		
	1	2	3
Politician I 1			
2			
3			

would be: Spend the first day in Bigtown; if the opponent also spends the first day in Bigtown, then spend the second day in Bigtown; however, if the opponent spends the first day in Megalopolis, then spend the second day in Megalopolis. There would be eight such strategies, one for each combination of the two first-day choices, the opponent's two first-day choices, and the two second-day choices.

Each entry in the payoff table for player I represents the *utility* to player I (or the negative utility to player II) of the outcome resulting from the corresponding strategies used by the two players. From the politician's viewpoint, the objective is to *win votes,* and each additional vote (before learning the outcome of the election) is of equal value to him. Therefore, the appropriate entries for the payoff table are the *total net votes won* from the opponent (i.e., the sum of the net vote changes in the two cities) resulting from these 2 days of campaigning. This formulation is summarized in Table 13.2.

However, we should also point out that this payoff table would *not* be appropriate if additional information were available to the politicians. In particular, if they knew exactly how the populace was planning to vote 2 days before the election, the only significance of the data prescribed by Table 13.2 would be to indicate which politician would win the election with each combination of strategies. Because the ultimate goal is to win the election, and because the size of the plurality is relatively inconsequential, the utility entries in the table then should be some positive constant (say, $+1$) when politician I would win and -1 when he would lose. Even if only a *probability* of winning can be determined for each combination of strategies, the appropriate entries would be *the probability of winning minus the probability of losing* because they then would represent *expected* utilities. However, sufficiently accurate data to make such determinations usually are not available.

Using the form given in Table 13.2, three alternative sets of data for the payoff table are given here to illustrate how to solve three different kinds of games.

Variation 1: Given that Table 13.3 is the payoff table for the two politicians (players), which strategy should each of them select?

Table 13.3 **Payoff Table for Variation 1 of the Political Campaign Problem**

		II		
		1	2	3
	1	1	2	4
I	*2*	1	0	5
	3	0	1	−1

This situation is a rather special one, where the answer can be obtained just by applying the concept of **dominated strategies** to rule out a succession of inferior strategies until only one choice remains. Specifically, a strategy can be eliminated from further consideration if it is *dominated* by another strategy, i.e., if there is another strategy that is *always at least as good* regardless of what the opponent does.

At the outset, Table 13.3 includes *no* dominated strategies for player II. However, for player I, strategy 3 is dominated by strategy 1 because the latter has larger payoffs ($1 \geq 0$, $2 \geq 1$, $4 \geq -1$) regardless of what player II does. Eliminating strategy 3 from further consideration yields the following reduced payoff table:

	1	2	3
1	1	2	4
2	1	0	5

Because both players are assumed to be rational, player II also can deduce that player I has only these two strategies remaining under consideration. Therefore, player II now *does* have a dominated strategy—strategy 3, which is dominated by both strategies 1 and 2 because they always have smaller losses (payoffs to player I) in this reduced payoff table (for strategy 1: $1 \leq 4$, $1 \leq 5$; for strategy 2: $2 \leq 4$, $0 \leq 5$). Eliminating this strategy yields

	1	2
1	1	2
2	1	0

At this point, strategy 2 for player I becomes dominated by strategy 1 because the latter is better in column 2 ($2 \geq 0$) and equally good in column 1 ($1 \geq 1$). Eliminating the dominated strategy leads to

	1	2
1	1	2

where strategy 2 for player II is dominated by strategy 1 ($1 \leq 2$). Consequently, both players should select their strategy 1. Player I then will receive a payoff of 1 from player II (i.e., politician I will gain 1,000 votes from politician II).

In general, the payoff to player I when both players play optimally is referred to as the **value of the game**. A game that has a value of *zero* is said to be a **fair game**. Since this particular game has a value of 1, it is *not* a fair game.

The concept of a *dominated strategy* is a very useful one for reducing the size of the payoff table that needs to be considered and, in unusual cases like this one, actually identifying the optimal solution for the game. However, most games require another approach to at least finish solving, as illustrated by the next two variations of the example.

Variation 2: Now suppose that the current data give Table 13.4 as the payoff table for the politicians (players). This game does not have dominated strategies, so it is not obvious what the players should do. What line of reasoning does game theory say they should use?

Consider player I. By selecting strategy 1, he could win 6 or he could lose as much as 3. However, because player II is rational and thus will protect himself from

Table 13.4 **Payoff Table for Variation 2 of the Political Campaign Problem**

		II 1	2	3	Minimum	
	1	−3	−2	6	−3	
I	2	2	0	2	0	← Maximin value
	3	5	−2	−4	−4	
	Maximum:	5	0	6		
			↑ Minimax value			

large payoffs to I, it seems probable that playing strategy 1 would result in a loss to player I. Similarly, by selecting strategy 3, player I could win 5, but more probably his rational opponent would avoid this loss and instead administer him a loss, which could be as large as 4. On the other hand, if player I selects strategy 2, he is guaranteed not to lose anything, and he could even win something. Therefore, because it provides the *best guarantee,* strategy 2 seems to be a ''rational'' choice for player I against his rational opponent.

Now consider player II. He could lose as much as 5 or 6 by using strategy 1 or 3, but is guaranteed at least breaking even with strategy 2. Therefore, using the same reasoning of seeking the *best guarantee* against a rational opponent, his apparent choice is strategy 2.

If both players choose their strategy 2, the result is that both break even. Thus, in this case, neither player improves upon his *best guarantee,* but both also are forcing the opponent into the same position. Even when the opponent deduces a player's strategy, the opponent cannot exploit this information to improve his position. Stalemate.

The end product of this line of reasoning is that each player should play in such a way as to *minimize his maximum losses* whenever the resulting choice of strategy cannot be exploited by the opponent to then improve his position. This so-called **minimax criterion** is a standard criterion proposed by game theory for selecting a strategy. In terms of the payoff table, it implies that *player I* should select the strategy whose *minimum payoff* is *largest,* whereas *player II* should choose the one whose *maximum payoff to player I* is the *smallest*. This criterion is illustrated in Table 13.4, where strategy 2 is identified as the ''maximin'' strategy for player I, and strategy 2 is the minimax strategy for player II. The resulting payoff of 0 is the *value of the game,* so this is a *fair game*.

Notice the interesting fact that the same entry in this payoff table yields both the maximin and minimax values. The reason is that this entry is both the minimum in its row and the maximum in its column. The position of any such entry is called a **saddle point**.

The fact that this game possesses a saddle point was actually crucial in determining how it should be played. Because of the saddle point, neither player can take advantage of the opponent's strategy to improve his own position. In particular, when player II predicts or learns that player I is using strategy 2, player II would only increase his losses if he were to change from his original plan of using his strategy 2. Similarly, player I would only worsen his position if he were to change his plan. Thus neither player has any motive to consider changing strategies, either to take advantage of his opponent or to prevent the opponent from taking advantage of him.

Table 13.5 **Payoff Table for Variation 3 of the Political Campaign Problem**

		II 1	II 2	II 3	Minimum	
I	1	0	−2	2	−2	← Maximin value
I	2	5	4	−3	−3	
I	3	2	3	−4	−4	
	Maximum:	5	4	2 ↑ Minimax value		

Therefore, since this is a **stable solution**, players I and II should exclusively use their maximin and minimax strategies, respectively.

As the next variation illustrates, some games do not possess a saddle point, in which case a more complicated analysis is required.

Variation 3: Late developments in the campaign result in the *final* payoff table for the two politicians (players) given by Table 13.5. How should this game be played?

Suppose that both players attempt to apply the *minimax criterion* in the same way as in variation 2. Player I can guarantee that he will lose no more than 2 by playing strategy 1. Similarly, player II can guarantee that he will lose no more than 2 by playing strategy 3.

However, notice that the maximin value (−2) and the minimax value (2) do not coincide in this case. The result is that there is *no saddle point*.

What are the resulting consequences if both players plan to use the strategies just derived? It can be seen that player I would win 2 from player II, which would make player II unhappy. Because player II is rational and can therefore foresee this outcome, he would then conclude that he can do much better, actually winning 2 rather than losing 2, by playing strategy 2 instead. Because player I is also rational, he would anticipate this switch and conclude that he can improve considerably, from −2 to 4, by changing to strategy 2. Realizing this, player II would then consider switching back to strategy 3 to convert a loss of 4 to a gain of 3. This possibility of a switch would cause player I to consider again using strategy 1, after which the whole cycle would start over again.

In short, the originally suggested solution (player I to play strategy 1 and player II to play strategy 3) is an **unstable solution**, so it is necessary to develop a more satisfactory solution. But what kind of solution should it be?

The key fact seems to be that whenever one player's strategy is predictable, the opponent here can take great advantage of this information to improve his position. Therefore, an essential feature of a rational plan for playing a game such as this one is that neither player should be able to deduce which strategy the other will use. Hence, rather than applying some known criterion for determining a single strategy that will definitely be used, it is necessary to choose among alternative acceptable strategies on some kind of random basis. By doing this, neither player knows in advance which of his own strategies will be used, let alone what his opponent will do.

This suggests, in very general terms, the kind of approach that is required for games lacking a saddle point. The next section discusses this approach more fully.

Given this foundation, we turn our attention to procedures for finding an optimal way of playing such games. This particular variation of the political campaign problem will continue to be used to illustrate these ideas as they are developed.

13.3 Games with Mixed Strategies

Whenever a game does not possess a saddle point, game theory advises each player to assign a probability distribution over his or her set of strategies. To express this mathematically, let

x_i = probability that player I will use strategy i ($i = 1, 2, \ldots, m$),

y_j = probability that player II will use strategy j ($j = 1, 2, \ldots, n$),

where m and n are the respective numbers of available strategies. Thus player I would specify his plan for playing the game by assigning values to $x_1, x_2, \ldots, x_m$. Because these values are probabilities, they would need to be nonnegative and add up to 1. Similarly, the plan for player II would be described by the values he assigns to his decision variables $y_1, y_2, \ldots, y_n$. These plans $(x_1, x_2, \ldots, x_m)$ and $(y_1, y_2, \ldots, y_n)$ are usually referred to as **mixed strategies**, and the original strategies would then be called *pure strategies*.

When actually playing the game, it is necessary for each player to use one of his pure strategies. However, this pure strategy would be chosen by using some random device to obtain a random observation from the probability distribution specified by the mixed strategy, where this observation would indicate which particular pure strategy to use.

To illustrate, suppose that players I and II in *variation* 3 of the political campaign problem (see Table 13.5) select the mixed strategies $(x_1, x_2, x_3) = (\frac{1}{2}, \frac{1}{2}, 0)$ and $(y_1, y_2, y_3) = (0, \frac{1}{2}, \frac{1}{2})$, respectively. This selection would say that player I is giving an equal chance (probability of $\frac{1}{2}$) to choosing either (pure) strategy 1 or 2, but he is discarding strategy 3 entirely. Similarly, player II is randomly choosing between his last two pure strategies. To play the game, each player could then flip a coin to determine which of his two acceptable pure strategies he will actually use.

Although no completely satisfactory measure of performance is available for evaluating mixed strategies, a very useful one is the *expected payoff*. Applying the probability theory definition of expected value, this quantity is

$$\text{Expected payoff} = \sum_{i=1}^{m} \sum_{j=1}^{n} p_{ij} x_i y_j,$$

where p_{ij} is the payoff if player I uses pure strategy i and player II uses pure strategy j. In the example of mixed strategies just given, there are four possible payoffs $(-2, 2, 4, -3)$, each occurring with a probability of $\frac{1}{4}$, so the expected payoff is $\frac{1}{4}(-2 + 2 + 4 - 3) = \frac{1}{4}$. Thus this measure of performance does not disclose anything about the risks involved in playing the game, but it does indicate what the average payoff will tend to be if the game is played many times.

Using this measure, game theory extends the concept of the minimax criterion to games that lack a saddle point and thus need mixed strategies. In this context, the **minimax criterion** says that a given player should select the mixed strategy that *minimizes* the *maximum expected loss* to himself. Equivalently, when focusing on payoffs (player I) rather than losses (player II), this criterion says to *maximin* instead, i.e., *maximize* the *minimum expected payoff* to the player. By the *minimum expected payoff* we mean the smallest possible expected payoff that can result from any mixed strategy with which the opponent can counter. Thus the mixed strategy for player I that is *optimal* according to this criterion is the one that provides the *guarantee* (minimum expected payoff) that is *best* (maximal). (The value of this best guarantee is the *maximin value,* denoted by $\underline{v}$.) Similarly, the *optimal* strategy for player II is the one that provides the *best guarantee,* where best now means *minimal* and guarantee refers to the *maximum expected loss* that can be administered by any of the opponent's mixed strategies. (This best guarantee is the *minimax value,* denoted by $\overline{v}$.)

Recall that when only pure strategies were used, games not having a saddle point turned out to be *unstable* (no stable solutions). The reason was essentially that $\underline{v} < \overline{v}$, so that the players would want to change their strategies to improve their positions. Similarly, for games with mixed strategies, it is necessary that $\underline{v} = \overline{v}$ for the optimal solution to be *stable*. Fortunately, according to the *minimax theorem* of game theory this condition always holds for such games.

> **Minimax theorem:** If mixed strategies are allowed, the pair of mixed strategies that is optimal according to the minimax criterion provides a *stable solution* with $\underline{v} = \overline{v} = v$ (the value of the game), so that neither player can do better by unilaterally changing his strategy.

One proof of this theorem is included in Sec. 13.5.

Although the concept of mixed strategies becomes quite intuitive if the game is played *repeatedly,* it requires some interpretation when the game is to be played just *once*. In this case, using a mixed strategy still involves selecting and using *one* pure strategy (randomly selected from the specified probability distribution), so it might seem more sensible to ignore this randomization process and just choose the one "best" pure strategy to be used. However, we have already illustrated for variation 3 in the preceding section that a player must *not* allow his opponent to deduce what his strategy will be (i.e., the solution procedure under the rules of game theory must not *definitely* identify which pure strategy will be used when the game is unstable). Furthermore, even if the opponent is able to use only his knowledge of the tendencies of the first player to deduce *probabilities* (for the pure strategy chosen) that are different from those for the *optimal* mixed strategy, then he still can take advantage of this knowledge to reduce the expected payoff to the first player. Therefore, the only way to *guarantee* attaining the optimal *expected* payoff v is to *randomly* select the pure strategy to be used from the probability distribution for the optimal mixed strategy. (Valid statistical procedures for making such a random selection are discussed in Sec. 15.2.)

Now we must show how to find the optimal mixed strategy for each player. There are several methods of doing this. One is a graphical procedure that may be used whenever one of the players has only two (undominated) pure strategies; this approach is described in the next section. When larger games are involved, the usual method is to transform the problem into a linear programming problem that would then be solved by the simplex method on a computer; Sec. 13.5 discusses this approach.

13.4 Graphical Solution Procedure

Consider any game with mixed strategies such that, after eliminating dominated strategies, one of the players has only two pure strategies. To be specific, let this player be player I. Because his mixed strategies are (x_1, x_2) and $x_2 = 1 - x_1$, it is necessary for him to solve only for the optimal value of x_1. However, it is straightforward to plot the expected payoff as a function of x_1 for each of his opponent's pure strategies. This graph can then be used to identify the point that maximizes the minimum expected payoff. The opponent's minimax mixed strategy can also be identified from the graph.

To illustrate this procedure, consider *variation 3* of the political campaign problem (see Table 13.5). Notice that the third pure strategy for player I is dominated by his second, so the payoff table can be reduced to the form given in Table 13.6. Therefore, for each of the pure strategies available to player II, the expected payoff for player I will be

(y_1, y_2, y_3)	Expected Payoff
(1, 0, 0)	$0x_1 + 5(1 - x_1) = 5 - 5x_1$
(0, 1, 0)	$-2x_1 + 4(1 - x_1) = 4 - 6x_1$
(0, 0, 1)	$2x_1 - 3(1 - x_1) = -3 + 5x_1$

Now plot these expected payoff lines on a graph, as shown in Fig. 13.1. For any given value of x_1 and of (y_1, y_2, y_3), the expected payoff will be the appropriate weighted average of the corresponding points on these three lines. In particular,

$$\text{Expected payoff} = y_1(5 - 5x_1) + y_2(4 - 6x_1) + y_3(-3 + 5x_1).$$

Thus, given x_1, the minimum expected payoff is given by the corresponding point on the "bottom" line. According to the minimax (or maximin) criterion, player I should select the value of x_1 giving the *largest* minimum expected payoff, so that

$$\underline{v} = v = \max_{0 \le x_1 \le 1} \{\min(-3 + 5x_1, 4 - 6x_1)\}.$$

Therefore, the optimal value of x_1 is the one at the intersection of the two lines $(-3 + 5x_1)$ and $(4 - 6x_1)$. Solving algebraically,

$$-3 + 5x_1 = 4 - 6x_1,$$

so that $x_1 = \frac{7}{11}$; thus $(x_1, x_2) = (\frac{7}{11}, \frac{4}{11})$ is the *optimal mixed strategy* for player I, and

$$\underline{v} = v = -3 + 5(\tfrac{7}{11}) = \tfrac{2}{11}$$

Table 13.6 **Reduced Payoff Table for Variation 3 of the Political Campaign Problem**

		Probability	y_1	*II* y_2	y_3
	Probability	Pure Strategy	1	2	3
I	x_1	1	0	−2	2
	$1 - x_1$	2	5	4	−3

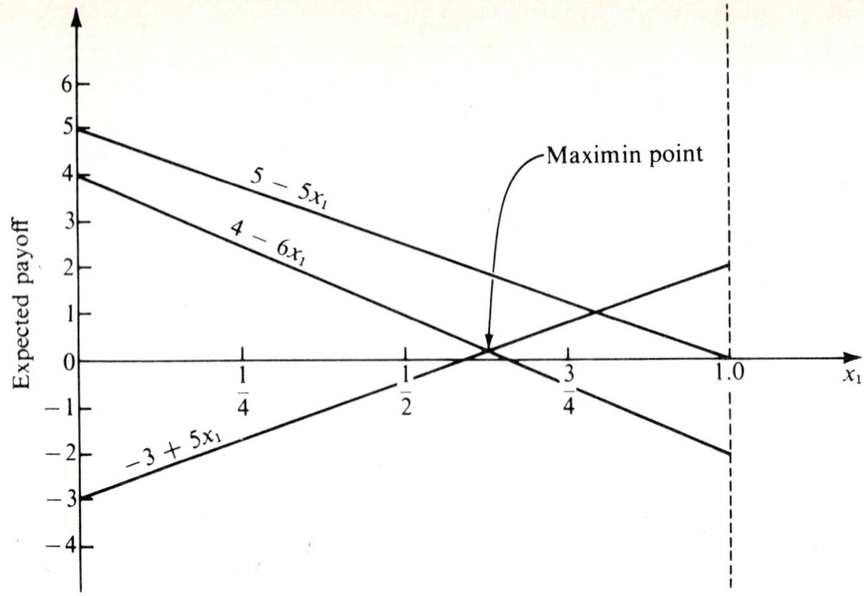

Figure 13.1 Graphical procedure for solving games.

is the value of the game.

To find the corresponding optimal mixed strategy for player II, we now reason as follows. According to the definition of the *minimax value* $\overline{v}$ and the minimax theorem, the expected payoff resulting from the optimal strategy $(y_1, y_2, y_3) = (y_1^*, y_2^*, y_3^*)$ will satisfy the condition

$$y_1^*(5 - 5x_1) + y_2^*(4 - 6x_1) + y_3^*(-3 + 5x_1) \le \overline{v} = v = \tfrac{2}{11}$$

for all values of x_1 $(0 \le x_1 \le 1)$. Furthermore, when player I is playing optimally (that is, $x_1 = \frac{7}{11}$), this inequality will be an equality, so that

$$\tfrac{20}{11}y_1^* + \tfrac{2}{11}y_2^* + \tfrac{2}{11}y_3^* = v = \tfrac{2}{11}.$$

Because (y_1, y_2, y_3) is a probability distribution, it is also known that

$$y_1^* + y_2^* + y_3^* = 1.$$

Therefore, $y_1^* = 0$ because $y_1^* > 0$ would violate the next-to-last equation; i.e., the expected payoff on the graph at $x_1 = \frac{7}{11}$ would be above the maximin point. (In general, any line that does not pass through the maximin point must be given a zero weight to avoid increasing the expected payoff above this point.)

Hence

$$y_2^*(4 - 6x_1) + y_3^*(-3 + 5x_1) \begin{cases} \le \frac{2}{11} & \text{for } 0 \le x_1 \le 1 \\ = \frac{2}{11} & \text{for } x_1 = \frac{7}{11}. \end{cases}$$

But y_2^* and y_3^* are numbers, so the left-hand side is the equation of a straight line, which is a fixed weighted average of the two "bottom" lines on the graph. Because the ordinate of this line must equal $\frac{2}{11}$ at $x_1 = \frac{7}{11}$, and because it must never exceed $\frac{2}{11}$, the line necessarily is horizontal. (This conclusion is always true unless the optimal value of x_1 is either zero or 1, in which case player II also should use a single pure strategy.) Therefore,

$$y_2^*(4 - 6x_1) + y_3^*(-3 + 5x_1) = \tfrac{2}{11}, \qquad \text{for } 0 \le x_1 \le 1.$$

Hence, to solve for y_2^* and y_3^*, select two values of x_1 (say, zero and 1), and solve the resulting two simultaneous equations. Thus

$$4y_2^* - 3y_3^* = \tfrac{2}{11},$$

$$-2y_2^* + 2y_3^* = \tfrac{2}{11},$$

so that $y_3^* = \frac{6}{11}$ and $y_2^* = \frac{5}{11}$. Therefore, the *optimal mixed strategy* for player II is $(y_1, y_2, y_3) = (0, \frac{5}{11}, \frac{6}{11})$.

If, in another problem, there should happen to be more than two lines passing through the maximin point, so that more than two of the y_j^* can be greater than zero, this condition would imply that there are many ties for the optimal mixed strategy for player II. One such strategy can then be identified by setting all but two of these y_j^* equal to zero and solving for the remaining two in the manner just described. For the remaining two, the associated lines must have positive slope in one case and negative slope in the other.

Although this graphical procedure has been illustrated for only one particular problem, essentially the same reasoning can be used to solve any game with mixed strategies that has only two undominated pure strategies for one of the players.

13.5 Solving by Linear Programming

Any game with mixed strategies can be solved by transforming the problem into a linear programming problem. As you will see, this transformation requires little more than applying the minimax theorem and using the definitions of the maximin value $\underline{v}$ and minimax value $\overline{v}$.

First, consider how to find the optimal mixed strategy for player I. As indicated in Sec. 13.3,

$$\text{Expected payoff} = \sum_{i=1}^{m} \sum_{j=1}^{n} p_{ij} x_i y_j$$

and the strategy $(x_1, x_2, \ldots, x_m)$ is optimal if

$$\sum_{i=1}^{m} \sum_{j=1}^{n} p_{ij} x_i y_j \ge \underline{v} = v$$

for all opposing strategies $(y_1, y_2, \ldots, y_n)$. Thus this inequality will need to hold, for example, for each of the pure strategies of player II, i.e., for each of the strategies $(y_1, y_2, \ldots, y_n)$ where one $y_j = 1$ and the rest equal zero. Substituting these values into the inequality yields

$$\sum_{i=1}^{m} p_{ij} x_i \ge v, \qquad \text{for } j = 1, 2, \ldots, n,$$

so that the inequality *implies* this set of n inequalities. Furthermore, this set of n inequalities *implies* the original inequality (rewritten),

$$\sum_{j=1}^{n} y_j \left(\sum_{i=1}^{m} p_{ij} x_i \right) \geq \sum_{j=1}^{n} y_j v = v,$$

since

$$\sum_{j=1}^{n} y_j = 1.$$

Because the implication goes in both directions, it follows that imposing this set of n linear inequalities is *equivalent* to requiring the original inequality to hold for all strategies $(y_1, y_2, \ldots, y_n)$. But these n inequalities are legitimate linear programming constraints, as are the additional constraints

$$x_1 + x_2 + \cdots + x_m = 1$$

$$x_i \geq 0, \qquad \text{for } i = 1, 2, \ldots, m$$

that are required to ensure that the x_i are probabilities. Therefore, any solution $(x_1, x_2, \ldots, x_m)$ that satisfies this entire set of linear programming constraints is the desired optimal mixed strategy.

Consequently, the problem of finding an optimal mixed strategy has been reduced to finding a *feasible solution* for a *linear programming* problem. The two remaining difficulties are (1) v is unknown and (2) the linear programming problem has no objective function. Fortunately, both these difficulties can be resolved at one stroke by replacing the unknown constant v by the variable x_{m+1} and then *maximizing* x_{m+1}, so that x_{m+1} automatically will equal v (by definition) at the *optimal* solution for the linear programming problem!

To summarize, player I would find his optimal mixed strategy by using the simplex method to solve the *linear programming* problem:

$$\text{Minimize} \qquad (-x_{m+1}),$$

subject to

$$p_{11}x_1 + p_{21}x_2 + \cdots + p_{m1}x_m - x_{m+1} \geq 0$$

$$p_{12}x_1 + p_{22}x_2 + \cdots + p_{m2}x_m - x_{m+1} \geq 0$$

$$\vdots$$

$$p_{1n}x_1 + p_{2n}x_2 + \cdots + p_{mn}x_m - x_{m+1} \geq 0$$

$$-(x_1 + x_2 + \cdots + x_m) = -1$$

and

$$x_i \geq 0, \qquad \text{for } i = 1, 2, \ldots, m.$$

(The objective function and equality constraint have been rewritten here in an equivalent way for later convenience.) Note that x_{m+1} is not restricted to be nonnegative, whereas the simplex method can be applied only after *all* the variables have nonnegativity constraints. However, this matter can be easily rectified, as will be discussed shortly.

Now consider player II. He could find his optimal mixed strategy by rewriting the payoff table as the payoff to himself rather than to player I and then by proceeding exactly as just described. However, it is enlightening to summarize his formulation in terms of the original payoff table. By proceeding in a way that is completely analogous to the one just described, player II would conclude that his optimal mixed strategy is given by the optimal solution to the *linear programming* problem:

$$\text{Maximize} \quad (-y_{n+1}),$$

subject to

$$p_{11}y_1 + p_{12}y_2 + \cdots + p_{1n}y_n - y_{n+1} \leq 0$$

$$p_{21}y_1 + p_{22}y_2 + \cdots + p_{2n}y_n - y_{n+1} \leq 0$$

$$\vdots$$

$$p_{m1}y_1 + p_{m2}y_2 + \cdots + p_{mn}y_n - y_{n+1} \leq 0$$

$$-(y_1 + y_2 + \cdots + y_n) = -1$$

and

$$y_j \geq 0, \qquad \text{for } j = 1, 2, \ldots, n.$$

Notice the key fact that this linear programming problem and the one given for player I are *dual* to each other. This fact has several important implications. One implication is that the optimal mixed strategies for both players can be found by solving only one of the linear programming problems because the optimal dual solution is an automatic by-product of the simplex method calculations to find the optimal primal solution. A second implication is that this brings all *duality theory* to bear upon the interpretation and analysis of games.

A related implication is that this provides a simple proof of the minimax theorem. Let x^*_{m+1} and y^*_{n+1} denote the value of x_{m+1} and y_{n+1} in the optimal solution of the respective linear programming problems. It is known from *duality theory* that $-x^*_{m+1} = -y^*_{n+1}$, so that $x^*_{m+1} = y^*_{n+1}$. However, it is evident from the definition of $\underline{v}$ and $\overline{v}$ that $\underline{v} = x^*_{m+1}$ and $\overline{v} = y^*_{n+1}$, so it follows that $\underline{v} = \overline{v}$ as claimed by the minimax theorem.

The objective functions of these two linear programming problems have been written as minimize $(-x_{m+1})$ and maximize $(-y_{n+1})$ simply to demonstrate that the two problems are *dual* to each other. Hereafter, these objective functions will be written in the more natural *equivalent* forms, maximize x_{m+1} and minimize y_{n+1}. For the same reason, the negative sign now should be deleted from both sides of the equality constraint in each problem.

One remaining loose end needs to be tied up, namely, what to do about x_{m+1} and y_{n+1} being unrestricted in sign in the linear programming formulations. If it is clear that $v \geq 0$ so that the optimal values of x_{m+1} and y_{n+1} are nonnegative, then it is safe to introduce nonnegativity constraints for these variables for purposes of applying the simplex method. However, if $v < 0$, then an adjustment needs to be made. One possibility is to replace v by the difference of two nonnegative variables. Another is to reverse players I and II so that the payoff table would be rewritten as the payoff to the original player II, which would make the corresponding value of v positive. A third, and the most commonly used, procedure is to add a sufficiently large fixed constant to all the entries in the payoff table so that the new value of the game will be positive. (For example, setting this constant equal to the absolute value of the largest negative entry will suffice.) Because this same constant is added to every entry, this adjustment cannot alter the optimal mixed strategies in any way, so they can now be obtained in the usual manner. The indicated value of the game would be increased by the amount of the constant, but this value can be readjusted after the solution has been obtained.

To illustrate this linear programming approach, consider again variation 3 of the political campaign problem after eliminating dominated strategy 3 for player I (see Table 13.6). Because there are some negative entries in the reduced payoff table, it is unclear at the outset whether the *value* of the game v is *nonnegative* (it turns out to be). For the moment, let us assume that $v \geq 0$ and proceed without making any of the adjustments discussed in the preceding paragraph.

To write out the linear programming model for player I for this example, note that p_{ij} in the general model is the entry in row i and column j of Table 13.6, for $i = 1, 2$ and $j = 1, 2, 3$. Using maximize x_{m+1} instead of the equivalent minimize $(-x_{m+1})$, with $m = 2$ and $n = 3$, the resulting model is

$$\text{Maximize} \quad x_3,$$

subject to

$$5x_2 - x_3 \geq 0$$
$$-2x_1 + 4x_2 - x_3 \geq 0$$
$$2x_1 - 3x_2 - x_3 \geq 0$$
$$x_1 + x_2 = 1$$

and

$$x_1 \geq 0, \quad x_2 \geq 0, \quad x_3 \geq 0.$$

Applying the simplex method to this linear programming problem yields $x_1^* = \frac{7}{11}$, $x_2^* = \frac{4}{11}$, $x_3^* = \frac{2}{11}$ as the optimal solution. (See Probs. 16 and 17.) Consequently, the optimal *mixed* strategy for player I according to the minimax criterion is $(x_1, x_2) = (\frac{7}{11}, \frac{4}{11})$, and the *value* of the game is $v = x_3^* = \frac{2}{11}$. The simplex method also yields the optimal solution to the *dual* (given next) of this problem, namely, $y_1^* = 0$, $y_2^* = \frac{5}{11}$, $y_3^* = \frac{6}{11}$, $y_4^* = \frac{2}{11}$, so the optimal *mixed* strategy for player II is $(y_1, y_2, y_3) = (0, \frac{5}{11}, \frac{6}{11})$.

The *dual* of the preceding problem is just the linear programming model for player II (the one with variables $y_1, y_2, \ldots, y_n, y_{n+1}$) shown earlier in this section. Plugging in the values of p_{ij} from Table 13.6, this model (in minimization form) is

$$\text{Minimize} \quad y_4,$$

subject to

$$-2y_2 + 2y_3 - y_4 \leq 0$$
$$5y_1 + 4y_2 - 3y_3 - y_4 \leq 0$$
$$y_1 + y_2 + y_3 = 1$$

and

$$y_1 \geq 0, \quad y_2 \geq 0, \quad y_3 \geq 0, \quad y_4 \geq 0.$$

Applying the simplex method directly to this model yields the optimal solution: $y_1^* = 0$, $y_2^* = \frac{5}{11}$, $y_3^* = \frac{6}{11}$, $y_4^* = \frac{2}{11}$ (as well as the optimal *dual* solution, $x_1^* = \frac{7}{11}$, $x_2^* = \frac{4}{11}$, $x_3^* = \frac{2}{11}$). Thus the optimal *mixed* strategy for player II is $(y_1, y_2, y_3) = (0, \frac{5}{11}, \frac{6}{11})$, and the *value* of the game is again seen to be $v = y_4^* = \frac{2}{11}$.

Because we already had found the optimal *mixed* strategy for player II while dealing with the first model, we did not have to solve the second one. In general, you always can find the optimal mixed strategies for *both* players by choosing just one of the models (either one) and then using the simplex method to solve for *both* the optimal solution and the optimal *dual* solution.

Both of these linear programming models assumed that $v \geq 0$. If this assumption were violated, what would happen is that both models would have *no feasible solu-*

tions, so the simplex method would stop quickly with this message. To avoid this risk, we could have added a positive constant, say 3 (the absolute value of the largest negative entry), to all of the entries in Table 13.6. This then would increase by 3 all of the coefficients of x_1, x_2, y_1, y_2, and y_3 in the inequality constraints of the two models. (See Prob. 13.)

13.6 Extensions

Although this chapter has considered only *two-person, zero-sum games* with a *finite* number of pure strategies, game theory extends far beyond this kind of game. In fact, extensive research has been done on a number of more complicated types of games, including the ones summarized in this section.

One such type is the *n-person game,* where more than two players may participate in the game. This generalization is particularly important because, in many kinds of competitive situations, there frequently are more than two competitors involved. This may occur, for example, in competition among business firms, in international diplomacy, and so forth. Unfortunately, the existing theory for such games is less satisfactory than it is for two person games.

Another generalization is the *nonzero-sum game,* where the sum of the payoffs to the players need not be zero (or any other fixed constant). This case reflects the fact that many competitive situations include noncompetitive aspects that contribute to the mutual advantage or mutual disadvantage of the players. For example, the advertising strategies of competing companies can affect not only how they will split the market but also the total size of the market for their competing products.

Because mutual gain is possible, nonzero-sum games are further classified in terms of the degree to which the players are permitted to cooperate. At one extreme is the *noncooperative game,* where there is no preplay communication between the players. At the other extreme is the *cooperative game,* where preplay discussions and binding agreements are permitted. For example, competitive situations involving trade regulations between countries, or collective bargaining between labor and management, might be formulated as cooperative games. When there are more than two players, cooperative games also allow some or all of the players to form coalitions.

Still another extension is to the class of *infinite games,* where the players have an infinite number of pure strategies available to them. These games are designed for the kind of situation where the strategy to be selected can be represented by a *continuous* decision variable. For example, this decision variable might be the time at which to take a certain action, or the proportion of one's resources to allocate to a certain activity, in a competitive situation. Much research has been concentrated on such games in recent years.

However, the analysis required in these extensions beyond the two-person, zero-sum, finite game is relatively complex and will not be pursued further here.

13.7 Conclusions

The general problem of how to make decisions in a competitive environment is a very common and important one. The fundamental contribution of game theory is that it provides a basic conceptual framework for formulating and analyzing such problems in simple situations. However, there is a considerable gap between what the theory can handle and the complexity of most competitive situations arising in practice. Therefore, the conceptual tools of game theory usually play just a supplementary role in dealing with these situations.

Because of the importance of the general problem, research is continuing with some success to extend the theory to more complex situations.

SELECTED REFERENCES

1. Begley, Sharon, with David Grant: "The Games Scholars Play," *Newsweek,* Sept. 6, 1982, p. 72.
2. Davis, M.: *Game Theory: An Introduction,* Basic Books, New York, 1983.
3. Kaplan, E. L.: *Mathematical Programming and Games,* Wiley, New York, 1982.
4. Luce, R. Duncan, and Howard Raiffa: *Games and Decisions,* Wiley, New York, 1957.
5. May, Francis B.: *Introduction to Games of Strategy,* Allyn & Bacon, Boston, 1970.
6. Moulin, Herve: *Game Theory for the Social Sciences,* 2d ed., New York University Press, New York, 1986.
7. Owen, Guillermo: *Game Theory,* 2d ed., Academic Press, New York, 1982.
8. Shubik, Martin: *Game Theory in the Social Sciences,* Vols. I (1982) and II (1987), MIT Press, Cambridge, Mass.
9. Szep, J., and F. Forgo: *Introduction to the Theory of Games,* D. Reidel, Boston, 1985.

PROBLEMS

1.* For each of the following payoff tables, determine the optimal strategy for each player by successively eliminating dominated strategies. (Indicate the order in which you eliminated strategies.)

(*a*)

		II		
		1	2	3
I	*1*	−3	1	2
	2	1	2	1
	3	1	0	−2

(*b*)

		II		
		1	2	3
I	*1*	1	2	0
	2	2	−3	−2
	3	0	3	−1

2. Consider the game having the following payoff table.

		II			
		1	2	3	4
I	*1*	2	−3	−1	1
	2	−1	1	−2	2
	3	−1	2	−1	3

Determine the optimal strategy for each player by successively eliminating dominated strategies. Give a list of the dominated strategies (and the corresponding dominating strategies) in the order in which you were able to eliminate them.

3. Find the saddle point for the game having the following payoff table.

		II		
		1	2	3
I	*1*	1	−1	1
	2	−2	0	3
	3	3	1	2

4. Find the saddle point for the game having the following payoff table.

		II			
		1	2	3	4
	1	3	−3	−2	−4
I	*2*	−4	−2	−1	1
	3	1	−1	2	0

5. Two companies share the bulk of the market for a particular kind of product. Each is now planning its new marketing plans for the next year in an attempt to wrest some sales away from the other company. (The total sales for the product are relatively fixed, so one company can only increase its sales by winning them away from the other.) Each company is considering three possibilities: (1) better packaging of the product, (2) increased advertising, and (3) a slight reduction in price. The costs of the three alternatives are quite comparable and sufficiently large that each company will select just one. The estimated effect of each combination of alternatives on the *increased percentage of the sales* for company I is

		II		
		1	2	3
	1	2	3	1
I	*2*	1	4	0
	3	3	−2	−1

Each company must make its selection before learning the decision of the other company.

(*a*) Without eliminating dominated strategies, use the minimax (or maximin) criterion to determine the best strategy for each side.

(*b*) Now identify and eliminate dominated strategies as far as possible. Make a list of the dominated strategies showing the order in which you were able to eliminate them. Then show the resulting reduced payoff table with no remaining dominated strategies.

6. The labor union and management of a particular company have been negotiating a new labor contract. However, negotiations have now come to an impasse, with management making a ''final'' offer of a wage increase of $1.10/hour and the union making a ''final'' demand of a $1.60/hour increase. Therefore, both sides have agreed to have an impartial arbitrator set the wage increase somewhere between $1.10/hour and $1.60/hour (inclusively).

The arbitrator has asked each side to submit to her a confidential proposal for a fair and economically reasonable wage increase (rounded to the nearest dime). From past experience, both sides know that this arbitrator normally accepts the proposal of the side that gives the most from its ''final'' figure. If neither side changes its final figure, or if they both give in the same amount, then the arbitrator normally compromises halfway between ($1.35 in this case). Each side now needs to determine what wage increase to propose for its own maximum advantage.

(*a*) Formulate this problem as a *two-person, zero-sum game.*

(*b*) Use the concept of dominated strategies to determine the best strategy for each side.

(*c*) Without eliminating dominated strategies, use the minimax criterion to determine the best strategy for each side.

7.* Two politicians soon will be starting their campaigns against each other for a certain political office. Each must now select the main issue he will emphasize as the theme of his campaign. Each has three advantageous issues from which to choose, but the relative effectiveness of each one would depend upon the issue chosen by his opponent. In particular, the estimated increase in the vote for politician I (expressed as a percentage of the total vote) resulting from each combination of issues is

		Issue for Politician II		
		1	2	3
Issue for politician I	*1*	7	−1	3
	2	1	0	2
	3	−5	−3	1

However, because considerable staff work is required to research and formulate the issue chosen, each politician must make his own choice before learning his opponent's choice. Which issue should he choose?

For each of the situations described here, formulate this problem as a *two-person, zero-sum game,* and then determine which issue should be chosen by each politician according to the specified criterion.

(*a*) The current preferences of the voters are very uncertain, so each additional percent of votes won by one of the politicians has the same value to him. Use the *minimax* criterion.

(*b*) A reliable poll has found that the percentage of the voters currently preferring politician I (before the issues have been raised) lies between 45 and 50 percent. (Assume a uniform distribution over this range.) Use the concept of *dominated strategies,* beginning with the strategies for politician I.

(*c*) Suppose that the percentage described in part (*b*) actually were 45 percent. Should politician I use the minimax criterion? Explain. Which issue would you recommend? Why?

8. Two manufacturers currently are competing for sales in two different but equally profitable product lines. In both cases the sales volume for manufacturer II is three times as large as that for manufacturer I. Because of a recent technological breakthrough, both manufacturers will be making a major improvement in both products. However, they are uncertain as to what development and marketing strategy they should follow.

If both product improvements are developed simultaneously, either manufacturer can have them ready for sale in 12 months. Another alternative is to have a "crash program" to develop only one product first to try to get it marketed ahead of the competition. By doing this, manufacturer II could have one product ready for sale in 9 months, whereas manufacturer I would require 10 months (because of previous commitments for its production facilities). For either manufacturer, the second product could then be ready for sale in an additional 9 months.

For either product line, if both manufacturers market their improved models simultaneously, it is estimated that manufacturer I would increase its share of the total future sales of this product by 8 percent of the total (from 25 to 33 percent). Similarly, manufacturer I would increase its share by 20, 30, and 40 percent of the total if it markets the product sooner than manufacturer II by 2, 6, and 8 months, respectively. On the other hand, manufacturer I would lose 4, 10, 12, and 14 percent of the total if manufacturer II markets it sooner by 1, 3, 7, and 10 months, respectively.

Formulate this problem as a two-person, zero-sum game, and then determine which strategy the respective manufacturers should use according to the minimax criterion.

9. Consider the following parlor game to be played between two players. Each player begins with three chips: one red, one white, and one blue. Each chip can be used only once.

To begin, each player selects one of his chips and places it on the table, concealed. Both players then uncover the chips and determine the payoff to the winning player. In particular, if both players play the same kind of chip, it is a draw; otherwise, the following table indicates the winner and how much he receives from the other player. Next, each player selects one of his two remaining chips and repeats the procedure, resulting in another payoff according to the following table. Finally, each player plays his one remaining chip, resulting in the third and final payoff.

Winning Chip	Payoff
Red beats White	\$50
White beats Blue	\$40
Blue beats Red	\$30
Matching colors	0

Formulate this problem as a two-person, zero-sum game by identifying the form of the strategies and payoffs.

10. Consider the game having the following payoff table.

		II 1	2
I	*1*	3	−2
	2	−1	2

Use the graphical procedure described in Sec. 13.4 to determine the value of the game and the optimal *mixed* strategy for each player according to the minimax criterion. Check your answer for player II by constructing *his* payoff table and applying the graphical procedure directly to this table.

11.* For each of the following payoff tables, use the graphical procedure described in Sec. 13.4 to determine the value of the game and the optimal *mixed* strategy for each player according to the minimax criterion:

(*a*)

		II 1	2	3
I	*1*	4	3	1
	2	0	1	2

(*b*)

		II 1	2	3
I	*1*	1	−1	3
	2	0	4	1
	3	3	−2	5
	4	−3	6	−2

12. Consider the following parlor game between two players. It begins when a referee flips a coin, notes whether it comes up heads or tails, and then shows this result to player I only. Player I may then either (1) pass and thereby pay \$5 to player II, or (2) he may bet. If player I passes, the game is terminated. However, if he bets, the game continues, in which case player II may then either (1) pass and thereby pay \$5 to player I, or (2) he may call. If player II calls, the referee then shows him the coin; if it came up heads, player II pays \$10 to player I; if it came up tails, player II receives \$10 from player I.

(*a*) Give the pure strategies for each player. (*Hint:* Player I will have four pure strategies, each one specifying how he would respond to each of the two results the referee can show him; player II will have two pure strategies, each one specifying how he will respond if player I bets.)

(*b*) Develop the payoff table for this game, using expected values for the entries when necessary. Determine whether it has a saddle point or not.

(*c*) Use the graphical procedure described in Sec. 13.4 to determine the optimal *mixed* strategy for each player according to the minimax criterion. Also give the corresponding value of the game.

13. Referring to the last paragraph of Sec. 13.5, suppose that 3 were added to all of the entries of Table 13.6 in order to ensure that the corresponding linear programming models for both players have feasible solutions with $x_3 \geq 0$ and $y_4 \geq 0$. Write out these two models. Based on the information given in Sec. 13.5, what are the optimal solutions for these two models? What is the relationship between x_3^* and y_4^*? What is the relationship between the *value* of the original game v and the values of x_3^* and y_4^*?

14.* Consider the game having the following payoff table.

		II			
		1	2	3	4
	1	5	0	3	1
I	*2*	2	4	3	2
	3	3	2	0	4

Use the approach described in Sec. 13.5 to formulate the problem of finding the optimal *mixed* strategies according to the minimax criterion as a *linear programming* problem.

15. For each of the following payoff tables, transform the problem of finding the minimax *mixed* strategies into an equivalent *linear programming* problem.

(*a*)

		II		
		1	2	3
	1	4	2	−3
I	*2*	−1	0	3
	3	2	3	−2

(*b*)

		II				
		1	2	3	4	5
	1	1	−3	2	−2	1
I	*2*	2	3	0	3	−2
	3	0	4	−1	−3	2
	4	−4	0	−2	2	−1

16. Consider variation 3 of the political campaign problem (see Table 13.6). Refer to the resulting *linear programming* model for player I given near the end of Sec. 13.5. Ignoring the objective function variable (x_3), plot the *feasible region* for x_1 and x_2 graphically. (*Hint:* This feasible region consists of a single line segment.) Next, write an algebraic expression for the maximizing value of x_3 for any point in this feasible region. Finally, use this expression to demonstrate that the optimal solution must, in fact, be the one given in Sec. 13.5.

17. Consider the *linear programming* model for player I given near the end of Sec. 13.5 for variation 3 of the political campaign problem (see Table 13.6). Verify the optimal *mixed* strategies for both players given in Sec. 13.5 by applying a computer code of the simplex method to this model to find *both* its optimal solution and its optimal *dual* solution.

18. The A. J. Swim Team soon will have an important swim meet with the G. N. Swim Team. Each team has a star swimmer (John and Mark, respectively) who can swim very well in the 100-yard butterfly, backstroke, and breaststroke events. However, the rules prevent them from being used in more than *two* of these events. Therefore, their coaches now need to decide how to use them to maximum advantage.

Each team will enter three swimmers per event (the maximum allowed). For each event, the following table gives the best time previously achieved by John and Mark as well as the best time for each of the other swimmers who will definitely enter that event. (Whichever event John or Mark does not swim, his team's third entry for that event will be slower than the two shown in the table.)

	A. J. Swim Team			G. N. Swim Team		
	Entry			*Entry*		
	1	2	John	Mark	1	2
Fly	1:01.6	59.1	57.5	58.4	1:03.2	59.8
Back	1:06.8	1:05.6	1:03.3	1:02.6	1:04.9	1:04.1
Breast	1:13.9	1:12.5	1:04.7	1:06.1	1:15.3	1:11.8

The points awarded are 5 points for first place, 3 for second place, 1 for third place, and none for lower places. Both coaches believe that all swimmers will essentially equal their best times in this meet. Thus John and Mark each will definitely be entered in two of these three events.

(*a*) The coaches must submit all their entries before the meet without knowing the entries for the other team, and no changes are permitted later. The outcome of the meet is very uncertain, so each additional point has equal value for the coaches. Formulate this problem as a *two-person, zero-sum* game. Eliminate dominated strategies, and then use the graphical procedure described in Sec. 13.4 to find the optimal *mixed* strategy for each team according to the minimax criterion.

(*b*) The situation and assignment are the same as in part (*a*), except that both coaches now believe that the A. J. Swim Team will win the swim meet if it can win 13 or more points in these three events, but will lose with less than 13 points. [Compare the resulting optimal mixed strategies with those obtained in part (*a*).]

(*c*) Now suppose that the coaches submit their entries during the meet one event at a time. When submitting his entries for an event, the coach does not know who will be swimming that event for the other team, but he does know who has swum *preceding* events. The three key events just discussed are swum in the order listed in the table. Once again, the A. J. Swim Team needs 13 points in these events to win the swim meet. Formulate this problem as a *two-person, zero-sum* game. Then use the concept of dominated strategies to determine the best strategy for the G. N. team that actually "guarantees" it will win under the assumptions being made.

(*d*) The situation is the same as in part (*c*). However, assume now that the coach for the G. N. team does not know about game theory and so may, in fact, choose any of his available strategies that have Mark swimming two events. Use the concept of dominated strategies to determine the best strategies from which the coach for the A. J. team should choose. If this coach knows that the other coach has a tendency to enter Mark in the butterfly and the backstroke more often than in the breaststroke, which strategy should he choose?

19. Consider the general $m \times n$, two-person, zero-sum game. Let p_{ij} denote the payoff to player I if he plays his strategy i ($i = 1, \ldots, m$) and player II plays his strategy j ($j = 1, \ldots, n$). Strategy 1 (say) for player I is said to be *weakly dominated* by strategy 2 (say) if $p_{1j} \leq p_{2j}$ for $j = 1, \ldots, n$ and $p_{1j} = p_{2j}$ for one or more values of j.

(*a*) Assume that the payoff table possesses one or more *saddle points*, so that the players have corresponding optimal pure strategies under the minimax criterion. Prove that eliminating *weakly dominated* strategies from the payoff table cannot eliminate all these saddle points and cannot produce any new ones.

(*b*) Assume that the payoff table does not possess any saddle points, so that the optimal strategies under the minimax criterion are *mixed* strategies. Prove that eliminating weakly dominated pure strategies from the payoff table cannot eliminate all optimal mixed strategies and cannot produce any new ones.

20. Briefly describe what you feel are the advantages and disadvantages of the minimax criterion.

14

Decision Analysis

14.1 Introduction

In recent years, decision analysis has become an important technique in business, industry, and government. Decision analysis provides a rational methodology for decision making in the face of uncertainty. It enables a manager to choose among alternatives in an optimal fashion, taking into account the worth of acquiring experimental data to reduce the uncertainty.

This chapter presents a framework for making decisions when (1) experimentation is infeasible and (2) experimentation is possible, resulting in the availability of sample data. The criterion of optimality used to select among alternatives will be the minimization of expected cost. Among the problems considered in this chapter are the following: What is the decision that minimizes expected cost, given the result of an experiment (if indeed an experiment is performed)? By following the optimal policy, what is the expected cost? If an experiment is performed, will it be worthwhile; that is, will the decrease in expected cost be more than the cost of the experiment?

Table 14.1 **Table of Profits for Oil Company**

	500,000-Barrel Well	200,000-Barrel Well	50,000-Barrel Well	Dry Well
Drill for oil	650,000	200,000	−25,000	−75,000
Unconditional lease	45,000	45,000	45,000	45,000
Conditional lease	250,000	100,000	0	0

Finally, what is the maximum amount of money that might be spent in order to eliminate all of the uncertainty?

EXAMPLE: Consider the following problem. An oil company owns some land that is purported to contain oil. The company classifies such land into four categories by the total number of barrels that are expected to be obtained from the well, i.e., a 500,000-barrel well, a 200,000-barrel well, a 50,000-barrel well, or a dry well. The company is faced with deciding whether to drill for oil, to unconditionally lease the land to an independent oil driller, or to conditionally lease the land at a rate depending upon the oil strike. The cost of drilling a producing well is \$100,000, and the cost of drilling a dry well is \$75,000. For producing wells, the profit per barrel of oil is \$1.50 (after deducting all production costs). Under the unconditional lease agreement, the company receives \$45,000 for the land, whereas under the conditional lease arrangement, the company receives 50 cents for each barrel of oil extracted, provided the land yields a 200,000- or 500,000-barrel strike; otherwise, it receives nothing.

The possible profits for the oil company are shown in Table 14.1.

14.2 Decision Making without Experimentation

General Framework

Before seeking a solution to the aforementioned problem, it is worthwhile to formulate a general framework for decision making. The decision maker must choose an action a from a set A of possible actions. In the oil-drilling example, the set A consists of three points, a_1, a_2, and a_3, that correspond to drilling for oil, unconditionally leasing the land, and conditionally leasing the land, respectively. In taking an action, the decision maker must be aware of its consequences, which will usually also be a function of the "state of nature." A state of nature θ is a representation of the actual real-world situation to which the action will apply. Generally, the states of nature are an enumeration, within the model according to some set of indices, of possible alternative representations of the physical phenomenon being studied. The set of possible values that θ can assume will be denoted by Θ. In the oil-drilling example, Θ consists of four points, θ_1, θ_2, θ_3, and θ_4, with θ_1 corresponding to the land yielding a 500,000-barrel well, θ_2 corresponding to a 200,000-barrel well, θ_3 corresponding to a 50,000-barrel well, and θ_4 corresponding to a dry well. Very often the states of nature are characterized by a parameter of a family of probability distributions. In the context of the oil-drilling example, the potential strikes might be viewed as the expected value of the random variable, oil yield, having some assumed form of probability distribution. Thus a representation of the model of this oil-drilling problem is that the oil yield in the site is a random variable with an unknown expected value. The company

Table 14.2 **Loss Function for Oil-Drilling Example**

Action \ State of Nature	θ_1: 500,000-Barrel Well	θ_2: 200,000-Barrel Well	θ_3: 50,000-Barrel Well	θ_4: Dry Well
a_1: drill for oil	−650,000	−200,000	25,000	75,000
a_2: unconditionally lease	−45,000	−45,000	−45,000	−45,000
a_3: conditionally lease	−250,000	−100,000	0	0

is willing to approximate this expected value by one of four values: 500,000 barrels, 200,000 barrels, 50,000 barrels, and no barrels (dry). Thus the states of nature become these possible values of the expected value of the random variable, oil yield.

To measure the consequences of a decision maker's action, we assume that there exists a *loss function* $l(a, \theta)$ that reflects the loss from taking action a when the state of nature is θ; it is defined for each combination of a and θ. If the problem is formulated in terms of gains, a gain can be termed as a negative loss. The loss function is generally measured in monetary terms, although other utility functions can be used. Note that $l(a, \theta)$ is assumed to be a function only of a and θ. The loss function for the oil-drilling example is easily obtained from Table 14.1 and is given in Table 14.2.[1] Although the loss function is easily obtained directly from the action and the state of nature in this example, occasionally the loss depends upon the outcome of a random variable whose probability distribution depends upon the true state of nature. For example, this situation would occur in the oil-drilling example if the profit were expressed directly in terms of the random variable, oil yield. The loss would then be a random variable, and $l(a, \theta)$ would then be interpreted as the expected value of the loss incurred when action a is taken and the true state of nature is θ. Hence even here the loss function depends upon only a and θ. In general, in formulating the problem, if the state of nature is defined so broadly that observing its value resolves all uncertainty relevant to the decision at hand, then the loss can always be expressed as a (deterministic) function of θ and the action a. If this case is not true, meaning that the observation of θ would still leave some uncertainty as to the ultimate consequence of a given action a, the loss function $l(a, \theta)$ is computed as the expected loss, given state θ and action a.

Minimax Criterion

If the true state of nature were known, it would be simple to choose the correct action, i.e., that action which has minimum loss. Unfortunately, the true state of nature is not generally known, and choosing a correct action is not simple. In the oil-drilling example, if $\theta = \theta_1$, a 500,000-barrel well, the best action is to drill for oil; whereas if $\theta = \theta_4$, a dry well, the best action is to lease unconditionally. This decision-theory formulation has the appearance of game theory as described in detail in Chap. 13, with the two players being the decision maker and nature. The actions correspond to the pure strategies of the decision maker, and the states of nature correspond to the pure strategies of nature. The payoff matrix in game theory is analogous to the loss

[1] In discussing this example throughout the chapter, many negative values appear. These are to be interpreted as gains or profit and should not cause the reader any trouble.

table. An approach for obtaining solutions to game theory problems is through the **minimax principle**. This principle tells the decision maker to find the maximum loss for each of his actions and to choose that action which has the smallest maximum loss. Similarly, the decision maker's opponent, nature in this case, should find the minimum loss to the decision maker for each one of her possible states of nature and present to the decision maker that state of nature which maximizes this minimum loss. If these loss values are equal, the game is said to have a **value**. If a game has a value, and if each player follows his optimal strategy, the decision maker can guarantee that his loss will never exceed the value. Furthermore, if the decision maker follows his optimal strategy and if nature deviates from hers, the loss to the decision maker can only be decreased. Unfortunately, in this context a value does not always exist. However, it does exist in the oil-drilling example. Using the minimax criterion, the decision maker should choose action a_2 and guarantee that his loss will not exceed $-45{,}000$. Similarly, nature should choose state $\theta = \theta_3$ or $\theta = \theta_4$ and guarantee that the decision maker's loss will be at least $-45{,}000$. Thus this "game" does indeed have a value, and the minimax strategy for the decision maker is to lease unconditionally.

A fundamental theorem in the theory of games states that if mixed strategies are allowed, and if the minimax principle is followed, the game always has a value. A **mixed strategy** for the decision maker is a probability distribution defined over the action space. The actual choice of strategy is dependent upon the outcome of a random device having a probability distribution associated with the action space. Thus choosing a mixed strategy is equivalent to choosing a probability distribution. Similarly, a mixed strategy for nature is a probability distribution defined over the possible states of nature. **Pure strategies** are just special cases of mixed strategies, where the probability assigned to the chosen action is 1 and the probability assigned to the other action is zero. Because both the action and the state of nature are random variables, the loss incurred is also a random variable, and again expected loss is the criterion.

However, even though the minimax principle has some attractive properties, it is seldom used in games against nature because it is an extremely conservative criterion in this context. The actions taken when this principle is used assume that nature is a conscious opponent that wants to inflict as much damage as possible on the decision maker. Generally, nature is not a malevolent opponent, and it is unlikely that the decision maker has to guard against such an occurrence.

Bayes' Criterion

The previous section pointed out that the minimax principle says to proceed as if nature will select a probability distribution, defined over the possible states of nature, which is least favorable to the decision maker. It was also noted that this approach is very conservative because there is no reason to expect nature to use this distribution. As a matter of fact, in some situations the decision maker will actually have some advance information about θ that contradicts this assumption about what nature will do. When the decision maker has such information, he certainly should take it into account. Such information can usually be translated into a probability distribution, acting as though the state of nature is a random variable, in which case this distribution is referred to as a **prior distribution**. Prior distributions are often subjective in that they may depend upon the experience or intuition of an individual.

For example, in the oil-drilling problem, the company has had some experience with wells in similar geographic areas and has concluded that about 10 percent of the strikes are 500,000-barrel wells, 15 percent are 200,000-barrel wells, 25 percent are 50,000-barrel wells, and 50 percent are dry wells. Hence these data can be translated into the prior distribution as follows:

$$P\{\theta = \theta_1\} = P_\theta(1) = 0.10$$

$$P\{\theta = \theta_2\} = P_\theta(2) = 0.15$$

$$P\{\theta = \theta_3\} = P_\theta(3) = 0.25$$

$$P\{\theta = \theta_4\} = P_\theta(4) = 0.50.$$

A procedure for using the prior distribution to aid in the selection of an action is Bayes' criterion. **Bayes' principle** tells the decision maker to select that action (called *Bayes' decision procedure*) which minimizes the expected loss. The expected loss $l(a)$ is evaluated with respect to the prior distribution, which is defined over the possible states of nature; i.e.,

$$l(a) = E[l(a, \theta)] = \begin{cases} \sum_{\text{all } k} l(a, k)P_\theta(k), & \text{if } \theta \text{ is discrete} \\ \int_{-\infty}^{\infty} l(a, y)P_\theta(y)\, dy, & \text{if } \theta \text{ is continuous.} \end{cases}$$

Thus, for the oil-drilling example, the expected loss $l(a)$ for each action is given by

$$\begin{aligned} l(a_1) = E[l(a_1, \theta)] &= -650{,}000(0.10) - 200{,}000(0.15) + 25{,}000(0.25) \\ &\quad + 75{,}000(0.50) \\ &= -\$51{,}250, \\ l(a_2) = E[l(a_2, \theta)] &= -45{,}000(0.10) - 45{,}000(0.15) - 45{,}000(0.25) \\ &\quad -45{,}000(0.50) \\ &= -\$45{,}000, \\ l(a_3) = E[l(a_3, \theta)] &= -250{,}000(0.10) - 100{,}000(0.15) \\ &= -\$40{,}000. \end{aligned}$$

Hence using Bayes' principle leads to selecting action a_1, that is, drill for oil, and the associated expected loss is $-\$51,250$ (profit). It is interesting to speculate as to whether the decision maker could have improved upon this expected loss by making use of a mixed strategy rather than a pure strategy (because nature is using the mixed strategy specified by the prior distribution). It can be shown that the decision maker cannot improve his position by using mixed strategies, so that it is sufficient for him to consider only pure strategies.

14.3 Decision Making with Experimentation

The previous sections assumed that the decision maker was to make his decision without experimentation. However, if some experimentation is possible (perhaps at a cost), the data derived from this experimentation should be incorporated into the decision-making process. For example, returning to the oil-drilling example, suppose that it is possible to obtain seismic soundings at a cost of $12,000. This information

Table 14.3 **Frequency of Seismic Classifications**

Seismic Classification	θ_1: 500,000-Barrel Well	θ_2: 200,000-Barrel Well	θ_3: 50,000-Barrel Well	θ_4: Dry Well
1	$7(\frac{7}{12})$	$9(\frac{9}{16})$	$11(\frac{11}{24})$	$9(\frac{9}{48})$
2	$4(\frac{4}{12})$	$3(\frac{3}{16})$	$6(\frac{6}{24})$	$13(\frac{13}{48})$
3	$1(\frac{1}{12})$	$2(\frac{2}{16})$	$3(\frac{3}{24})$	$15(\frac{15}{48})$
4	$0(\frac{0}{12})$	$2(\frac{2}{16})$	$4(\frac{4}{24})$	$11(\frac{11}{48})$

leads to four possible seismic classifications, denoted by (1), (2), (3), and (4). Classification (1) denotes that there is definitely a closed geologic structure to the site (a very favorable condition if the presence of oil is desired); classification (2) denotes that there is probably a closed structure to the site; classification (3) denotes that there is a nonclosed structure to the site (a relatively unfavorable condition); and classification (4) denotes that there is no structure to the site (an unfavorable condition). Based upon past examination of similar geologic areas (100 such examinations), the company obtains the data presented in Table 14.3.[1] The values in parentheses in each cell can be interpreted as conditional probabilities, given the state of nature: e.g., if the well is a 200,000-barrel well, then $\frac{9}{16}$ can be interpreted as the conditional probability that the seismic reading is classified as (2) (probably a closed structure to the site); if the well is dry, then $\frac{15}{48}$ can be interpreted as the conditional probability that the seismic reading is classified as (3) (a nonclosed structure to the site); and so on. Before proceeding with the example, we shall discuss a general method for incorporating these data.

Let X denote the information made available by experimentation obtained from a random sample. X is then a random variable and may be viewed as a function of the sample data; for example, X may denote a sample mean, the maximum of the sample, a vector of the sample observations, the third observation in a sample, and so forth. The decision maker is to choose a **decision procedure**, rule, or strategy, which tells him the form and amount of experimentation and what action to take for each possible value that X may take on. Denote this function to be chosen as $d[x]$, so that if the random variable X takes on the value x, then $a = d[x]$ would be the action to be taken. The decision maker, then, is interested in choosing a function d, from among the many possible decision functions, that is, in some sense, optimal. (Indeed, part of the problem here is to choose a good working definition for the term *optimal*.) To evaluate a decision function, we must explore its consequences. Because the action taken, a, is a function of the outcome of the random variable X, then $d[X]$ is also a random variable, and the loss associated with that action also depends upon the outcome of this random variable. An appropriate measure of the consequences of taking action $a = d[X]$, when the true state of nature is θ, is then given by the expected value of the loss. This quantity will be known as the **risk function** $R(d, \theta)$; that is,

$$R(d, \theta) = E[l(d[X], \theta)],$$

where the expectation is taken with respect to the probability distribution of the random variable X, and the loss function includes the cost of experimentation.

[1] Although the actual fraction of wells falling historically into the four categories differs slightly from the prior distribution, the prior probabilities given under Bayes' criterion in Sec. 14.2 are thought to be more representative of what to expect for this particular site and will be used subsequently.

Now consider how to apply this approach to the oil-drilling example. Suppose the following decision rule, d_1, is to be evaluated. If the seismic reading is classified as (1), take action a_1; if the seismic reading is classified as (2) or (3), take action a_3; and if the seismic reading is classified as (4), take action a_2; that is,

$$\begin{aligned} d_1[x] &= a_1, \qquad \text{for } x = 1 \\ &= a_2, \qquad \text{for } x = 4 \\ &= a_3, \qquad \text{for } x = 2 \text{ or } x = 3. \end{aligned}$$

Therefore,

$$\begin{aligned} R(d_1, \theta_1) &= -650{,}000(\tfrac{7}{12}) - 45{,}000(0) - 250{,}000(\tfrac{4}{12} + \tfrac{1}{12}) + 12{,}000 \\ &= -\$471{,}333, \\ R(d_1, \theta_2) &= -200{,}000(\tfrac{9}{16}) - 45{,}000(\tfrac{2}{16}) - 100{,}000(\tfrac{3}{16} + \tfrac{2}{16}) + 12{,}000 \\ &= -\$137{,}375, \\ R(d_1, \theta_3) &= 25{,}000(\tfrac{11}{24}) - 45{,}000(\tfrac{4}{24}) + 0(\tfrac{6}{24} + \tfrac{3}{24}) + 12{,}000 \\ &= \$15{,}958, \\ R(d_1, \theta_4) &= 75{,}000(\tfrac{9}{48}) - 45{,}000(\tfrac{11}{48}) + 0(\tfrac{13}{48} + \tfrac{15}{48}) + 12{,}000 \\ &= \$15{,}750. \end{aligned}$$

Note that the 12,000 represents the cost of obtaining the seismic data.

Thus it is evident how the risk function for a given decision procedure is evaluated. The risk function provides a means for defining *optimality*. An optimal decision function might be defined as one that will minimize the risk for every value of θ. However, it is evident that an optimal decision function (in this sense) may not always exist and, in fact, does not exist in most cases. Thus the preceding definition is inadequate. Hence another definition of optimality is considered in the next section.

Bayes' Procedures

Even when data are available, there is no best definition of *optimal procedures*. With data, it is still possible to use a minimax criterion or a minimax decision function, but it too suffers from the same disadvantages as it does when no data are available; i.e., it assumes that nature will act as a conscious opponent and confront the decision maker with the least favorable distribution of θ.

If the decision maker has some advance information about the states of nature that can be described in terms of a prior distribution, then *Bayes' principle* can be applied to the *risk function*. If the states of nature are discrete, **Bayes' risk** corresponding to a decision function d and a prior probability distribution of θ, $P_\theta(k)$, is given by

$$B(d) = \sum_{\text{all } k} R(d, k)P_\theta(k).$$

If the states of nature are continuous, Bayes' risk corresponding to a prior probability density function of θ, $P_\theta(y)$, is given by

$$B(d) = \int_{-\infty}^{\infty} R(d, y)P_\theta(y)\, dy.$$

Bayes' risk provides another means for defining optimality for decision rules using *Bayes' principle*. *Bayes' principle* tells the decision maker to select that function d (called the *Bayes' decision procedure*) which minimizes $B(d)$. A method for finding Bayes' decision procedure follows.

When no data were available, using Bayes' procedure led us to select that action which minimized the expected loss; this expectation was evaluated with respect to the prior distribution of θ. Now that data are available, additional information is available about the state of nature. For example, if the seismic data are classified as (4), they are evidence that the strike will not be a 500,000-barrel well and probably not a 200,000-barrel well. Hence, after observing the experimental data, we should update the prior distribution by using more timely information about the probability distribution of the state of nature. Such updated information is called the *posterior distribution* of θ, given the prior distribution and the data $X = x$. The **posterior distribution** of θ is just the conditional distribution of θ, given $X = x$. If θ is discrete, the posterior distribution will be denoted by $h_{\theta|X=x}(k)$, and, if θ is continuous, then the posterior distribution will be denoted by $h_{\theta|X=x}(y)$. The method for calculating the posterior distribution is given later. However, if the method used for calculating Bayes' procedure when no data are available is followed (selecting that action which minimizes the *expected loss*), with this expectation now evaluated with respect to the *posterior distribution* of θ, given $X = x$, this decision procedure minimizes $B(d)$. Hence it is Bayes' procedure. This statement is not obvious, but it can easily be proved. Thus, to find Bayes' procedure, the decision maker computes the posterior distribution of θ, given $X = x$. He then chooses that action which minimizes the expected loss[1] $l_h(a)$ (including the cost of experimentation),[2] with this expectation evaluated with respect to the *posterior* distribution of θ, given $X = x$, where

$$l_h(a) = E[l(a, \theta)] = \begin{cases} \sum_{\text{all } k} l(a, k)h_{\theta|X=x}(k), & \text{if } \theta \text{ is discrete} \\ \int_{-\infty}^{\infty} l(a, y)h_{\theta|X=x}(y)\,dy, & \text{if } \theta \text{ is continuous.} \end{cases}$$

In the oil-drilling example, the *posterior distribution* can be calculated by methods to be discussed later in this section and is given in Table 14.4.

Suppose that the seismic reading is classified as (3) (the geologic site has a nonclosed structure). To obtain Bayes' procedure, we compute the expected loss with respect to the posterior distribution of θ, given $X = 3$, for each of the actions as

Table 14.4 **Values of $h_{\theta|X=x}(k)$, the Posterior Distribution of θ**

x \ k	θ_1	θ_2	θ_3	θ_4
1	0.166	0.240	0.327	0.267
2	0.129	0.108	0.241	0.522
3	0.039	0.087	0.146	0.728
4	0	0.107	0.238	0.655

[1] Note that loss is used rather than risk.

[2] Although the notation $l_h(a)$ does not show it, one should remember that this quantity does indeed depend upon the experimental outcome, x.

follows:

$$l_h(a_1) = E[l(a_1, \theta)] = -650{,}000(0.039) - 200{,}000(0.087) + 25{,}000(0.146) + 75{,}000(0.728) + 12{,}000$$

$$= \$27{,}500,$$

$$l_h(a_2) = E[l(a_2, \theta)] = -45{,}000 + 12{,}000 = -\$33{,}000,$$

$$l_h(a_3) = E[l(a_3, \theta)] = -250{,}000(0.039) - 100{,}000(0.087) + 12{,}000$$

$$= -\$6{,}450.$$

Bayes' procedure selects action a_2 (because this action minimizes the expected loss), which implies that the company should unconditionally lease the land. Thus we see that the experimental data change the action of the decision maker. Without experimentation, using Bayes' procedure recommended drilling for oil, whereas the information obtained from the seismic data recommends that the company unconditionally lease the land. Incidentally, although Table 14.4 was obtained for all values of x, it was necessary to obtain the values for only $x = x_3$. In fact, this method of computing Bayes' procedure has the important advantage that it is necessary to compute only the optimal $d[x]$ for the single point that corresponds to the outcome of the experiment. Using the basic formula for $B(d)$ to find Bayes' procedure requires the determination of the entire optimal decision function, which is generally more difficult.

Calculation of the Posterior Distribution

Denote by (θ, X) a bivariate random variable having a joint probability distribution. Consider the case where (θ, X) is a discrete bivariate random variable, with joint probability distribution given by $P_{\theta X}(k, j)$. Each random variable θ and X has a marginal distribution. In fact, $P_\theta(k)$, the prior distribution of θ, is the marginal distribution of θ. The usual expression given as the probability distribution of the random variable X actually corresponds to the conditional probability distribution of X, given θ. For example, if X has a Poisson distribution with parameter $\theta = 24$, then $e^{-24}24^j/j!$ is just the conditional probability distribution function of X, given $\theta = 24$.[1] To indicate that this distribution is a conditional distribution, we introduce the notation

$$Q_{X|\theta=k}(j) = P\{X = j|\theta = k\}.$$

Thus, in the Poisson example,

$$Q_{X|\theta=24}(j) = P\{X = j|\theta = 24\} = \frac{e^{-24}24^j}{j!}$$

represents the conditional probability distribution of X, given that $\theta = 24$, and it has the form of a Poisson distribution with parameter $\theta = 24$.

If the joint distribution of (θ, X) is of interest, the expression

$$P_{\theta X}(k, j) = Q_{X|\theta=k}(j)P_\theta(k)$$

[1] In order to be consistent with the notation introduced in this chapter, θ is used instead of the more usual symbol λ to denote the parameter of the Poisson distribution.

can be used to evaluate it. $Q_X(j)$, the marginal distribution of X, can also be obtained; i.e.,

$$Q_X(j) = \sum_{\text{all } k} P_{\theta X}(k, j) = \sum_{\text{all } k} Q_{X|\theta=k}(j)P_\theta(k).$$

Finally, the only remaining probability distribution that has not been discussed is the conditional distribution of θ, given $X = j$, that is, the posterior distribution of θ, given $X = j$, $h_{\theta|X=j}(k)$. An alternative expression for the joint probability distribution of (θ, X) is given by

$$P_{\theta X}(k, j) = h_{\theta|X=j}(k)Q_X(j).$$

Equating the two expressions for $P_{\theta X}(k, j)$, and letting $j = x$ (the outcome of the experiment), leads to the important result from which the posterior distribution can be calculated; i.e.,

$$h_{\theta|X=x}(k) = \frac{Q_{X|\theta=k}(x)P_\theta(k)}{Q_X(x)}.$$

Thus, in summary, the posterior distribution can be calculated by using the preceding expression. $P_\theta(k)$ is the prior distribution. $Q_{X|\theta=k}(x)$ is the ordinary expression for the probability distribution of the random variable X evaluated at $X = x$, but it is written in this form to show its dependence upon the value of the parameter θ. The function $Q_X(x)$ is the marginal distribution of the random variable X, evaluated at $X = x$, and is obtained from

$$Q_X(x) = \sum_{\text{all } k} Q_{X|\theta=k}(x)P_\theta(k).$$

Returning to the oil-drilling example, suppose that the seismic reading is classified as (3) (the geologic site has a nonclosed structure). Recall that the prior distribution of the classification of the land is assumed to be

$$P\{\theta = \theta_1\} = P_\theta(1) = 0.10$$

$$P\{\theta = \theta_2\} = P_\theta(2) = 0.15$$

$$P\{\theta = \theta_3\} = P_\theta(3) = 0.25$$

$$P\{\theta = \theta_4\} = P_\theta(4) = 0.50.$$

It is necessary to evaluate the expressions $h_{\theta|X=3}(1)$, $h_{\theta|X=3}(2)$, $h_{\theta|X=3}(3)$, and $h_{\theta|X=3}(4)$, where

$$h_{\theta|X=3}(k) = \frac{Q_{X|\theta=k}(3)P_\theta(k)}{Q_X(3)}, \qquad \text{for } k = 1, 2, 3, \text{ and } 4.$$

In this case, $Q_{X|\theta=k}(3) = P\{X = 3|\theta = k\}$ is just the probability that the seismic reading will be classified as (3), given that the well is a θ_k-barrel well, and these values can be obtained directly from Table 14.3. Hence

$$Q_{X|\theta=1}(3) = \tfrac{1}{12}, \qquad Q_{X|\theta=2}(3) = \tfrac{1}{8}, \qquad Q_{X|\theta=3}(3) = \tfrac{1}{8}, \qquad Q_{X|\theta=4}(3) = \tfrac{5}{16}.$$

The marginal distribution of the random variable X evaluated at $X = 3$ can now be obtained; that is,

$$\begin{aligned} Q_X(3) &= Q_{X|\theta=1}(3)P_\theta(1) + Q_{X|\theta=2}(3)P_\theta(2) + Q_{X|\theta=3}(3)P_\theta(3) + Q_{X|\theta=4}(3)P_\theta(4) \\ &= \tfrac{1}{12}(0.1) + \tfrac{1}{8}(0.15) + \tfrac{1}{8}(0.25) + \tfrac{5}{16}(0.50) = 0.2147, \end{aligned}$$

so that the posterior distribution can now be calculated; i.e.,

$$h_{\theta|X=3}(1) = \frac{(\frac{1}{12})(0.1)}{0.2147} = 0.039$$

$$h_{\theta|X=3}(2) = \frac{(\frac{1}{8})(0.15)}{0.2147} = 0.087$$

$$h_{\theta|X=3}(3) = \frac{(\frac{1}{8})(0.25)}{0.2147} = 0.146$$

$$h_{\theta|X=3}(4) = \frac{(\frac{5}{16})(0.50)}{0.2147} = 0.728.$$

There exists a simple tabular algorithm that will yield Table 14.4, i.e, the posterior distribution of θ. Table 14.5 is such a table. The entries in the top row of Table 14.5*a* are simply the prior probabilities of the well sizes, whereas the entries in the bottom four rows are the conditional probabilities of a seismic classification given the size of a well. The entries in the first four columns of Table 14.5*b* are the entries of the corresponding elements of Table 14.5*a*, multiplied by the appropriate prior probability [e.g., $0.0583 = (\frac{7}{12})(0.10)$]. The entries in the last column of Table 14.5*b* are simply the sum of the four elements in the corresponding row (e.g., $0.3511 = 0.0583 + 0.0844 + 0.1146 + 0.0938$). The entries in Table 14.5*c*, the conditional probabilities of obtaining a particular well size given a seismic classification, are the entries of the corresponding elements of Table 14.5*b*, divided by the appropriate element $Q_X(x)$, in the last column of Table 14.5*b* (e.g., $0.166 \doteq 0.0583/0.3511$).

The expression for the posterior distribution has been given, where θ and X are both discrete random variables. If θ is discrete and X is continuous, the posterior distribution $h_{\theta|X=x}(k)$ is given by

$$h_{\theta|X=x}(k) = \frac{f_{X|\theta=k}(x)P_\theta(k)}{f_X(x)},$$

where $P_\theta(k)$ is the prior distribution of θ, and $f_{X|\theta=k}(x)$ is the ordinary expression for the density function of the random variable X, but it is written in this form to show its dependence upon the value of the parameter θ. The function $f_X(x)$ is the marginal

Table 14.5 **Tabular Algorithm for Computation of Posterior Distribution—Oil-Drilling Example**

(*a*)

Prior Dist. ***$P_\theta(k)$***	***0.10***	***0.15***	***0.25***	***0.50***
	$Q_{X\|\theta=k}(x)$			
x \ k	θ_1	θ_2	θ_3	θ_4
Seismic classification 1	$\frac{7}{12}$	$\frac{9}{16}$	$\frac{11}{24}$	$\frac{9}{48}$
2	$\frac{4}{12}$	$\frac{3}{16}$	$\frac{6}{24}$	$\frac{13}{48}$
3	$\frac{1}{12}$	$\frac{2}{16}$	$\frac{3}{24}$	$\frac{15}{48}$
4	$\frac{0}{12}$	$\frac{2}{16}$	$\frac{4}{24}$	$\frac{11}{48}$

(*b*)

	$Q_{X\|\theta=k}(x)P_\theta(k)$				$Q_X(x)$
x \ k	θ_1	θ_2	θ_3	θ_4	
Seismic classification 1	0.0583	0.0844	0.1146	0.0938	0.3511
2	0.0333	0.0281	0.0625	0.1354	0.2593
3	0.0083	0.0188	0.0313	0.1563	0.2147
4	0.0	0.0188	0.0417	0.1146	0.1751

(*c*)

Posterior Distribution of θ	$h_{\theta\|X=x}(k)$			
x \ k	θ_1	θ_2	θ_3	θ_4
Seismic classification 1	0.166	0.240	0.327	0.267
2	0.129	0.108	0.241	0.522
3	0.039	0.087	0.146	0.728
4	0.0	0.107	0.238	0.655

density of the random variable X, and it is obtained from

$$f_X(x) = \sum_{\text{all } k} f_{X|\theta=k}(x)P_\theta(k).$$

If both θ and X are continuous, the posterior distribution $h_{\theta|X=x}(y)$ is given by

$$h_{\theta|X=x}(y) = \frac{f_{X|\theta=y}(x)P_\theta(y)}{f_X(x)},$$

where $P_\theta(y)$ is the prior density function of θ. The function $f_{X|\theta=y}(x)$ has been defined previously. $f_X(x)$ is the marginal density of the random variable X, and it is obtained from

$$f_X(x) = \int_{-\infty}^{\infty} f_{X|\theta=y}(x)P_\theta(y)\, dy.$$

Finally, if θ is continuous and X is discrete, the posterior distribution $h_{\theta|X=x}(y)$ is given by

$$h_{\theta|X=x}(y) = \frac{Q_{X|\theta=y}(x)P_\theta(y)}{Q_X(x)},$$

where $P_\theta(y)$ is the prior density function of θ. The function $Q_{X|\theta=y}(x)$ is the ordinary expression for the probability distribution of the random variable X, that is, $P\{X = x|\theta = y\}$, but it is written in this form to show its dependence upon the value of the parameter θ. $Q_X(x)$ is the marginal distribution of the random variable X, and it is obtained from

$$Q_X(x) = \int_{-\infty}^{\infty} Q_{X|\theta=y}(x)P_\theta(y)\, dy.$$

Value of Experimentation

Before performing any experiment, we should determine its potential value. Suppose the experiment can lead to perfect information about the state of nature. What is this perfect information worth? In the oil-drilling example, seismic information that is *imperfect* costs \$12,000. If perfect information saves, say, only \$10,000, seismic information should be forgone because it is too expensive. If we knew that the strike would be a 500,000-barrel well (state of nature is θ_1), then the best action to take clearly would be a_1, drill for oil. We can see from Table 14.2 that this action would lead to a loss of $-$\$650,000. Similarly, if we knew that the strike would be a 200,000-barrel well (state of nature is θ_2), the best action again would be a_1, with a corresponding loss of $-$\$200,000. However, if we knew that the strike would be a 50,000-barrel well (state of nature is θ_3), the best action would be a_2, unconditionally lease the land, with a corresponding loss of $-$\$45,000. Finally, if we knew that the well was dry (state of nature is θ_4), the best action again would be a_2, also with a corresponding loss of $-$\$45,000. Because the (prior) probabilities of each of these states are known, the expected loss with perfect information available about the state of nature $E(PI)$ is given by

$$\begin{aligned} E(PI) &= -650{,}000(0.1) - 200{,}000(0.15) - 45{,}000(0.25) - 45{,}000(0.50) \\ &= -\$128{,}750. \end{aligned}$$

Bayes' solution (without any data) provided for an expected loss of $-$\$51,250, which is substantially more than the expected loss with perfect information, so that experi-

Table 14.6 **Bayes' Actions, Expected Losses, and Marginal Distribution for Oil Example**

x	Bayes' Action	Expected Loss	Marginal Distribution $Q_X(x)$
1	a_1	−115,700	0.351
2	a_1	− 48,275	0.259
3	a_2	− 33,000	0.215
4	a_2	− 33,000	0.175

mentation can lead to potential savings. In fact, the decision maker should be willing to pay a cost of up to $-51{,}250 + 128{,}750 = \$77{,}500$ for perfect information.

Now that experimentation may be desirable, it is useful to determine the value of obtaining seismic data. It has been shown that if the seismic data are classified as (3), the optimal (Bayes) action is a_2, with a corresponding loss of $-\$33{,}000$. Using the same techniques, the optimal (Bayes) actions can be obtained if the seismic data are classified as (1), (2), or (4). These results are summarized in Table 14.6. Because the entries for the expected loss depend upon the outcome of the experiment, x, the overall measure of the effectiveness of the experiment requires obtaining a weighted sum that is weighted with respect to the marginal probability distribution of the random variable, X, of the Bayes' losses. The calculations required for obtaining this marginal distribution, $Q_X(x)$, are described earlier in the section dealing with the calculation of the posterior distribution, where we obtained $Q_X(3)$. Other values are given in Table 14.6. Hence the weighted sum of the Bayes' losses, called the *unconditional expected loss with experimentation*, is given by

$$-115{,}700(0.351) - 48{,}275(0.259) - 33{,}000(0.215) - 33{,}000(0.175) = -\$65{,}984.$$

The value of the experiment (beyond its cost of \$12,000) is then given by the difference between this weighted sum and the Bayes' loss without data; i.e.,

$$-65{,}984 + 51{,}250 = -\$14{,}734,$$

indicating an expected savings of \$14,734 due to following an optimal decision procedure with experimentation. Hence obtaining seismic soundings does reduce the total expected cost.

In fact, if the nonoptimal procedure, d_1, presented earlier in this section, is followed, its weighted risk is given by

$$-471{,}333(0.1) - 137{,}375(0.15) + 15{,}958(0.25) + 15{,}750(0.50) = -\$55{,}875.$$

Thus an expected savings of \$10,109 ($= -65{,}984 + 55{,}875$) is obtained by using the (optimal) Bayes' procedure rather than the procedure d_1.

14.4 Decision Trees

An alternative method to the analysis presented in this chapter is the use of *decision trees*. A **decision tree** is a graphical method of expressing, in chronological order, the alternative actions that are available to the decision maker and the choices determined by chance. Decision trees consist of forks (nodes) and branches. There are two types of forks: decision forks, represented by squares, □; and chance forks, represented by circles, ○. Branches are straight lines that emanate from decision forks or chance forks. When a decision maker encounters a decision fork, she must choose one of the alternative branches to travel on. When a decision maker encounters a chance fork, she has no control over which branch to travel on. Instead, her path is determined by chance events whose probabilities are those associated with the branches that emanate from the chance fork.

For example, the decision tree for the oil-drilling problem is given in Fig. 14.1, and you are urged to refer to this figure throughout the ensuing discussion. Initially the decision maker has a choice of not using seismic soundings or using seismic soundings. Either action triggers some consequences. If the decision not to use seismic soundings is made, the decision maker is led down the appropriate path, arriving at a fork (node) with branches marked: drill, unconditionally lease, and conditionally lease. She must choose one of these branches on which to continue. If she chooses to drill, she is led down the appropriate path, arriving at a fork with branches marked: 500,000-barrel well, 200,000-barrel well, 50,000-barrel well, and dry well. The choice of the branch on which to continue is a chance event.

Depending upon the outcome of this chance event, she reaches a terminating point. Similarly, if the initial decision is to use seismic soundings, the decision maker is led down the appropriate path, arriving at a fork with branches marked: definitely closed structure, probably closed structure, nonclosed structure, no structure. The choice of the branch on which to continue is a chance event. If by chance the data reveal a nonclosed structure, then this branch is chosen, and she arrives at a fork with branches marked: drill, unconditionally lease, and conditionally lease. The decision maker must choose one of these branches on which to continue. If she chooses to drill, she is led down the appropriate branch, arriving at a fork with branches marked: 500,000-barrel well, 200,000-barrel well, 50,000-barrel well, and dry well. The choice of the branch on which to continue is again a chance event. Depending upon the outcome of this chance event, she reaches a terminating point. The entire tree can be completed in this fashion.[1]

The previous discussion presents a graphical method for representing the decision problem. However, nothing was said about how to choose the optimal path to travel on. Basically, the calculations described in the earlier sections of this chapter are required, namely, finding posterior probabilities, marginal probabilities, and Bayes' risks. For each possible path that can be followed, the loss is specified at the terminal point. Working *backward* from each terminal point to the nearest fork (a chance fork), we place a loss at that fork, this cost being the expected cost taken with respect to the probabilities associated with the branches. These probabilities represent the probability of a certain state of nature, indicated by the terminal branch, being chosen, given the path followed to the last fork. For example, for the no-seismic data drill path, the probability of the 500,000-barrel branch is just the prior probability that the well is a 500,000-barrel well—that is, 0.1. For the seismic data, nonclosed structure drill path, the probability of the 500,000-barrel branch is just the posterior probability that the well is a 500,000-barrel well, given that the seismic reading was classified as a nonclosed structure—that is, 0.039. Again, working backward, we find that the next fork is a decision fork. The loss associated with this fork is the minimum loss over the branches associated with that node. On the no-seismic data path, −51,250 represents the minimum of −51,250, −45,000, and −40,000 and is as-

[1] Only part of the tree is presented in Fig. 14.1, but all other branches can be easily drawn.

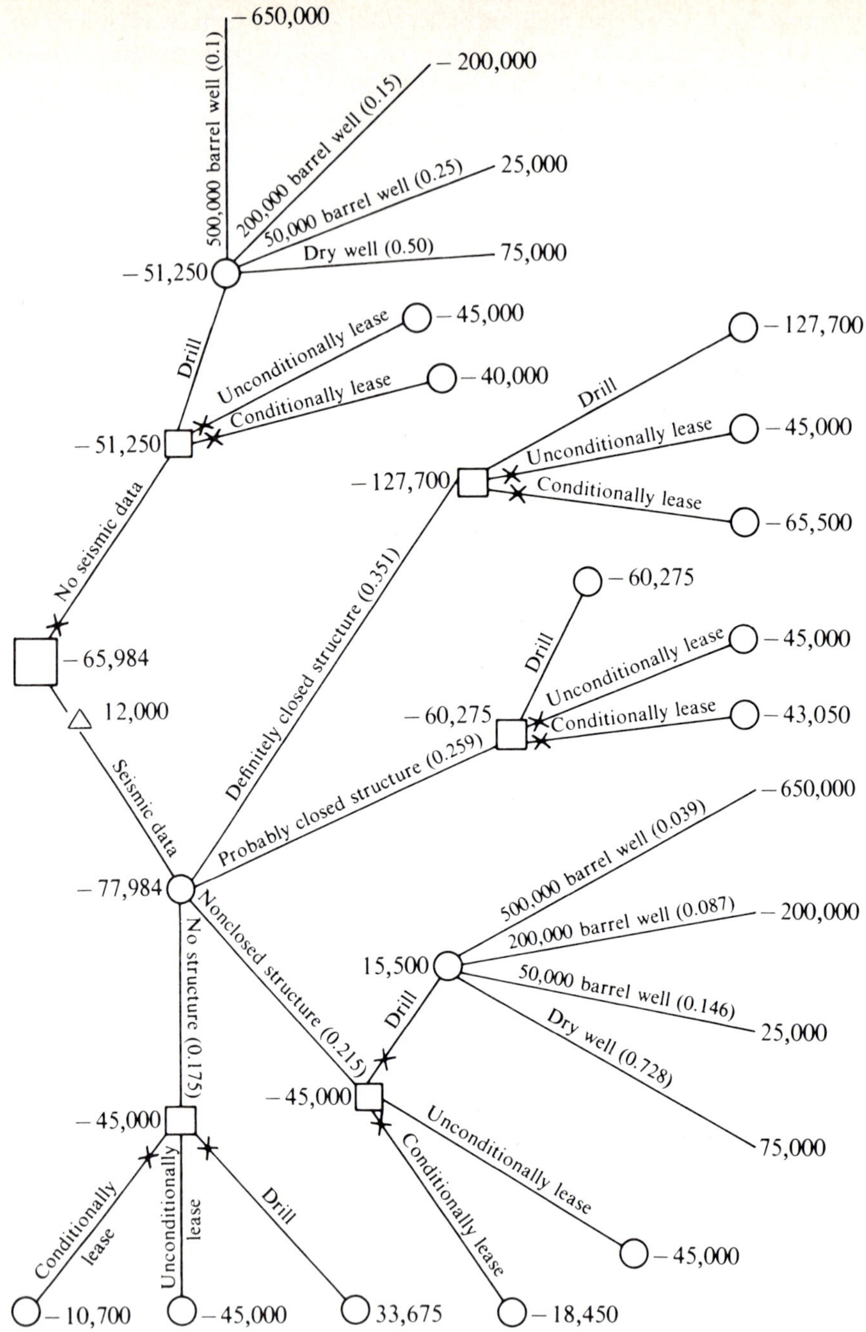

Figure 14.1 Decision tree for oil-drilling example.

sociated with the action *drill*. Hence, this action is the best one to take, given that the decision maker is at that fork. The symbol (×) through the other two branches eliminates the actions *unconditionally lease* and *conditionally lease* from further consideration on that path. Similarly, on the seismic data, nonclosed structure path, −45,000 represents the minimum of 15,500, −45,000, and −18,450 and is associated with the action *unconditionally lease*. Hence, this action is the best one to take, given that the decision maker is at that fork. The symbol (×) through the other two branches eliminates the other two actions from further consideration on that path. The next fork on the seismic data path is a chance fork. The loss associated with this fork is the expected cost taken with respect to the probabilities associated with the branches. These probabilities represent the (unconditional) probability that the seismic reading, indicated by the branch, is obtained, given the path followed to this fork. For the seismic data path, the probability that the seismic reading will lead to the nonclosed structure branch is just the unconditional probability that it will be classified as a nonclosed structure—that is, 0.215. Finally, the beginning fork has a loss associated with it of −65,984. This loss is the minimum of −51,250 that is associated with the no-seismic data branch and −65,984, which is associated with the seismic data branch (and obtained by adding the cost of taking seismic soundings, 12,000, to the −77,984 attached to the branch). Note that the cost of taking seismic soundings is denoted by the symbol △ on the branch. Hence the no-seismic data branch is eliminated, and the optimal procedure is to follow the seismic data path, leading to an expected profit of 65,984, which is, of course, the solution obtained earlier. Again, it is worthwhile to note that the calculations required by using decision tree analysis are identical to those required by using the previously described analytical methods.

14.5 Utility Function

The oil-drilling example assumed that an expected loss (profit) in monetary terms was the appropriate measure of the consequences of taking an action, given a state of nature. However, there are many situations where this assumption is inappropriate. For example, suppose that an individual was offered the choice of (1) accepting a 50–50 chance of winning $10,000 or nothing or (2) receiving $4,000 with certainty. Many people would prefer the $4,000 even though the expected payoff on the 50–50 chance of winning $10,000 is $5,000. A company may be unwilling to invest a large sum of money in a new product even if the expected profit is substantial if there is a risk of losing their investment and thereby becoming bankrupt. People buy insurance even though it is a poor investment; the insurance company must pay expenses *and* make a profit. Do these examples invalidate the previous material? Fortunately, the answer is no, because there is a way of transforming monetary values into an appropriate scale that reflects the decision makers' preferences. This scale is called the *utility scale*, and it becomes the appropriate measure of the consequences of taking an action, given a state of nature. A detailed discussion of *utilities* can be found in the references at the end of this chapter.

14.6 Carnival Example

A carnival is scheduled to appear in a city on a given date. The profits that will be obtained depend heavily upon the weather. In particular, if the weather is rainy, the carnival loses $15,000; if it is cloudy, the carnival loses $5,000; and if it is sunny, the carnival makes a profit of $10,000. The carnival has to set up equipment for its show, but it can cancel the show prior to setting up its equipment. This action results in a loss of $1,000. Furthermore, by incurring an additional cost of $1,000, the

Table 14.7 **Weather Bureau Data for Carnival Example**

Probability that Forecast Is	*Actual Weather*		
	Rain	Clouds	Sun
Rain (*RF*)	0.7	0.2	0.1
Clouds (*CF*)	0.2	0.6	0.2
Sun (*SF*)	0.1	0.2	0.7

carnival can postpone its setup decision until the day before the scheduled performance. At this time, the carnival can obtain the local weather report. The Weather Bureau has compiled data based upon its predictions; these data are given in Table 14.7. Furthermore, the Weather Bureau has compiled a prior distribution of the weather. In particular, the probabilities of rain, clouds, and sun are 0.1, 0.3, and 0.6, respectively.

We shall analyze this example by first using decision tree analysis. The unevaluated decision tree is shown in Fig. 14.2, and it is a graphical representation of the decision problem.[1] Note that the lower part of the tree represents the *no-data case*, while the upper part uses additional information from *experimentation*. The first decision fork to confront the decision maker requires making a choice between using or not using the Weather Bureau's report (fork 1). If the choice is not to use the Weather Bureau's information, the decision maker is led down the path that arrives at decision fork 2 with branches marked *setup* and *no setup* (cancel). The selection of one of these branches results in a flow into a chance fork (either fork 3 or fork 4). The choice of the branch on which to continue is a chance event, and, depending upon the outcome of this chance event, the decision maker reaches one of the terminating points: rain, clouds, or sun. Associated with each of these terminating points is a monetary consequence.

If, at initial decision fork 1, the decision maker chooses to use the Weather Bureau's information, he is led down the path that arrives at fork 5, with branches marked: weather forecast rain, weather forecast cloudy, weather forecast sunny. The choice of the branch on which to continue is a chance event. If (by chance) the Weather Bureau forecasts a sunny day, then the decision maker chooses this branch, and it leads to decision fork 11, which has subsequent branches identical to those described for fork 2.

This discussion has been concerned with a graphical method for representing the decision problem, and it has not been concerned with how to choose the optimal path to travel on. Determining the optimal path requires calculations similar to those described for the oil-drilling example. In particular, the posterior distribution of the states of nature (rain, clouds, or sun) is required, given the weather forecast. This posterior distribution is given in Table 14.7, and the details will be presented later in this section. These posterior probabilities are necessary for the evaluation of the decision tree.

Figure 14.3 is the evaluated decision tree for the carnival example. For each possible path that can be followed, the loss is specified at the terminal point. If we work *backward* from each terminal point to the nearest fork (a chance fork), the loss placed at that fork is the expected cost taken with respect to the probabilities associated

[1] The forks are numbered arbitrarily for discussion purposes.

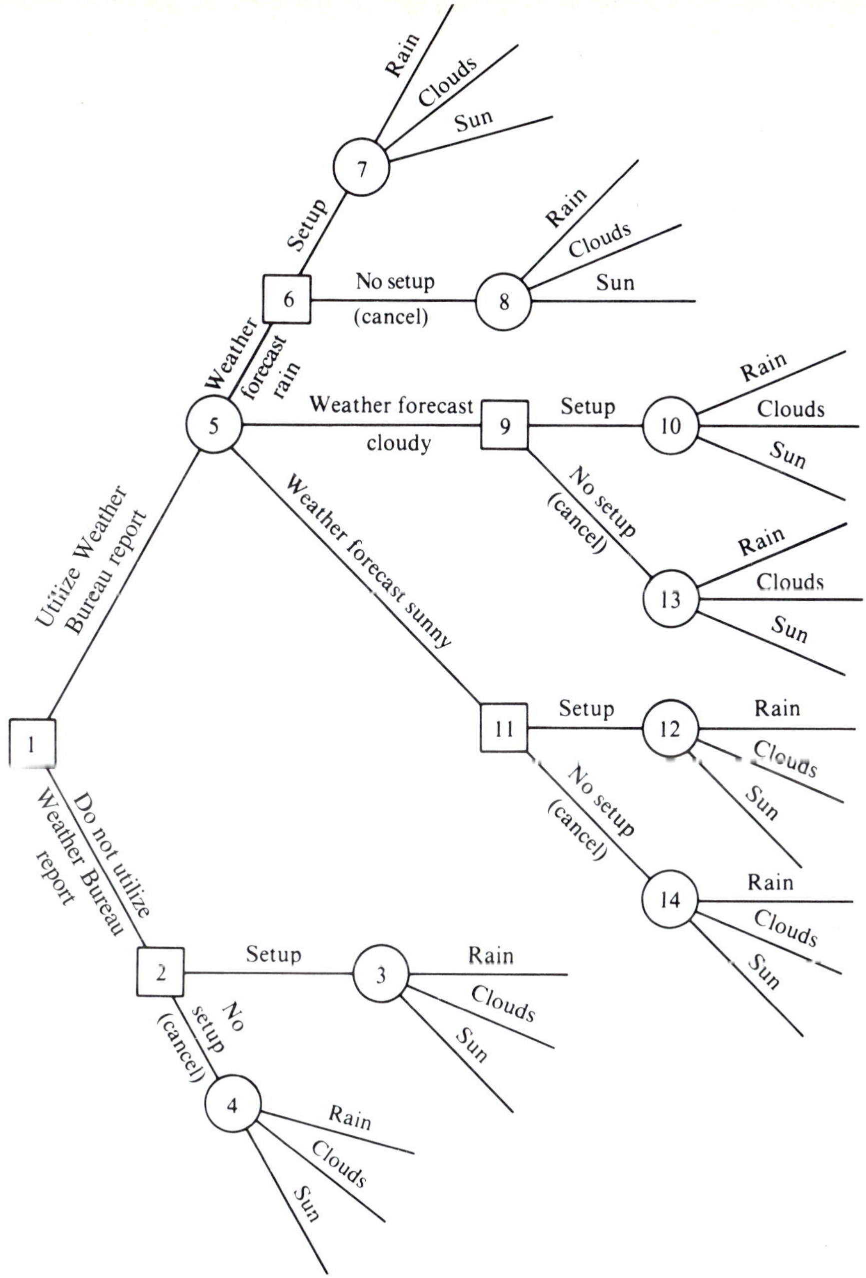

Figure 14.2 Unevaluated decision tree for carnival example.

with the branches. These probabilities represent the probability of the state of nature (indicated by the terminal branch) being chosen, given the path followed to the last fork. For example, for the *Do not utilize Weather Bureau report, Setup path*, the probability of reaching the sun branch is just the prior probability that a day will be sunny—that is, 0.6 (this probability is shown on the branch). The expected cost at fork 3 is −\$3,000 (profit). For the *Utilize Weather Bureau report, Weather forecast*

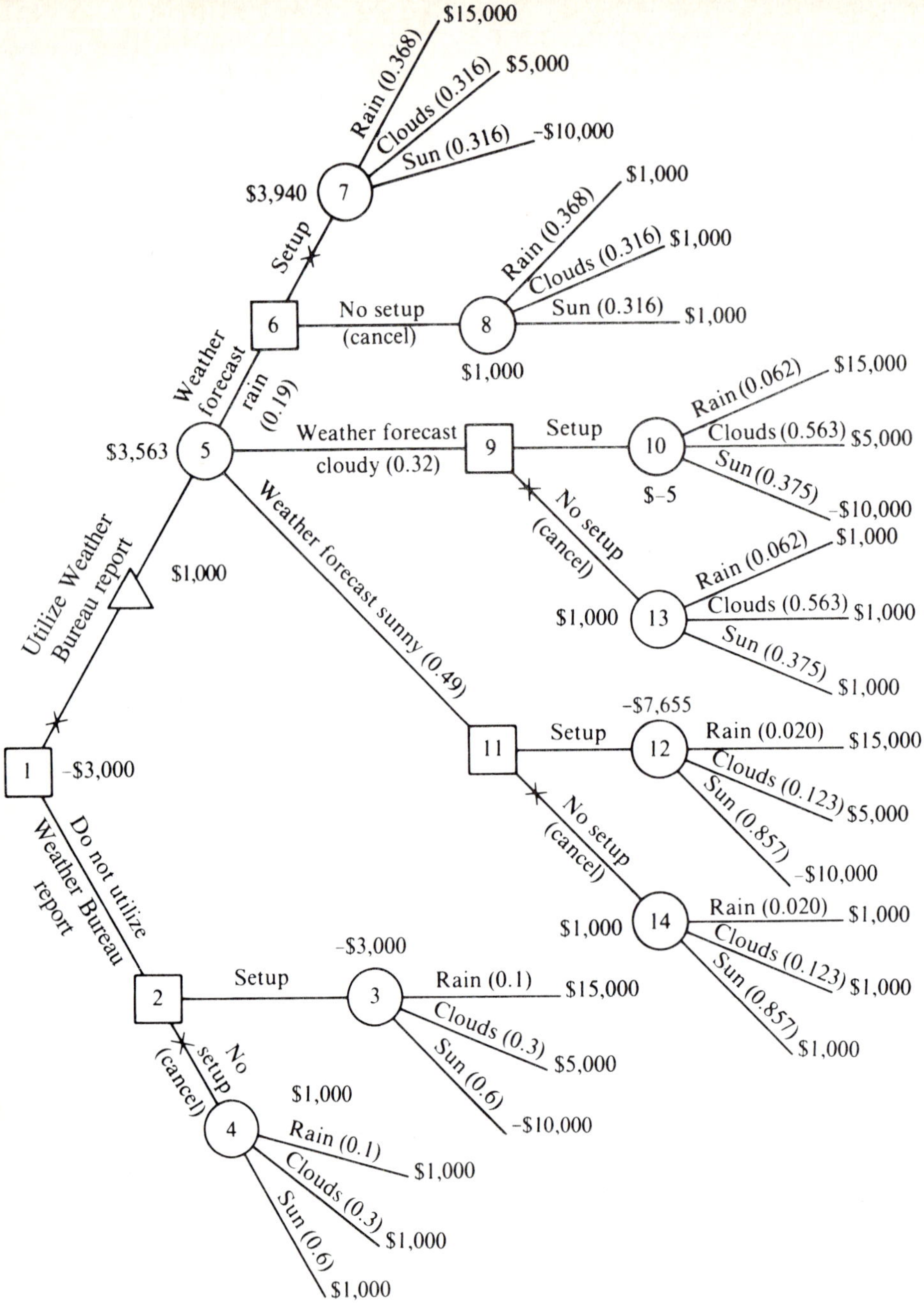

Figure 14.3 Evaluated decision tree for carnival example.

sunny, Setup path, the probability of reaching the sun branch is just the posterior probability that a day will be sunny, given that the weather forecast is for a sunny day—that is, 0.857 (this posterior probability is shown on the branch). The expected cost at fork 12 is −$7,655 (profit). Working backward from fork 12 leads to fork 11, a decision fork. Fork 14 also leads into fork 11. The loss associated with fork 11 is the minimum loss over the two branches that emanated from that fork. The loss of −$7,655 represents the minimum of −$7,655 and $1,000, and it is associated with the action *Setup*. Hence this action is the best one to take, given that the decision maker is at fork 11. The symbol (×) through the other branch eliminates that branch from further consideration on that path. Again, working *backward* from fork 11 leads to fork 5, a chance fork. The loss associated with this fork is the expected cost taken with respect to the probabilities associated with the branches. These probabilities represent the (unconditional) probability that the Weather Bureau's forecast, indicated by the branch, is obtained, given the path followed to this fork. For the *Utilize Weather Bureau report path*, the probability that the weather forecast will be sunny is simply the *unconditional* probability that the forecast will call for sun—that is, 0.49. This probability appears on the appropriate branch in Fig. 14.3, and its calculation will be discussed later in this section.[1] The expected cost at fork 5 is −$3,563 [−7,655(0.49) + −5(0.32) + 1,000(0.19)]. Continuing to work backward from fork 5 leads to fork 1, the initial decision fork. Fork 2 also leads into this initial decision fork. The loss associated with fork 1 is the minimum loss over the (two) branches that emanate from that fork. The loss of −$3,000 represents the minimum of −3,000 and −2,563 (after adding the $1,000 cost of using the Weather Bureau's forecast), and it is associated with the action that calls for not using the Weather Bureau's forecast. Indeed, the optimal decision is not to use the Weather Bureau's forecast and to set up for the carnival, leading to an expected profit of $3,000.

Although incurring a cost of $1,000 to use the Weather Bureau's report was not worthwhile, a cost of $563 or less would be worthwhile. Finally, what is the most that the carnival should be willing to pay for any type of information about the weather? The expected loss with perfect information is given by

$$E(PI) = 1{,}000(0.1) + 1{,}000(0.3) - 10{,}000(0.6) = -\$5{,}600,$$

so that some type of experimentation may lead to potential savings, with a decision maker being willing to pay a cost for perfect information up to

$$-3{,}000 + 5{,}600 = \$2{,}600.$$

Calculation of the Posterior Distribution for the Carnival Example

To calculate the posterior distribution, we need to use the same tabular algorithm that was presented for the oil-drilling example. The results are given in Table 14.8.

The entries in the top row of Table 14.8*a* are simply the prior probabilities of the weather conditions, whereas the entries in the bottom three rows are the conditional probabilities of a forecast type, given a particular weather condition. The entries in the first three columns of Table 14.8*b* are the entries of the corresponding elements of Table 14.8*a* multiplied by the appropriate prior probability [e.g., 0.07 = (0.70)(0.10)]. The entries in the last column of Table 14.8*b* are simply the sum of the three elements in the corresponding row (e.g., 0.19 = 0.07 + 0.06 + 0.06). The entries in Table 14.8*c*, the conditional probabilities of obtaining a particular weather condition given a particular forecast type, are the entries of the corresponding elements of Table 14.8*b* divided by the appropriate element, $Q_X(x)$, in the last column of Table 14.8*b* (e.g., 0.368 = 0.07/0.19).

[1] These probabilities are also obtained from the tabular algorithm for the computation of the posterior distribution (the last column of Table 14.8*b*).

Table 14.8 **Tabular Algorithm for Computation of Posterior Distribution for the Carnival Example**

	(a)			(b)					(c)			
Prior Distribution $P_\theta(k)$	**0.10**	**0.30**	**0.60**						***Posterior Distribution of*** θ			
	$Q_{X\|\theta=k}(x)$			$Q_{X\|\theta=k}(x)P_\theta(k)$				$Q_X(x)$	$h_{\theta\|X=x}(k)$			
x \ k	Rain θ_1	Clouds θ_2	Sun θ_3	x \ k	Rain θ_1	Clouds θ_2	Sun θ_3		x \ k	Rain θ_1	Clouds θ_2	Sun θ_3
Rain forecast (*RF*)	0.70	0.20	0.10	*RF*	0.07	0.06	0.06	0.19	*RF*	0.368	0.316	0.316
Cloudy forecast (*CF*)	0.20	0.60	0.20	*CF*	0.02	0.18	0.12	0.32	*CF*	0.062	0.563	0.375
Sunny forecast (*SF*)	0.10	0.20	0.70	*SF*	0.01	0.06	0.42	0.49	*SF*	0.020	0.123	0.857

Non-Decision Tree Analysis of the Carnival Example

An alternative (but equivalent) method for solving the carnival example is to use the techniques given in Secs. 14.2 and 14.3. There are two potential actions (other than deciding whether or not to use the Weather Bureau's data), namely, *setup* (a_1) and *no setup* (a_2). There are three states of nature: rain (R), clouds (C), and sun (S). The loss function is presented in Table 14.9. The problem will be solved by first assuming the local Weather Bureau's report is not available (no data). The expected loss $l(a)$ for each action is given by

$$l(a_1) = 15{,}000(0.1) + 5{,}000(0.3) - 10{,}000(0.6) = -\$3{,}000,$$

$$l(a_2) = 1{,}000(0.1) + 1{,}000(0.3) + 1{,}000(0.6) = \$1{,}000.$$

Hence Bayes' principle leads to selecting action a_1, that is, setting up, and the associated expected loss is $-\$3{,}000$ (profit).

Now, the Weather Bureau's report will be assumed to be available but at a cost of \$1,000. The posterior distribution can be obtained and is given in Table 14.8*c*. This table was obtained by using the tabular algorithm. Entries can also be obtained directly from the usual expressions for posterior probabilities. For example, the posterior distribution that the weather will be sunny, given that the forecast is for rain, is given by

$$\frac{(0.1)(0.6)}{(0.7)(0.1) + (0.2)(0.3) + (0.1)(0.6)} = 0.316.$$

The optimal Bayes' actions, given the various forecasts, are presented in Table 14.10. A typical entry in Table 14.10 will be obtained next. Given that the forecast

Table 14.9 **Loss Function for Carnival Example**

	State of Nature		
Action	Rain	Clouds	Sun
a_1: setup	15,000	5,000	−10,000
a_2: no setup	1,000	1,000	1,000

Table 14.10 Table of Bayes' Actions, Expected Losses, and Marginal Distribution for Carnival Example

Forecast	Bayes' Action	Expected Loss	Marginal Distribution
RF	a_2	2,000	0.19
CF	a_1	995	0.32
SF	a_1	−6,655	0.49

is for rain,

$$l_h(a_1) = (0.368)(15{,}000) + (0.316)(5{,}000) - (0.316)(10{,}000) + 1{,}000$$

$$= \$4{,}940,$$

$$l_h(a_2) = \$1{,}000 + 1{,}000 = \$2{,}000.$$

Hence the recommended Bayes' action is a_2, and the corresponding expected loss is \$2,000. Similarly, the marginal probability that the forecast will be sun is given by

$$(0.1)(0.1) + (0.2)(0.3) + (0.7)(0.6) = 0.49.$$

This result is also obtained in the last column of Table 14.8*b*. Hence, if the Weather Bureau's reports are to be used, the Bayes' actions call for *setting up* if the forecast is for clouds or sun but for *not setting up* if the forecast is for rain.

Is it desirable to use the Weather Bureau's data? The answer to this question calls for computing the unconditional expected loss with experimentation (weighted sum of the Bayes' losses); i.e.,

$$2{,}000(0.19) + 995(0.32) - 6{,}655(0.49) = -2{,}563.$$

Thus the value of the Weather Bureau's data is given by

$$-2{,}563 + 3{,}000 = 437,$$

(which is an actual loss relative to the no-data case), so that the Weather Bureau's data are not worth the \$1,000 cost.

14.7 Conclusions

Decision analysis has become an important technique in the solution of business problems. It can be applied to broad problems facing management, such as determining whether to enter a new product field, or it can be used to solve smaller problems, such as the one illustrated by the oil-drilling example. It is characterized by the decision maker enumerating all the available courses of action, expressing the utilities, and quantifying the subjective probabilities. When these data are available, decision analysis becomes a powerful tool in determining an optimal course of action.

SELECTED REFERENCES

1. Brown, R. B., A. S. Kahr, and C. Peterson: *Decision Analysis for the Manager*, Holt, New York, 1974.
2. Holloway, C. A.: *Decision Making under Uncertainty*, Prentice-Hall, Englewood Cliffs, N. J., 1979.

3. Keeney, R. L., and H. Raiffa: *Decisions with Multiple Objectives*, Wiley, New York, 1976.
4. LaValle, I. H.: *Fundamentals of Decision Analysis*, Holt, New York, 1978.
5. Raiffa, H.: *Decision Analysis*, Addison-Wesley, Reading, Mass., 1968.
6. Schlaifer, R. O.: *Analysis of Decisions under Uncertainty*, McGraw-Hill, New York, 1969.
7. Winkler, R. L.: *Introduction to Bayesian Inference and Decision*, Holt, New York, 1972.

PROBLEMS

1.* A new type of airplane is to be purchased by the Air Force, and the number of *spare* engines to be ordered must be determined. The Air Force must order these spare engines in batches of five, and it can choose among only 15, 20, or 25 spares. The supplier of these engines has two plants, and the Air Force must make its decision prior to knowing which plant will be used. From past experience, the Air Force knows that the number of spare engines required when production takes place at Plant A is approximated by a Poisson distribution with parameter $\theta = 21$, whereas the number of spare engines required when production takes place at Plant B is approximated by a Poisson distribution with parameter $\theta = 24$. The cost of a spare engine purchased now is \$400,000, whereas the cost of a spare engine purchased at a later date is \$900,000. Holding costs and interest are to be neglected. Spares must always be supplied if they are demanded, and unused engines will be scrapped when the airplanes become obsolete. From these data, the loss function can be computed as

Action \ State of Nature	θ_1: $\theta = 21$	θ_2: $\theta = 24$
a_1: order 15	1.155×10^7	1.414×10^7
a_2: order 20	1.012×10^7	1.207×10^7
a_3: order 25	1.047×10^7	1.135×10^7

The Air Force knows from past experience that two-thirds of all types of airplane engines are produced in Plant A, and only one-third are produced in Plant B. Furthermore, it knows that a similar type of engine was produced for an earlier version of the current airplane under consideration. The order size for this earlier type was the same as for the current model. Furthermore, its nonobsolete life is identical with that planned for the present version. The engine for the current order will be produced in the same plant as the previous model, although the Air Force does not know which of the two plants this is. The Air Force does have access to the data on the number of spares actually required for the older version (which had a Poisson distribution), but it does not have time to determine the production location.

(*a*) What action does the Bayes' procedure recommend, assuming that the information on the old airplane model is not available?

(*b*) How much money is it worthwhile to pay for "perfect information"?

(*c*) Assuming that cost of data on the old airplane model is free and that 30 spares were required, determine the Bayes' action.

2. A large mill is faced with the problem of extending \$100,000 credit to a new customer, a dress manufacturer. The mill classifies typical companies into the following categories: poor risk, average risk, and good risk. Their experience indicates that 20 percent of similar companies are poor risks, 50 percent are average risks, and 30 percent are good risks. If credit is extended, the expected profit for poor risks is −\$15,000, for average risks \$10,000, and for good risks \$20,000. If credit is not extended, the dress manufacturer will turn to another mill. The mill is able to consult a credit-rating organization for a fee of \$2,000. Their experience with this credit-rating company is as follows:

Credit Company Evaluation	Actual Credit Rating, %		
	Poor	Average	Good
Poor	50	40	20
Average	40	50	40
Good	10	10	40

(*a*) What action does Bayes' procedure recommend, assuming the credit-rating company is not used?
(*b*) How much money is it worthwhile to pay for "perfect information"?
(*c*) What is the optimal expected loss if the credit-rating company data are used? Does it pay to use these data?
(*d*) What action does Bayes' procedure recommend if the credit-rating company determines the dress manufacturer to be a poor risk?

3. Use the scenario given in Prob. 2.
(*a*) Draw and properly label the decision tree.
(*b*) Evaluate the decision tree.
(*c*) Determine the optimal policy.
(*d*) How much money is it worthwhile to pay for "perfect information"?

4. A manufacturer produces items that have a probability p of being defective. These items are formed into lots of 150. Past experience indicates that p is either 0.05 or 0.25, and, furthermore, in 80 percent of the lots produced, p equals 0.05 (and, in 20 percent of the lots, p equals 0.25). These items are then used in an assembly, and ultimately their quality is determined before the final assembly leaves the plant. Initially the manufacturer can *either* screen each item in a lot at a cost of $15 per item and replace defective items *or* use the items directly without screening. If the latter action is chosen, the cost of rework is ultimately $100/defective item. For these data, the costs per lot can be calculated as follows:

	$p = 0.05$	$p = 0.25$
Screen	1,500	1,500
Do not screen	750	3,750

Because screening requires scheduling of inspectors and equipment, the decision to screen or not screen must be made 2 days before the potential screening takes place. However, one item can be taken from the lot and sent to a laboratory, and its quality (defective or nondefective) can be reported before the screen/no-screen decision must be made. The cost of this initial inspection is $125.
(*a*) What action does Bayes' procedure recommend without looking at the single item?
(*b*) How much money is it worthwhile to pay for "perfect information"?
(*c*) What is the optimal expected cost if the quality of items is determined before the screen/no-screen decision is made?
(*d*) What action does Bayes' procedure recommend if the quality of one of the items is determined and found to be defective?

5. Use the scenario given in Prob. 4.
(*a*) Draw and properly label the decision tree.
(*b*) Evaluate the decision tree.
(*c*) Determine the optimal policy.
(*d*) How much money is it worthwhile to pay for "perfect information"?

6.* Assume that there are two weighted coins. Coin 1 has a probability of 0.3 of turning up heads, and coin 2 has a probability of 0.6 of turning up heads. A coin is tossed once. The decision maker must decide which coin was tossed. The probability that coin 1 was tossed is 0.6, and the probability that coin 2 was tossed is 0.4. The loss matrix is as follows:

	Coin 1 Tossed	Coin 2 Tossed
a_1: say coin 1 tossed	0	1
a_2: say coin 2 tossed	1	0

(*a*) What is the Bayes' procedure (action) before the coin is tossed?
(*b*) What is the Bayes' procedure if the outcome is heads? What if it is tails?

7. Use the scenario given in Prob. 6.
(*a*) Draw and properly label the decision tree.
(*b*) Evaluate the decision tree.
(*c*) Determine the optimal policy.

8.* A company has developed a new chip that will enable it to enter the microcomputer field if it so desires. Alternatively, it can sell its rights for $800,000. If it chooses to build computers, the profitability of the venture depends upon the company's ability to market the microcomputer during the first year. It has sufficient access to retail outlets so that it can guarantee sales of 1,000 computers. On the other hand, if this computer catches on, it can sell 10,000 machines. The company believes that both sales alternatives are equally likely and that all other alternatives are negligible. The cost of setting up the assembly line is $600,000. The difference between the selling price and the variable cost is $600. Market research can be performed at a cost of $400,000 to determine which of the two levels of demand is more realistic. Previous experience indicates that such market research is correct two-thirds of the time.

(*a*) What action does Bayes' procedure recommend, assuming market research is not used?
(*b*) How much money is it worthwhile to pay for "perfect information"?
(*c*) What is the optimal expected loss if market research is used? Does it pay to use market research?
(*d*) What action does Bayes' procedure recommend if market research determined that only 1,000 computers will be sold?

9. Use the scenario given in Prob. 8.
(*a*) Draw and properly label the decision tree.
(*b*) Evaluate the decision tree.
(*c*) Determine the optimal policy.
(*d*) How much money is it worthwhile to pay for "perfect information"?

10. A new type of camera film has been developed. It is packaged in sets of five sheets, each sheet providing an instantaneous snapshot. Because this process is new, the manufacturer has attached an additional sheet to the package, so that the store may test 1 sheet before it sells the package of 5. In promoting the film, the manufacturer offers to refund the entire purchase price of the film if 1 of the 5 is defective. This refund must be paid by the camera store, and the selling price has been fixed at $1 if this guarantee is to be valid. The camera store may sell the film for 50 cents if the preceding guarantee is replaced by one that pays 10 cents for each defective sheet. The cost of the film to the camera store is 20 cents, and the film is not returnable. The store may take three actions:

a_1: scrap the film,
a_2: sell the film for $1,
a_3: sell the film for 50 cents.

(*a*) If the six states of nature correspond to 0, 1, 2, 3, 4, 5 defective sheets in the package, complete the following loss table:

a \ θ	0	1	2	3	4	5
a_1	0.20					
a_2	−0.80		0.20			
a_3	−0.30	−0.20		0.00		

(*b*) The store has accumulated the following information on sales of 60 such packages:

Quality of Attached Sheet	Defectives in Box					
	0	1	2	3	4	5
Good	10	8	6	4	2	0
Bad	0	2	4	6	8	10
Total	10	10	10	10	10	10

These data indicate that each state of nature is equally likely, so that this prior can be assumed. What is Bayes' procedure (before testing the attached sheet) for a package of film?

(*c*) What is the optimal expected loss for a package of film if the attached sheet is tested? What action does Bayes' procedure recommend if the sheet is good? If it is bad?

11. There are two biased coins with probabilities of landing heads 0.8 and 0.4, respectively. One of the coins is chosen at random (each with probability $\frac{1}{2}$), and you are to receive \$100 if you correctly predict how many heads will occur in two coin tosses.

(*a*) Using Bayes' criterion, what would you predict, and what is the corresponding expected gain?

(*b*) Suppose now that you may observe a practice coin toss before predicting. Using Bayes' criterion, what would you predict based on each of the possible outcomes of the practice toss?

(*c*) What is your expected gain with the practice toss?

12. A private university is considering whether or not to hold an extensive centennial campaign next fall to raise funds for a new athletic field. The response to the campaign depends heavily upon the success of this spring's varsity baseball team. If the baseball team has a winning season (W), many of the alumni will contribute, and the campaign will raise \$3 million. If the team has a losing season (L), few will contribute, and the campaign will lose \$2 million. If no campaign is undertaken, no costs are incurred.

(*a*) Based upon past performance, the baseball team has had winning seasons 60 percent of the time. How much should the president of the university be willing to pay for "perfect information"?

(*b*) Now there is the possibility of hiring a professional talent scout to look at the baseball team before March 25, the date when both the baseball season begins and the decision of whether or not to hold the campaign must be made. For \$100,000, this talent scout correctly predicts, three-fourths of the time, what kind of season, W or L, a team will have. Find the posterior distribution of the outcome of the baseball team's season, given that the scout's prediction is for (1) a winning season, W_p, and for (2) a losing season, L_p.

(*c*) Evaluate the decision tree. Should the president pay \$100,000 for the scout's prediction?

13. An emerging presidential candidate is considering whether or not to run in the high-stakes Super Tuesday primaries. If he enters the Super Tuesday (S.T.) primaries, he and his advisors believe that he will either do well (finish first or second) or do poorly (finish third or worse) with probabilities 0.4 and 0.6, respectively. Doing well on Super Tuesday will net the candidate's campaign approximately $4 million, whereas a poor showing will mean a loss of $2.5 million after paying for numerous TV ads. Alternatively, he may choose not to run at all on Super Tuesday and incur no costs.

The candidate's advisors realize that his chances of success on Super Tuesday may be affected by the outcome of the smaller New Hampshire (N.H.) primary occurring three weeks before Super Tuesday. Political analysts feel that the results of New Hampshire's primary are correct two-thirds of the time in predicting the results of the Super Tuesday primaries. Among the advisors is a decision analysis expert who uses this information to calculate the following probabilities:

$$P\{\text{candidate does well in S.T. primaries, candidate does well in N.H.}\} = \tfrac{4}{7}$$

$$P\{\text{candidate does well in S.T. primaries, candidate does poorly in N.H.}\} = \tfrac{1}{4}$$

$$P\{\text{candidate does well in N.H. primary}\} = \tfrac{7}{15}$$

The cost of entering and campaigning in the New Hampshire primary is estimated to be $400,000.

(*a*) Draw and properly label the decision tree.
(*b*) Evaluate the decision tree.
(*c*) Determine the strategy that minimizes the candidate's expected losses.
(*d*) How much money is it worthwhile to pay for "perfect information"?

14. A doctor has a seriously ill patient but has had trouble diagnosing the specific cause of the illness. The doctor now has narrowed the cause down to two alternatives: disease *A* or disease *B*. Based on the evidence so far, she feels that the two alternatives are equally likely.

Beyond the testing already done, there is no test available to determine if the cause is disease *B*. One test is available for disease *A*, but it has two major problems. First, it is very expensive. Second, it is somewhat unreliable, giving an accurate result only 80 percent of the time. Thus, it will give a positive result (indicating disease *A*) for only 80 percent of patients who have disease *A*, whereas it will give a positive result for 20 percent of patients who actually have disease *B* instead.

Disease *B* is a very serious disease with no known treatment. It is sometimes fatal, and those who survive remain in poor health with a poor quality of life thereafter. The prognosis is similar for victims of disease *A* if it is left untreated. However, there is a fairly expensive treatment available that eliminates the danger of death for those with disease *A*, and it may return them to good health. Unfortunately, it is a relatively radical treatment that always leads to death if the patient actually has disease *B* instead.

The probability distribution for the prognosis for the current patient is given for each case in the following table, where the column headings (after the first one) indicate the disease for this patient.

	Outcome Probabilities			
	No Treatment		*Receive Treatment for Disease A*	
Outcome	*A*	*B*	*A*	*B*
Die	0.2	0.5	0	1.0
Survive with poor health	0.8	0.5	0.5	0
Return to good health	0	0	0.5	0

The patient has assigned the following utilities to the possible outcomes.

Outcome	Utility
Die	0
Survive with poor health	10
Return to good health	30

In addition, these utilities should be incremented by -2 if the patient incurs the cost of the test for disease A and by -1 if the patient (or the patient's estate) incurs the cost of the treatment for disease A.

Use decision analysis with a complete decision tree to determine if the patient should undergo the test for disease A and then how to proceed (receive the treatment for disease A?) in order to maximize the patient's expected utility.

15. Solve the oil-drilling example if the profit per barrel of oil is increased to $4.

16.* Solve the oil-drilling example by using the prior distribution formed from the data in Table 14.3.

17. Solve the carnival example if the cost of using the Weather Bureau's data is reduced to $400.

18. Refer to the scenarios in the appropriate problems and assume that no experimental data are available; draw, properly label, and evaluate the decision tree:

(*a*) For Prob. 2;
(*b*) For Prob. 4;
(*c*) For Prob. 6;
(*d*) For Prob. 8.

15

Simulation

The technique of **simulation** has long been an important tool of the designer. For example, simulating airplane flight in a wind tunnel is standard practice when designing a new airplane. Theoretically, the laws of physics could be used to obtain the same information about how the performance of the airplane changes as design parameters are altered, but, as a practical matter, the analysis would be too complicated to do it all. Another alternative would be to build real airplanes with alternative designs and test them in actual flight to choose the final design, but this would be far too expensive (as well as unsafe). Therefore, after performing some preliminary theoretical analysis to develop a *rough* design, simulating flight in a wind tunnel is a vital tool for experimenting with *specific* designs. This simulation amounts to *imitating* the performance of a real airplane in a controlled environment in order to *estimate* what its actual performance would be. After developing a detailed design in this way, a prototype model can then be built and tested in actual flight in order to fine-tune the final design.

Simulation plays essentially this same role in many operations research studies. However, rather than designing an airplane, the operations research team is concerned with developing a design (or operating policy) for some *stochastic system* (a system that evolves *probabilistically* over time). Some of these stochastic systems resemble the examples of Markov chains and queueing systems described in Chaps. 5–7, and others are more complicated. Rather than using a wind tunnel, the performance of the real system is *imitated* by using probability distributions to *randomly generate* the various events that occur in the system. Therefore, a simulation model *synthesizes* the system by building it up component by component and event by event. It then *runs* the simulated system to obtain *statistical observations* of the performance of the system that result from the various randomly generated events. Because the *simulation runs* typically require generating and processing a vast amount of data, these simulated statistical experiments normally are performed on a computer.

When simulation needs to be used as part of an operations research study, it commonly is preceded and followed by the same steps as for the airplane design. In particular, some preliminary theoretical analysis is done first to develop a *rough* design of the system. Then simulation is used to experiment with *specific* designs in order to *estimate* what the actual performance of the system would be. After developing a detailed design in this way, the system is tested in actual use in order to fine-tune the final design.

Operations research teams typically use simulation when the stochastic system involved is too complex to be analyzed satisfactorily by the kinds of analytical models described in the preceding chapters. One of the main strengths of the analytical approach is that it abstracts the essence of the problem and reveals its underlying structure, thereby providing insight into the cause-and-effect relationships within the system. Therefore, if it is possible to construct an analytical model that is both a reasonable idealization of the problem and amenable to solution, this approach usually is superior to simulation. However, many problems are so complex that they cannot be solved analytically. Thus, even though simulation tends to be a relatively expensive procedure, it often provides the only practical approach to a problem.

In essence, then, the operations research view of simulation is that it is a *controlled statistical sampling technique* for *estimating* the performance of complex stochastic systems when analytical models do not suffice. Rather than describing the overall behavior of the system directly, the *simulation model* describes the operation of the system in terms of *individual events* of the individual components of the system. In particular, the system is divided into elements whose behavior can be predicted, at least in terms of probability distributions, for each of the various possible *states of the system* and its inputs. The interrelationships among the elements also are built into the model.

Thus simulation provides a means of dividing the model-building job into smaller component parts that can be formulated more readily (e.g., a component part might be a simple queueing system) and then combining these component parts in their natural order. After constructing the model, we can then activate it by using random numbers to generate simulated events over time according to the appropriate probability distributions. The result is a simulation of the actual operation of the system over time, and we can record its aggregate behavior. By repeating this process for the various alternative configurations for the design and operating policies of the system, and by comparing their performance, we can identify the most promising configurations. Because of statistical error, it is impossible to guarantee that the configuration yielding the best simulated performance is indeed the optimal one, but it should be at least near optimal if the simulated experiment was designed properly.

Thus simulation typically is nothing more or less than the technique of performing *sampling experiments* on the model of the system. The experiments are done on the model rather than on the real system itself only because the latter would be too inconvenient, expensive, and time consuming. Otherwise, simulated experiments should be viewed as virtually indistinguishable from ordinary statistical experiments, so they also should be based upon sound statistical theory.

This chapter focuses on *discrete event simulations* (as opposed to *continuous simulations*), i.e., simulations where changes in the *state of the system* occur at random points in time (as opposed to continuously) as a result of the occurrence of discrete events. The basic building blocks of a model for a discrete event simulation are the possible states and events, a simulation clock for recording the passage of (simulated) time, a mechanism for randomly generating the different kinds of events, and a mechanism for then generating state transitions.

These building blocks are described in Sec. 15.1 in the context of illustrative examples. The second section elaborates on the formulation and implementation of simulation models. The next two sections then focus on the design and analysis of the program of statistical experimentation inherent in a simulation study.

15.1 Illustrative Examples

This section uses some relatively simple stochastic systems to introduce and illustrate some basic concepts of simulation. The first system is so simple, in fact, that the simulation does not even need to be performed on a computer. The second system incorporates more of the normal features of a simulation, although it too is simple enough to be solved analytically. Following these two examples, we then survey some more typical applications of simulation.

Example 1—A Coin-Flipping Game

Suppose you were offered a chance to play a game whereby you would repeatedly flip an unbiased coin until the *difference* between the number of heads tossed and the number of tails tossed is *three*. You would be required to pay $1 for each flip of the coin, but you would receive $8 at the end of each play of the game. You are not allowed to quit during a play of the game. Thus you win money if the number of flips required is fewer than eight, but you lose money if more than eight flips are required. How would you decide whether or not to play this game?

Many people would base this decision on simulation, although they probably would not call it by that name. (There is also an analytical solution for this game, but it is not a particularly elementary one.) In this case, simulation amounts to nothing more than playing the game alone many times until it becomes clear whether it is worthwhile playing for money. Half an hour spent in repeatedly flipping a coin and recording the earnings or losses that would have resulted might be sufficient. This is a true simulation because you are *imitating* the actual play of the game *without* actually winning or losing any money.

How would this simulated experiment be executed on a computer? Although the computer cannot flip coins, it can generate numbers. Therefore, it would generate (or be given) a sequence of random digits, each of which would correspond to a flip of a coin. (The generation of random numbers is discussed in Sec. 15.2.) The probability distribution for the outcome of a flip is that the probability of a head is $\frac{1}{2}$ and the probability of a tail is $\frac{1}{2}$, whereas there are 10 possible values of a random digit, each having a probability of $\frac{1}{10}$. Therefore, five of these values (say, 0, 1, 2, 3, 4) would be assigned an association with a *head* and the other five (say, 5, 6, 7, 8, 9) with a *tail*. Thus the computer would simulate the playing of the game by examining each new random digit generated and labeling it a *head* or a *tail*, according to its value. It would continue doing this, recording the outcome of each simulated play of the game, as long as desired.

To illustrate the computer approach to this simulated experiment, we suppose that the computer generated the following sequence of random digits:

8, 1, 3, 7, 2, 7, 1, 6, 5, 5, 7, 9, 0, 0, 3, 4, 3, 5, 6, 8, 5,
8, 9, 4, 8, 0, 4, 8, 6, 5, 3, 5, 9, 2, 5, 7, 9, 7, 2, 9, 3, 9,
8, 5, 8, 9, 2, 5, 7, 6, 9, 7, 6, 0, 7, 3, 9, 8, 2, 7, 1, 0, 3,
2, 6, 2, 7, 1, 3, 7, 0, 4, 4, 1, 8, 3, 2, 1, 3, 9, 5, 9, 0, 5,
0, 3, 8, 7, 8, 9, 5, 4, 0, 8, 3, 8, 0, 1.

Thus, denoting a head by H and a tail by T, the first simulated play of the game is *THHTHTHTTTT*, requiring 11 simulated flips of a coin. The subsequent simulated plays of the game require 5, 5, 9, 7, 7, 5, 3, 17, 5, 5, 3, 9, and 7 simulated flips, respectively. This experiment has a sample size of 14 (14 simulated plays of the game), where the individual observations are the number of flips required for a play of the game. One useful statistic is

$$\text{Sample average} = \frac{11 + 5 + \cdots + 7}{14} = 7,$$

because the sample average provides an *estimate* of the true *mean* of the underlying probability distribution.

This sample average of 7 would seem to indicate that, on the average, you should win about \$1 each time you play the game. Therefore, if you do not have a relatively high aversion to risk, it appears that you should choose to play this game, preferably a large number of times.

However, beware! One of the common errors in the use of simulation is that conclusions are based on overly small samples, because statistical analysis was inadequate or totally lacking. In this case, the *sample standard deviation* is 3.67, so that the estimated *standard deviation* of the *sample average* is $3.67/\sqrt{14} \approx 0.98$. Therefore, even if it is assumed that the probability distribution of the number of flips required for a play of the game is a *normal distribution* (which is a gross assumption because the true distribution is *skewed*), any reasonable *confidence interval* for the true *mean* of this distribution would extend far above 8. Hence a much larger sample size is required before we can draw a valid conclusion at a reasonable level of statistical significance. Unfortunately, because the standard deviation of a sample average is inversely proportional to the *square root* of the sample size, a large increase in the sample size is required to yield a relatively small increase in the precision of the estimate of the true mean. In this case, it appears that an additional 100 simulated plays of the game might be adequate.

It so happens that the true *mean* of the number of flips required for a play of this game is 9. Thus, in the long run, you actually would lose about \$1 each time you played the game.

Although formally constructing a full-fledged *simulation model* is not really necessary for this simple simulation, we now shall do so for illustrative purposes. The *stochastic system* being simulated is the successive flipping of the coin for a play of the game. The *simulation clock* records the number of (simulated) flips t that have occurred so far. The information about the system that defines its current status, i.e., the *state of the system*, is

$$N(t) = \text{number of heads minus number of tails after } t \text{ flips.}$$

The *events* that change the state of the system are the flipping of a *head* or the flipping of a *tail*. The *event generation mechanism* is the generation of a *random digit* where

$$0\text{–}4 \Rightarrow \text{a head,}$$

$$5\text{–}9 \Rightarrow \text{a tail.}$$

The *state transition mechanism* is to set

$$N(t) = \begin{cases} N(t-1) + 1, & \text{if flip } t \text{ is a head} \\ N(t-1) - 1, & \text{if flip } t \text{ is a tail.} \end{cases}$$

The simulated game then ends at the first value of t where $N(t) = \pm 3$, where the resulting sampling *observation* for the simulated experiment is $(8 - t)$, the amount won (positive or negative) for that play of the game.

The next example will illustrate these building blocks of a simulation model for a more typical stochastic system.

Example 2—An M/M/1 Queueing System

Consider the M/M/1 queueing theory model (Poisson input, exponential service times, and single server) that was discussed at the beginning of Sec. 6.6. Although this model already has been solved analytically, it will be instructive to consider how to study it using simulation. To be specific, suppose that the values of the *arrival rate* λ and *service rate* μ are

$$\lambda = 5 \text{ per hour}, \qquad \mu = 3 \text{ per hour.}$$

To summarize the physical operation of the system, arriving customers enter the queue, eventually are serviced by the server, and then leave. Thus it is necessary for the *simulation model* to describe and synchronize the arrival of customers and the servicing of customers.

Starting at time 0, the *simulation clock* records the amount of (simulated) time t that has transpired so far during the simulation run. The information about the queueing system that defines its current status, i.e., the *state of the system*, is

$$N(t) = \text{number of customers in the system at time } t.$$

The *events* that change the state of the system are the *arrival* of a customer or a *service completion* for the customer currently in service (if any). We shall describe the *event generation mechanism* a little later. The *state transition mechanism* is to

$$\text{Reset } N(t) = \begin{cases} N(t) + 1, & \text{if an arrival occurs at time } t \\ N(t) - 1, & \text{if a service completion occurs at time } t. \end{cases}$$

There are two basic methods used for advancing the simulation clock and recording the operation of the system. We did not distinguish between these methods for Example 1 because they actually coincide for that simple situation. However, we now shall describe and illustrate these two **time advance mechanisms** (*fixed-time incrementing* and *next-event incrementing*) in turn.

With the **fixed-time incrementing** time advance mechanism, the following two-step procedure is used repeatedly.

Summary of Fixed-Time Incrementing

1. *Advance time* by a small *fixed amount*.
2. *Update the system* by determining what events occurred during the elapsed time interval and what the resulting state of the system is. Also record desired information about the performance of the system.

For the queueing theory model under consideration, only two types of events can occur during each of these elapsed time intervals, namely, one or more *arrivals* and one or more *service completions*. Furthermore, the probability of two or more arrivals or of two or more service completions during an interval is negligible for this model if the interval is relatively short. Thus the only two possible events during such an interval that need to be investigated are the arrival of one customer and the service completion for one customer. Each of these events has a known probability.

To illustrate, let us use 0.1 hour (6 minutes) as the small fixed amount by which the clock is advanced each time. (Normally, a considerably smaller time interval would be used to make negligible the probability of multiple arrivals or multiple service completions, but this choice will create more action for illustrative purposes.) Because both interarrival times and service times have an exponential distribution, the probability P_A that a time interval of 0.1 hour will include an *arrival* is

$$P_A = 1 - e^{-3/10} = 0.259,$$

and the probability P_D that it will include a *departure* (service completion), given that a customer was being served at the beginning of the interval, is

$$P_D = 1 - e^{-5/10} = 0.393.$$

To randomly generate either kind of event according to these probabilities, the approach is similar to that in Example 1. The computer again is used to generate a *random number*, but this time with multiple digits rather than one. Placing a decimal point in front of the number then makes it a *uniform random number* on (0, 1), that is, a random observation from the *uniform distribution* between 0 and 1. Denoting this uniform random number by r_A,

$$r_A < 0.259 \Rightarrow \text{an arrival occurred},$$

$$r_A \geq 0.259 \Rightarrow \text{an arrival did not occur}.$$

Similarly, with *another* uniform random number r_D,

$$r_D < 0.393 \Rightarrow \text{a departure occurred},$$

$$r_D \geq 0.393 \Rightarrow \text{a departure did not occur},$$

given that a customer was being served at the beginning of the time interval. With no customer in service then (i.e., no customers in the system), it is assumed that no departure can occur during the interval even if an arrival does occur.

Table 15.1 shows the result of using this approach for 10 iterations of the *fixed-time incrementing* procedure, starting with no customers in the system and using time units of *minutes*.

Step 2 of the procedure (updating the system) includes recording the desired measures of performance about the aggregate behavior of the system during this time interval. For example, it could record the *number of customers* in the queueing system and the *waiting time* of any customer who just completed his or her wait. If it is sufficient to estimate only the mean rather than the probability distribution of each of these random variables, the computer would merely add the value (if any) at the end of the current time interval to a cumulative sum. The sample averages would be obtained after the simulation run was completed by dividing these sums by the sample sizes involved, namely, the total number of time intervals and the total number of customers, respectively.

Next-event incrementing differs from fixed-time incrementing in that the simulation clock is incremented by a *variable* amount rather than by a fixed amount each time. This variable amount is the time from the event that has just occurred until the *next event* of any kind occurs; i.e., the clock jumps from event to event. A summary follows.

Summary of Next-Event Incrementing

1. *Advance time* to the time of the *next event* of any kind.
2. *Update the system* by determining its new state that results from this event and by randomly generating the time until the next occurrence of any event type that can occur from this state (if not previously generated). Also record desired information about the performance of the system.

For this example the computer needs to keep track of two future events, namely, the next arrival and the next service completion (if a customer currently is being served). These times are obtained by taking a random observation from the probability distribution of interarrival and service times, respectively. As before, the computer takes such a random observation by generating and using a random number. (This technique will be discussed in Sec. 15.2.) Thus, each time an arrival or service completion occurs, the computer determines how long it will be until the next time this event will occur, adds this time to the current clock time, and then stores this sum in a computer file. (If the service completion leaves no customers in the system,

Table 15.1 **Applying Fixed-Time Incrementing to Example 2**

t (min)	$N(t)$	r_A	Arrival in Interval?	r_D	Departure in Interval?
0	0				
6	1	0.096	Yes	—	
12	1	0.569	No	0.665	No
18	1	0.764	No	0.842	No
24	0	0.492	No	0.224	Yes
30	0	0.950	No	—	
36	0	0.610	No	—	
42	1	0.145	Yes	—	
48	1	0.484	No	0.552	No
54	1	0.350	No	0.590	No
60	0	0.430	No	0.041	Yes

then the generation of the time until the next service completion is postponed until the next arrival occurs.) To determine which event will occur next, the computer finds the minimum of the clock times stored in the file. To expedite the bookkeeping involved, simulation programming languages provide a "timing routine" that determines the occurrence time and type of the next event, advances time, and transfers control to the appropriate subprogram for the event type.

Table 15.2 shows the result of applying this approach through five iterations of the *next-event incrementing* procedure, starting with no customers in the system and using time units of *minutes*. For later reference, we include the *uniform random numbers* r_A and r_D used to generate the interarrival times and service times, respectively, by the method to be described in Sec. 15.2. These r_A and r_D are the same as used in Table 15.1 in order to provide a truer comparison between the two *time advance mechanisms*.

The next-event incrementing procedure is considerably better suited for this example and similar stochastic systems than the fixed-time incrementing procedure. Next-event incrementing requires fewer iterations to cover the same amount of simulated time, and it generates a precise schedule for the evolution of the system rather than a rough approximation.

The next-event incrementing procedure will be illustrated again in Sec. 15.4 (see Table 15.11) in the context of a full statistical experiment for estimating certain measures of performance for another queueing system.

Several pertinent questions about how to conduct a simulation study of this type still remain to be answered. These answers are presented in a broader context in subsequent sections.

Typical Applications

During the early 1980s, a survey was made of many leading American firms to learn more about their use of simulation, as reported by David Christie and Hugh Watson in Selected Reference 4. One major finding was the identification of the functional areas of the company where simulation was being applied. The results are shown in Table 15.3, where production (manufacturing) leads the list, closely followed by corporate planning, engineering, finance, and research and development. As the percentages indicate, many of the companies are applying simulation in most of these areas.

This survey also found that the development of simulation models had spread far beyond centralized operations research (or management science) departments. In fact, in 54 percent of the companies, simulation models were being created in func-

Table 15.2 **Applying Next-Event Incrementing to Example 2**

t (min)	$N(t)$	r_A	Next Interarrival Time	r_D	Next Service Time	Next Arrival	Next Departure	Next Event
0	0	0.096	2.019	—	—	2.019	—	Arrival
2.019	1	0.569	16.833	0.665	13.123	18.852	15.142	Departure
15.142	0	—	—	—	—	18.852	—	Arrival
18.852	1	0.764	28.878	0.842	22.142	47.730	40.994	Departure
40.994	0	—	—	—	—	47.730	—	Arrival
47.730	1							

Table 15.3 **Percentage of Surveyed Companies Using Simulation in Certain Functional Areas**

Functional Area	Percentage
Production	59%
Corporate planning	53%
Engineering	46%
Finance	41%
Research and development	37%
Marketing	24%
Data processing	16%
Personnel	10%

tional area departments, and corporate planning departments were doing so in 30 percent of the companies. The clear majority of both kinds of departments, as well as OR departments, reported that they perceived the results of their simulation applications to be "good," as opposed to "fair" or "poor."

More recent reports (e.g., see Selected Reference 9) indicate that the use of simulation continues to grow rapidly in the production (manufacturing) area, as well as in the corporate planning and finance areas. Growth in the latter areas has been aided by the recent development of specialized simulation programming languages for financial planning.

Another important recent development has been the increasing use of *computer graphics* to generate animated displays of the movement of entities through the simulated system. For example, in the simulation of manufacturing systems, the moving entities can represent components being manufactured and various materials-handling devices such as automated guided vehicles. The computer graphics provide greater insight into the performance of the system for any given design, and they also add credibility to the results of the simulation study.

There have been numerous applications of simulation in a wide variety of contexts. Some examples are listed here to illustrate the great versatility of this technique:

1. Simulation of the operations at a large airport by an airline company to test changes in company policies and practices (e.g., amounts of maintenance capacity, berthing facilities, spare aircraft, and so on).
2. Simulation of the passage of traffic across a junction with time-sequenced traffic lights to determine the best time sequences.
3. Simulation of a maintenance operation to determine the optimal size of repair crews.
4. Simulation of the flux of uncharged particles through a radiation shield to determine the intensity of the radiation that penetrates the shield.
5. Simulation of steel-making operations to evaluate changes in operating practices and the capacity and configuration of the facilities.
6. Simulation of the U.S. economy to predict the effect of economic policy decisions.
7. Simulation of large-scale military battles to evaluate defensive and offensive weapons systems.
8. Simulation of large-scale distribution and inventory control systems to improve the design of these systems.

9. Simulation of the overall operation of an entire business firm to evaluate broad changes in the policies and operation of the firm and also to provide a business game for training executives.
10. Simulation of a telephone communications system to determine the capacity of the respective components required to provide satisfactory service at the most economic level.
11. Simulation of the operation of a developed river basin to determine the best configuration of dams, power plants, and irrigation works to provide the desired level of flood control and water-resource development.
12. Simulation of the operation of a production line to determine the amount of in-process storage space that should be provided.

15.2 Formulating and Implementing a Simulation Model

Constructing the Model

The first step in a simulation study is to develop a model representing the system to be investigated. This step requires the analyst to become thoroughly familiar with the operating realities of the system and the objectives of the study. Given this requirement, the analyst probably would attempt to reduce the real system to a logical flow diagram. The system is thereby broken down into a set of components linked together by a master flow diagram, where the components themselves may be broken down into subcomponents, and so on. Ultimately the system is decomposed into a set of elements for which operating rules may be given. These operating rules predict the events that will be generated by the corresponding elements, perhaps in terms of probability distributions. After specifying these elements, rules, and logical linkages, the analyst needs to test the model thoroughly piece by piece. This testing can be done partially by performing a gross version of the simulation on a calculator and checking whether each input is received from the appropriate source and whether each output is acceptable to the next submodel. However, the individual components of the model also should be tested alone to verify that their internal performance is reasonably consistent with reality.

It should be emphasized that, like any operations research model, the simulation model need not be a completely realistic representation of the real system. In fact, it appears that most simulation models err on the side of being overly realistic rather than overly idealized. With the former approach, the model easily degenerates into a mass of trivia and meandering details, so that a great deal of programming and computer time is required to obtain a small amount of information. Furthermore, failing to strip away trivial factors to get down to the core of the system may obscure the significance of those results that are obtained.

If the behavior of an element cannot be predicted exactly, given the state of the system, it is better to take random observations from the probability distributions involved than to use averages to simulate the performance of this element. This statement is true even when one is interested only in the average aggregate performance of the system, because combining average performances for the individual elements may result in something far from average for the overall system.

One question that may arise when choosing probability distributions for the model is whether to use frequency distributions of historical data or to seek the theoretical probability distribution that best fits these data. The latter alternative usually is preferable because it avoids reproducing the idiosyncrasies of a certain period in the past.

When constructing the building blocks of a simulation model described and illustrated in the preceding section, one key step is the definition of the *state of the system*. The state must include the relevant information about the current status of the system so that generating the simulated evolution of the system based upon the state provides an accurate representation of the behavior of the real system. The state must also allow measuring and combining quantities that yield meaningful estimates for measures of performance of the system. Frequently, the state must be represented by a *vector* (a set of state variables) rather than a single variable. For complex stochastic systems, there sometimes are alternative reasonable definitions for the state of the system.

Generating Random Numbers

As the examples in Sec. 15.1 demonstrated, implementing a simulation model requires random numbers to obtain random observations from probability distributions. One method for generating such random numbers is to use a physical device such as a spinning disk or an electronic randomizer. Several tables of random numbers have been generated in this way, including one containing 1 million random digits, published by the Rand Corporation. An excerpt from the Rand table is given in Table 15.4.

When performing a simulation on a computer, the needed random numbers normally are generated directly by the computer by using a ''random number generator.'' A **random number generator** is an algorithm that produces sequences of

Table 15.4 **Table of Random Digits**

09656	96657	64842	49222	49506	10145	48455	23505	90430	04180
24712	55799	60857	73479	33581	17360	30406	05842	72044	90764
07202	96341	23699	76171	79126	04512	15426	15980	88898	06358
84575	46820	54083	43918	46989	05379	70682	43081	66171	38942
38144	87037	46626	70529	27918	34191	98668	33482	43998	75733
48048	56349	01986	29814	69800	91609	65374	22928	09704	59343
41936	58566	31276	19952	01352	18834	99596	09302	20087	19063
73391	94006	03822	81845	76158	41352	40596	14325	27020	17546
57580	08954	73554	28698	29022	11568	35668	59906	39557	27217
92646	41113	91411	56215	69302	86419	61224	41936	56939	27816
07118	12707	35622	81485	73354	49800	60805	05648	28898	60933
57842	57831	24130	75408	83784	64307	91620	40810	06539	70387
65078	44981	81009	33697	98324	46928	34198	96032	98426	77488
04294	96120	67629	55265	26248	40602	25566	12520	89785	93932
48381	06807	43775	09708	73199	53406	02910	83292	59249	18597
00459	62045	19249	67095	22752	24636	16965	91836	00582	46721
38824	81681	33323	64086	55970	04849	24819	20749	51711	86173
91465	22232	02907	01050	07121	53536	71070	26916	47620	01619
50874	00807	77751	73952	03073	69063	16894	85570	81746	07568
26644	75871	15618	50310	72610	66205	82640	86205	73453	90232

Source: Reproduced with permission from The Rand Corporation: *A Million Random Digits with 100,000 Normal Deviates*. Copyright, The Free Press, Glencoe, Ill., 1955, top of p. 182.

numbers that follow a specified probability distribution and possess the appearance of randomness. The reference to "sequence of numbers" means that the algorithm produces many random numbers in a serial manner. Although an individual user may need only relatively few of the numbers, it is generally required that the algorithm be capable of producing many numbers. "Probability distribution" implies that a probability statement can be associated with the occurrence of each number produced by the algorithm. The probability distribution is usually taken to be the *uniform distribution* between 0 and 1, in which case the numbers generated by the algorithm may be called *uniform random numbers* or simply random numbers.

Strictly speaking, the numbers generated by the computer should not be called random numbers because they are predictable and reproducible (which sometimes is advantageous), given the random number generator being used. Therefore, they are sometimes given the name **pseudo-random numbers**. However, the important point is that they satisfactorily play the role of random numbers in the simulation if the method used to generate them is valid.

Various relatively sophisticated statistical procedures have been proposed for testing whether a generated sequence of numbers has an acceptable appearance of randomness.[1] Basically the requirements are that each successive number in the sequence must have an equal probability of taking on any one of the possible values, and it must be statistically independent of the other numbers in the sequence. In other words, the numbers need to be random observations from a *uniform distribution*.

There are a number of *random number generators* available, of which the most popular are the *congruential methods* (additive, multiplicative, and mixed). The *mixed congruential method* includes features of the other two, so we shall discuss it first.

The **mixed congruential method** generates a *sequence* of random numbers by always calculating the next random number from the last one obtained, given an initial random number x_0 (called the **seed**), which may be obtained from some published source such as the Rand table. In particular, it calculates the $(n + 1)$st random number, x_{n+1}, from the nth random number, x_n, by using the recurrence relation

$$x_{n+1} \equiv (ax_n + c)(\text{modulo } m),$$

where a, c, and m are positive integers ($a < m$, $c < m$). This mathematical notation signifies that x_{n+1} is the *remainder* when $(ax_n + c)$ is divided by m. Thus the *possible* values of x_{n+1} are $0, 1, \ldots, m - 1$, so that m represents the desired number of *different* values that could be generated for the random numbers.

To illustrate, suppose that $m = 8$, $a = 5$, $c = 7$, and $x_0 = 4$. The resulting sequence of random numbers is calculated in Table 15.5. (The sequence cannot be continued further because it would just begin repeating the numbers in the same order.) Note that this sequence includes each of the eight possible numbers exactly once. This property is desirable, but it does not occur with some choices of a and c. (Try $a = 4$, $c = 7$, $x_0 = 3$.) Fortunately, there are rules available for choosing values of a and c that will guarantee this property. (There are no restrictions on the seed, x_0, because it affects only where the sequence begins and not the progression of numbers.)

For a binary computer with a word size of b bits, the usual choice for m is $m = 2^b$; this is the total number of nonnegative integers that can be expressed within the capacity of the word size. (Any undesired integers that arise in the sequence of

[1] See Selected References 6 and 14 for further information about these tests and the generation of random numbers.

Table 15.5 **Illustration of Mixed Congruential Method**

n	x_n	$5x_n + 7$	$(5x_n + 7)/8$	x_{n+1}
0	4	27	$3 + \frac{3}{8}$	3
1	3	22	$2 + \frac{6}{8}$	6
2	6	37	$4 + \frac{5}{8}$	5
3	5	32	$4 + \frac{0}{8}$	0
4	0	7	$0 + \frac{7}{8}$	7
5	7	42	$5 + \frac{2}{8}$	2
6	2	17	$2 + \frac{1}{8}$	1
7	1	12	$1 + \frac{4}{8}$	4

random numbers are just not used.) With this choice of m, we can ensure that each possible number occurs exactly once before any number is repeated by selecting any of the values $a = 1, 5, 9, 13, \ldots$ and $c = 1, 3, 5, 7, \ldots$. For a decimal computer with a word size of d digits, the usual choice for m is $m = 10^d$, and the same property is ensured by selecting any of the values $a = 1, 21, 41, 61, \ldots$ and $c = 1, 3, 7, 9, 11, 13, 17, 19, \ldots$ (that is, all positive *odd* integers *except* those ending with the digit 5). The specific selection can be made on the basis of the *serial correlation* between successively generated numbers, which differs considerably among these alternatives.[1]

Occasionally, random numbers with only a relatively small number of digits are desired. For example, suppose that only three digits are desired, so that the possible values can be expressed as 000, 001, . . . , 999. In such a case, the usual procedure still is to use $m = 2^b$ or $m = 10^d$, so that an extremely large number of random numbers can be generated before the sequence starts repeating itself. However, except for purposes of calculating the next random number, all but three digits of each random number would be discarded. One convention is to take the *last* three digits (i.e., the three trailing digits).

The **multiplicative congruential method** is just the special case of the mixed congruential method where $c = 0$. The **additive congruential method** also is similar, but it sets $a = 1$ and replaces c by some random number preceding x_n in the sequence, for example, x_{n-1} (so that more than one seed is required to start calculating the sequence).

Among the possible random number generators (choices of a and m) based on the multiplicative congruential method, perhaps the most widely used is the *Learmouth-Lewis generator,*

$$x_{n+1} \equiv 7^5 x_n \text{ (modulo } 2^{31} - 1).$$

This generator has been tested extensively, and the results of the statistical tests indicate that it is very satisfactory. Versions of this generator are used, e.g., in IBM versions of APL, in the International Mathematics and Statistics Library (IMSL) package, and in the random number generator package LLRANDOM. Tables of suitable seeds also are available.

[1] See Coveyou, R. R.: "Serial Correlation in the Generation of Pseudo-Random Numbers," *Journal of the Association of Computing Machinery,* **7**:72–74, 1960.

Given a sequence of random numbers, how can one generate a sequence of random observations from a given probability distribution?

For simple discrete distributions, one answer is quite evident, as demonstrated by Example 1 in Sec. 15.1. Merely allocate the possible values of a random number to the various numbers in the probability distribution in direct proportion to the respective probabilities of those numbers. For example, consider the probability distribution of the outcome of a throw of two dice. It is known that the probability of throwing a 2 is $\frac{1}{36}$ (as is the probability of throwing a 12), the probability of throwing a 3 is $\frac{2}{36}$, and so on. Therefore, $\frac{1}{36}$ of the possible values of a random number should be associated with throwing a 2, $\frac{2}{36}$ of the values with throwing a 3, and so forth. Thus, if two-digit random numbers are being used, 72 of the 100 values would be selected for consideration, so that a random number would be rejected if it took on any one of the other 28 values. Then two of the 72 possible values (say, 00 and 01) would be assigned an association with throwing a 2, four of them (say, 02, 03, 04, and 05) would be assigned to throwing a 3, and so on.

For more complicated distributions, whether discrete or continuous, a generalization of this approach called the **inverse transformation method** can sometimes be used to generate random observations. Letting X be the random variable involved, denote the cumulative distribution function by

$$F(x) = P\{X \leq x\}.$$

Generating each observation then requires the following two steps.

Summary of Inverse Transformation Method

1. Generate a *uniform random number* r between 0 and 1.
2. Set $F(x) = r$ and solve for x, which then is the desired random observation from the probability distribution.

This procedure is illustrated in Fig. 15.1 for the case where $F(x)$ is plotted graphically and the uniform random number r happens to be 0.5269.

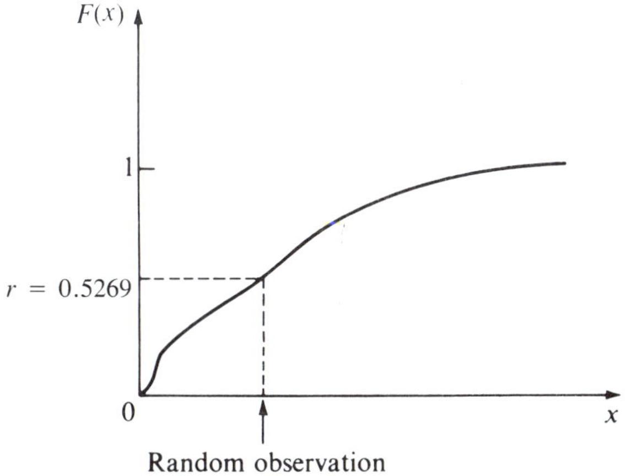

Figure 15.1 Illustration of the *inverse transformation method* for obtaining a random observation from a given probability distribution.

Although the graphical procedure illustrated by Fig. 15.1 is convenient if the simulation is done manually, the computer must revert to some alternative approach. For discrete distributions, a *table look-up approach* can be used by constructing a table that gives the "range" (jump) in the value of $F(x)$ for each value of $X = x$. For certain *continuous* distributions, the $F(x) = r$ equation can be solved *analytically* for x, as will now be illustrated.

Consider the *exponential distribution* (see Sec. 6.4) that has the cumulative distribution function

$$F(x) = 1 - e^{-\alpha x}, \qquad \text{for } x \geq 0,$$

where $1/\alpha$ is the mean of the distribution. Setting $F(x) = r$ thereby yields

$$1 - e^{-\alpha x} = r,$$

so that

$$e^{-\alpha x} = 1 - r.$$

Therefore, taking the natural logarithm of both sides,

$$\ln (e^{-\alpha x}) = \ln (1 - r),$$

so that

$$-\alpha x = \ln (1 - r),$$

which yields

$$x = \frac{\ln (1 - r)}{-\alpha}$$

as the desired random observation from the exponential distribution. (It should be noted that other, more complicated, techniques have also been developed for the exponential distribution[1] that are faster for a computer than calculating a logarithm.)

Note that $(1 - r)$ is itself a *uniform random number*. Therefore, to save a subtraction, it is common in practice simply to use the *original* uniform random number r directly in place of $(1 - r)$.

A natural extension of this procedure for the exponential distribution also can be used to generate a random observation from an *Erlang* (gamma) *distribution* (see Sec. 6.7). The sum of k independent exponential random variables, each with mean $1/k\alpha$, has the Erlang distribution with shape parameter k and mean $1/\alpha$. Therefore, given a sequence of k random numbers between 0 and 1, say, $r_1, r_2, \ldots, r_k$, the desired random observation from the Erlang distribution is

$$x = \sum_{i=1}^{k} \frac{\ln (1 - r_i)}{-k\alpha},$$

which reduces to

$$x = -\frac{1}{k\alpha} \ln \left\{ \prod_{i=1}^{k} (1 - r_i) \right\},$$

where Π denotes multiplication. Once again, the subtractions may be eliminated simply by using the r_i directly in place of the $(1 - r_i)$.

A particularly simple technique for generating a random observation from a *normal distribution* is obtained by applying the central limit theorem. Because a uniform random number has a *uniform distribution* from 0 to 1, it has mean $\frac{1}{2}$ and standard deviation $1/\sqrt{12}$. Therefore, this theorem implies that the sum of n uniform

[1] For example, see Ahrens, J. H., and V. Dieter: "Efficient Table-Free Sampling Methods for Exponential, Cauchy, and Normal Distributions," *Communications of the ACM*, **31**:1330–1337, 1988.

random numbers has approximately a normal distribution with mean $n/2$ and standard deviation $\sqrt{n/12}$. Thus, if $r_1, r_2, \ldots, r_n$ are a sample of uniform random numbers, then

$$x = \frac{\sigma}{\sqrt{n/12}} \sum_{i=1}^{n} r_i + \left(\mu - \frac{n}{2} \frac{\sigma}{\sqrt{n/12}}\right)$$

is a random observation from an approximately normal distribution with mean μ and standard deviation σ. This approximation is an excellent one (except in the tails of the distribution), even with small values of n. Thus values of n from 5 to 10 may be adequate; $n = 12$ also is a convenient value, because it eliminates the square root terms from the preceding expression.

Various *exact* techniques for generating random observations from a normal distribution have also been developed.[1] These exact techniques are sufficiently fast that, in practice, they generally are used instead of the approximate method described above.

A simple method for handling the *chi-square distribution* is to use the fact that it is obtained by summing squares of standardized normal random variables. Thus, if $y_1, y_2, \ldots, y_n$ are a sample of random observations from a normal distribution with mean 0 and standard deviation 1, such as could be obtained (approximately) by the technique just described, then

$$x = \sum_{i=1}^{n} y_i^2$$

is a random observation from a chi-square distribution with n degrees of freedom.

For many continuous distributions, including the normal and chi-square distributions, it is not feasible to apply the *inverse transformation method* because $x = F^{-1}(r)$ cannot be computed (or at least computed efficiently). Therefore, several other types of methods have been developed to generate random observations from such distributions. Frequently, these methods are considerably faster than the inverse transformation method even when the latter method can be used. To provide some notion of the approach for these alternative methods, we now shall illustrate one called the **acceptance-rejection method** on a simple example.

Consider the *triangular distribution* having the probability density function

$$f(x) = \begin{cases} x, & \text{if } 0 \le x \le 1 \\ 1 - (x - 1), & \text{if } 1 \le x \le 2 \\ 0, & \text{otherwise.} \end{cases}$$

The acceptance-rejection method uses the following two steps (perhaps repeatedly) to generate a random observation.

1. Generate a uniform random number r_1 between 0 and 1, and set $x = 2r_1$.
2. *Accept* x with

$$\text{Probability} = \begin{cases} x, & \text{if } 0 \le x \le 1 \\ 1 - (x - 1), & \text{if } 1 \le x \le 2 \end{cases}$$

to be the desired random observation. Otherwise, *reject* x and repeat the two steps.

[1] Ibid.

To randomly generate the event of accepting (or rejecting) x according to this probability, the method implements step 2 as follows:

2. Generate a uniform random number r_2 between 0 and 1.

$$\text{Accept } x \quad \text{if } r_2 \le f(x).$$

$$\text{Reject } x \quad \text{if } r_2 > f(x).$$

If x is rejected, repeat the two steps.

Because $x = 2r_1$ is being accepted with a probability $= f(x)$, the probability distribution of *accepted values* has $f(x)$ as its density function, so accepted values are valid *random observations* from $f(x)$.

We were fortunate in this example that the *largest* value of $f(x)$ for any x was exactly one. If this largest value were $L \neq 1$ instead, then r_2 would be multiplied by L in step 2. With this adjustment, the method is easily extended to other probability density functions over a finite interval, and similar concepts can be used over an infinite interval as well.

Preparing a Simulation Program

A number of detailed decisions confront the person who must write the computer program for executing a simulation. Although an extensive discussion of these issues is beyond the scope of this book, we shall mention several major considerations.

The basic purpose of most simulation studies is to compare alternatives. Therefore, the simulation program must be flexible enough to accommodate readily the alternatives that will be considered. Because it often is impossible to predict exactly what interesting alternatives will be uncovered during the course of the study, it is essential that flexibility and provision for rapid, simple modifications be built into the program.

Most of the instructions in a simulation program are logical operations, whereas the relatively little actual arithmetic work required is usually of a very simple type. This consideration should be reflected in the choice of computer equipment and programming language to be used.

The considerations just mentioned actually provided part of the motivation for the development of general *simulation programming languages*. For example, GPSS and SIMSCRIPT are two such languages that are widely used. These languages are designed especially to expedite the type of programming (and reprogramming) unique to simulation. Their specific purposes include the following. One objective is to provide a convenient means of describing the elements that commonly appear in simulation models. A second is to expedite changing the design and operating policies of the system being simulated, so that a large number of configurations (including some suggested during the course of the study) can be considered easily. Another service provided by the simulation languages is some type of internal timing and control mechanism, with related commands, to assist in the kind of bookkeeping that is required when executing a simulation run. They also are designed to obtain data and statistics conveniently on the aggregate behavior of the system being simulated. Finally, these languages provide simple operational procedures, such as introducing changes into the simulation model, initializing the state of the model, altering the kind of output data to be generated, and stacking a series of simulation runs.

For all these reasons a simulation program often should be written in one of these simulation languages rather than in a general programming language. The tremendous savings in programming time ordinarily provided by the simulation languages usually compensate for any slight loss in computer running time.

Finally, it should be emphasized that the strategy of the simulation study should be planned carefully before finishing the simulation program. Merely letting the computer compile masses of data in a blind search for attractive alternatives is far from adequate. Simulation basically is a means for conducting an experimental investigation. Therefore, just as with a physical experiment, careful attention should be given to the construction of a theory of formal hypotheses to be tested and to the skillful design of a statistical experiment that will yield valid conclusions. This subject is discussed in Secs. 15.3 and 15.4.

Validating the Model

The typical simulation model consists of a high number of elements, rules, and logical linkages. Therefore, even when the individual components have been carefully tested, numerous small approximations can still cumulate into gross distortions in the output of the overall model. Consequently, after writing and debugging the computer program, it is important to test the *validity* of the model for reasonably predicting the aggregate behavior of the system being simulated.

When some form of the real system has already been in operation, its performance data should be compared with the corresponding output data from the model. Standard statistical tests can sometimes be used to determine whether the differences in the means, variances, and probability distributions generating the two sets of data are statistically significant. The time-dependent behavior of the data might also be compared statistically. If the data are not amenable to statistical analysis, another approach is to ask personnel familiar with the behavior of the real system if they can discriminate between the two sets of data.

If the model is intended to simulate alternative design configurations or operating policies for a proposed system for which no actual data are available, it may be worthwhile to conduct a *field test* to collect some real data to compare with the output of the model. Conducting such a test might involve constructing a small prototype of some version of the proposed system and placing it into operation. Another possibility might be to temporarily alter an existing system to correspond to one of the proposals.

However, field tests frequently are too expensive and time consuming to be used. Without any real data as a standard of comparison, the only way to validate the overall model is to have knowledgeable people carefully check the credibility of output data for a variety of situations. Even when no basis exists for checking the reasonableness of the data for a *single* situation, some conclusions usually can be drawn about how the *relative* performance of the system should change as various parameters are changed. It is especially important to convince the *decision makers* of the credibility of the model, so they will be willing to use it at least to *aid* their decisions. If the model may be used again in the future, careful records of its predictions and of actual results should be kept to continue the validation process.

15.3 Experimental Design for Simulation

Selecting a Statistical Procedure[1]

The underlying statistical theory applicable to *simulated* experimentation is essentially indistinguishable from that for *physical* experimentation. Thus the design of a simulated experiment should be based upon the large body of knowledge comprising the science of statistics.

There are, however, differences between physical and simulated experimentation regarding the emphasis placed on using the various types of statistical procedures. Physical experiments frequently involve testing hypotheses about the value of a population parameter or about the equality of several population means. Simulated experiments typically place more emphasis on *optimization*. It probably is taken for granted that alternative design configurations have different population means for the measure of performance of the system. Instead, the objective of the simulation study often is to find the alternative yielding the largest mean for the measure of performance.[2] Hence *multiple decision tests* and complete or partial *ordering procedures* frequently are appropriate for simulated experiments. Furthermore, *sequential procedures* tend to be useful, both because the evolution of the experiments may be difficult to predict and because a simulated experiment often can be resumed relatively easily.

Another difference between these two types of experiments is the degree to which the experimental conditions can be held constant when comparing alternatives. Only simulated experiments can control the variability in the behavior of the elements of the system during the course of the experiment. By reproducing the same sequence of random numbers for each alternative simulated, it often is possible to reproduce an identical sequence of events. This reproduction sharpens the contrast between alternatives by reducing the residual variation in the differences in the aggregate performance of the system, so that much smaller sample sizes are required to detect statistically significant differences. Therefore, this approach usually is far superior to generating new random numbers for each alternative.

The fact that reproducing the same random numbers does not yield statistically independent results should not be of great concern. The correct procedure for comparing only two alternatives is to pair the results regarding the aggregate performance of the system that were produced by the same events. Because these pairs of results are obtained under the same experimental conditions, the *differences* between them become the relevant sample observations. This sample would be used to test the hypothesis that the mean of these differences is zero and to obtain a confidence interval for this mean. This result would thereby indicate whether there is a statistically significant difference between the means of the measure of performance for the two alternatives. If more than two alternatives need to be compared, the *Bonferroni in-*

[1] This subsection assumes some knowledge of statistical procedures, as reviewed in Chap. 4.

[2] In some cases, however, the objective is just to *describe* the performance of proposed systems or policies for management's evaluation and decision making, so *point estimates* and *confidence intervals* probably would be obtained. Simulated experiments also are occasionally conducted to determine which factors significantly influence the performance of the system (perhaps to guide subsequent experimentation), in which case *analysis of variance* probably would be used.

equality can be used to construct *simultaneous confidence intervals* on the means of the differences for the various pairs of alternatives.[1]

Often it is possible to express the alternatives in terms of the values of one or more continuous design variables. In these cases, there actually is an infinite number of alternatives (although the differences among some of them are minute). Because it would be impossible to simulate all of them, it is necessary to take a selective sample of these alternatives and then estimate the value of the design variables that will maximize some expected measure of performance for the system. There exists considerable literature that gives efficient procedures for experimentally determining the maximum of a mathematical function to within a specified accuracy.[2]

Variance-Reducing Techniques

Because considerable computer time usually is required for simulation runs, it is important to obtain as much and as precise information as possible from the amount of simulation that can be done. Unfortunately, there has been a tendency in practice to apply simulation uncritically without giving adequate thought to the efficiency of the experimental design. This tendency has occurred despite the fact that considerable progress has been made in developing special techniques for increasing the precision (i.e., decreasing the variance) of sample estimators.

These variance-reducing techniques often are called **Monte Carlo techniques** (a term sometimes applied to simulation in general). Because they tend to be rather sophisticated, it is not possible to explore them deeply here. However, we shall attempt to impart the flavor of these techniques and the great increase in precision they sometimes provide by presenting two of them in the following example.

Consider the *exponential distribution* whose parameter has a value of 1. Thus its probability density function is $f(x) = e^{-x}$, as shown in Fig. 15.2, and its cumulative distribution function is $F(x) = 1 - e^{-x}$. It is known that the mean of this distribution is 1. However, suppose that this mean were not known and that we want to estimate this mean by using simulation.

To provide a standard of comparison for the two variance-reducing techniques, we consider first the straightforward simulation approach, sometimes called the *crude Monte Carlo technique*. This approach involves generating some *random observations* from the exponential distribution under consideration and then using the *average* of these observations to estimate the mean. As described in Sec. 15.2, these random observations would be

$$x_i = -\ln(1 - r_i), \qquad \text{for } i = 1, 2, \ldots, n,$$

where $r_1, r_2, \ldots, r_n$ are uniform random numbers between 0 and 1. If we use a portion of Table 15.4 to obtain 10 such uniform random numbers, the resulting random observations are shown in Table 15.6. (These same random numbers also are used to illustrate the variance-reducing techniques to sharpen the comparison.)

Notice that the sample average in Table 15.6 is 0.779, as opposed to the true mean of 1.000. However, because the standard deviation of the sample average hap-

[1] See Bowker, Albert H., and Gerald J. Lieberman: *Engineering Statistics*, 2d ed., pp. 304–308, Prentice-Hall, Englewood Cliffs, N.J., 1972.

[2] A survey of the procedures available for this problem is given by Wilde, Douglass J.: *Optimum Seeking Methods*, Prentice-Hall, Englewood Cliffs, N.J., 1964.

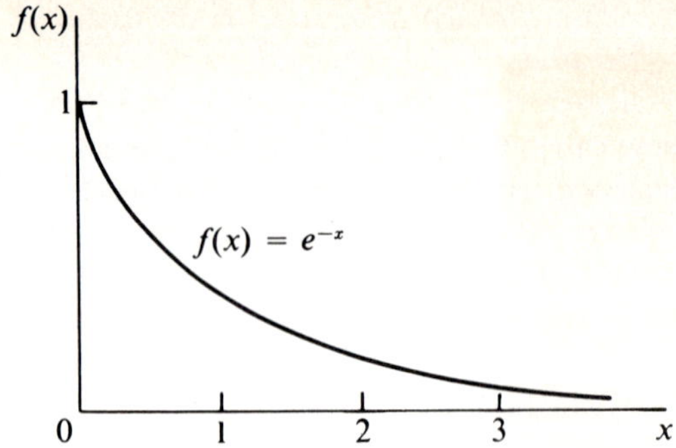

Figure 15.2 Probability density function for example.

pens to be $1/\sqrt{n}$, or $1/\sqrt{10}$ in this case (as could be estimated from the sample), an error of this amount or larger would occur approximately one-half of the time. Furthermore, because the standard deviation of a sample average is always inversely proportional to $\sqrt{n}$, this sample size would need to be quadrupled to reduce this standard deviation by one-half. These somewhat disheartening facts suggest the need for other techniques that would obtain such estimates more precisely and more efficiently.

A relatively simple Monte Carlo technique for obtaining better estimates is **stratified sampling**. There are two shortcomings of the crude Monte Carlo approach that are rectified by stratified sampling. First, by the very nature of randomness, a random sample may not provide a particularly uniform cross section of the distribution. For example, the random sample given in Table 15.6 has no observations between 0.014 and 0.328, even though the probability that a random observation will fall inside this interval is greater than $\frac{1}{4}$. Second, certain portions of a distribution may be more critical than others for obtaining a precise estimate, but random sampling gives

Table 15.6 **Example for Crude Monte Carlo Technique**

i	Random Number* r_i	Random Observation $x_i = -\ln(1 - r_i)$
1	0.495	0.684
2	0.335	0.408
3	0.791	1.568
4	0.469	0.633
5	0.279	0.328
6	0.698	1.199
7	0.013	0.014
8	0.761	1.433
9	0.290	0.343
10	0.693	1.183
	Total =	7.793
	Estimate of mean =	0.779

* Actually, 0.0005 was added to the indicated value for each of the r_i, so that the range of their possible values would be from 0.0005 to 0.9995 rather than from 0.000 to 0.999.

no special priority to obtaining observations from these portions. For example, the tail of an exponential distribution is especially critical in determining its mean. However, the rando[illegible] than 1.568, even though the[illegible] explanation is the basic one [illegible] true mean. Stratified sampli[illegible] on into portions called *stra*[illegible] with disproportionately hea[illegible]

To illustra[illegible] strata in the manner shown i[illegible] observations approximately fr[illegible] . To ensure that the random [illegible] at portion of the distribution, [illegible] he indicated range for $F(x)$, [illegible] er of observations to be ge[illegible] n.[1] The last column then sho[illegible] the *ratio* of the *sampling pro*[illegible] the stratum) to the *distributio*[illegible] alling inside the stratum). The[illegible] tance of the respective strata i[illegible]

Given the f[illegible] Table 15.7, the same random [illegible] n in the fifth column in Table 1[illegible] hted average of these observati[illegible] distribution have been sample[illegible] e, divide the observations from [illegible] to give proportionate weighti[illegible] vn in the last column of Table 1[illegible] s the desired estimate of the me[illegible]

The second [illegible] e method of **complementary r**[illegible] hat the "luck of the draw" on t[illegible] e average of

Table 15.7 **Formulati**[illegible]

Stratum	Portio[illegible] Distrib[illegible]
1	$0 \leq F(x)$[illegible]
2	$0.64 \leq F(x)$[illegible]
3	$0.96 \leq F(x)$[illegible]

[1] These sample sizes are roughly based on a recommended guideline that they be proportional to the *product* of the *probability* of a random observation falling inside the corresponding stratum *times* the *standard deviation* within this stratum.

[2] This method is a special case of the method of *antithetic variates,* which attempts to generate *pairs* of random observations having a high *negative* correlation, so that the combined average will tend to be closer to the mean.

Table 15.8 **Example for Stratified Sampling**

Stratum	i	Random Number r_i	Stratum Random No. r'_i	Stratum Random Observation $x'_i = -\ln(1 - r'_i)$	Sampling Weight w_i	x'_i/w_i
1	1	0.495	0.317	0.381	$\frac{5}{8}$	0.610
	2	0.335	0.215	0.242	$\frac{5}{8}$	0.387
	3	0.791	0.507	0.707	$\frac{5}{8}$	1.131
	4	0.469	0.300	0.357	$\frac{5}{8}$	0.571
2	5	0.279	0.729	1.306	$\frac{5}{4}$	1.045
	6	0.698	0.864	1.995	$\frac{5}{4}$	1.596
	7	0.013	0.644	1.033	$\frac{5}{4}$	0.826
	8	0.761	0.884	2.154	$\frac{5}{4}$	1.723
3	9	0.290	0.9716	3.561	5	0.712
	10	0.693	0.9877	4.398	5	0.880
					Total =	9.481
					Estimate of mean =	0.948

the resulting random observations to be substantially on one side of the true mean, whereas the *complements* of those uniform random numbers (which are themselves uniform random numbers) would have tended to yield a nearly opposite result. (For example, the uniform random numbers in Table 15.6 average less than 0.5, and none are as large as 0.8, which led to an estimate substantially below the true mean.) Therefore, using *both* the original uniform random numbers *and* their complements to generate random observations and then calculating the *combined* sample average should provide a more precise estimator of the mean. This approach is illustrated in Table 15.9,[1] where the first three columns come from Table 15.6 and the last two columns use the complementary uniform random numbers, which results in a combined sample average of 0.920.

This example has suggested that the variance-reducing techniques provide a much more precise estimator of the mean than does straightforward simulation. These results definitely were not a coincidence, as a derivation of the variance of the estimators would show. In comparison with straightforward simulation, these techniques (including several more complicated ones not presented here) do indeed provide a much more precise estimator with the same amount of computer time, or they provide as precise an estimator with much less computer time. Despite the fact that additional analysis may be required to incorporate one or more of these techniques into the simulation study, the rewards should not be forgone readily.

Although this example was a particularly simple one, it is often possible, though more difficult, to apply these techniques to much more complex problems. For example, suppose that the objective of the simulation study is to estimate the mean waiting time of customers in a queueing system (such as those described in Sec. 7.1). Because both the probability distribution of interarrival times and the probability distribution of service times are involved, and because consecutive waiting times are not statistically independent, this problem may appear to be beyond the capabilities of the variance-reducing techniques. However, as has been described in detail else-

[1] It should be noted that 20 calculations of a logarithm were required in this case, in contrast to the 10 that were required by each of the preceding techniques.

Table 15.9 **Example for Method of Complementary Random Numbers**

i	Random Number r_i	Random Observation $x_i = -\ln(1 - r_i)$	Complementary Random Number $r'_i = 1 - r_i$	Random Observation $x'_i = -\ln(1 - r'_i)$
1	0.495	0.684	0.505	0.702
2	0.335	0.408	0.665	1.092
3	0.791	1.568	0.209	0.234
4	0.469	0.633	0.531	0.756
5	0.279	0.328	0.721	1.275
6	0.698	1.199	0.302	0.359
7	0.013	0.014	0.987	4.305
8	0.761	1.433	0.239	0.272
9	0.290	0.343	0.710	1.236
10	0.693	1.183	0.307	0.366
		Total: 7.793		10.597

Estimate of mean $= \frac{1}{2}(0.779 + 1.060) = 0.920$

where,[1] these techniques and others can indeed be applied to this type of problem very advantageously. For example, the method of *complementary random numbers* can be applied simply by repeating the original simulation run, substituting the complements of the original uniform random numbers to generate the corresponding random observations.

Tactical Problems

There are several special *tactical* issues that arise in connection with gathering the data from simulated experiments. We shall briefly describe these here and then subsequently elaborate on certain ways of dealing with them.

Many simulation studies are concerned with investigating systems that operate continually in a steady-state condition. Unfortunately, a simulation model cannot be operated this way; it must be started and stopped. Because of the artificiality introduced by the abrupt beginning of operation, the performance of the simulated system does not become representative of the corresponding real-world system until it too has essentially reached a steady-state condition (i.e., until the probability distribution of the state of the simulated system has essentially reached a limiting *equilibrium* distribution). Thus one tactical problem is how to obtain data that are relevant for predicting the *steady-state* behavior of the real system.

The traditional way of dealing with this problem is to run the simulation model for some time without collecting data until it is believed that the simulated system has essentially reached a steady-state condition. Unfortunately, it is difficult to estimate just how long this *stabilization period* needs to be. Furthermore, available analytical results suggest that a surprisingly long period is required, so that a great deal of unproductive computer time must be expended. Section 15.4 presents a statistical approach that eliminates these difficulties.

A related tactical issue is the selection of the *starting conditions* for the simulated system. The traditional recommendation is that the simulated system should be started

[1] Ehrenfeld, S., and S. Ben-Tuvia: "The Efficiency of Statistical Simulation Procedures," *Technometrics*, **4**(2):257–275, 1962. Also see Selected References 7, 8, and 14 for a general discussion of the theory and application of various variance-reducing techniques.

in a state as representative of steady-state conditions as possible to minimize the required length of the stabilization period. However, the underlying objective of the simulated experiment is to *estimate* these conditions, so little advance information may be available to guide the selection in this way. The procedure in Sec. 15.4 also eliminates this difficulty.

Most statistical sampling procedures assume that the experimental output data are in the form of a collection of distinct and statistically independent random observations from some underlying probability distribution. By contrast, because of the nature of the problems for which simulation is used, the observations from a simulated experiment are likely to be highly correlated. For example, there is a high correlation between the waiting times of consecutive customers in a queueing system. Furthermore, many measures of performance are such that the simulated experiment yields this measure continuously as a function of time rather than as a sequence of separate observations. Thus another tactical problem is how to collect the data so as to circumvent these difficulties.

One traditional method is to execute a series of completely separate and independent simulation runs of equal length and to use the average measure of performance for each run (excluding the initial stabilization period) as an individual observation. The main disadvantage is that each run requires an initial stabilization period for approaching a steady-state condition, so that much of the simulation time is unproductive. The second traditional method eliminates this disadvantage by making the runs consecutively, using the ending condition of one run as the steady-state starting condition for the next run. In other words, one continuous overall simulation run (except for the one initial stabilization period) is divided for bookkeeping purposes into a series of equal portions (runs). The average measure of performance for each portion is then treated as an individual observation. The disadvantage of this method is that it does not eliminate the correlation between observations entirely, even though it may reduce it considerably by making the portions sufficiently long.

Once again, these difficulties are eliminated by the statistical approach described in the next section.

15.4 The Regenerative Method of Statistical Analysis

We have just described several difficult *tactical* problems in gathering data from simulated experiments and the shortcomings of traditional statistical procedures in dealing with these problems. We now present an innovative statistical approach that is especially designed to eliminate these problems.

The basic concept underlying this approach is that for many systems a simulation run can be divided into a series of **cycles** such that the evolution of the system in a cycle is a probabilistic replica of the evolution in any other cycle. Thus, if we calculate an appropriate measure of the length of the cycle along with some *statistic* to summarize the behavior of interest within each cycle, these statistics for the respective cycles constitute a series of independent and identically distributed observations that can be analyzed by standard statistical procedures. Because the system keeps going through these independent and identically distributed cycles whether or not it is in a steady-state condition, these observations are directly applicable from the outset for estimating the steady-state behavior of the system.

For cycles to possess these properties, they must each *begin* at the same **regeneration point**, i.e., at the point where the system probabilistically restarts, and can proceed without any knowledge of its past history. The system can be viewed as *regenerating* itself at this point in the sense that the probabilistic structure of the future behavior of the system depends upon being at this point and not on anything that happened previously. (This property is the *Markovian property* described in Sec. 5.3 for Markov chains.) A cycle *ends* when the system again reaches the regeneration point (when the next cycle begins). Thus the **length of a cycle** is just the elapsed time between consecutive occurrences of the regeneration point, which is a random variable that depends upon the evolution of the system.

When *next-event incrementing* is used, a typical regeneration point is a point at which an event has just occurred but no future events have yet been scheduled. Thus nothing needs to be known about the history of previous schedulings, and the simulation can start from scratch in scheduling future events. When *fixed-time incrementing* is used, a regeneration point is a point at which the probabilities of possible events occurring during the next unit of time do not depend upon when any past events occurred, but only on the current state of the system.

Not every system possesses regeneration points, so this **regenerative method** of collecting data cannot always be used. Furthermore, even when there are regeneration points, the one chosen to define the beginning and ending points of the cycles must recur frequently enough so that a substantial number of cycles will be obtained with a reasonable amount of computer time.[1] Thus some care must be taken to choose a suitable regeneration point.

Perhaps the most important application of the regenerative method to date has been to the simulation of *queueing systems*, including *queueing networks* (see Sec. 6.9) such as the ones that arise in computer modeling.[2]

Example

Suppose that information needs to be obtained about the steady-state behavior of a system that can be formulated as a *single-server queueing system* (see Sec. 6.2). However, both the interarrival and service times have a *discrete uniform distribution* with a probability of $\frac{1}{10}$ of the values of 6, 8, . . . , 24 and the values of 1, 3, . . . , 19, respectively. Because analytical results are not available, simulation with *next-event incrementing* is to be used to obtain the desired results.

Except for the distributions involved, the general approach is the same as described in Sec. 15.1 for Example 2. In particular, the building blocks of the simulation model are the same as specified there, including defining the *state of the system* as the number of customers in the system. Suppose that *one-digit* random numbers are used to generate the random observations from the distributions, as shown in Table 15.10. Beginning the simulation run with *zero* customers in the system then yields

[1] The theoretical requirements for the method are that the expected cycle length be *finite* and that the number of cycles would go to infinity if the system continued operating indefinitely.

[2] See, for example, Iglehart, Donald L., and Gerald S. Shedler: *Regenerative Simulation of Passage Times in Networks of Queues*, Lecture Notes in Control and Information Sciences, Vol. 4, Springer-Verlag, New York, 1980. For another exposition that emphasizes applications to computer system modeling, see Shedler, G. S.: *Regeneration and Networks of Queues*, Springer-Verlag, New York, 1987.

Table 15.10 **Correspondence between Random Numbers and Random Observations for Queueing System Example**

Random Number	Interarrival Time	Service Time
0	6	1
1	8	3
⋮	⋮	⋮
9	24	19

the results summarized in Table 15.11 and Fig. 15.3, where the random numbers are obtained sequentially as needed from the tenth row of Table 15.4.[1]

For this system, one *regeneration point* is where an *arrival* occurs with *no* previous customers left. At this point, the process probabilistically restarts, so the probabilistic structure of when future arrivals and service completions will occur is completely independent of any previous history. The only relevant information is that the system has just entered the special state of having had *zero* customers *and* having the time until the next arrival reach *zero*. The simulation run would not previously have scheduled any future events but would now generate *both* the next interarrival time and the service time for the customer that just arrived.

The only other regeneration points for this system are where an arrival and a service completion occur simultaneously, with a prespecified number of customers in the system. However, the regeneration point described in the preceding paragraph occurs much more frequently and thus is a better choice for defining a cycle. With this selection, the first five complete cycles of the simulation run are those shown in Fig. 15.3. (In most cases, you should have a considerably larger number of cycles in the entire simulation run in order to have sufficient precision in the statistical analysis.)

Various types of information about the steady-state behavior of the system can be obtained from this simulation run, including *point estimates* and *confidence intervals* for the expected number of customers in the system, the expected waiting time, and so on. In each case, it is necessary to use only the corresponding statistics from the respective cycles and the lengths of the cycles. We shall first present the general statistical expressions for the regenerative method and then apply them to this example.

Statistical Formulas

Formally speaking, the statistical problem for the regenerative method is to obtain estimates of the expected value of some random variable X of interest. This estimate is to be obtained by calculating a statistic Y for each cycle and an appropriate measure Z of the *size* of the cycle such that

$$E(X) = \frac{E(Y)}{E(Z)}.$$

[1] When both an interarrival time and a service time need to be generated at the same time, the interarrival time is obtained first.

Table 15.11 **Simulation Run for Queueing System Example**

Time	Number of Customers	Random Number	Next Arrival	Next Service Completion
0	0	9	24	—
24	1	2, 6	34	37
34	2	4	48	37
37	1	6	48	50
48	2	4	62	50
50	1	1	62	53
53	0	—	62	—
62	1	1, 1	70	65
65	0	—	70	—
70	1	3, 9	82	89
82	2	1	90	89
89	1	4	90	98
90	2	1	98	98
98	2	1, 5	106	109
106	3	6	124	109
109	2	2	124	114
114	1	1	124	117
117	0	—	124	—
124	1	5, 6	140	137
137	0	—	140	—
140	1	9, 3	164	147
147	0	—	164	—
164	1			

(The regenerative property ensures that such a *ratio formula* holds for many steady-state random variables X.) Thus, if n complete cycles are generated during the simulation run, the data gathered are $Y_1, Y_2, \ldots, Y_n$ and $Z_1, Z_2, \ldots, Z_n$ for the respective cycles.

Letting $\overline{Y}$ and $\overline{Z}$, respectively, denote the sample averages for these two sets of data, the corresponding *point estimate* of $E(X)$ would be obtained from the formula

$$\text{Est}\ \{E(X)\} = \frac{\overline{Y}}{\overline{Z}}.$$

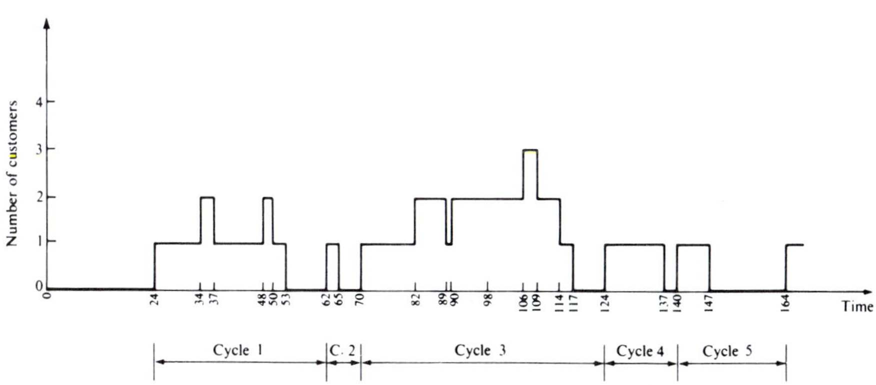

Figure 15.3 Outcome of the simulation run for the queueing system example.

To obtain a *confidence interval* for $E(X)$, we must first calculate several quantities from the data. These quantities include the *sample variances*

$$s_{11}^2 = \frac{1}{n-1}\sum_{i=1}^{n}(Y_i - \bar{Y})^2 = \frac{1}{n-1}\sum_{i=1}^{n}Y_i^2 - \frac{1}{n(n-1)}\left(\sum_{i=1}^{n}Y_i\right)^2,$$

$$s_{22}^2 = \frac{1}{n-1}\sum_{i=1}^{n}(Z_i - \bar{Z})^2 = \frac{1}{n-1}\sum_{i=1}^{n}Z_i^2 - \frac{1}{n(n-1)}\left(\sum_{i=1}^{n}Z_i\right)^2,$$

and the combined *sample covariance*

$$s_{12}^2 = \frac{1}{n-1}\sum_{i=1}^{n}(Y_i - \bar{Y})(Z_i - \bar{Z})$$

$$= \frac{1}{n-1}\sum_{i=1}^{n}Y_iZ_i - \frac{1}{n(n-1)}\left(\sum_{i=1}^{n}Y_i\right)\left(\sum_{i=1}^{n}Z_i\right).$$

Also let

$$s^2 = s_{11}^2 - 2\left(\frac{\bar{Y}}{\bar{Z}}\right)s_{12}^2 + \left(\frac{\bar{Y}}{\bar{Z}}\right)^2 s_{22}^2.$$

Finally, let α be the constant such that $(1 - 2\alpha)$ is the desired *confidence coefficient* for the confidence interval, and look up K_α in Table A5.1 (see Appendix 5) for the normal distribution. If n is not too small, an *asymptotic confidence interval* for $E(X)$ is then given by

$$\frac{\bar{Y}}{\bar{Z}} - \frac{K_\alpha s}{\bar{Z}\sqrt{n}} \leq E(X) \leq \frac{\bar{Y}}{\bar{Z}} + \frac{K_\alpha s}{\bar{Z}\sqrt{n}};$$

that is, the probability is approximately $(1 - 2\alpha)$ that the endpoints of an interval generated in this way will surround the actual value of $E(X)$.

Application of the Statistical Formulas to the Example

Consider first how to estimate the *expected waiting time* for a customer *before* beginning service (denoted by W_q in Chap. 6). Thus the random variable X now would represent a customer's waiting time excluding service, so that

$$W_q = E(X).$$

The corresponding information gathered during the simulation run is the *actual* waiting time (excluding service) incurred by the respective customers. Therefore, for each cycle, the summary statistic Y would be the *sum of the waiting times*, and the size of the cycle Z would be the *number of customers*, so that

$$W_q = \frac{E(Y)}{E(Z)}.$$

For cycle 1, a total of three customers are processed, so $Z_1 = 3$. The first customer incurs no waiting before beginning service, the second waits 3 units of time (from 34 to 37), and the third waits 2 units of time (from 48 to 50), so $Y_1 = 5$. We proceed similarly for the other cycles. The data for the problem are

$$\begin{aligned} Y_1 &= 5, & Z_1 &= 3 \\ Y_2 &= 0, & Z_2 &= 1 \\ Y_3 &= 34, & Z_3 &= 5 \\ Y_4 &= 0, & Z_4 &= 1 \\ Y_5 &= 0, & Z_5 &= 1 \\ \overline{Y} &= 7.8, & \overline{Z} &= 2.2. \end{aligned}$$

Therefore, the *point estimate* of W_q is

$$\text{Est}\,\{W_q\} = \frac{\overline{Y}}{\overline{Z}} = \frac{7.8}{2.2} = 3\,\frac{6}{11}.$$

To obtain a 95 percent *confidence interval* for W_q, the preceding formulas are first used to calculate

$$s_{11}^2 = 219.20, \qquad s_{22}^2 = 3.20, \qquad s_{12}^2 = 24.80, \qquad s = 9.14.$$

Because $(1 - 2\alpha) = 0.95$, then $\alpha = 0.025$, so that $K_\alpha = 1.96$ from Table A5.1. The resulting confidence interval is

$$-0.09 < W_q \leq 7.19;$$

that is,

$$W_q \leq 7.19.$$

The reason that this confidence interval is so wide (even including impossible negative values) is that the number of sample observations (cycles), $n = 5$, is so small. Note in the general formula that the width of the confidence interval is *inversely proportional* to the *square root* of n, so that, e.g., quadrupling n reduces the width by half (assuming no change in s or $\overline{Z}$). Given preliminary values of s and $\overline{Z}$ from a short preliminary simulation run (such as the run in Table 15.11), this relationship makes it possible to estimate in advance the width of the confidence interval that would result from any given choice of n for the full simulation run. The final choice of n can then be made based on the trade-off between computer time and the precision of the statistical analysis.

Now suppose that this simulation run is to be used to estimate P_0, the probability of having *zero* customers in the system. [The theoretical value is known to be $P_0 = 1 - \lambda/\mu = 1 - (\frac{1}{15})/(\frac{1}{10}) = \frac{1}{3}$.] The corresponding information obtained during the simulation run is the fraction of time during which the system is empty. Therefore, the summary statistic Y for each cycle would be the *total time* during which *no* customers are present, and the size Z would be the *length* of the cycle, so that

$$P_0 = \frac{E(Y)}{E(Z)}.$$

The length of cycle 1 is 38 (from 24 to 62), so that $Z_1 = 38$. During this time, the system is empty from 53 to 62, so that $Y_1 = 9$. Proceeding in this manner for the other cycles, the following data are obtained for the problem:

$$\begin{aligned} Y_1 &= 9, & Z_1 &= 38 \\ Y_2 &= 5, & Z_2 &= 8 \\ Y_3 &= 7, & Z_3 &= 54 \\ Y_4 &= 3, & Z_4 &= 16 \\ Y_5 &= 17, & Z_5 &= 24 \\ \bar{Y} &= 8.2, & \bar{Z} &= 28. \end{aligned}$$

Thus the *point estimate* of P_0 is

$$\text{Est}\,\{P_0\} = \frac{8.2}{28} = 0.293.$$

By calculating

$$s_{11}^2 = 29.20, \qquad s_{22}^2 = 334, \qquad s_{12}^2 = 17, \qquad s = 6.92,$$

a 95 percent *confidence interval* for P_0 is found to be

$$0.076 \le P_0 \le 0.510.$$

(The wide range of this interval indicates that a much longer simulation run would be needed to obtain a relatively precise estimate of P_0.)

If we redefine Y appropriately, the same approach also can be used to estimate other probabilities involving the number of customers in the system. However, because this number never exceeded 3 during this simulation run, a much longer run will be needed if the probability involves larger numbers.

The other basic queueing theory expected values defined in Sec. 6.2 (W, L_q, L) can be estimated from the estimate of W_q by using the relationships among these four expected values given near the end of Sec. 6.2. However, they can also be estimated directly from the results of the simulation run. For example, because the *expected number of customers waiting to be served* is

$$L_q = \sum_{n=2}^{\infty} (n - 1)P_n,$$

it can be estimated by defining

$$Y = \sum_{n=2}^{\infty} (n - 1)T_n,$$

where T_n is the *total time* that exactly n customers are in the system during the cycle. (This definition of Y actually is *equivalent* to the definition used when estimating W_q.) In this case, Z would be defined as it would be when estimating any P_n, namely, the *length* of the cycle. The resulting *point estimate* of L_q then turns out to be simply the *point estimate* of W_q *multiplied by* the actual *average arrival rate* for the complete cycles observed.

It is also possible to estimate *higher moments* of these probability distributions by redefining Y accordingly. For example, the *second moment* about the origin of the *number of customers waiting to be served* (N_q),

$$E(N_q^2) = \sum_{n=2}^{\infty} (n - 1)^2 P_n,$$

can be estimated by redefining

$$Y = \sum_{n=2}^{\infty} (n - 1)^2 T_n.$$

This point estimate, along with the point estimate of L_q (the first moment of N_q) just described, can then be used to estimate the *variance of* N_q. Specifically, because of the general relationship between variance and moments, this variance is

$$\text{Var}(N_q) = E(N_q^2) - L_q^2.$$

Therefore, its point estimate is obtained by substituting in the point estimates of the quantities on the right-hand side of this relationship.

Finally, we should mention that it was unnecessary to generate the first *inter-arrival* time (24) for the simulation run summarized in Table 15.11 and Fig. 15.3 because this time played no role in the statistical analysis. It is more efficient with the regenerative method just to start the run at the regeneration point.

Selected References 5 and 14 (Chap. 6) provide considerably more information about the regenerative method, including how it can be applied to more complicated kinds of problems than those considered here.

15.5 Conclusions

Simulation is a widely used tool for estimating the performance of complex stochastic systems if contemplated designs or operating policies were to be used.

We have focused in this chapter on the use of simulation for predicting the *steady-state* behavior of systems whose states change only at discrete points in time. However, by having a series of runs begin with the prescribed *starting conditions*, we can also use simulation to describe the *transient* behavior of a proposed system. Furthermore, if we use differential equations, simulation can be applied to systems whose states change *continuously* with time.

Simulation is indeed a very versatile tool. However, it is by no means a panacea. Simulation is inherently an imprecise technique. It provides only *statistical estimates* rather than exact results, and it *compares alternatives* rather than generating an optimal one. Furthermore, simulation is a *slow and costly* way to study a problem. It usually requires a large amount of time and expense for analysis and programming, in addition to considerable computer running time. Simulation models tend to become unwieldy, so that the number of cases that can be run and the accuracy of the results obtained often turn out to be very inadequate. Finally, simulation yields only *numerical data* about the performance of the system, so that it provides no additional insight into the cause-and-effect relationships within the system except for the clues that can be gleaned from these numbers (and from the analysis required to construct the simulation model). Therefore, it is very expensive to conduct a sensitivity analysis of the parameter values assumed by the model. The only possible way would be to conduct new series of simulation runs with different parameter values, which would tend to provide relatively little information at a relatively high cost.

Simulation provides a way of *experimenting* with proposed systems or policies without actually implementing them. Sound statistical theory should be used in designing these experiments. Surprisingly long simulation runs often are needed to obtain *statistically significant* results. However, *variance-reducing techniques* can be very helpful in reducing the length of the runs needed.

Several *tactical* problems arise when we apply traditional statistical estimation procedures to simulated experiments. These problems include prescribing appropriate *starting conditions*, determining when a *steady-state condition* has essentially been reached, and dealing with *statistically dependent* observations. These problems can be eliminated by using the *regenerative method* of statistical analysis. However, there are some restrictions on when this method can be applied.

Simulation unquestionably has an important place in the theory and practice of operations research. It is an invaluable tool for use on those problems where analytical techniques are inadequate.

SELECTED REFERENCES

1. Aburdene, Maurice F.: *Computer Simulation of Dynamic Systems,* Wm. C. Brown, Dubuque, Iowa, 1988.
2. Banks, Jerry, and John S. Carson, II: *Discrete-Event System Simulation,* Prentice-Hall, Englewood Cliffs, N.J., 1984.
3. Bratley, Paul, Bennett L. Fox, and Linus Schrage: *A Guide to Simulation,* Springer-Verlag, New York, and Secaucus, N.J., 1983.
4. Christy, D. P., and H. J. Watson: "The Application of Simulation: A Survey of Industry Practice," *Interfaces,* **13**(5):47–52, October 1983.
5. Crane, M. A., and A. J. Lemoine: *An Introduction to the Regenerative Method for Simulation Analysis*, Springer-Verlag, Berlin, 1977.
6. Fishman, George S.: *Principles of Discrete Event Simulation*, Wiley, New York, 1978.
7. Hammersley, J. M., and D. C. Handscombe: *Monte Carlo Methods*, Methuen, London, 1964.
8. Kleijnen, J. P. C.: *Statistical Techniques in Simulation*, Marcel Dekker, New York, *Part I,* 1974; *Part II*, 1975.
9. Law, Averill M.: "Introduction to Simulation: A Powerful Tool for Complex Manufacturing Systems," *Industrial Engineering*, **18**(5):46–63, May 1986.
10. Law, Averill M., and W. David Kelton: *Simulation Modeling and Analysis,* McGraw-Hill, New York, 1982.
11. MacDougall, M. H.: *Simulating Computer Systems: Techniques and Tools,* MIT Press, Cambridge, Mass., 1987.
12. Pritsker, A. Alan B.: *Introduction to Simulation and SLAM II*, 3d ed., Halsted Press–Wiley, New York, 1986.
13. Pritsker, A. Alan B., and C. Elliott Sigal: *Management Decision Making: A Network Simulation Approach,* Prentice-Hall, Englewood Cliffs, N.J., 1983.
14. Rubenstein, Reuven Y.: *Simulation and the Monte Carlo Method*, Wiley, New York, 1981.
15. Taha, Hamdy A.: *Simulation Modeling and SIMNET*, Prentice-Hall, Englewood Cliffs, N.J., 1987.
16. Watson, Hugh J.: *Computer Simulation in Business*, Wiley, New York, 1981.

(Random numbers needed to do these problems manually should be obtained from Table 15.4. For each part, use the digits *consecutively* starting from the front of the top row to form *three-digit* random numbers 096, 569, 665, and so on.)

1.* Use the *mixed congruential method* to generate the following sequences of random numbers:

(*a*) A sequence of 10 *one-digit* random numbers such that $x_{n+1} \equiv (x_n + 3)(\text{modulo } 10)$ and $x_0 = 2$.

(*b*) A sequence of eight random numbers between 0 and 7 such that $x_{n+1} \equiv (5x_n + 1)$ (modulo 8) and $x_0 = 1$.

(*c*) A sequence of five *two-digit* random numbers such that $x_{n+1} \equiv (61x_n + 27)(\text{modulo } 100)$ and $x_0 = 10$.

2. Use the *mixed congruential method* to generate a sequence of five *two-digit* random numbers such that $x_{n+1} \equiv (41x_n + 33)(\text{modulo } 100)$ and $x_0 = 48$.

3. Use the *mixed congruential method* to generate the following sequences of random numbers:

(*a*) A sequence of five random numbers between 0 and 31 such that $x_{n+1} \equiv (13x_n + 15)(\text{modulo } 32)$ and $x_0 = 14$.

(*b*) A sequence of three *three-digit* random numbers such that $x_{n+1} \equiv (201x_n + 503)(\text{modulo } 1{,}000)$ and $x_0 = 485$.

4.* Use the one-digit random numbers—5, 2, 4, 9, 7—to generate random observations for each of the following situations:

(*a*) Throwing an unbiased coin.

(*b*) Throwing a die.

(*c*) The color of a traffic light found by a randomly arriving car when it is green 40 percent of the time, yellow 10 percent of the time, and red 50 percent of the time.

5. Generate five random observations from a *uniform distribution* between −10 and +40.

6.* Suppose that random observations are needed from the *triangular distribution* whose probability density function is

$$f(x) = \begin{cases} 2x, & \text{if } 0 \le x \le 1 \\ 0, & \text{otherwise.} \end{cases}$$

(*a*) Derive an expression for each random observation as a function of the uniform random number r.

(*b*) Generate five random observations.

7. Generate three random observations from each of the following probability distributions:

(*a*) The *uniform distribution* from 25 to 75.

(*b*) The distribution whose probability density function is

$$f(x) = \begin{cases} \frac{1}{4}(x + 1)^3, & \text{if } -1 \le x \le 1 \\ 0, & \text{otherwise.} \end{cases}$$

(*c*) The distribution whose probability density function is

$$f(x) = \begin{cases} \frac{1}{200}(x - 40), & \text{if } 40 \le x \le 60 \\ 0, & \text{otherwise.} \end{cases}$$

8. Generate three random observations from each of the following probability distributions:

(*a*) The random variable X has $P\{X = 0\} = \frac{1}{2}$. Given $X \neq 0$, it has a uniform distribution between -5 and 15.

(*b*) The distribution whose probability density function is

$$f(x) = \begin{cases} x - 1, & \text{if } 1 \le x \le 2 \\ 3 - x, & \text{if } 2 \le x \le 3. \end{cases}$$

(*c*) The *geometric distribution* with parameter $p = \frac{1}{3}$, so that

$$P\{X = k\} = \begin{cases} \frac{1}{3}\left(\frac{2}{3}\right)^{k-1}, & \text{if } k = 1, 2, \ldots \\ 0, & \text{otherwise.} \end{cases}$$

9. Generate three random observations from a *normal distribution* with mean $= 10$ and standard deviation $= 5$. (Use $n = 3$ for each observation.)

10. Generate four random observations from a *normal distribution* with mean $= 0$ and standard deviation $= 1$. (Use $n = 3$ for each observation.) Then use these four observations to generate two random observations from a *chi-square distribution* with 2 degrees of freedom.

11.* Generate two random observations from each of the following probability distributions:

(*a*) The exponential distribution with mean $= 4$.

(*b*) The *Erlang* distribution with mean $= 4$ and shape parameter $k = 2$ (that is, standard deviation $= 2\sqrt{2}$).

(*c*) The normal distribution with mean $= 4$ and standard deviation $= 2\sqrt{2}$. (Use $n = 6$ for each observation.)

12. Generate four random observations from an *exponential distribution* with mean $=$ 1. Then use these four observations to generate one random observation from an *Erlang distribution* with mean $= 4$ and shape parameter $k = 4$.

13. Use the *acceptance-rejection method* to generate three random observations from the triangular distribution used to illustrate this method in Sec. 15.2.

14. Use the *acceptance-rejection method* to generate three random observations from the probability density function

$$f(x) = \begin{cases} \frac{1}{50}(x - 10), & \text{if } 10 \le x \le 20 \\ 0, & \text{otherwise.} \end{cases}$$

15. The weather can be considered a stochastic system, because it evolves in a probabilistic manner from one day to the next. Suppose for a certain location that this probabilistic evolution satisfies the following description:

The probability of rain tomorrow is 0.6 if it is raining today.

The probability of being clear (no rain) tomorrow is 0.8 if it is clear today.

Simulate the evolution of the weather for 10 days, beginning the day after a clear day.

16. The game of craps requires the player to throw two dice one or more times until a decision has been reached as to whether he wins or loses. He wins if the first throw results in a sum of 7 or 11, or, alternatively, if the first sum is 4, 5, 6, 8, 9, or 10 and the same sum reappears before a sum of 7 has appeared. Conversely, he loses if the first throw results in a sum of 2, 3, or 12, or, alternatively, if the first sum is 4, 5, 6, 8, 9, or 10 and a sum of 7 appears before the first sum reappears.

(*a*) Simulate five plays of this game to start the process of estimating the probability of winning.

(*b*) For a large number of plays of the game, the proportion of wins has *approximately* a *normal distribution* with mean $= 0.493$ and standard deviation $= 0.5/\sqrt{n}$. Use this information to calculate the number of simulated plays that would be required to have a probability of at least 0.95 that the proportion of wins will be less than 0.5.

17. Consider the M/M/1 queueing theory model that was discussed in Sec. 6.6 and Example 2, Sec. 15.1. Suppose that the mean arrival rate is 5 per hour, the mean service rate is 10 per hour, and you are required to estimate the expected waiting time before service begins by using simulation.

(*a*) Starting with the system empty, use *next-event incrementing* to perform the simulation until two service completions have occurred.
(*b*) Starting with the system empty, use *fixed-time incrementing* (with 2 minutes as the time unit) to perform the simulation until two service completions have occurred.
(*c*) Write a computer simulation program with *next-event incrementing* for this problem. Use the *regenerative method* with 100 cycles to obtain a point estimate and 95 percent confidence interval for the steady-state expected waiting time before service begins. Compare these results with the theoretical value.
(*d*) Write a computer simulation program with *fixed-time incrementing* and 0.1 minute as the time unit. Use the *regenerative method* with 100 cycles to obtain a point estimate and 95 percent confidence interval for the steady-state expected waiting time before service begins. Compare these results with the theoretical value.

18.* Consider the probability distribution whose probability density function is

$$f(x) = \begin{cases} \dfrac{1}{x^2}, & \text{if } 1 \le x \le \infty \\ 0, & \text{otherwise.} \end{cases}$$

The problem is to perform a simulated experiment, with the help of variance-reducing techniques, for estimating the mean of this distribution. To provide a standard of comparison, also derive the mean analytically.

For each of the following cases, generate 10 observations and calculate the resulting estimate of the mean:

(*a*) Use the *crude Monte Carlo method.*
(*b*) Use *stratified sampling* with *three* strata: $0 < F(x) \le 0.6$, $0.6 < F(x) \le 0.9$, $0.9 < F(x) \le 1$, with 3, 3, and 4 observations, respectively.
(*c*) Use the *method of complementary random numbers.*

19. One product produced by a certain company requires that bushings be drilled into a metal block and that cylindrical shafts be inserted into the bushings. The shafts are required to have a radius of at least 1.0000 inch, but the radius should be as little larger than this as possible. In actuality, the probability distribution of what the radius of a shaft will be (in inches) has the probability density function

$$f_s(x) = \begin{cases} 400e^{-400(x-1.0000)}, & \text{if } x \ge 1.0000 \\ 0, & \text{otherwise.} \end{cases}$$

Similarly, the probability distribution of what the radius of a bushing will be (in inches) has the probability density function

$$f_B(x) = \begin{cases} 100, & \text{if } 1.0000 \le x \le 1.0100 \\ 0, & \text{otherwise.} \end{cases}$$

The clearance between a bushing and a shaft is the difference in their radii. Because they are selected at random, there occasionally is interference (i.e., negative clearance) between a bushing and a shaft that were to be mated. The objective is to determine how frequently this interference will happen under the current probability distributions.

Perform a simulated experiment for estimating the probability of interference. Notice that almost all cases of interference will occur when the radius of the bushing is much closer to 1.0000 inch than to 1.0100 inches. Therefore, it appears that an efficient experiment would generate most of the simulated bushings from this critical portion of the distribution. Take this observation into account in part (*b*). For each of the following cases, generate 10 observations and calculate the resulting estimate of the probability of interference:

(*a*) Use the *crude Monte Carlo method.*
(*b*) Develop and apply a *stratified sampling* approach to this problem.
(*c*) Use the *method of complementary random numbers*.

20. Simulation is being used to study a system whose measure of performance X will be partially determined by the outcome of a certain external factor. This factor has three possible outcomes (unfavorable, neutral, and favorable) that will occur with equal probability ($\frac{1}{3}$). Because the favorable outcome would greatly increase the spread of possible values of X, this outcome is more critical than the others for estimating the mean and variance of X. Therefore, a *stratified sampling* approach has been adopted, with six random observations of the value of X generated under the favorable outcome, three generated under the neutral outcome, and one generated under the unfavorable outcome—as follows:

Outcome of External Factor	Simulated Values of X
Favorable	8, 5, 1, 6, 3, 7
Neutral	3, 5, 2
Unfavorable	2

(*a*) Develop the resulting estimate of $E(X)$.
(*b*) Develop the resulting estimate of $E(X^2)$.

21.* A certain single-server system has been simulated, with the following sequence of waiting times before service for the respective customers. Use the *regenerative method* to obtain a point estimate and 90 percent confidence interval for the steady-state expected waiting time before service.

(*a*) 0, 5, 4, 0, 2, 0, 3, 1, 6, 0.
(*b*) 0, 3, 2, 0, 3, 1, 5, 0, 0, 2, 4, 0, 3, 5, 2, 0.

22. Consider the queueing system example presented in Sec. 15.4 for the regenerative method. Explain why the point where a *service completion* occurs with *no* other customers left is *not* a regeneration point.

23. A company has been having a maintenance problem with a certain complex piece of equipment. This equipment contains four identical vacuum tubes that have been the cause of the trouble. The problem is that the tubes fail fairly frequently, thereby forcing the equipment to be shut down while a replacement is made. The current practice is to replace tubes only when they fail. However, a proposal has been made to replace all four tubes whenever any one of them fails to reduce the frequency with which the equipment must be shut down. The objective is to compare these two alternatives on a cost basis.

The pertinent data are the following. For each tube, the operating time until failure has approximately a *uniform distribution* from 1,000 to 2,000 hours. The equipment must be shut down for 1 hour to replace one tube or for 2 hours to replace all four tubes. The total cost associated with shutting down the equipment and replacing tubes is \$100/hour plus \$20 for each new tube.

(*a*) Starting with four new tubes, simulate the operation of the two alternative policies for 5,000 hours of simulated time.
(*b*) Use the data from part (*a*) to make a preliminary comparison of the two alternatives on a cost basis.

(*c*) For the *proposed* policy, describe an appropriate *regeneration point* for defining cycles that will permit applying the *regenerative method* of statistical analysis. Explain why the regenerative method cannot be applied to the *current* policy.

(*d*) For the *proposed* policy, use the *regenerative method* to obtain a point estimate and 95 percent confidence interval for the steady-state *expected cost per hour* from the data obtained in part (*a*).

(*e*) Write a computer simulation program for the two alternative policies. Then repeat parts (*a*), (*b*), and (*d*) on the computer, with 100 cycles for the *proposed* policy and 55,000 hours of simulated time (including a stabilization period of 5,000 hours) for the current policy.

24. A manufacturing company has two planers for cutting flat surfaces in large work pieces of two different types. The time required to perform each job varies somewhat, depending largely upon the number of passes that must be made. In particular, for both types of work pieces, the time required by a planer has approximately the following probability distribution:

Time (min)	Probability
10	0.30
20	0.25
30	0.18
40	0.12
50	0.08
60	0.045
70	0.015
80	0.007
90	0.003

Every half hour one work piece of *both* types is brought to the planer department.

Unfortunately, the planer department has had a difficult time keeping up with its workload. Frequently there are a number of work pieces waiting for a free planer. This waiting has seriously disrupted the production schedule for the subsequent operations, thereby greatly increasing the cost of in-process inventory as well as the cost of idle equipment and resulting lost production. Therefore, a proposal has been made to obtain one additional planer to relieve this bottleneck.

It is estimated that the total incremental cost (including capital recovery cost) associated with obtaining and operating another planer would be $30/hour. (This estimate takes into account the fact that, even with an additional planer, the total running time for all the planers will remain the same.) It is also estimated that the total cost associated with work pieces having to wait to be processed is $200 per work piece per hour and $100 per work piece per hour for work pieces of the first and second types, respectively. Because of this difference in costs, work pieces of the first type always are given priority over those of the second type. In other words, if a planer becomes free when work pieces of both types are waiting, a work piece of the first type always is chosen to be processed next.

(*a*) Starting with all planers idle waiting for work pieces to arrive momentarily, use *next-event incrementing* to simulate the operation of the two alternative policies (the status quo or obtaining one additional planer) for 3 hours of simulated time.

(*b*) Describe an appropriate *regeneration point* for defining cycles that will permit applying the *regenerative method* of statistical analysis to this problem.

(*c*) Write a computer simulation program for the two alternative policies. Use the *regenerative method* with 100 cycles each to compare the two alternatives on a cost basis.

25. Select any of the typical applications of simulation listed at the end of Sec. 15.2 and develop a simulation model for this type of problem.

Appendixes

Appendix 1

Convexity

The concept of *convexity* is frequently used in operations research work. Therefore, we introduce the properties of convex (or concave) functions and convex sets.

Definition: A *function* of a single variable, $f(x)$, is a **convex function** if, for each pair of values of x, say, x' and x'',

$$f(\lambda x'' + (1 - \lambda)x') \leq \lambda f(x'') + (1 - \lambda)f(x')$$

for all values of λ such that $0 \leq \lambda \leq 1$. It is a **strictly convex function** if $\leq$ can be replaced by $<$. It is a **concave function** (or a **strictly concave function**) if this statement holds when $\leq$ is replaced by $\geq$ (or by $>$).

This definition has an enlightening geometric interpretation. Consider the graph of the function $f(x)$ drawn as a function of x. Then $[x', f(x')]$ and $[x'', f(x'')]$ are two points on the graph of $f(x)$, and $[\lambda x'' + (1 - \lambda)x', \lambda f(x'') + (1 - \lambda)f(x')]$ represents the various points on the line segment between these two points when $0 \leq \lambda \leq 1$. Thus the original inequality in the definition indicates that this line segment lies entirely above or on the graph of the function. Therefore, $f(x)$ is convex if, for each pair of points on the graph of $f(x)$, the line segment joining these two points lies entirely above or on the graph of $f(x)$. In other words, $f(x)$ is convex if it is "always bending upward." (This condition is sometimes referred to as "concave upward," as opposed to "concave downward" for a concave function.) To be more precise, if $f(x)$ possesses a second derivative everywhere, then $f(x)$ is convex if and only if $d^2f(x)/dx^2 \geq 0$ for all values of x [for which $f(x)$ is defined]. Similarly, $f(x)$ is strictly convex when $d^2f(x)/dx^2 > 0$, concave when $d^2f(x)/dx^2 \leq 0$, and strictly concave when $d^2f(x)/dx^2 < 0$. Some examples are given in Figs. A1.1 to A1.4.

The concept of a convex function also generalizes to functions of more than one variable. Thus if $f(x)$ is replaced by $f(x_1, x_2, \ldots, x_n)$, the definition just given still applies if x is replaced everywhere by $(x_1, x_2, \ldots, x_n)$. Similarly, the corresponding geometric interpretation is still valid after generalizing the concepts of *points* and *line segments*. Thus, just as a particular value of (x, y) is interpreted as a point in two-dimensional space, each possible value of $(x_1, x_2, \ldots, x_m)$ may be thought of as a point in m-dimensional (Euclidean) space. By letting $m = n + 1$, the points on the graph of $f(x_1, x_2, \ldots, x_n)$ become the possible values of $[x_1, x_2, \ldots, x_n, f(x_1, x_2, \ldots, x_n)]$. Another point, $(x_1, x_2, \ldots, x_n, x_{n+1})$, is said to lie above, on, or below the graph of $f(x_1, x_2, \ldots, x_n)$, according to whether x_{n+1} is larger, equal to, or smaller than $f(x_1, x_2, \ldots, x_n)$, respectively.

Definition: The **line segment** joining any two points $(x_1', x_2', \ldots, x_m')$ and $(x_1'', x_2'', \ldots, x_m'')$ is the collection of points.

$$(x_1, x_2, \ldots, x_m) = [\lambda x_1'' + (1 - \lambda)x_1', \lambda x_2'' + (1 - \lambda)x_2', \ldots, \lambda x_m'' + (1 - \lambda)x_m']$$

such that $0 \leq \lambda \leq 1$.

Thus a line segment in m-dimensional space is a direct generalization of a line segment in two-dimensional space. For example, if

$$(x_1', x_2') = (2, 6), \qquad (x_1'', x_2'') = (3, 4),$$

then the line segment joining them is the collection of points,

$$(x_1, x_2) = [3\lambda + 2(1 - \lambda), 4\lambda + 6(1 - \lambda)]$$

where $0 \leq \lambda \leq 1$.

Definition: $f(x_1, x_2, \ldots, x_n)$ is a **convex function** if, for each pair of points on the graph of $f(x_1, x_2, \ldots, x_n)$, the line segment joining these two points lies entirely above or on the graph of $f(x_1, x_2, \ldots, x_n)$. It is a **strictly convex function** if this line segment actually lies entirely above this graph except at the endpoints of the line segment. **Concave functions** and **strictly concave functions** are defined in exactly the same way, except that *above* is replaced by *below*.

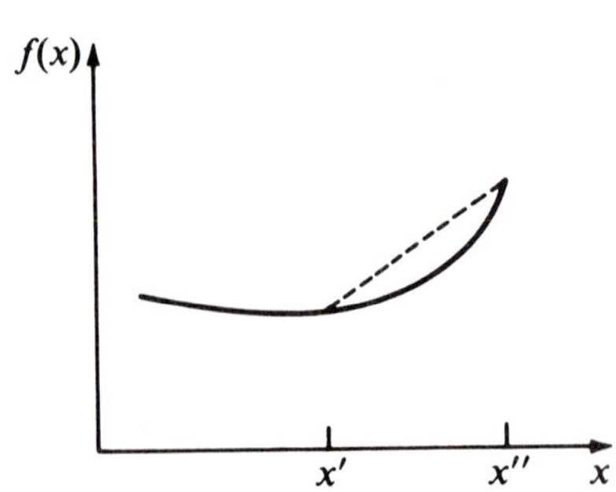

Figure A1.1 A convex function.

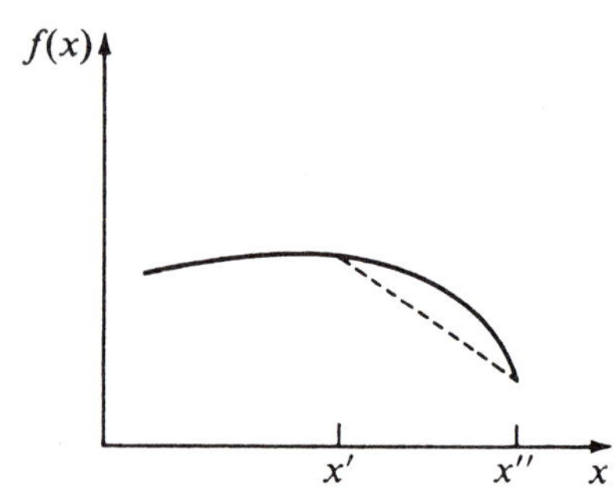

Figure A1.2 A concave function.

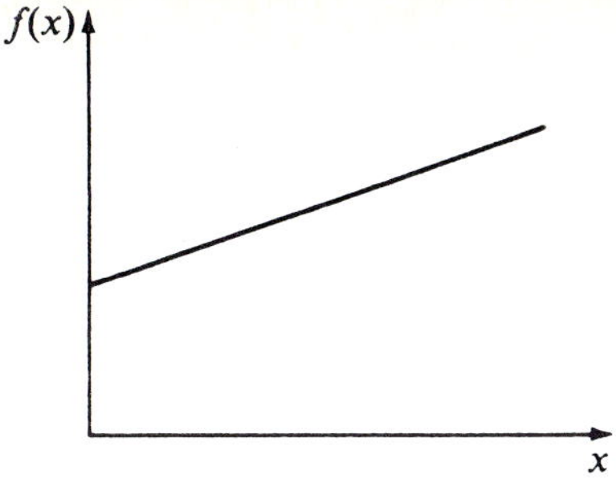

Figure A1.3 A function that is both convex and concave.

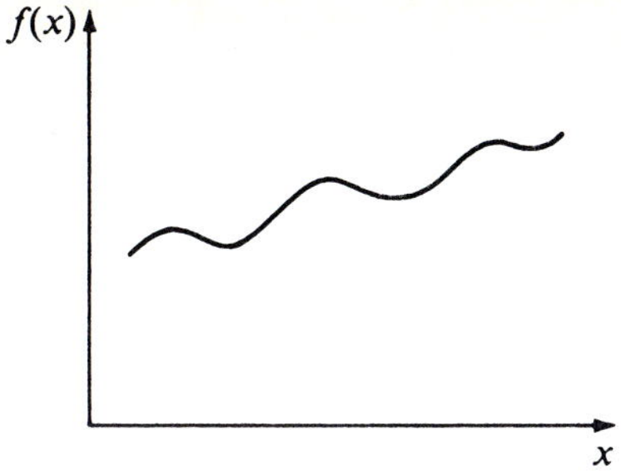

Figure A1.4 A function that is neither convex nor concave.

Just as the second derivative can be used (when it exists everywhere) to check whether a function of a single variable is convex, so second partial derivatives can be used to check functions of several variables, although in a more complicated way. For example, if there are two variables, then $f(x_1, x_2)$ is convex if and only if

(1) $$\frac{\partial^2 f(x_1, x_2)}{\partial x_1^2} \frac{\partial^2 f(x_1, x_2)}{\partial x_2^2} - \left[\frac{\partial^2 f(x_1, x_2)}{\partial x_1 \partial x_2}\right]^2 \geq 0,$$

(2) $$\frac{\partial^2 f(x_1, x_2)}{\partial x_1^2} \geq 0,$$

and

(3) $$\frac{\partial^2 f(x_1, x_2)}{\partial x_2^2} \geq 0,$$

for all possible values of (x_1, x_2), assuming that these partial derivatives exist everywhere. It is strictly convex if $\geq$ can be replaced by $>$ in all three conditions [but now condition (3) is superfluous and can be omitted because it is implied by the other two conditions], whereas $f(x_1, x_2)$ is concave if $\geq$ can be replaced by $\leq$ in conditions (2) and (3). When there are more than two variables, the conditions for convexity are a generalization of the ones just shown. In mathematical terminology, $f(x_1, x_2, \ldots, x_n)$ is convex if and only if its $n \times n$ Hessian matrix is positive semidefinite for all possible values of $(x_1, x_2, \ldots, x_n)$.

Thus far convexity has been treated as a general property of a function. However, many nonconvex functions do satisfy the conditions for convexity over certain intervals for the respective variables. Therefore, it is meaningful to talk about a function being convex over a certain region. For example, a function is said to be convex within a neighborhood of a specified point if its second derivative or partial derivatives satisfy the conditions for convexity at that point. This concept is useful in Appendix 2.

Finally, two particularly important properties of convex functions should be mentioned. First, if $f(x_1, x_2, \ldots, x_n)$ is a convex function, then $g(x_1, x_2, \ldots, x_n) = -f(x_1, x_2, \ldots, x_n)$ is a concave function, and vice versa. Second, the sum of convex functions is a convex function. To illustrate,

$$f_1(x_1) = x_1^4 + 2x_1^2 - 5x_1$$

and

$$f_2(x_1, x_2) = x_1^2 + 2x_1x_2 + x_2^2$$

are both convex functions, as you can verify by calculating their second derivatives. Therefore, the sum of these functions,

$$f(x_1, x_2) = x_1^4 + 3x_1^2 - 5x_1 + 2x_1x_2 + x_2^2,$$

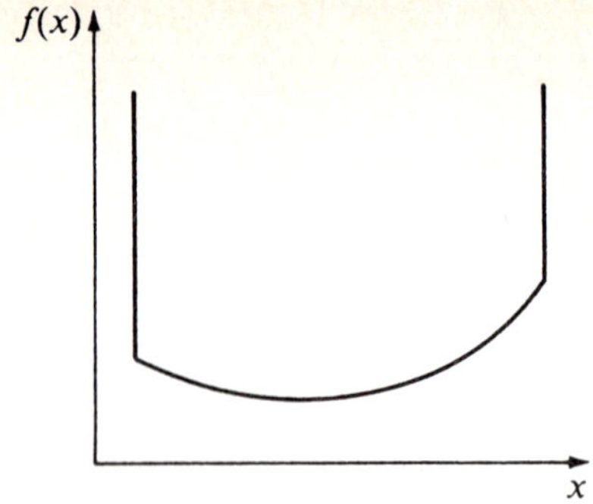

Figure A1.5 Example of a convex set determined by a convex function.

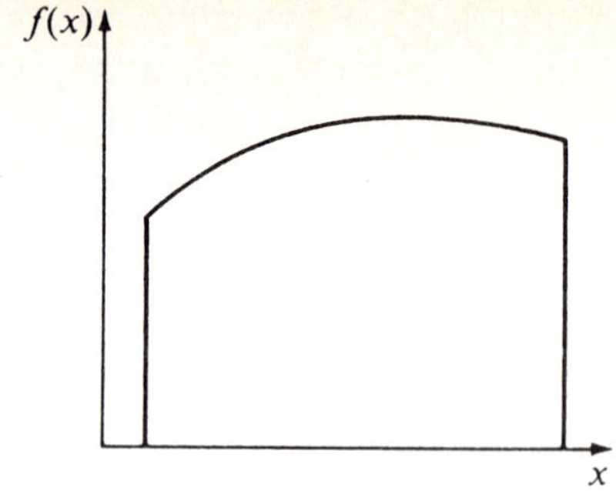

Figure A1.6 Example of a convex set determined by a concave function.

is a convex function, whereas its negative,

$$g(x_1, x_2) = -x_1^4 - 3x_1^2 + 5x_1 - 2x_1x_2 - x_2^2,$$

is a concave function.

The concept of a convex function leads quite naturally to the related concept of a **convex set**. Thus, if $f(x_1, x_2, \ldots, x_n)$ is a convex function, then the collection of points that lie above or on the graph of $f(x_1, x_2, \ldots, x_n)$ forms a convex set. Similarly, the collection of points that lie below or on the graph of a concave function is a convex set. These cases are illustrated in Figs. A1.5 and A1.6 for the case of a single independent variable. Furthermore, convex sets have the important property that, for any given group of convex sets, the collection of points that lie in all of them (i.e., the intersection of these convex sets) is also a convex set. Therefore, the collection of points that lie both above or on a convex function and below or on a concave function is a convex set, as illustrated in Fig. A1.7. Thus convex sets may be viewed intuitively as a collection of points whose bottom boundary is a convex function and whose top boundary is a concave function. To be a bit more precise, a convex set may be defined as follows:

Definition: A **convex set** is a collection of points such that, for each pair of points in the collection, the entire line segment joining these two points is also in the collection.

The distinction between nonconvex sets and convex sets is illustrated in Figs. A1.8 and A1.9. Thus the set of points shown in Fig. A1.8 is not a convex set because there exist many pairs of these points, for example, (1, 2) and (2, 1), such that the line segment between them

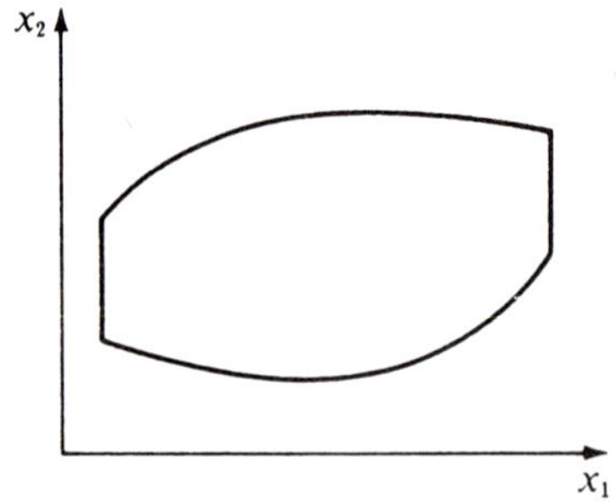

Figure A1.7 Example of a convex set determined by both convex and concave functions.

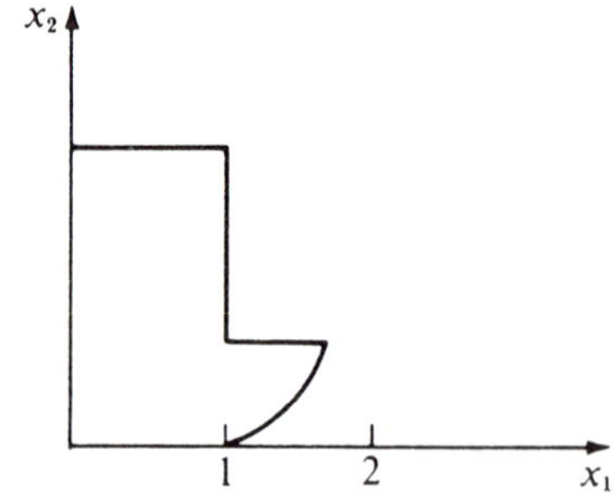

Figure A1.8 Example of a set that is not convex.

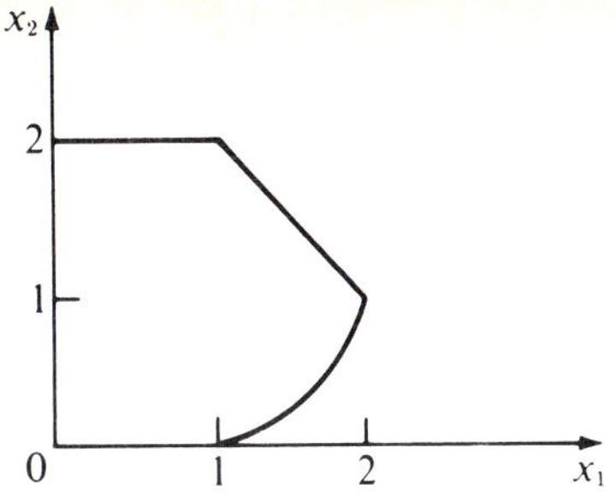

Figure A1.9 Example of a convex set.

does not lie entirely within the set. This is not the case for the set in Fig. A1.9, which is convex.

In conclusion, the useful concept of an *extreme point* of a convex set needs to be introduced.

Definition: An **extreme point** of a convex set is a point in the set that does not lie on any line segment that joins two other points in the set.

Thus the extreme points of the convex set in Fig. A1.9 are (0, 0), (0, 2), (1, 2), (2, 1), (1, 0), and all the infinite number of points on the boundary between (2, 1) and (1, 0). If this particular boundary were a line segment instead, then the set would have only the five listed extreme points.

Appendix 2

Classical Optimization Methods

This appendix reviews the classical methods of calculus for finding a solution that maximizes or minimizes (1) a function of a single variable, (2) a function of several variables, and (3) a function of several variables subject to constraints on the values of these variables. It is assumed that the functions considered possess continuous first and second derivatives and partial derivatives everywhere. Some of the concepts discussed next have been introduced briefly in Secs. 14.2 and 14.3.

Consider a function of a single variable, such as that shown in Fig. A2.1. A necessary condition for a particular solution, $x = x^*$, to be either a minimum or a maximum is that

$$\frac{df(x)}{dx} = 0 \qquad \text{at } x = x^*.$$

Thus in Fig. A2.1 there are five solutions satisfying these conditions. To obtain more information about these five so-called *critical points,* it is necessary to examine the second derivative. Thus, if

$$\frac{d^2f(x)}{dx^2} > 0 \qquad \text{at } x = x^*,$$

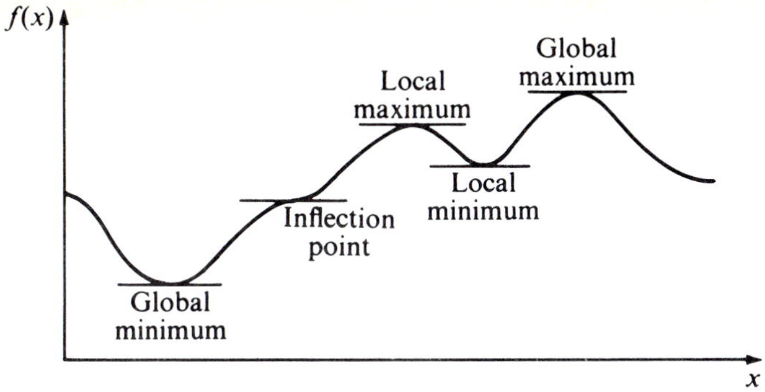

Figure A2.1 A function having several maxima and minima.

then x^* must be at least a *local minimum* [that is, $f(x^*) \leq f(x)$ for all x sufficiently close to x^*]. Using the language introduced in Appendix 1, we can say that x^* must be a local minimum if $f(x)$ is *strictly convex* within a neighborhood of x^*. Similarly, a sufficient condition for x^* to be a *local maximum* (given that it satisfies the necessary condition) is that $f(x)$ is *strictly concave* within a neighborhood of x^* (that is, the second derivative is *negative* at x^*). If the second derivative is zero, the issue is not resolved (the point may even be an *inflection point*), and it is necessary to examine higher derivatives.

To find a *global minimum* [i.e., a solution x^* such that $f(x^*) \leq f(x)$ for all x], it is necessary to compare the local minima and identify the one that yields the smallest value of $f(x)$. If this value is less than $f(x)$ as $x \to -\infty$ and as $x \to +\infty$ (or at the endpoints of the function, if it is defined only over a finite interval), then this point is a global minimum. Such a point is shown in Fig. A2.1, along with the global maximum, which is identified in an analogous way.

However, if $f(x)$ is known to be either a convex or concave function (see Appendix 1 for a description of such functions), the analysis becomes much simpler. In particular, if $f(x)$ is a *convex* function, such as the one shown in Fig. A1.1, then any solution x^* such that

$$\frac{df(x)}{dx} = 0 \qquad \text{at } x = x^*$$

is known automatically to be a *global minimum*. In other words, this condition is not only a *necessary* but a *sufficient* condition for a global minimum of a convex function. If this function actually is strictly convex, then this solution must be the only global minimum. (However, if the function is either always decreasing or always increasing, so the derivative is nonzero for all values of x, then there will be no global minimum at a finite value of x.) Otherwise, there could be a tie for the global minimum over a single interval where the derivative is zero. Similarly, if $f(x)$ is a *concave* function, then having

$$\frac{df(x)}{dx} = 0 \qquad \text{at } x = x^*$$

becomes both a *necessary* and *sufficient* condition for x^* to be a *global maximum*.

The analysis for an unconstrained function of several variables $f(\mathbf{x})$, where $\mathbf{x} = (x_1, x_2, \ldots, x_n)$, is similar. Thus a *necessary* condition for a solution $\mathbf{x} = \mathbf{x}^*$ to be either a minimum or a maximum is that

$$\frac{\partial f(\mathbf{x})}{\partial x_j} = 0 \qquad \text{at } \mathbf{x} = \mathbf{x}^*, \qquad \text{for } j = 1, 2, \ldots, n.$$

After identifying the critical points that satisfy this condition, each such point is then classified as a local minimum or maximum if the function is *strictly convex* or *strictly concave*, respectively, within a neighborhood of the point. (Additional analysis is required if the function is neither.) The *global minimum* and *maximum* would be found by comparing the local minima and maxima and then checking the value of the function as some of the variables approach $-\infty$ or $+\infty$. However, if the function is known to be *convex* or *concave*, then a critical point must be a *global minimum* or a *global maximum*, respectively.

Now consider the problem of finding the *minimum* or *maximum* of the function $f(\mathbf{x})$, subject to the restriction that $\mathbf{x}$ must satisfy all the equations

$$g_1(\mathbf{x}) = b_1$$
$$g_2(\mathbf{x}) = b_2$$
$$\vdots$$
$$g_m(\mathbf{x}) = b_m,$$

where $m < n$. For example, if $n = 2$ and $m = 1$, the problem might be

$$\text{Maximize} \quad f(x_1, x_2) = x_1^2 + 2x_2,$$

subject to

$$g(x_1, x_2) = x_1^2 + x_2^2 = 1.$$

In this case (x_1, x_2) is restricted to be on the circle of radius 1, whose center is at the origin, so that the goal is to find the point on this circle that yields the largest value of $f(x_1, x_2)$. This example is soon solved after a general approach to the problem is outlined.

A classical method of dealing with this problem is the **method of Lagrange multipliers.** This procedure begins by formulating the **Lagrangian function,**

$$h(\mathbf{x}, \boldsymbol{\lambda}) = f(\mathbf{x}) - \sum_{i=1}^{m} \lambda_i[g_i(\mathbf{x}) - b_i],$$

where the new variables $\boldsymbol{\lambda} = (\lambda_1, \lambda_2, \ldots, \lambda_m)$ are called *Lagrange multipliers*. Notice the key fact that for the *feasible* values of $\mathbf{x}$,

$$g_i(\mathbf{x}) - b_i = 0, \qquad \text{for all } i,$$

so $h(\mathbf{x}, \boldsymbol{\lambda}) = f(\mathbf{x})$. Therefore, it can be shown that if $(\mathbf{x}, \boldsymbol{\lambda}) = (\mathbf{x}^*, \boldsymbol{\lambda}^*)$ is a *local* or *global minimum* or *maximum* for the unconstrained function $h(\mathbf{x}, \boldsymbol{\lambda})$, then $\mathbf{x}^*$ is a corresponding *critical point* for the original problem. As a result, the method now reduces to analyzing $h(\mathbf{x}, \boldsymbol{\lambda})$ by the procedure just described for unconstrained functions. Thus the $(n + m)$ partial derivatives would be set equal to zero; that is,

$$\frac{\partial h}{\partial x_j} = \frac{\partial f}{\partial x_j} - \sum_{i=1}^{m} \lambda_i \frac{\partial g_i}{\partial x_j} = 0, \qquad \text{for } j = 1, 2, \ldots, n,$$

$$\frac{\partial h}{\partial \lambda_i} = -g_i(\mathbf{x}) + b_i = 0, \qquad \text{for } i = 1, 2, \ldots, m,$$

and then the critical points would be obtained by solving these equations for $(\mathbf{x}, \boldsymbol{\lambda})$. Notice that the last m equations are equivalent to the constraints in the original problem, so only feasible solutions are considered. After further analysis to identify the *global minimum* or *maximum* of $h(\cdot)$, the resulting value of $\mathbf{x}$ is then the desired solution to the original problem.

It should be pointed out that from a practical computational viewpoint, the method of Lagrange multipliers is not a particularly powerful procedure. It is often essentially impossible to solve the equations to obtain the critical points. Furthermore, even when they can be obtained, the number of critical points may be so large (often infinite) that it is impractical to attempt to identify a global minimum or maximum. However, for certain types of small problems, this method can sometimes be used successfully. To illustrate, consider the example introduced earlier. In this case,

$$h(x_1, x_2) = x_1^2 + 2x_2 - \lambda[x_1^2 + x_2^2 - 1],$$

so that

$$\frac{\partial h}{\partial x_1} = 2x_1 - 2\lambda x_1 = 0,$$

$$\frac{\partial h}{\partial x_2} = 2 - 2\lambda x_2 = 0,$$

$$\frac{\partial h}{\partial \lambda} = -[x_1^2 + x_2^2 - 1] = 0.$$

The first equation implies that either $\lambda = 1$ or $x_1 = 0$. If $\lambda = 1$, then the other two equations imply that $x_2 = 1$ and $x_1 = 0$. If $x_1 = 0$, then the third equation implies that $x_2 = \pm 1$. Therefore, the two critical points for the original problem are $(x_1, x_2) = (0, 1)$ and $(0, -1)$. Thus it is apparent that these points are the *global maximum* and *minimum*, respectively.

In presenting the classical optimization methods just described, we have assumed that you are already familiar with derivatives and how to obtain them. However, there is a special case of importance in operations research work that warrants additional explanation, namely, the derivative of an integral. In particular, consider how to find the derivative of the function

$$F(y) = \int_{g(y)}^{h(y)} f(x, y)\, dx,$$

where $g(y)$ and $h(y)$ are the limits of integration expressed as functions of y. To begin, suppose that these limits of integration are constants, so that $g(y) = a$ and $h(y) = b$, respectively. For this special case, it can be shown that, given the regularity conditions assumed at the beginning of this appendix, the derivative is simply

$$\frac{d}{dy} \int_a^b f(x, y)\, dx = \int_a^b \frac{\partial f(x, y)}{\partial y}\, dx.$$

For example, if $f(x, y) = e^{-xy}$, $a = 0$, and $b = \infty$, then

$$\frac{d}{dy} \int_0^\infty e^{-xy}\, dx = \int_0^\infty (-x)e^{-xy}\, dx = -\frac{1}{y^2}$$

at any positive value of y. Thus the intuitive procedure of interchanging the order of differentiation and integration is valid for this case. However, finding the derivative becomes a little more complicated than this when the limits of integration are functions. In particular,

$$\frac{d}{dy} \int_{g(y)}^{h(y)} f(x, y)\, dx = \int_{g(y)}^{h(y)} \frac{\partial f(x, y)}{\partial y}\, dx + f(h(y), y)\frac{dh(y)}{dy} - f(g(y), y)\frac{dg(y)}{dy},$$

where $f(h(y), y)$ is obtained by writing out $f(x, y)$ and then replacing x by $h(y)$ wherever it appears, and similarly for $f(g(y), y)$. To illustrate, if $f(x, y) = x^2y^3$, $g(y) = y$, and $h(y) = 2y$, then

$$\frac{d}{dy} \int_y^{2y} x^2y^3\, dx = \int_y^{2y} 3x^2y^2\, dx + (2y)^2y^3(2) - y^2y^3(1) = 14y^5$$

at any positive value of y.

Appendix 3

Matrices and Matrix Operations

A **matrix** is defined to be a *rectangular array of numbers*. For example,

$$\mathbf{A} = \begin{bmatrix} 2 & 5 \\ 3 & 0 \\ 1 & 1 \end{bmatrix}$$

is a 3×2 matrix (where 3×2 denotes "3 by 2") because it is a rectangular array of numbers with three rows and two columns. (Matrices are denoted in this book by **boldface capital letters**.) The numbers in the rectangular array are called the **elements** of the matrix. For example,

$$\mathbf{B} = \begin{bmatrix} 1 & 2.4 & 0 & \sqrt{3} \\ -4 & 2 & -1 & 15 \end{bmatrix}$$

is a 2 × 4 matrix whose elements are 1, 2.4, 0, $\sqrt{3}$, −4, 2, −1, and 15. Thus, in more general terms,

$$\mathbf{A} = \begin{bmatrix} a_{11} & a_{12} & \cdots & a_{1n} \\ a_{21} & a_{22} & \cdots & a_{2n} \\ \vdots & \vdots & & \vdots \\ a_{m1} & a_{m2} & & a_{mn} \end{bmatrix} = \|a_{ij}\|$$

is an $m \times n$ matrix, where $a_{11}, \ldots, a_{mn}$ represent the numbers that are the elements of this matrix; $\|a_{ij}\|$ is shorthand notation for identifying the matrix whose element in row i and column j is a_{ij} for every $i = 1, 2, \ldots, m$ and $j = 1, 2, \ldots, n$.

Because matrices do not possess a numerical value, they cannot be added, multiplied, and so on as if they were individual numbers. However, it is sometimes desirable to perform certain manipulations on arrays of numbers. Therefore, rules have been developed for performing operations on matrices that are analogous to arithmetic operations. To describe these, let $\mathbf{A} = \|a_{ij}\|$ and $\mathbf{B} = \|b_{ij}\|$ be two matrices having the same number of rows and the same number of columns. Then **A** and **B** are said to be equal (**A** = **B**) if and only if *all* of the corresponding elements are equal ($a_{ij} = b_{ij}$ for all i and j). The operation of multiplying a matrix by a number (denote this number by k) is performed by multiplying each element of the matrix by k, so that

$$k\mathbf{A} = \|ka_{ij}\|.$$

For example,

$$3\begin{bmatrix} 1 & \frac{1}{3} & 2 \\ 5 & 0 & -3 \end{bmatrix} = \begin{bmatrix} 3 & 1 & 6 \\ 15 & 0 & -9 \end{bmatrix}.$$

To add **A** and **B**, simply add the corresponding elements, so that

$$\mathbf{A} + \mathbf{B} = \|a_{ij} + b_{ij}\|.$$

To illustrate,

$$\begin{bmatrix} 5 & 3 \\ 1 & 6 \end{bmatrix} + \begin{bmatrix} 2 & 0 \\ 3 & 1 \end{bmatrix} = \begin{bmatrix} 7 & 3 \\ 4 & 7 \end{bmatrix}.$$

Similarly, subtraction is done as

$$\mathbf{A} - \mathbf{B} = \mathbf{A} + (-1)\mathbf{B},$$

so that

$$\mathbf{A} - \mathbf{B} = \|a_{ij} - b_{ij}\|.$$

For example,

$$\begin{bmatrix} 5 & 3 \\ 1 & 6 \end{bmatrix} - \begin{bmatrix} 2 & 0 \\ 3 & 1 \end{bmatrix} = \begin{bmatrix} 3 & 3 \\ -2 & 5 \end{bmatrix}.$$

Note that, with the exception of multiplication by a number, all the preceding operations are defined only when the two matrices involved are of the same size. However, all of them are straightforward because they involve performing only the same comparison or arithmetic operation on the corresponding elements of the matrices.

There exists one additional elementary operation that has not been defined, **matrix multiplication**, but it is considerably more complicated. To find the element in row i, column j of the matrix resulting from multiplying **A** times **B**, it is necessary to multiply each element in row i of **A** by the corresponding element in column j of **B** and then to add these products. Therefore, the matrix multiplication is defined if and only if the number of columns of **A** equals the number of rows of **B**, because this condition is required if we are to perform the specified element-by-element multiplication. Thus, if **A** is an $m \times n$ matrix and **B** is an $n \times r$ matrix, then their product is

$$\mathbf{AB} = \left\| \sum_{k=1}^{n} a_{ik}b_{kj} \right\|.$$

To illustrate,
$$\begin{bmatrix} 1 & 2 \\ 4 & 0 \\ 2 & 3 \end{bmatrix}\begin{bmatrix} 3 & 1 \\ 2 & 5 \end{bmatrix} = \begin{bmatrix} 1(3)+2(2) & 1(1)+2(5) \\ 4(3)+0(2) & 4(1)+0(5) \\ 2(3)+3(2) & 2(1)+3(5) \end{bmatrix} = \begin{bmatrix} 7 & 11 \\ 12 & 4 \\ 12 & 17 \end{bmatrix}.$$

On the other hand, if one attempts to multiply these matrices in the reverse order, the resulting product,

$$\begin{bmatrix} 3 & 1 \\ 2 & 5 \end{bmatrix}\begin{bmatrix} 1 & 2 \\ 4 & 0 \\ 2 & 3 \end{bmatrix},$$

is not even defined. Even when both **AB** and **BA** are defined,

$$\mathbf{AB} \neq \mathbf{BA}$$

in general. Thus *matrix multiplication* should be viewed as a specially designed operation whose properties are quite different from those of *arithmetic multiplication*. To understand why this special definition was adopted, consider the following system of equations:

$$\begin{aligned} 2x_1 - x_2 + 5x_3 + x_4 &= 20 \\ x_1 + 5x_2 + 4x_3 + 5x_4 &= 30 \\ 3x_1 + x_2 - 6x_3 + 2x_4 &= 20. \end{aligned}$$

Rather than writing these equations out as shown here, they can be written much more concisely in matrix form as

$$\mathbf{Ax} = \mathbf{b},$$

where
$$\mathbf{A} = \begin{bmatrix} 2 & -1 & 5 & 1 \\ 1 & 5 & 4 & 5 \\ 3 & 1 & -6 & 2 \end{bmatrix}, \quad \mathbf{x} = \begin{bmatrix} x_1 \\ x_2 \\ x_3 \\ x_4 \end{bmatrix}, \quad \mathbf{b} = \begin{bmatrix} 20 \\ 30 \\ 20 \end{bmatrix}.$$

It is this kind of multiplication for which matrix multiplication is designed.

Carefully note that *matrix division* is not defined.

Although the matrix operations described here do not possess certain of the properties of arithmetic operations, they do satisfy the following laws:

$$\begin{aligned} \mathbf{A} + \mathbf{B} &= \mathbf{B} + \mathbf{A}, \\ (\mathbf{A} + \mathbf{B}) + \mathbf{C} &= \mathbf{A} + (\mathbf{B} + \mathbf{C}), \\ \mathbf{A}(\mathbf{B} + \mathbf{C}) &= \mathbf{AB} + \mathbf{AC}, \\ \mathbf{A}(\mathbf{BC}) &= (\mathbf{AB})\mathbf{C}, \end{aligned}$$

when the relative sizes of these matrices are such that the indicated operations are defined.

Another type of matrix operation, which has no arithmetic analog, is the **transpose operation**. This operation involves nothing more than interchanging the rows and columns of the matrix, which is frequently useful for performing the multiplication operation in the desired way. Thus, for any matrix $\mathbf{A} = \|a_{ij}\|$, its transpose $\mathbf{A}^T$ is

$$\mathbf{A}^T = \|a_{ji}\|.$$

For example, if
$$\mathbf{A} = \begin{bmatrix} 2 & 5 \\ 1 & 3 \\ 4 & 0 \end{bmatrix},$$

then
$$\mathbf{A}^T = \begin{bmatrix} 2 & 1 & 4 \\ 5 & 3 & 0 \end{bmatrix}.$$

Zero and 1 are numbers that play a special role in arithmetic. There also exist special matrices that play a similar role in matrix theory. In particular, the matrix that is analogous to 1 is the **identity matrix I**, which is a square matrix whose elements are zeroes except for ones along the main diagonal. Thus

$$\mathbf{I} = \begin{bmatrix} 1 & 0 & 0 & \cdots & 0 \\ 0 & 1 & 0 & \cdots & 0 \\ 0 & 0 & 1 & \cdots & 0 \\ \vdots & \vdots & \vdots & & \vdots \\ 0 & 0 & 0 & \cdots & 1 \end{bmatrix}.$$

The number of rows or columns of **I** can be specified as desired. The analogy of **I** to 1 follows from the fact that for any matrix **A**,

$$\mathbf{IA} = \mathbf{A} = \mathbf{AI},$$

where **I** is assigned the appropriate number of rows and columns in each case for the multiplication operation to be defined. Similarly, the matrix that is analogous to zero is the so-called **null matrix 0**, which is a matrix of any size whose elements are *all zeroes*. Thus

$$\mathbf{0} = \begin{bmatrix} 0 & 0 & \cdots & 0 \\ 0 & 0 & \cdots & 0 \\ \vdots & \vdots & & \vdots \\ 0 & 0 & \cdots & 0 \end{bmatrix}.$$

Therefore, for any matrix **A**,

$$\mathbf{A} + \mathbf{0} = \mathbf{A}, \qquad \mathbf{A} - \mathbf{A} = \mathbf{0}, \qquad \text{and} \qquad \mathbf{0A} = \mathbf{0} = \mathbf{A0},$$

where **0** is the appropriate size in each case for the operations to be defined.

On certain occasions, it is useful to partition a matrix into several smaller matrices called **submatrices**. For example, one possible way of partitioning a 3 × 4 matrix would be

$$\mathbf{A} = \left[\begin{array}{c|ccc} a_{11} & a_{12} & a_{13} & a_{14} \\ \hline a_{21} & a_{22} & a_{23} & a_{24} \\ a_{31} & a_{32} & a_{33} & a_{34} \end{array}\right] = \begin{bmatrix} a_{11} & \mathbf{A}_{12} \\ \mathbf{A}_{21} & \mathbf{A}_{22} \end{bmatrix},$$

where $\mathbf{A}_{12} = [a_{12}, \; a_{13}, \; a_{14}]$, $\quad \mathbf{A}_{21} = \begin{bmatrix} a_{21} \\ a_{31} \end{bmatrix}$, $\quad \mathbf{A}_{22} = \begin{bmatrix} a_{22} & a_{23} & a_{24} \\ a_{32} & a_{33} & a_{34} \end{bmatrix}$

all are submatrices. Rather than perform operations element by element on such partitioned matrices, we can instead do them in terms of the submatrices, provided the partitionings are such that the operations are defined. For example, if **B** is a partitioned 4 × 1 matrix such that

$$\mathbf{B} = \left[\begin{array}{c} b_1 \\ \hline b_2 \\ b_3 \\ b_4 \end{array}\right] = \begin{bmatrix} b_1 \\ \mathbf{B}_2 \end{bmatrix},$$

then

$$\mathbf{AB} = \left[\begin{array}{c} a_{11}b_1 + \mathbf{A}_{12}\mathbf{B}_2 \\ \hline \mathbf{A}_{21}b_1 + \mathbf{A}_{22}\mathbf{B}_2 \end{array}\right].$$

A special kind of matrix that plays an important role in matrix theory is the kind that has either a *single row* or a *single column*. Such matrices are often referred to as **vectors**. Thus

$$\mathbf{x} = [x_1, x_2, \ldots, x_n]$$

is a **row vector**, and

$$\mathbf{x} = \begin{bmatrix} x_1 \\ x_2 \\ \vdots \\ x_n \end{bmatrix}$$

is a **column vector**. (Vectors are denoted in this book by **boldface lowercase letters.**) These vectors also are sometimes called *n-vectors* to indicate that they have n elements. For example,

$$\mathbf{x} = [1, 4, -2, \tfrac{1}{3}, 7]$$

is a 5-vector. A **null vector 0** is either a row vector or a column vector whose elements are *all zeroes,* i.e.,

$$\mathbf{0} = [0, 0, \ldots, 0], \qquad \mathbf{0} = \begin{bmatrix} 0 \\ 0 \\ \vdots \\ 0 \end{bmatrix}$$

(Although the same symbol **0** is used for either kind of *null vector,* as well as for a *null matrix,* the context normally will identify which it is.)

One reason vectors play an important role in matrix theory is that any $m \times n$ matrix can be partitioned into either m row vectors or n column vectors, and important properties of the matrix can be analyzed in terms of these vectors. To amplify, consider a set of n-vectors, $\mathbf{x}_1, \mathbf{x}_2, \ldots, \mathbf{x}_m$, of the same type (i.e., they are either all row vectors or all column vectors).

Definition: A set of vectors $\mathbf{x}_1, \mathbf{x}_2, \ldots, \mathbf{x}_m$ is said to be **linearly dependent** if there exist m numbers (denoted by $c_1, c_2, \ldots, c_m$), some of which are not zero, such that

$$c_1\mathbf{x}_1 + c_2\mathbf{x}_2 + \cdots + c_m\mathbf{x}_m = \mathbf{0}.$$

Otherwise, the set is said to be **linearly independent**.

To illustrate, if $m = 3$ and

$$\mathbf{x}_1 = [1, 1, 1]$$

$$\mathbf{x}_2 = [0, 1, 1]$$

$$\mathbf{x}_3 = [2, 5, 5],$$

then

$$2\mathbf{x}_1 + 3\mathbf{x}_2 - \mathbf{x}_3 = \mathbf{0},$$

so that

$$\mathbf{x}_3 = 2\mathbf{x}_1 + 3\mathbf{x}_2.$$

Thus $\mathbf{x}_1, \mathbf{x}_2, \mathbf{x}_3$ would be linearly dependent because one of them is a linear combination of the others. However, if $\mathbf{x}_3$ were changed to

$$\mathbf{x}_3 = [2, 5, 6]$$

instead, then $\mathbf{x}_1, \mathbf{x}_2, \mathbf{x}_3$ would be linearly independent.

Definition: The **rank** of a *set* of vectors is the largest number of *linearly independent vectors* that can be chosen from the set.

Continuing the preceding example, the rank of the set of vectors $\mathbf{x}_1, \mathbf{x}_2, \mathbf{x}_3$ was 2, but it became 3 after changing $\mathbf{x}_3$.

Definition: A **basis** for a *set* of vectors is a *collection* of linearly independent vectors taken from the set such that every vector in the set is a linear combination of the vectors in the collection (i.e., every vector in the set equals the sum of certain multiples of the vectors in the collection).

To illustrate, $\mathbf{x}_1$ and $\mathbf{x}_2$ constituted a basis for $\mathbf{x}_1$, $\mathbf{x}_2$, $\mathbf{x}_3$ in the preceding example before $\mathbf{x}_3$ was changed.

Theorem A3.1: A *collection* of r linearly independent vectors chosen from a set of vectors is a *basis* for the set if and only if the set has rank r.

Given the preceding results regarding vectors, it is now possible to present certain important concepts regarding matrices.

Definition: The **row rank** of a matrix is the *rank* of its set of *row vectors*. The **column rank** of a matrix is the *rank* of its *column vectors*.

For example, if the matrix **A** is

$$\mathbf{A} = \begin{bmatrix} 1 & 1 & 1 \\ 0 & 1 & 1 \\ 2 & 5 & 5 \end{bmatrix},$$

then its row rank was shown to be 2. Note that the column rank of **A** is also 2. This fact is no coincidence, as the following general theorem indicates.

Theorem A3.2: The *row rank* and *column rank* of a matrix are *equal*.

Thus it is only necessary to speak of *the rank* of a matrix.

The final concept to be discussed is that of the **inverse of a matrix**. For any nonzero number k, there exists a reciprocal or inverse, $k^{-1} = 1/k$, such that

$$kk^{-1} = k^{-1}k = 1.$$

Is there an analogous concept that is valid in matrix theory? In other words, for a given matrix **A** other than the null matrix, does there exist a matrix $\mathbf{A}^{-1}$ such that

$$\mathbf{AA}^{-1} = \mathbf{A}^{-1}\mathbf{A} = \mathbf{I}?$$

If **A** is not a square matrix (i.e., if the number of rows and columns of **A** differ), the answer is *never*, because these matrix products would necessarily have a different number of rows for the multiplication to be defined (so that the equality operation would not be defined). However, if **A** is square, then the answer is *under certain circumstances*, as indicated in Theorem A3.3.

Definition: A matrix is called **nonsingular** if its rank equals both the number of rows and the number of columns. Otherwise, it is called **singular**.

Thus only square matrices can be *nonsingular*. A useful way of testing for nonsingularity is provided by the fact that a square matrix is nonsingular if and only if *its determinant is nonzero*.

Theorem A3.3: (*a*) If **A** is *nonsingular*, there is a unique nonsingular matrix $\mathbf{A}^{-1}$, called the **inverse** of **A**, such that $\mathbf{AA}^{-1} = \mathbf{I} = \mathbf{A}^{-1}\mathbf{A}$.
(*b*) If **A** is *nonsingular* and **B** is a matrix for which either $\mathbf{AB} = \mathbf{I}$ or $\mathbf{BA} = \mathbf{I}$, then $\mathbf{B} = \mathbf{A}^{-1}$.
(*c*) *Only nonsingular* matrices have *inverses*.

To illustrate, consider the matrix

$$\mathbf{A} = \begin{bmatrix} 5 & -4 \\ 1 & -1 \end{bmatrix}.$$

Notice that the rank of **A** is 2, so it is *nonsingular*. Therefore, **A** must have an *inverse,* which happens to be

$$\mathbf{A}^{-1} = \begin{bmatrix} 1 & -4 \\ 1 & -5 \end{bmatrix}.$$

Hence,

$$\mathbf{A}\mathbf{A}^{-1} = \begin{bmatrix} 5 & -4 \\ 1 & -1 \end{bmatrix}\begin{bmatrix} 1 & -4 \\ 1 & -5 \end{bmatrix} = \begin{bmatrix} 1 & 0 \\ 0 & 1 \end{bmatrix},$$

and

$$\mathbf{A}^{-1}\mathbf{A} = \begin{bmatrix} 1 & -4 \\ 1 & -5 \end{bmatrix}\begin{bmatrix} 5 & -4 \\ 1 & -1 \end{bmatrix} = \begin{bmatrix} 1 & 0 \\ 0 & 1 \end{bmatrix}.$$

Appendix 4

Simultaneous Linear Equations

Consider the system of simultaneous linear equations

$$
\begin{aligned}
a_{11}x_1 + a_{12}x_2 + \cdots + a_{1n}x_n &= b_1, \\
a_{21}x_1 + a_{22}x_2 + \cdots + a_{2n}x_n &= b_2, \\
&\vdots \\
a_{m1}x_1 + a_{m2}x_2 + \cdots + a_{mn}x_n &= b_m.
\end{aligned}
$$

It is commonly assumed that this system has a solution, and a unique solution, if and only if $m = n$. However, this assumption is an oversimplification. It raises the questions: Under what conditions will these equations have a simultaneous solution? Given that they do, when will there be only one such solution? If there is a unique solution, how can it be identified in a systematic way? These questions are the ones we explore in this appendix. The discussion of the first two questions assumes that you are familiar with the basic information about matrices in Appendix 3.

The preceding system of equations can also be written in matrix form as

$$\mathbf{Ax} = \mathbf{b},$$

where
$$\mathbf{A} = \begin{bmatrix} a_{11} & a_{12} & \cdots & a_{1n} \\ a_{21} & a_{22} & \cdots & a_{2n} \\ & & \vdots & \\ a_{m1} & a_{m2} & \cdots & a_{mn} \end{bmatrix}, \quad \mathbf{x} = \begin{bmatrix} x_1 \\ x_2 \\ \vdots \\ x_n \end{bmatrix}, \quad \mathbf{b} = \begin{bmatrix} b_1 \\ b_2 \\ \vdots \\ b_m \end{bmatrix}.$$

The first two questions can be answered immediately in terms of the properties of these matrices. First, the system of equations possesses at least one solution if and only if the *rank* of **A** equals the *rank* of [**A**, **b**]. (Notice that equality is guaranteed if the *rank* of **A** equals m.) This result follows immediately from the definitions of *rank* and *linear independence* given in Appendix 3, because if the *rank* of [**A**, **b**] exceeds the *rank* of **A** by 1 (the only other possibility), then **b** is *linearly independent* of the column vectors of **A** (that is, **b** cannot equal any linear combination **Ax** of these vectors).

Second, given that these ranks are equal, there are then two possibilities. If the *rank* of **A** is n (its maximum possible value), then the system of equations will possess exactly *one solution*. [This result follows from Theorem A3.1, the definition of a *basis,* and part (*b*) of Theorem A3.3.] If the *rank* of **A** is *less* than n, then there will exist an *infinite number of solutions*. (This result follows from the fact that for any *basis* of the column vectors of **A**, the $\mathbf{x}_j$ corresponding to column vectors not in this basis can be assigned any value, and there will still exist a solution for the other variables as before.)

Finally, it should be noted that if **A** and [**A**, **b**] have a *common rank* r such that $r < m$, then $(m - r)$ of the equations must be linear combinations of the other ones, so that these $(m - r)$ *redundant* equations can be deleted without affecting the solution(s). It then follows from the preceding results that this system of equations (with or without the redundant equations) possesses at least one solution, where the number of solutions is *one* if $r = n$ or *infinite* if $r < n$.

Now consider how to find a solution to the system of equations. Assume for the moment that $m = n$ and **A** is nonsingular, so that a unique solution exists. This solution can be obtained by the **Gauss-Jordan method of elimination** (commonly called **Gaussian elimination**), which proceeds as follows. To begin, eliminate the first variable from all but one (say, the first) of the equations by adding an appropriate multiple (positive or negative) of this equation to each of the others. (For convenience, this one equation would be divided by the coefficient of this variable, so that the final value of this coefficient is 1.) Next, proceed in the same way to eliminate the second variable from all equations except one new one (say, the second). Then repeat this procedure for the third variable, the fourth variable, and so on, until each of the n variables remains in only one of the equations and each of the n equations contains exactly one of these variables. The desired solution can then be read from the equations directly.

To illustrate the *Gauss-Jordan method of elimination,* we consider the following system of linear equations:

(1) $$x_1 - x_2 + 4x_3 = 10$$

(2) $$-x_1 + 3x_2 = 10$$

(3) $$2x_2 + 5x_3 = 22.$$

The method begins by eliminating x_1 from all but the first equation. This first step is executed simply by adding Eq. (1) to Eq. (2), which yields

(1) $$x_1 - x_2 + 4x_3 = 10$$

(2) $$2x_2 + 4x_3 = 20$$

(3) $$2x_2 + 5x_3 = 22.$$

The next step is to eliminate x_2 from all but the second equation. Begin this step by dividing Eq. (2) by 2, so that x_2 will have a coefficient of +1, as follows:

$$(1) \qquad x_1 - x_2 + 4x_3 = 10$$

$$(2) \qquad x_2 + 2x_3 = 10$$

$$(3) \qquad 2x_2 + 5x_3 = 22.$$

Then add Eq. (2) to Eq. (1), and subtract two times Eq. (2) from Eq. (3), which yields

$$(1) \qquad x_1 + 6x_3 = 20$$

$$(2) \qquad x_2 + 2x_3 = 10$$

$$(3) \qquad x_3 = 2.$$

The final step is to eliminate x_3 from all but the third equation. This step requires subtracting six times Eq. (3) from Eq. (1) and subtracting two times Eq. (3) from Eq. (2), which yields

$$(1) \qquad x_1 = 8$$

$$(2) \qquad x_2 = 6$$

$$(3) \qquad x_3 = 2.$$

Thus the desired solution is $(x_1, x_2, x_3) = (8, 6, 2)$, and the procedure is completed.

Now consider briefly what happens if the Gauss-Jordan method of elimination is applied when $m \neq n$ and/or **A** is singular. As we discussed earlier, there are three possible cases to consider. First, if the rank of [**A**, **b**] exceeds the rank of **A** by 1, then *no solution* to the system of equations will exist. In this case, the Gauss-Jordan method obtains an equation where the left-hand side has vanished (i.e., all the coefficients of the variables are zero), whereas the right-hand side is nonzero. This signpost indicates that no solution exists, so there is no reason to proceed further.

The second case is where both of these ranks are equal to n, so that a *unique solution* exists. This case implies that $m \geq n$. If $m = n$, then the previous assumptions must hold and no difficulty arises. Therefore, suppose that $m > n$, so that there are $(m - n)$ redundant equations. In this case, all these redundant equations are eliminated (i.e., both the left-hand and right-hand sides would become zero) during the process of executing the Gauss-Jordan method, so the unique solution is identified just as it was before.

The final case is where both the ranks are equal to r, where $r < n$, so that the system of equations possesses an *infinite number of solutions*. In this case, at the completion of the Gauss-Jordan method, each of the r variables remains in only one of the equations, and each of the r equations (any additional equations have vanished) contains exactly one of these variables. However, each of the other $(n - r)$ variables either vanishes or remains in some of the equations. Therefore, any solution obtained by assigning arbitrary values to the $(n - r)$ variables, and then identifying the respective values of the r variables from the single final equation in which each one appears, is a solution to the system of simultaneous equations. Equivalently, the transfer of these $(n - r)$ variables to the right-hand side of the equations (either before or after the method is executed) identifies the solutions for the r variables as a function of these extra variables.

Appendix 5 Tables

Table A5.1 **Areas Under the Normal Curve from K_α to ∞**

$$P\{\text{normal} \geq K_\alpha\} = \int_{K_\alpha}^{\infty} \frac{1}{\sqrt{2\pi}} e^{-x^2/2}\, dx = \alpha$$

K_α	.00	.01	.02	.03	.04	.05	.06	.07	.08	.09
0.0	.5000	.4960	.4920	.4880	.4840	.4801	.4761	.4721	.4681	.4641
0.1	.4602	.4562	.4522	.4483	.4443	.4404	.4364	.4325	.4286	.4247
0.2	.4207	.4168	.4129	.4090	.4052	.4013	.3974	.3936	.3897	.3859
0.3	.3821	.3783	.3745	.3707	.3669	.3632	.3594	.3557	.3520	.3483
0.4	.3446	.3409	.3372	.3336	.3300	.3264	.3228	.3192	.3156	.3121
0.5	.3085	.3050	.3015	.2981	.2946	.2912	.2877	.2843	.2810	.2776
0.6	.2743	.2709	.2676	.2643	.2611	.2578	.2546	.2514	.2483	.2451
0.7	.2420	.2389	.2358	.2327	.2296	.2266	.2236	.2206	.2177	.2148
0.8	.2119	.2090	.2061	.2033	.2005	.1977	.1949	.1922	.1894	.1867
0.9	.1841	.1814	.1788	.1762	.1736	.1711	.1685	.1660	.1635	.1611
1.0	.1587	.1562	.1539	.1515	.1492	.1469	.1446	.1423	.1401	.1379
1.1	.1357	.1335	.1314	.1292	.1271	.1251	.1230	.1210	.1190	.1170
1.2	.1151	.1131	.1112	.1093	.1075	.1056	.1038	.1020	.1003	.0985
1.3	.0968	.0951	.0934	.0918	.0901	.0885	.0869	.0853	.0838	.0823
1.4	.0808	.0793	.0778	.0764	.0749	.0735	.0721	.0708	.0694	.0681
1.5	.0668	.0655	.0643	.0630	.0618	.0606	.0594	.0582	.0571	.0559
1.6	.0548	.0537	.0526	.0516	.0505	.0495	.0485	.0475	.0465	.0455
1.7	.0446	.0436	.0427	.0418	.0409	.0401	.0392	.0384	.0375	.0367
1.8	.0359	.0351	.0344	.0336	.0329	.0322	.0314	.0307	.0301	.0294
1.9	.0287	.0281	.0274	.0268	.0262	.0256	.0250	.0244	.0239	.0233
2.0	.0228	.0222	.0217	.0212	.0207	.0202	.0197	.0192	.0188	.0183
2.1	.0179	.0174	.0170	.0166	.0162	.0158	.0154	.0150	.0146	.0143
2.2	.0139	.0136	.0132	.0129	.0125	.0122	.0119	.0116	.0113	.0110
2.3	.0107	.0104	.0102	.00990	.00964	.00939	.00914	.00889	.00866	.00842
2.4	.00820	.00798	.00776	.00755	.00734	.00714	.00695	.00676	.00657	.00639
2.5	.00621	.00604	.00587	.00570	.00554	.00539	.00523	.00508	.00494	.00480
2.6	.00466	.00453	.00440	.00427	.00415	.00402	.00391	.00379	.00368	.00357
2.7	.00347	.00336	.00326	.00317	.00307	.00298	.00289	.00280	.00272	.00264
2.8	.00256	.00248	.00240	.00233	.00226	.00219	.00212	.00205	.00199	.00193
2.9	.00187	.00181	.00175	.00169	.00164	.00159	.00154	.00149	.00144	.00139

K_α	.0	.1	.2	.3	.4	.5	.6	.7	.8	.9
3	.00135	$.0^{3}968$	$.0^{3}687$	$.0^{3}483$	$.0^{3}337$	$.0^{3}233$	$.0^{3}159$	$.0^{3}108$	$.0^{4}723$	$.0^{4}481$
4	$.0^{4}317$	$.0^{4}207$	$.0^{4}133$	$.0^{5}854$	$.0^{5}541$	$.0^{5}340$	$.0^{5}211$	$.0^{5}130$	$.0^{6}793$	$.0^{6}479$
5	$.0^{6}287$	$.0^{6}170$	$.0^{7}996$	$.0^{7}579$	$.0^{7}333$	$.0^{7}190$	$.0^{7}107$	$.0^{8}599$	$.0^{8}332$	$.0^{8}182$
6	$.0^{9}987$	$.0^{9}530$	$.0^{9}282$	$.0^{9}149$	$.0^{10}777$	$.0^{10}402$	$.0^{10}206$	$.0^{10}104$	$.0^{11}523$	$.0^{11}260$

Source: Croxton, F. E.: *Tables of Areas in Two Tails and in One Tail of the Normal Curve*. Copyright 1949 by Prentice-Hall, Inc., Englewood Cliffs, N.J.

Table A5.2 **100 α Percentage Points of Student's *t* Distribution**
***P*{Student's *t* with *v* Degrees of Freedom ≥ Tabled Value} = α**

v \ α	0.40	0.25	0.10	0.05	0.025	0.01	0.005	0.0025	0.001	0.0005
1	0.325	1.000	3.078	6.314	12.706	31.821	63.657	127.32	318.31	636.62
2	0.289	0.816	1.886	2.920	4.303	6.965	9.925	14.089	22.327	31.598
3	0.277	0.765	1.638	2.353	3.182	4.541	5.841	7.453	10.214	12.924
4	0.271	0.741	1.533	2.132	2.776	3.747	4.604	5.598	7.173	8.610
5	0.267	0.727	1.476	2.015	2.571	3.365	4.032	4.773	5.893	6.869
6	0.265	0.718	1.440	1.943	2.447	3.143	3.707	4.317	5.208	5.959
7	0.263	0.711	1.415	1.895	2.365	2.998	3.499	4.029	4.785	5.408
8	0.262	0.706	1.397	1.860	2.306	2.896	3.355	3.833	4.501	5.041
9	0.261	0.703	1.383	1.833	2.262	2.821	3.250	3.690	4.297	4.781
10	0.260	0.700	1.372	1.812	2.228	2.764	3.169	3.581	4.144	4.587
11	0.260	0.697	1.363	1.796	2.201	2.718	3.106	3.497	4.025	4.437
12	0.259	0.695	1.356	1.782	2.179	2.681	3.055	3.428	3.930	4.318
13	0.259	0.694	1.350	1.771	2.160	2.650	3.012	3.372	3.852	4.221
14	0.258	0.692	1.345	1.761	2.145	2.624	2.977	3.326	3.787	4.140
15	0.258	0.691	1.341	1.753	2.131	2.602	2.947	3.286	3.733	4.073
16	0.258	0.690	1.337	1.746	2.120	2.583	2.921	3.252	3.686	4.015
17	0.257	0.689	1.333	1.740	2.110	2.567	2.898	3.222	3.646	3.965
18	0.257	0.688	1.330	1.734	2.101	2.552	2.878	3.197	3.610	3.922
19	0.257	0.688	1.328	1.729	2.093	2.539	2.861	3.174	3.579	3.883
20	0.257	0.687	1.325	1.725	2.086	2.528	2.845	3.153	3.552	3.850
21	0.257	0.686	1.323	1.721	2.080	2.518	2.831	3.135	3.527	3.819
22	0.256	0.686	1.321	1.717	2.074	2.508	2.819	3.119	3.505	3.792
23	0.256	0.685	1.319	1.714	2.069	2.500	2.807	3.104	3.485	3.767
24	0.256	0.685	1.318	1.711	2.064	2.492	2.797	3.091	3.467	3.745
25	0.256	0.684	1.316	1.708	2.060	2.485	2.787	3.078	3.450	3.725
26	0.256	0.684	1.315	1.706	2.056	2.479	2.779	3.067	3.435	3.707
27	0.256	0.684	1.314	1.703	2.052	2.473	2.771	3.057	3.421	3.690
28	0.256	0.683	1.313	1.701	2.048	2.467	2.763	3.047	3.408	3.674
29	0.256	0.683	1.311	1.699	2.045	2.462	2.756	3.038	3.396	3.659
30	0.256	0.683	1.310	1.697	2.042	2.457	2.750	3.030	3.385	3.646
40	0.255	0.681	1.303	1.684	2.021	2.423	2.704	2.971	3.307	3.551
60	0.254	0.679	1.296	1.671	2.000	2.390	2.660	2.915	3.232	3.460
120	0.254	0.677	1.289	1.658	1.980	2.358	2.617	2.860	3.160	3.373
∞	0.253	0.674	1.282	1.645	1.960	2.326	2.576	2.807	3.090	3.291

Source: Table 12 of *Biometrika Tables for Statisticians,* vol. I, 3d ed., 1966, by permission of the Biometrika Trustees.

Table A5.3 **100 α Percentage Points of Chi-Square Distribution**
***P*{Chi Square with *v* Degrees of Freedom $\geq$ Tabled Value} = α**

v \ α	0.995	0.99	0.975	0.95	0.90	0.75	0.50
1	0.0^4393	0.0^3157	0.0^3982	0.00393	0.0158	0.102	0.455
2	0.0100	0.0201	0.0506	0.103	0.211	0.575	1.386
3	0.0717	0.115	0.216	0.352	0.584	1.213	2.366
4	0.207	0.297	0.484	0.711	1.064	1.923	3.357
5	0.412	0.554	0.831	1.145	1.610	2.675	4.351
6	0.676	0.872	1.237	1.635	2.204	3.455	5.348
7	0.989	1.239	1.690	2.167	2.833	4.255	6.346
8	1.344	1.646	2.180	2.733	3.490	5.071	7.344
9	1.735	2.088	2.700	3.325	4.168	5.899	8.343
10	2.156	2.558	3.247	3.940	4.865	6.737	9.342
11	2.603	3.053	3.816	4.575	5.578	7.584	10.341
12	3.074	3.571	4.404	5.226	6.304	8.438	11.340
13	3.565	4.107	5.009	5.892	7.042	9.299	12.340
14	4.075	4.660	5.629	6.571	7.790	10.165	13.339
15	4.601	5.229	6.262	7.261	8.547	11.036	14.339
16	5.142	5.812	6.908	7.962	9.312	11.912	15.338
17	5.697	6.408	7.564	8.672	10.085	12.792	16.338
18	6.265	7.015	8.231	9.390	10.865	13.675	17.338
19	6.844	7.633	8.907	10.117	11.651	14.562	18.338
20	7.434	8.260	9.591	10.851	12.443	15.452	19.337
21	8.034	8.897	10.283	11.591	13.240	16.344	20.337
22	8.643	9.542	10.982	12.338	14.041	17.240	21.337
23	9.260	10.196	11.688	13.091	14.848	18.137	22.337
24	9.886	10.856	12.401	13.848	15.659	19.037	23.337
25	10.520	11.524	13.120	14.611	16.473	19.939	24.337
26	11.160	12.198	13.844	15.379	17.292	20.843	25.336
27	11.808	12.879	14.573	16.151	18.114	21.749	26.336
28	12.461	13.565	15.308	16.928	18.939	22.657	27.336
29	13.121	14.256	16.047	17.708	19.768	23.567	28.336
30	13.787	14.953	16.791	18.493	20.599	24.478	29.336
40	20.707	22.164	24.433	26.509	29.051	33.660	39.335
50	27.991	29.707	32.357	34.764	37.689	42.942	49.335
60	35.535	37.485	40.482	43.188	46.459	52.294	59.335
70	43.275	45.442	48.758	51.739	55.329	61.698	69.334
80	51.172	53.540	57.153	60.391	64.278	71.145	79.334
90	59.196	61.754	65.647	69.126	73.291	80.625	89.334
100	67.328	70.065	74.222	77.929	82.358	90.133	99.334
K_α	−2.576	−2.326	−1.960	−1.645	−1.282	−0.6745	0.000

Source: Abridged from Table 8 of *Biometrika Tables for Statisticians,* vol. I, 3d ed., 1966, by permission of the Biometrika Trustees.

Table A5.3 ***(continued)***

0.25	0.10	0.05	0.025	0.01	0.005	0.001	α / v
1.323	2.706	3.841	5.024	6.635	7.879	10.828	1
2.773	4.605	5.991	7.378	9.210	10.597	13.816	2
4.108	6.251	7.815	9.348	11.345	12.838	16.266	3
5.385	7.779	9.488	11.143	13.277	14.860	18.467	4
6.626	9.236	11.070	12.832	15.086	16.750	20.515	5
7.841	10.645	12.592	14.449	16.812	18.548	22.458	6
9.037	12.017	14.067	16.013	18.475	20.278	24.322	7
10.219	13.362	15.507	17.535	20.090	21.955	26.125	8
11.389	14.684	16.919	19.023	21.666	23.589	27.877	9
12.549	15.987	18.307	20.483	23.209	25.188	29.588	10
13.701	17.275	19.675	21.920	24.725	26.757	31.264	11
14.845	18.549	21.026	23.337	26.217	28.300	32.909	12
15.984	19.812	22.362	24.736	27.688	29.819	34.528	13
17.117	21.064	23.685	26.119	29.141	31.319	36.123	14
18.245	22.307	24.996	27.488	30.578	32.801	37.697	15
19.369	23.542	26.296	28.845	32.000	34.267	39.252	16
20.489	24.769	27.587	30.191	33.409	35.718	40.790	17
21.605	25.989	28.869	31.526	34.805	37.156	43.312	18
22.718	27.204	30.144	32.852	36.191	38.582	43.820	19
23.828	28.412	31.410	34.170	37.566	39.997	45.315	20
24.935	29.615	32.671	35.479	38.932	41.401	46.797	21
26.039	30.813	33.924	36.781	40.289	42.796	48.268	22
27.141	32.007	35.172	38.076	41.638	44.181	49.728	23
28.241	33.196	36.415	39.364	42.980	45.558	51.179	24
29.339	34.382	37.652	40.646	44.314	46.928	52.620	25
30.434	35.563	38.885	41.923	45.642	48.290	54.052	26
31.528	36.741	40.113	43.194	46.963	49.645	55.476	27
32.620	37.916	41.337	44.461	48.278	50.993	56.892	28
33.711	39.087	42.557	45.722	49.588	52.336	58.302	29
34.800	40.256	43.773	46.979	50.892	53.672	59.703	30
45.616	51.805	55.758	59.342	63.691	66.766	73.402	40
56.334	63.167	67.505	71.420	76.154	79.490	86.661	50
66.981	74.397	79.082	83.298	88.379	91.952	99.607	60
77.577	85.527	90.531	95.023	100.425	104.215	112.317	70
88.130	96.578	101.879	106.629	112.329	116.321	124.839	80
98.650	107.565	113.145	118.136	124.116	128.299	137.208	90
109.141	118.498	124.342	129.561	135.807	140.169	149.449	100
+0.6745	+1.282	+1.645	+1.960	+2.326	+2.576	+3.090	K_α

For $v > 100$ take

$$\chi^2 = v\left(1 - \frac{2}{9v} + K_\alpha\sqrt{\frac{2}{9v}}\right)^2 \quad \text{or} \quad \chi^2 = \frac{1}{2}\left[K_\alpha + \sqrt{(2v-1)}\right]^2,$$

according to the degree of accuracy required. K_α is the standardized normal deviate corresponding to α and is shown in the bottom line of the table.

Table A5.4 Summation of Terms of the Poisson Distribution: 1,000$P\{$Poisson with Parameter $\lambda \leq c\}$

c \ λ	0.01	0.02	0.03	0.04	0.05	0.06	0.07	0.08	0.09
0	990	980	970	961	951	942	932	923	914
1	1,000	1,000	1,000	999	999	998	998	997	996
2				1,000	1,000	1,000	1,000	1,000	1,000

c \ λ	0.10	0.15	0.20	0.25	0.30	0.35	0.40	0.45	0.50
0	905	861	819	779	741	705	670	638	607
1	995	990	982	974	963	951	938	925	910
2	1,000	999	999	998	996	994	992	989	986
3		1,000	1,000	1,000	1,000	1,000	999	999	998
4							1,000	1,000	1,000

c \ λ	0.55	0.60	0.65	0.70	0.75	0.80	0.85	0.90	0.95	1.00
0	577	549	522	497	472	449	427	407	387	368
1	894	878	861	844	827	809	791	772	754	736
2	982	977	972	966	959	953	945	937	929	920
3	998	997	996	994	993	991	989	987	984	981
4	1,000	1,000	999	999	999	999	998	998	997	996
5			1,000	1,000	1,000	1,000	1,000	1,000	1,000	999
6										1,000

c \ λ	1.05	1.10	1.15	1.20	1.25	1.30	1.35	1.40	1.45	1.50
0	350	333	317	301	287	273	259	247	235	223
1	717	699	681	663	645	627	609	592	575	558
2	910	900	890	879	868	857	845	833	821	809
3	978	974	970	966	962	957	952	946	940	934
4	996	995	993	992	991	989	988	986	984	981
5	999	999	999	998	998	998	997	997	996	996
6	1,000	1,000	1,000	1,000	1,000	1,000	999	999	999	999
7							1,000	1,000	1,000	1,000

c \ λ	1.55	1.60	1.65	1.70	1.75	1.80	1.85	1.90	1.95	2.00
0	212	202	192	183	174	165	157	150	142	135
1	541	525	509	493	478	463	448	434	420	406
2	796	783	770	757	744	731	717	704	690	677
3	928	921	914	907	899	891	883	875	866	857
4	979	976	973	970	967	964	960	956	952	947
5	995	994	993	992	991	990	988	987	985	983
6	999	999	998	998	998	997	997	997	996	995
7	1,000	1,000	1,000	1,000	1,000	999	999	999	999	999
8						1,000	1,000	1,000	1,000	1,000

Source: Reproduced by permission from Hillier, F. S., and F. D. Lo: *Tables for Multiple-Server Queueing Systems Involving Erlang Distributions,* Technical Report #14, NSF GK-2925, Department of Operations Research, Stanford University, Stanford, Calif., December 28, 1971.

c \ λ	2.10	2.20	2.30	2.40	2.50	2.60	2.70	2.80	2.90	3.00
0	122	111	100	091	082	074	067	061	055	050
1	380	355	331	308	287	267	249	231	215	199
2	650	623	596	570	544	518	494	469	446	423
3	839	819	799	779	758	736	714	692	670	647
4	938	928	916	904	891	877	863	848	832	815
5	980	975	970	964	958	951	943	935	926	916
6	994	993	991	988	986	983	979	976	971	966
7	999	998	997	997	996	995	993	992	990	988
8	1,000	1,000	999	999	999	999	998	998	997	996
9			1,000	1,000	1,000	1,000	999	999	999	999
10							1,000	1,000	1,000	1,000

c \ λ	3.10	3.20	3.30	3.40	3.50	3.60	3.70	3.80	3.90	4.00
0	045	041	037	033	030	027	025	022	020	018
1	185	171	159	147	136	126	116	107	099	092
2	401	380	359	340	321	303	285	269	253	238
3	625	603	580	558	537	515	494	473	453	433
4	798	781	763	744	725	706	687	668	648	629
5	906	895	883	871	858	844	830	816	801	785
6	961	955	949	942	935	927	918	909	899	889
7	986	983	980	977	973	969	965	960	955	949
8	995	994	993	992	990	988	986	984	981	979
9	999	998	998	997	997	996	995	994	993	992
10	1,000	1,000	999	999	999	999	998	998	998	997
11			1,000	1,000	1,000	1,000	1,000	999	999	999
12								1,000	1,000	1,000

c \ λ	4.10	4.20	4.30	4.40	4.50	4.60	4.70	4.80	4.90	5.00
0	017	015	014	012	011	010	009	008	007	007
1	085	078	072	066	061	056	052	048	044	040
2	224	210	197	185	174	163	152	143	133	125
3	414	395	377	359	342	326	310	294	279	265
4	609	590	570	551	532	513	495	476	458	440
5	769	753	737	720	703	686	668	651	634	616
6	879	867	856	844	831	818	805	791	777	762
7	943	936	929	921	913	905	896	887	877	867
8	976	972	968	964	960	955	950	944	938	932
9	990	989	987	985	983	980	978	975	972	968
10	997	996	995	994	993	992	991	990	988	986
11	999	999	998	998	998	997	997	996	995	995
12	1,000	1,000	999	999	999	999	999	999	998	998
13			1,000	1,000	1,000	1,000	1,000	1,000	999	999
14									1,000	1,000

Table A5.4 (continued)

c \ λ	5.10	5.20	5.30	5.40	5.50	5.60	5.70	5.80	5.90	6.00
0	006	006	005	005	004	004	003	003	003	002
1	037	034	031	029	027	024	022	021	019	017
2	116	109	102	095	088	082	077	072	067	062
3	251	238	225	213	202	191	180	170	160	151
4	423	406	390	373	358	342	327	313	299	285
5	598	581	563	546	529	512	495	478	462	446
6	747	732	717	702	686	670	654	638	622	606
7	856	845	833	822	809	797	784	771	758	744
8	925	918	911	903	894	886	877	867	857	847
9	964	960	956	951	946	941	935	929	923	916
10	984	982	980	977	975	972	969	965	961	957
11	994	993	992	990	989	988	986	984	982	980
12	998	997	997	996	996	995	994	993	992	991
13	999	999	999	999	998	998	998	997	997	996
14	1,000	1,000	1,000	1,000	999	999	999	999	999	999
15					1,000	1,000	1,000	1,000	1,000	1,000

c \ λ	6.10	6.20	6.30	6.40	6.50	6.60	6.70	6.80	6.90	7.00
0	002	002	002	002	002	001	001	001	001	001
1	016	015	013	012	011	010	009	009	008	007
2	058	054	050	046	043	040	037	034	032	030
3	143	134	126	119	112	105	099	093	087	082
4	272	259	247	235	224	213	202	192	182	173
5	430	414	399	384	369	355	341	327	314	301
6	590	574	558	542	527	511	495	480	465	450
7	730	716	702	687	673	658	643	628	614	599
8	837	826	815	803	792	780	767	755	742	729
9	909	902	894	886	877	869	860	850	840	830
10	953	949	944	939	933	927	921	915	908	901
11	978	975	972	969	966	963	959	955	951	947
12	990	989	987	986	984	982	980	978	976	973
13	996	995	995	994	993	992	991	990	989	987
14	998	998	998	997	997	997	996	996	995	994
15	999	999	999	999	999	999	998	998	998	998
16	1,000	1,000	1,000	1,000	1,000	999	999	999	999	999
17						1,000	1,000	1,000	1,000	1,000

c \ λ	7.10	7.20	7.30	7.40	7.50	8.00	8.50	9.00	9.50	10.00
0	001	001	001	001	001	000	000	000	000	000
1	007	006	006	005	005	003	002	001	001	000
2	027	025	024	022	020	014	009	006	004	003
3	077	072	067	063	059	042	030	021	015	010
4	164	156	147	140	132	100	074	055	040	029
5	288	276	264	253	241	191	150	116	089	067
6	435	420	406	392	378	313	256	207	165	130
7	584	569	554	539	525	453	386	324	269	220
8	716	703	689	676	662	593	523	456	392	333
9	820	810	799	788	776	717	653	587	522	458
10	894	887	879	871	862	816	763	706	645	583
11	942	937	932	926	921	888	849	803	752	697
12	970	967	964	961	957	936	909	876	836	792
13	986	984	982	980	978	966	949	926	898	864
14	994	993	992	991	990	983	973	959	940	917
15	997	997	996	996	995	992	986	978	967	951
16	999	999	999	998	998	996	993	989	982	973
17	1,000	1,000	999	999	999	998	997	995	991	986
18			1,000	1,000	1,000	999	999	998	996	993
19						1,000	999	999	998	997
20							1,000	1,000	999	998
21									1,000	999
22										1,000

c \ λ	10.5	11.0	11.5	12.0	12.5	13.0	13.5	14.0	14.5	15.0
0	000	000	000	000	000	000	000	000	000	000
1	000	000	000	000	000	000	000	000	000	000
2	002	001	001	001	000	000	000	000	000	000
3	007	005	003	002	002	001	001	000	000	000
4	021	015	011	008	005	004	003	002	001	001
5	050	038	028	020	015	011	008	006	004	003
6	102	079	060	046	035	026	019	014	010	008
7	179	143	114	090	070	054	041	032	024	018
8	279	232	191	155	125	100	079	062	048	037
9	397	341	289	242	201	166	135	109	088	070
10	521	460	402	347	297	252	211	176	145	118
11	639	579	520	462	406	353	304	260	220	185
12	742	689	633	576	519	463	409	358	311	268
13	825	781	733	682	628	573	518	464	413	363
14	888	854	815	772	725	675	623	570	518	466
15	932	907	878	844	806	764	718	669	619	568
16	960	944	924	899	869	835	798	756	711	664
17	978	968	954	937	916	890	861	827	790	749
18	988	982	974	963	948	930	908	883	853	819
19	994	991	986	979	969	957	942	923	901	875
20	997	995	992	988	983	975	965	952	936	917
21	999	998	996	994	991	986	980	971	960	947
22	999	999	998	997	995	992	989	983	976	967
23	1,000	1,000	999	999	998	996	994	991	986	981
24			1,000	999	999	998	997	995	992	989
25				1,000	999	999	998	997	996	994
26					1,000	1,000	999	999	998	997
27							1,000	999	999	998
28								1,000	999	999
29									1,000	1,000

Table A5.4 (continued)

c \ λ	16	17	18	19	20	21	22	23	24	25
1	000	000	000	000	000	000	000	000	000	000
2	000	000	000	000	000	000	000	000	000	000
3	000	000	000	000	000	000	000	000	000	000
4	000	000	000	000	000	000	000	000	000	000
5	001	001	000	000	000	000	000	000	000	000
6	004	002	001	001	000	000	000	000	000	000
7	010	005	003	002	001	000	000	000	000	000
8	022	013	007	004	002	001	001	000	000	000
9	043	026	015	009	005	003	002	001	000	000
10	077	049	030	018	011	006	004	002	001	001
11	127	085	055	035	021	013	008	004	003	001
12	193	135	092	061	039	025	015	009	005	003
13	275	201	143	098	066	043	028	017	011	006
14	368	281	208	150	105	072	048	031	020	012
15	467	371	287	215	157	111	077	052	034	022
16	566	468	375	292	221	163	117	082	056	038
17	659	564	469	378	297	227	169	123	087	060
18	742	655	562	469	381	302	232	175	128	092
19	812	736	651	561	470	384	306	238	180	134
20	868	805	731	647	559	471	387	310	243	185
21	911	861	799	725	644	558	472	389	314	247
22	942	905	855	793	721	640	556	472	392	318
23	963	937	899	849	787	716	637	555	473	394
24	978	959	932	893	843	782	712	635	554	473
25	987	975	955	927	888	838	777	708	632	553
26	993	985	972	951	922	883	832	772	704	629
27	996	991	983	969	948	917	877	827	768	700
28	998	995	990	980	966	944	913	873	823	763
29	999	997	994	988	978	963	940	908	868	818
30	999	999	997	993	987	976	959	936	904	863
31	1,000	999	998	996	992	985	973	956	932	900
32		1,000	999	998	995	991	983	971	953	929
33			1,000	999	997	994	989	981	969	950
34				999	999	997	994	988	979	966
35				1,000	999	998	996	993	987	978
36					1,000	999	998	996	992	985
37						999	999	997	995	991
38						1,000	999	999	997	994
39							1,000	999	998	997
40								1,000	999	998
41									999	999
42									1,000	999
43										1,000

Table A5.5† **1 Percentage Points of the *F* Distribution: Table of $F_{0.01;v_1,v_2}$**

P **{*F* with v_1 and v_2 Degrees of Freedom ≤ Tabled Value} = α**

Degrees of Freedom for the Numerator (v_1)

v_2 \ v_1	1	2	3	4	5	6	7	8	9	10	12	15	20	24	30	40	60	120	∞
1	4052	4999.5	5403	5625	5764	5859	5928	5981	6022	6056	6106	6157	6209	6235	6261	6287	6313	6339	6366
2	98.50	99.00	99.17	99.25	99.30	99.33	99.36	99.37	99.39	99.40	99.42	99.43	99.45	99.46	99.47	99.47	99.48	99.49	99.50
3	34.12	30.82	29.46	28.71	28.24	27.91	27.67	27.49	27.35	27.23	27.05	26.87	26.69	26.60	26.50	26.41	26.32	26.22	26.13
4	21.20	18.00	16.69	15.98	15.52	15.21	14.98	14.80	14.66	14.55	14.37	14.20	14.02	13.93	13.84	13.75	13.65	13.56	13.46
5	16.26	13.27	12.06	11.39	10.97	10.67	10.46	10.29	10.16	10.05	9.89	9.72	9.55	9.47	9.38	9.29	9.20	9.11	9.02
6	13.75	10.92	9.78	9.15	8.75	8.47	8.26	8.10	7.98	7.87	7.72	7.56	7.40	7.31	7.23	7.14	7.06	6.97	6.88
7	12.25	9.55	8.45	7.85	7.46	7.19	6.99	6.84	6.72	6.62	6.47	6.31	6.16	6.07	5.99	5.91	5.82	5.74	5.65
8	11.26	8.65	7.59	7.01	6.63	6.37	6.18	6.03	5.91	5.81	5.67	5.52	5.36	5.28	5.20	5.12	5.03	4.95	4.86
9	10.56	8.02	6.99	6.42	6.60	5.80	5.61	5.47	5.35	5.26	5.11	4.96	4.81	4.73	4.65	4.57	4.48	4.40	4.31
10	10.04	7.56	6.55	5.99	5.64	5.39	5.20	5.06	4.94	4.85	4.71	4.56	4.41	4.33	4.25	4.17	4.08	4.00	3.91
11	9.65	7.21	6.22	5.67	5.32	5.07	4.89	4.74	4.63	4.54	4.40	4.25	4.10	4.02	3.94	3.86	3.78	3.69	3.60
12	9.33	6.93	5.95	5.41	5.06	4.82	4.64	4.50	4.39	4.30	4.16	4.01	3.86	3.78	3.70	3.62	3.54	3.45	3.36
13	9.07	6.70	5.74	5.21	4.86	4.62	4.44	4.30	4.19	4.10	3.96	3.82	3.66	3.59	3.51	3.43	3.34	3.25	3.17
14	8.86	6.51	5.56	5.04	4.69	4.46	4.28	4.14	4.03	3.94	3.80	3.66	3.51	3.43	3.35	3.27	3.18	3.09	3.00
15	8.68	6.36	5.42	4.89	4.56	4.32	4.14	4.00	3.89	3.80	3.67	3.52	3.37	3.29	3.21	3.13	3.05	2.96	2.87
16	8.53	6.23	5.29	4.77	4.44	4.20	4.03	3.89	3.78	3.69	3.55	3.41	3.26	3.18	3.10	3.02	2.93	2.84	2.75
17	8.40	6.11	5.18	4.67	4.34	4.10	3.93	3.79	3.68	3.59	3.46	3.31	3.16	3.08	3.00	2.92	2.83	2.75	2.65
18	8.29	6.01	5.09	4.58	4.25	4.01	3.84	3.71	3.60	3.51	3.37	3.23	3.08	3.00	2.92	2.84	2.75	2.66	2.57
19	8.18	5.93	5.01	4.50	4.17	3.94	3.77	3.63	3.52	3.43	3.03	3.15	3.00	2.92	2.84	2.76	2.67	2.58	2.49
20	8.10	5.85	4.94	4.43	4.10	3.87	3.70	3.56	3.46	3.37	3.23	3.09	2.94	2.86	2.78	2.69	2.61	2.52	2.42
21	8.02	5.78	4.87	4.37	4.04	3.81	3.64	3.51	3.40	3.31	3.17	3.03	2.88	2.80	2.72	2.64	2.55	2.46	2.36
22	7.95	5.72	4.82	4.31	3.99	3.76	3.59	3.45	3.35	3.26	3.12	2.98	2.83	2.75	2.67	2.58	2.50	2.40	2.31
23	7.88	5.66	4.76	4.26	3.94	3.71	3.54	3.41	3.30	3.21	3.07	2.93	2.78	2.70	2.62	2.54	2.45	2.35	2.26
24	7.82	5.61	4.72	4.22	3.90	3.67	3.50	3.36	3.26	3.17	3.03	2.89	2.74	2.66	2.58	2.49	2.40	2.31	2.21
25	7.77	5.57	4.68	4.18	3.85	3.63	3.46	3.32	3.22	3.13	2.99	2.85	2.70	2.62	2.54	2.45	2.36	2.27	2.17
26	7.72	5.53	4.64	4.14	3.82	3.59	3.42	3.29	3.18	3.09	2.96	2.81	2.66	2.58	2.50	2.42	2.33	2.23	2.13
27	7.68	5.49	4.60	4.11	3.78	3.56	3.39	3.26	3.15	3.06	2.93	2.78	2.63	2.55	2.47	2.38	2.29	2.20	2.10
28	7.64	5.45	4.57	4.07	3.75	3.53	3.36	3.23	3.12	3.03	2.90	2.75	2.60	2.52	2.44	2.35	2.26	2.17	2.06
29	7.60	5.42	4.54	4.04	3.73	3.50	3.33	3.20	3.09	3.00	2.87	2.73	2.57	2.49	2.41	2.33	2.23	2.14	2.03
30	7.56	5.39	4.51	4.02	3.70	3.47	3.30	3.17	3.07	2.98	2.84	2.70	2.55	2.47	2.39	2.30	2.21	2.11	2.01
40	7.31	5.18	4.31	3.83	3.51	3.29	3.12	2.99	2.89	2.80	2.66	2.52	2.37	2.29	2.20	2.11	2.02	1.92	1.80
60	7.08	4.98	4.13	3.65	3.34	3.12	2.95	2.82	2.72	2.63	2.50	2.35	2.20	2.12	2.03	1.94	1.84	1.73	1.60
120	6.85	4.79	3.95	3.48	3.17	2.96	2.79	2.66	2.56	2.47	2.34	2.19	2.03	1.95	1.86	1.76	1.66	1.53	1.38
∞	6.63	4.61	3.78	3.32	3.02	2.80	2.64	2.51	2.41	2.32	2.18	2.04	1.88	1.79	1.70	1.59	1.47	1.32	1.00

Degrees of freedom for the denominator (v_2)

† Reproduced from Table 18 of Biometrika *Tables for Statisticians*, vol. I, 3d ed., 1966, by permission of the Biometrika Trustees.

$F_{0.99,v_1,v_2} = 1/F_{0.01,v_2,v_1}$

Table A5.5† *(continued)*

v_2 \ v_1 (Degrees of freedom for the denominator (v_2))	Degrees of Freedom for the Numerator (v_1) 1	2	3	4	5	6	7	8	9	10	12	15	20	24	30	40	60	120	∞
1	647.8	799.5	864.2	899.6	921.8	937.1	948.2	956.7	963.3	968.6	976.7	984.9	993.1	997.2	1001	1006	1010	1014	1018
2	38.51	39.00	39.17	39.25	39.30	39.33	39.36	39.37	39.39	39.40	39.41	39.43	39.45	39.46	39.46	39.47	39.48	39.49	39.50
3	17.44	16.04	15.44	15.10	14.88	14.73	14.62	14.54	14.47	14.42	14.34	14.25	14.17	14.12	14.08	14.04	13.99	13.95	13.90
4	12.22	10.65	9.98	9.60	9.36	9.20	9.07	8.98	8.90	8.84	8.75	8.66	8.56	8.51	8.46	8.41	8.36	8.31	8.26
5	10.01	8.43	7.76	7.39	7.15	6.98	6.85	6.76	6.68	6.62	6.52	6.43	6.33	6.28	6.23	6.18	6.12	6.07	6.02
6	8.81	7.26	6.60	6.23	5.99	5.82	5.70	5.60	5.52	5.46	5.37	5.27	5.17	5.12	5.07	5.01	4.96	4.90	4.85
7	8.07	6.54	5.89	5.52	5.29	5.12	4.99	4.90	4.82	4.76	4.67	4.57	4.47	4.42	4.36	4.31	4.25	4.20	4.14
8	7.57	6.06	5.42	5.05	4.82	4.65	4.53	4.43	4.36	4.30	4.20	4.10	4.00	3.95	3.89	3.84	3.78	3.73	3.67
9	7.21	5.71	5.08	4.72	4.48	4.32	4.20	4.10	4.03	3.96	3.87	3.77	3.67	3.61	3.56	3.51	3.45	3.39	3.33
10	6.94	5.46	4.83	4.47	4.24	4.07	3.95	3.85	3.78	3.72	3.62	3.52	3.42	3.37	3.31	3.26	3.20	3.14	3.08
11	6.72	5.26	4.63	4.28	4.04	3.88	3.76	3.66	3.59	3.53	3.43	3.33	3.23	3.17	3.12	3.06	3.00	2.94	2.88
12	6.55	5.10	4.47	4.12	3.89	3.73	3.61	3.51	3.44	3.37	3.28	3.18	3.07	3.02	2.96	2.91	2.85	2.79	2.72
13	6.41	4.97	4.35	4.00	3.77	3.60	3.48	3.39	3.31	3.25	3.15	3.05	2.95	2.89	2.84	2.78	2.72	2.66	2.60
14	6.30	4.86	4.24	3.89	3.66	3.50	3.38	3.29	3.21	3.15	3.05	2.95	2.84	2.79	2.73	2.67	2.61	2.55	2.49
15	6.20	4.77	4.15	3.80	3.58	3.41	3.29	3.20	3.12	3.06	2.96	2.86	2.76	2.70	2.64	2.59	2.52	2.46	2.40
16	6.12	4.69	4.08	3.73	3.50	3.34	3.22	3.12	3.05	2.99	2.89	2.79	2.68	2.63	2.57	2.51	2.45	2.38	2.32
17	6.04	4.62	4.01	3.66	3.44	3.28	3.16	3.06	2.98	2.92	2.82	2.72	2.62	2.56	2.50	2.44	2.38	2.32	2.25
18	5.98	4.56	3.95	3.61	3.38	3.22	3.10	3.01	2.93	2.87	2.77	2.67	2.56	2.50	2.44	2.38	2.32	2.26	2.19
19	5.92	4.51	3.90	3.56	3.33	3.17	3.05	2.96	2.88	2.82	2.72	2.62	2.51	2.45	2.39	2.33	2.27	2.20	2.13
20	5.87	4.46	3.86	3.51	3.29	3.13	3.01	2.91	2.84	2.77	2.68	2.57	2.46	2.41	2.35	2.29	2.22	2.16	2.09
21	5.83	4.42	3.82	3.48	3.25	3.09	2.97	2.87	2.80	2.73	2.64	2.53	2.42	2.37	2.31	2.25	2.18	2.11	2.04
22	5.79	4.38	3.78	3.44	3.22	3.05	2.93	2.84	2.76	2.70	2.60	2.50	2.39	2.33	2.27	2.21	2.14	2.08	2.00
23	5.75	4.35	3.75	3.41	3.18	3.02	2.90	2.81	2.73	2.67	2.57	2.47	2.36	2.30	2.24	2.18	2.11	2.04	1.97
24	5.72	4.32	3.72	3.38	3.15	2.99	2.87	2.78	2.70	2.64	2.54	2.44	2.33	2.27	2.21	2.15	2.08	2.01	1.94
25	5.69	4.29	3.69	3.35	3.13	2.97	2.85	2.75	2.68	2.61	2.51	2.41	2.30	2.24	2.18	2.12	2.05	1.98	1.91
26	5.66	4.27	3.67	3.33	3.10	2.94	2.82	2.73	2.65	2.59	2.49	2.39	2.28	2.22	2.16	2.09	2.03	1.95	1.88
27	5.63	4.24	3.65	3.31	3.08	2.92	2.80	2.71	2.63	2.57	2.47	2.36	2.25	2.19	2.13	2.07	2.00	1.93	1.85
28	5.61	4.22	3.63	3.29	3.06	2.90	2.78	2.69	2.61	2.55	2.45	2.34	2.23	2.17	2.11	2.05	1.98	1.91	1.83
29	5.59	4.20	3.61	3.27	3.04	2.88	2.76	2.67	2.59	2.53	2.43	2.32	2.21	2.15	2.09	2.03	1.96	1.89	1.81
30	5.57	4.18	3.59	3.25	3.03	2.87	2.75	2.65	2.57	2.51	2.41	2.31	2.20	2.14	2.07	2.01	1.94	1.87	1.79
40	5.42	4.05	3.46	3.13	2.90	2.74	2.62	2.53	2.45	2.39	2.29	2.18	2.07	2.01	1.94	1.88	1.80	1.72	1.64
60	5.29	3.93	3.34	3.01	2.79	2.63	2.51	2.41	2.33	2.27	2.17	2.06	1.94	1.88	1.82	1.74	1.67	1.58	1.48
120	5.15	3.80	3.23	2.89	2.67	2.52	2.39	2.30	2.22	2.16	2.05	1.94	1.82	1.76	1.69	1.61	1.53	1.43	1.31
∞	5.02	3.69	3.12	2.79	2.57	2.41	2.29	2.19	2.11	2.05	1.94	1.83	1.71	1.64	1.57	1.48	1.39	1.27	1.00

† $F_{0.975,v_1,v_2} = 1/F_{0.025,v_2,v_1}$

Table A5.5† (continued)

v_2 \ v_1	1	2	3	4	5	6	7	8	9	10	12	15	20	24	30	40	60	120	∞
1	161.4	199.5	215.7	224.6	230.2	234.0	236.8	238.9	240.5	241.9	243.9	245.9	248.0	249.1	250.1	251.1	252.2	253.3	254.3
2	18.51	19.00	19.16	19.25	19.30	19.33	19.35	19.37	19.38	19.40	19.41	19.43	19.45	19.45	19.46	19.47	19.48	19.49	19.50
3	10.13	9.55	9.28	9.12	9.01	8.94	8.89	8.85	8.81	8.79	8.74	8.70	8.66	8.64	8.62	8.59	8.57	8.55	8.53
4	7.71	6.94	6.59	6.39	6.26	6.16	6.09	6.04	6.00	5.96	5.91	5.86	5.80	5.77	5.75	5.72	5.69	5.66	5.63
5	6.61	5.79	5.41	5.19	5.05	4.95	4.88	4.82	4.77	4.74	4.68	4.62	4.56	4.53	4.50	4.46	4.43	4.40	4.36
6	5.99	5.14	4.76	4.53	4.39	4.28	4.21	4.15	4.10	4.06	4.00	3.94	3.87	3.84	3.81	3.77	3.74	3.70	3.67
7	5.59	4.74	4.35	4.12	3.97	3.87	3.79	3.73	3.68	3.64	3.57	3.51	3.44	3.41	3.38	3.34	3.30	3.27	3.23
8	5.32	4.46	4.07	3.84	3.69	3.58	3.50	3.44	3.39	3.35	3.28	3.22	3.15	3.12	3.08	3.04	3.01	2.97	2.93
9	5.12	4.26	3.86	3.63	3.48	3.37	3.29	3.23	3.18	3.14	3.07	3.01	2.94	2.90	2.86	2.83	2.79	2.75	2.71
10	4.96	4.10	3.71	3.48	3.33	3.22	3.14	3.07	3.02	2.98	2.91	2.85	2.77	2.74	2.70	2.66	2.62	2.58	2.54
11	4.84	3.98	3.59	3.36	3.20	3.09	3.01	2.95	2.90	2.85	2.79	2.72	2.65	2.61	2.57	2.53	2.49	2.45	2.40
12	4.75	3.89	3.49	3.26	3.11	3.00	2.91	2.85	2.80	2.75	2.69	2.62	2.54	2.51	2.47	2.43	2.38	2.34	2.30
13	4.67	3.81	3.41	3.18	3.03	2.92	2.83	2.77	2.71	2.67	2.60	2.53	2.46	2.42	2.38	2.34	2.30	2.25	2.21
14	4.60	3.74	3.34	3.11	2.96	2.85	2.76	2.70	2.65	2.60	2.53	2.46	2.39	2.35	2.31	2.27	2.22	2.18	2.13
15	4.54	3.68	3.29	3.06	2.90	2.79	2.71	2.64	2.59	2.54	2.48	2.40	2.33	2.29	2.25	2.20	2.16	2.11	2.07
16	4.49	3.63	3.24	3.01	2.85	2.74	2.66	2.59	2.54	2.49	2.42	2.35	2.28	2.24	2.19	2.15	2.11	2.06	2.01
17	4.45	3.59	3.20	2.96	2.81	2.70	2.61	2.55	2.49	2.45	2.38	2.31	2.23	2.19	2.15	2.10	2.06	2.01	1.96
18	4.41	3.55	3.16	2.93	2.77	2.66	2.58	2.51	2.46	2.41	2.34	2.27	2.19	2.15	2.11	2.06	2.02	1.97	1.92
19	4.38	3.52	3.13	2.90	2.74	2.63	2.54	2.48	2.42	2.38	2.31	2.23	2.16	2.11	2.07	2.03	1.98	1.93	1.88
20	4.35	3.49	3.10	2.87	2.71	2.60	2.51	2.45	2.39	2.35	2.28	2.20	2.12	2.08	2.04	1.99	1.95	1.90	1.84
21	4.32	3.47	3.07	2.84	2.68	2.57	2.49	2.42	2.37	2.32	2.25	2.18	2.10	2.05	2.01	1.96	1.92	1.87	1.81
22	4.30	3.44	3.05	2.82	2.66	2.55	2.46	2.40	2.34	2.30	2.23	2.15	2.07	2.03	1.98	1.94	1.89	1.84	1.78
23	4.28	3.42	3.03	2.80	2.64	2.53	2.44	2.37	2.32	2.27	2.20	2.13	2.05	2.01	1.96	1.91	1.86	1.81	1.76
24	4.26	3.40	3.01	2.78	2.62	2.51	2.42	2.36	2.30	2.25	2.18	2.11	2.03	1.98	1.94	1.89	1.84	1.79	1.73
25	4.24	3.39	2.99	2.76	2.60	2.49	2.40	2.34	2.28	2.24	2.16	2.09	2.01	1.96	1.92	1.87	1.82	1.77	1.71
26	4.23	3.37	2.98	2.74	2.59	2.47	2.39	2.32	2.27	2.22	2.15	2.07	1.99	1.95	1.90	1.85	1.80	1.75	1.69
27	4.21	3.35	2.96	2.73	2.57	2.46	2.37	2.31	2.25	2.20	2.13	2.06	1.97	1.93	1.88	1.84	1.79	1.73	1.67
28	4.20	3.34	2.95	2.71	2.56	2.45	2.36	2.29	2.24	2.19	2.12	2.04	1.96	1.91	1.87	1.82	1.77	1.71	1.65
29	4.18	3.33	2.93	2.70	2.55	2.43	2.35	2.28	2.22	2.18	2.10	2.03	1.94	1.90	1.85	1.81	1.75	1.70	1.64
30	4.17	3.32	2.92	2.69	2.53	2.42	2.33	2.27	2.21	2.16	2.09	2.01	1.93	1.89	1.84	1.79	1.74	1.68	1.62
40	4.08	3.23	2.84	2.61	2.45	2.34	2.25	2.18	2.12	2.08	2.00	1.92	1.84	1.79	1.74	1.69	1.64	1.58	1.51
60	4.00	3.15	2.76	2.53	2.37	2.25	2.17	2.10	2.04	1.99	1.92	1.84	1.75	1.70	1.65	1.59	1.53	1.47	1.39
120	3.92	3.07	2.68	2.45	2.29	2.17	2.09	2.02	1.96	1.91	1.83	1.75	1.66	1.61	1.55	1.50	1.43	1.35	1.25
∞	3.84	3.00	2.60	2.37	2.21	2.10	2.01	1.94	1.88	1.83	1.75	1.67	1.57	1.52	1.46	1.39	1.32	1.22	1.00

Columns: *Degrees of Freedom for the Numerator* (v_1); rows: *Degrees of freedom for the denominator* (v_2)

† $F_{0.95,v_1,v_2} = 1/F_{0.05,v_2,v_1}$

Answers to Selected Problems

Chapter 3

1. (*a*) The six colored sides.
(*b*) $P\{X = 0\} = P\{X = 1\} = P\{X = 2\} = \frac{1}{3}$.
(*c*) $E[Y] = \frac{14}{3}$.

4. (*a*) $k = \theta + 1$.
(*b*) $F_x(b) = b\theta$ for $b \leq \theta$ and $Fx(b) = b(1 + \theta) - \theta$ for $\theta \leq b \leq 1$.
(*c*) $E[X] = \frac{1}{2}(1 + \theta - \theta^2)$;
(*d*) no.

7. (*a*)

k	-2	-1	0	1	2
$P\{X = k\}$	$\frac{1}{10}$	$\frac{2}{10}$	$\frac{3}{10}$	$\frac{1}{10}$	$\frac{3}{10}$

(*c*) $E[X] = \frac{3}{10}$.

10. (*a*) $P\{X \geq 25\} = 0.527$ and $P\{X = 20\} = 0.051$.
(b) $P\{\text{shortage}\} = 0.022$.

13. (*a*) $E[V] = 45$ and $\sigma_Z = 0.139$; (*b*) $P\{Z \geq 45.2\} = 0.0749$.

16. $n = 20$.

19.

	k	0	1	2	3
(a)	$P\{X_1 = k\}$	$\frac{1}{2}$	$\frac{1}{2}$	0	0
	$P\{X_2 = k\}$	$\frac{1}{8}$	$\frac{3}{8}$	$\frac{1}{2}$	0
(b)	$P\{X_1 = k \mid X_2 = 1\}$	$\frac{2}{3}$	$\frac{1}{3}$	0	0
(e)	$P\{X_1 + X_2 = k\}$	$\frac{1}{8}$	$\frac{1}{4}$	$\frac{1}{4}$	$\frac{3}{8}$

(c) No.
(d) $E[X_1] = \frac{1}{2}$; var $[X_1] = \frac{1}{4}$.
$E[X_2] = \frac{11}{8}$; var $[X_2] = \frac{31}{64}$.

Chapter 4

1. $E(\overline{X} - \mu)^2 = \sigma^2/2$.
$E(T_2 - \mu)^2 = 5\sigma^2/9$.
Therefore, $\overline{X} = T_1$ is better estimate.
4. (a) All values.
(b) $\lambda = \frac{1}{2}$.
9. (a) $\hat{y} = 121.04 - 1.0346x$.
(b) (46.64, 81.64).
(c) (42.13, 86.14).
(d) (37.1, 91.2).

Chapter 5

5. (a) All states belong to the same recurrent class.
10. (b) $\pi_0 = \pi_1 = \pi_2 = \pi_3 = \pi_4 = \frac{1}{5}$.
20. (a) $\pi_0 = 0.182$, $\pi_1 = 0.285$, $\pi_2 = 0.368$, $\pi_3 = 0.165$.
(b) 31.42.

Chapter 6

1. Input source: population having hair; customers: customers needing haircuts; queue: customers waiting for a barber; queue discipline: first-come-first-served; service mechanism: barber(s).
4. (a) 0.135.
(b) 0.270.
(c) 0.0527.
7. (b) $P_0 = \frac{2}{5}$, $P_n = (\frac{3}{5})(\frac{1}{2})^n$.
(c) $L = \frac{6}{5}$, $L_q = \frac{3}{5}$, $W = \frac{1}{25}$, $W_q = \frac{1}{50}$.
12. $\frac{31}{32}$.
21. (a) 0.429.
(b) 0.154.
(c) 0.072.

23.

Part	P_0	P_1	P_2	P_3	E (Not Running)
(a)	0.493	0.329	0.146	0.032	0.718
(b) (i)	0.333	0.222	0.148	0.099	2.000
(b) (ii)	0.415	0.277	0.185	0.123	1.015
(c)	0.546	0.364	0.081	0.009	0.553

27. (*a*) W_q (exponential) $= 2W_q$ (constant) $= \frac{8}{5}W_q$ (Erlang).
(*b*) W_q (new) $= \frac{1}{2}W_q$ (old) and L_q (new) $= L_q$ (old) for all distributions.
29. Current policy: $L = 1$; proposed policy: $L = \frac{13}{16}$.
36. (*a*) $W = \frac{1}{2}$.
(*b*) $W_1 = 0.20$, $W_2 = 0.35$, $W_3 = 1.10$.
(*c*) $W_1 = 0.125$, $W_2 = 0.3125$, $W_3 = 1.250$.

41.

Service Distribution	P_0	P_1	P_2	L
Erlang	0.561	0.316	0.123	0.561
Exponential	0.571	0.286	0.143	0.571

Chapter 7

2. (*a*) $E(WC) = 16$.
(*c*) $E(WC) = 26\frac{1}{2}$.
8. Status quo: $E(TC) = 50$; proposal: $E(TC) = 75.75$; keep status quo.
11. (*a*) Crew size = 2.
(*b*) Crew size = 3.
13. $\mu = 1.15$ minimizes $E(TC)$.
19. (*a*) $E(T) = \dfrac{6.8r}{v}$.
(*c*) $E(T) = \dfrac{67r}{21v}$.
22. One doctor: $E(TC) = 624.80$; two doctors: 92.95; have two doctors.

Chapter 8

3.

	Store		
	1	2	3
Allocations	1	2	2
	3	2	0

9.

Phase	(*a*)	(*b*)
1	2 *M*	2.945 *M*
2	1 *M*	1.055 *M*
3	1 *M*	0
Market share	6%	6.302%

12. $x_1 = -2 + \sqrt{13} \approx 1.6056$, $x_2 = 5 - \sqrt{13} \approx 1.3944$; $Z = 98.233$.
22. Produce 2 on first production run; if none acceptable, produce 2 on second run. Expected cost = \$573.

Chapter 9

1. (*a*) $t = 1.83$, $Q = 54.77$.
(*b*) $t = 1.91$, $Q = 57.45$, $S = 52.22$.
4. $t = 3.26$, $Q = 26{,}046$, $S = 24{,}572$.

15. Produce 7 units in period 1 and 7 units in period 3.
18. Produce 3 units in period 1 and 4 units in period 3.
22. Produce 1,857 loaves.
24. $(s, S) = (1, 5)$.
28. (*a*) $G(y) = (\frac{3}{10})y + 70e^{-y/25} - \frac{15}{2}$.
(*b*) $(k, Q) = (21, 100)$ policy.
31. If $x \le 46$, order $46 - x$ units; otherwise, do not order.
34. If $x \le 2$, order $2 - x$ units; otherwise, do not order.
39. If $x \le y^0$, order $y^0 - x$ units; otherwise, do not order.
$y^0 = \mu - c(1 - \alpha)/2$.
42. $s = 24$, $Q = 58$, $(s, S) = (24, 82)$.

Chapter 10

1. 2,090.5.
3. 783.2.
10. (*a*)

10/80	7.167						
1/81	7.233	1/82	7.267	1/83	9.633	1/84	9.267
4/81	7.733	4/82	8.067	4/83	10.367	4/84	8.867
7/81	7.433	7/82	8.700	7/83	10.433	7/84	8.267
10/81	7.500	10/82	9.467	10/83	10.267	10/84	7.967

(*b*) 1.184.
24. (*a*) $410.333 + 17.630t$.
(*b*) 604.267.
(*c*)

3	431.6	6	445.220	9	468.880
4	434.84	7	452.098	10	478.992
5	439.356	8	460.089	11	490.193

(*d*)

3	465.52	6	517.405	9	567.864
4	483.075	7	534.236	10	585.671
5	500.196	8	551.212	11	604.329

Chapter 11

1. (*b*) Use slow service when no customers or one customer is present and fast service when two customers are present.

2.
$$\text{Minimize } 3y_{01} + 9y_{02} + 3y_{11} + 9y_{12} + 28y_{21} + 34y_{22},$$
subject to
$$y_{01} + y_{02} - (\tfrac{1}{2}y_{01} + \tfrac{3}{10}y_{11} + \tfrac{1}{2}y_{02} + \tfrac{2}{5}y_{12}) = 0$$
$$y_{11} + y_{12} - (\tfrac{1}{2}y_{01} + \tfrac{1}{2}y_{11} + \tfrac{3}{5}y_{21} + \tfrac{1}{2}y_{02} + \tfrac{1}{2}y_{12} + \tfrac{4}{5}y_{22}) = 0$$
$$y_{21} + y_{22} - (\tfrac{2}{10}y_{11} + \tfrac{2}{5}y_{21} + \tfrac{1}{10}y_{12} + \tfrac{1}{5}y_{22}) = 0$$
$$y_{01} + y_{02} + y_{11} + y_{12} + y_{21} + y_{22} = 1$$
and
$$y_{ik} \ge 0 \quad \text{for } i = 0, 1, 2 \text{ and } k = 1, 2.$$

7. State 1: attempt ace; state 2: attempt lob.

8.
$$\text{Minimize } -\tfrac{1}{8}y_{01} + \tfrac{7}{24}y_{02} + \tfrac{1}{2}y_{11} + \tfrac{5}{12}y_{12},$$
subject to
$$y_{01} + y_{02} - (\tfrac{3}{8}y_{01} + y_{11} + \tfrac{7}{8}y_{02} + y_{12}) = 0$$
$$y_{11} + y_{12} - (\tfrac{5}{8}y_{01} + \tfrac{1}{8}y_{02}) = 0$$
$$y_{01} + y_{02} + y_{11} + y_{12} = 1$$
and
$$y_{ik} \ge 0 \quad \text{for } i = 0, 1 \text{ and } k = 1, 2.$$

22. Reject \$600 offer, accept any of the other two.

23. Minimize $60(y_{01} + y_{11} + y_{21}) - 600y_{02} - 800y_{12} - 1{,}000y_{22}$,

subject to
$$y_{01} + y_{02} - (0.95)(\tfrac{5}{8})(y_{01} + y_{11} + y_{21}) = \tfrac{5}{8}$$
$$y_{11} + y_{12} - (0.95)(\tfrac{1}{4})(y_{01} + y_{11} + y_{21}) = \tfrac{1}{4}$$
$$y_{21} + y_{22} - (0.95)(\tfrac{1}{8})(y_{01} + y_{11} + y_{21}) = \tfrac{1}{8}$$
and
$$y_{ik} \geq 0 \quad \text{for } i = 0, 1, 2 \text{ and } k = 1, 2.$$

24. After three iterations, approximation is, in fact, the optimal policy given in Prob. 22.

31. Use fertilizer B regardless of crop quality.

32. Minimize $-5400(y_{01} + y_{11}) - 6200(y_{02} + y_{12})$,

subject to
$$y_{01} + y_{02} - (\tfrac{1}{2})(\tfrac{3}{5}y_{01} + \tfrac{3}{5}y_{11} + \tfrac{4}{5}y_{02} + \tfrac{4}{5}y_{12}) = \tfrac{1}{2}$$
$$y_{11} + y_{12} - (\tfrac{1}{2})(\tfrac{2}{5}y_{01} + \tfrac{2}{5}y_{11} + \tfrac{1}{5}y_{02} + \tfrac{1}{5}y_{12}) = \tfrac{1}{2}$$
$$y_{ik} \geq 0 \quad \text{for } i = 0, 1 \text{ and } k = 1, 2.$$

33. After three iterations, approximation is optimal policy given in Prob. 31.

41. Use fertilizer B in all four periods regardless of crop quality.

48. In periods 1 to 3 do nothing when the machine is in states 0 or 1, overhaul when machine is in state 2, and replace when in state 3. In period 4 do nothing when machine is in state 0, 1, or 2, and replace when in state 3.

Chapter 12

1. Paths are $\{x_1, x_2\}$ and $\{x_1, x_3\}$.
$\Phi(x_1, x_2, x_3) = \max(x_1 x_2, x_1 x_3) = x_1 \max(x_2, x_3) = x_1[1 - (1 - x_2)(1 - x_3)]$.

3. $R(p_1, p_2, p_3) = p_1[1 - (1 - p_2)(1 - p_3)]$.

8. (*a*) Minimal paths are $\{x_1, x_3\}$ and $\{x_2, x_4\}$.
Minimal cuts are $\{x_1, x_2\}$, $\{x_1, x_4\}$, $\{x_2, x_3\}$, and $\{x_3, x_4\}$.
(*b*) $R(p_1, p_2, p_3) = 1 - (1 - p_1p_3)(1 - p_2p_4) = 0.9639$ when $p_i = 0.90$.
(*c*) Upper bound = exact system reliability.
Lower bound = $(1 - q_1q_2)(1 - q_1q_4)(1 - q_2q_3)(1 - q_3q_4) = 0.9606$ when $p_i = 0.90$, $q_i = 0.10$.

12. (*a*) $0.659 \leq R(t) \leq 1$
(*b*) $0 \leq R(1) \leq 0.324$.

Chapter 13

1. (*a*) Player I: strategy 2; player II: strategy 1.

7. (*a*) Politician I: issue 2; politician II: issue 2.
(*b*) Politician I: issue 1; politician II: issue 2.
(*c*) Minimax criterion says politician I can use any issue, but issue 1 offers politician I the only chance of winning if politician II is not "smart."

11. (*a*) $(x_1, x_2) = (\tfrac{2}{5}, \tfrac{3}{5})$; $(y_1, y_2, y_3) = (\tfrac{1}{5}, 0, \tfrac{4}{5})$; $v = \tfrac{8}{5}$.

14. Minimize $-x_4$,

subject to
$$5x_1 + 2x_2 + 3x_3 - x_4 \geq 0$$
$$4x_2 + 2x_3 - x_4 \geq 0$$
$$3x_1 + 3x_2 \qquad - x_4 \geq 0$$
$$x_1 + 2x_2 + 4x_3 - x_4 \geq 0$$
$$-x_1 - x_2 - x_3 = -1$$
and
$$x_1 \geq 0, \quad x_2 \geq 0, \quad x_3 \geq 0, \quad x_4 \geq 0.$$

Chapter 14

1. (*a*) a_3.
   (*b*) Up to \$230,000.
   (*c*) a_3.
6. (*a*) Guess coin 1.
   (*b*) Heads: coin 2; tails: coin 1.
8. (*a*) Produce the chip.
   (*b*) Up to \$400,000.
   (*c*) −\$2,300,000. It doesn't pay to use market research.
   (*d*) Produce the chip.
16. Bayes' procedure without seismic soundings is a_1 with expected loss of −\$68,000.

Seismic Sounding	Bayes' Action	*E*(Loss)	$Q_x(x)$
1	a_1	−137,700	0.360
2	a_1	−67,925	0.260
3	a_2	−33,000	0.210
4	a_2	−33,000	0.170

Value of seismic soundings is \$781, so they should not be used.

Chapter 15

1. (*a*) 5, 8, 1, 4, 7, 0, 3, 6, 9, 2.
4. (*a*) Assigning numbers 0, 1, 2, 3, 4 to heads and 5, 6, 7, 8, 9 to tails gives the sequence *THHTT*.
6. (*a*) $x = \sqrt{r}$.
11. (*a*) $x = -4 \ln(1 - r)$.
    (*b*) $x = -2 \ln[(1 - r_1)(1 - r_2)]$.
    (*c*) $x = 4 \sum_{i=1}^{6} r_i - 8$.
18. Use first 10 three-digit decimals from Table 15.4 and generate observations from

$$x_i = \frac{1}{1 - r_i}.$$

Method:	Analytic	Monte Carlo	Stratified Sampling	Complementary Numbers
Mean:	∞	4.3969	8.7661	3.812

21. (*a*) Est$\{W_q\} = 2\frac{1}{3}$ and $P\{1.572 \leq W_q \leq 3.094\} = 0.90$.

Author Index

Subject Index

Instructions for Using OR COURSEWARE

IBM Compatible Computers

1. Insert diskette in drive A.
2. Make drive A the current drive.
3. Type the name of the program (ProbMod) and press <ENTER>.
4. Programs may be copied to and run from a hard disk.

Macintosh Computers

1. Insert the diskette and double click on the ProbMod program icon
2. If the computer you are using does not have arrow keys, use the following substitutes:

UP
{

LEFT [ ] RIGHT

}
DOWN

3. The program may be copied to and run from a hard disk.